RESEARCH ON EVOLUTION EQUATION COMPENDIUM. VOLUME 2

RESEARCH ON EVOLUTION EQUATION COMPENDIUM. VOLUME 2

GASTON M. N'GUÉRÉKATA
EDITOR

Nova Science Publishers, Inc.
New York

LIBRARY OF CONGRESS CATALOGING-IN-PUBLICATION DATA

ISBN: 978-1-60456-752-6
Available upon request

Published by Nova Science Publishers, Inc. ✦ New York

CONTENTS

PREFACE

Chapter 1 - The behavior of inhomogeneous material, where the inhomogeneities are on a small scale, is of considerable interest in material sciences, mechanical engineering and in many technological applications, especially those involving composite materials. The theory of homogenization was created to model and predict the behavior of such material, when the inhomogeneities are on such a scale much smaller than the linear dimension of the system. In these situations, continuum models, in which rapid oscillation of the material properties is averaged, are sufficient to describe the behavior of the system. Moreover, they have the advantage of avoiding extensive numerical computations needed when the small scale behavior is taken into account. There have been numerous publications on homogenization theory and its applications in the past decades, the authors refer to the books by Bensoussan et al., Sanchez-Palencia and Jikov et al.

For homogenization of nonlinear equations, a standard approach is using energy methods by Tartar and Murat; however, in some cases, it is difficult to work out energy method sin was first introduced by Nguetseng, later elaborated by Allaire; the advantage of this method is that the homogenized equations and proof of convergence are obtained in a single step. Recently Caffarelli proposed an alternative approach in which the homogenization be considered as a viscosity process. Following the idea by Caffarelli, Garcia-Azorero et al. proved that the uniform limit of the solutions at scale e to quasilinear parabolic equations is a viscosity solution to the homogenized equation. The use of viscosity solution methods was first considered by Evansor divergence structure PDE.

In this paper, the authors shall use the method of double-scale convergence to derive the homogerized equation of the quasilinear parabolic equations considered in the references they site.

Chapter 2 - The authors consider a modified $2d$ second grade incompressible fluid with a higher order viscosity term, and show that this nonlinear evolution system possesses an exponential attractor. This object gives a more precise description of the asymptotic behaviour of dynamical systems than the global attractor.

Chapter 3 - In this work, the authors study the existence of periodic solutions for some partial functional differential equations with infinite delay. The authors assume that the linear part is not necessarily densely defined and satisfies the Hille-Yosida condition, and the phase space is chosen to be C for some decreasing function g from (-¥,0] to [1,¥). The authors also present a related Massera type result, namely the existence of a bounded solution on $\mathbb{R}^+$ implies the existence of a periodic solution.

Chapter 4 - The so called Lions sentinel method is applied to some time varying system with two time scale variables and with missing data. As an example, some population dynamics problem with missing data is considered. The two time scale variables represent the running time variable and the age of individuals. This insight seems to be new. Building the object sentinel leads to some null-controllability problem with restricted control variables and two time scale variables. Chapter 5 - In this work, the authors prove the existence of a center manifold for some partial functional differential equations, whose linear part is not necessarily densely defined but satisfies the Hille-Yosida condition. The attractiveness of the center manifold is also shown when the unstable space is reduced to zero. The authors prove that the flow on the center manifold is completely determined by an ordinary differential equation in a finite dimensional space. In some critical cases, when the exponential stability is not possible, the authors prove that the uniform asymptotic stability of the equilibrium is completely determined by the uniform asymptotic stability of the reduced system on the center manifold.

Chapter 6 - In this paper the authors study stability radii of positive linear functional differential systems in Banach spaces under multi-perturbations and multi-affine perturbations. The authors prove that for the class of positive systems, complex stability radius, real stability radius and positive stability radius of positive systems under multi-perturbations (or multi-affine perturbations) coincide and they are computed via a simple formula. The authors illustrated the obtained results by an example.

Chapter 7 - This paper considers Levinson's Theorem for nonlocal potentials. The effect of continuum bound states is clarified, and the presence of spurious states is introduced. Differences between definitions of the phase shift are discussed, with special attention given to discontinuities in the phase shift. Examples from nuclear physics emphasize the utility of this theorem and illustrate some of the more delicate aspects of the derivation. A summary is provided which analyzes certain crucial points in the discussion as they relate to previousattempts to obtain a generalized Levinson's theorem. NUCLEAR REACTIONS Levinson's Theorem, nonlocal potentialscontinuum bound states, spurious states, Jost functions, Fredholm determinants, definition of the phase shift.

Chapter 8 - For the FitzHugh-Nagumo system defined on *IR*, the authors prove the existence of a compact global attractor in a weighted Sobolev space which contains bounded solutions, in particular, traveling wave and spatially periodic solutions. The authors also study the behavior of the global attractors as a parameter goes to zero. Although the limiting system for =0does not possess a bounded attracting set, the authors show that there exists a constant 0 such that global attractors for 0 <= are all contained in a compact subset of the phase space. Furthermore, the authors construct a compact local attractor for the limiting system and establish the upper semicontinuity of the global attractors of perturbed system and the local attractor of the limiting system.

Chapter 9 - In this paper, the authors consider partial differential equations with nonautonomous past delay in L^1-phase space. Here, the authors consider general delay operators given by Stieltjes integrals. Using the Miyadera-Voigt perturbation result, the authors show the wellposedness of these equations. The authors study also the robustness of some asymptotic properties, as asymptotic stability, almost periodicity and almost automorphy, under the nonautonomous past delay effect. As an application, the authors give a dynamical population equation.

Chapter 10 - In this paper, the authors solve the problem of constructing the kernels of transmutation operator and give the generalized transmutation operator for a singular problem.

Chapter 11 - In this paper, the authors show that under very general conditions, elliptic operator with zero boundary condition can generate fractional resolvent families.

Chapter 12 - This paper is a survey devoted to the author's recent results related to the existence and uniqueness of pseudo almost periodic solutions to some abstract differential equations and partial differential equations. Some of those results will be slightly modified and applied to some concrete problems. As main tools, the authors will make extensive use of the method of analytic semigroups, and both the Banach and Zima's fixed-point principles. Applications include the study of pseudo almost periodic solutions to some boundary-value problems such as the heat equation with delay as well as the logistic differential equation.

Chapter 13 - In the 1970s, Cowin and Hegedus introduced an adaptive elasticity model for bone deposition and reabsorption (remodeling). In 1998, Figueiredo and Trabucho published a paper on an asymptotic model of an adaptive elastic rod. Afterwards, Monnier and Trabucho proved the existence and uniqueness of a solution for this model; Trabucho himself extended this result to a model that contains non-linear strain terms along with linear strain terms in a rate remodeling equation. Here, it is proposed to introduce a *Kelvin-Voigt* model for the plate that includes non-linear terms rate remodelling terms with respect to strain and the time derivative of strain. As many biomechanists consider cancellous bone to be structure consisting of both rode and plates, adding the study of the plate permitsre a more realistic modeling of the trabeculae of spongy bone. Recent studies have suggested that the dissipation of acoustic energy in cancellous bone interrogation via ultrasound is due to the viscous properties of the trabeculae, rather than that of interstitial blood and marrow. This implies use of a Kelvin-Voigt model for both the plate-like and rod-like trabeculae.

Chapter 14 – The authors investigate some inner bilateral obstacle problems for a class of strongly degenerate parabolic-hyperbolic quasilinear operators associated with homogeneous Dirichlet data in a multidimensional bounded domain. The authors first introduce the concept of an entropy process solution, more convenient and generalizing the notion of an entropy solution. Moreover, the boundary conditions are expressed by using the background of *Divergence Measure Fields*. The authors ensure that proposed definition warrants uniqueness. The existence of an entropy process solution is obtained through the vanishing viscosity and penalization methods.

Chapter 15 - Using the Perov's fixed point theorem and the fiber contractions theorem are obtained sufficient conditions for the smooth dependence by the end points of the solution of the two point boundary value problem for second order ODE's. Finally, an application in the metabolism control is presented.

Chapter 16 - In this paper the authors use variational methods to study the nonexistence of positive solutions for the following doubly nonlinear degenerate parabolic equations on Carnot groups:

$$\begin{cases} \frac{\partial u}{\partial t} = \nabla_G \cdot \left(u^{m-1} |\nabla_G u|^{p-2} \nabla_G u\right) + V u^{m+p-2} & \text{in } \Omega \times (0,T), \\ u(x,0) = u_0(x) \geq 0 & \text{in } \Omega, \\ u(x,t) = 0 & \text{on } \partial\Omega \times (0,T). \end{cases}$$

Here Ω is a bounded domain with smooth boundary in a Carnot group G ∇_G is the horizontal gradient on G, $T > 0$, $V \in L^1_{loc}(\Omega)$, u_0 is not identically zero, $m \in R$, $1 < p < Q$ and $m + p - 2 > 0$.

Chapter 17 - The authors deal with the scalar conservation law in a one dimensional bounded domain Ω: $\partial_t u + \partial x(k(x)\, g(u)) = 0$, associated with a bounded initial value u_0. The function k is supposed to be bounded, discontinuous at $\{x_0 = 0\}$, and with bounded variation. A weak entropy formulation for the Cauchy problem has been introduced by J.D Towers. The existence and the uniqueness is proved by N. Seguin and J. Vovelle through a regularization of the function k. The authors generalize the definition of J.D Towers and the authors adapt the method developed in to establish an existence and uniqueness property in the case of the homogeneous Dirichlet boundary conditions.

Chapter 18 - A new analytical solution to the one-dimensional delayed diffusion equation is derived in terms of the Lambert W-function. A new critical value of the thermal lag time parameter is also noted. The effects of varying this parameter are examined using numerical methods.

Chapter 19 - The authors are concerned with the shape of the level sets of solution of reaction diffusion equations. Using the maximum principle the authors find sufficient conditions for their concavity.

Chapter 20 - In this note the authors present some theorems on the superposition of a function which satisfies the Hölder condition and an (NS^p)-almost periodic (a.p.) function, as well an S^p-a.p. function. Moreover, the authors prove a theorem on the superposition of a continuous function and an (NH)-a.p. function. In the following, the authors give a theorem on the superposition of a differentiable function and a V_p-a.p. function. Finally, the authors prove a theorem on the superposition of a differentiable function and an $(NC^{(1)})$-a.p. function.

Chapter 21 - In this article, the authors study the quasi geostrophic equation, which is a prototypical geophysical fluid model. The authors will show the existence of time-periodic solutions for any the Coriolis parameter and the Ekman dissipation constant with nonhomogeneous boundary conditions.

Chapter 22 - The authors investigate the loop stability conditions of differential systems, that is, the conditions of invariance of the Jacobian spectrum under any variation of entries that leaves unchanged the loop structure. The authors use the dynamical properties of Jacobian Loops described by the products of the matrix entries under cyclic permutations of the indices. It appears that all k-order Feedback Loops given by the union of disjoint simple loops involving k variables must be positive for any asymptotic stable behavior. The authors also conjecture that the Loop structure requires a negative Feedback Loop of the system dimension for the onset of chaotic behavior.

Chapter 23 - In this paper sufficient conditions for oscillation of all solutions are given for the equations

$$\dot{x}(t) + \sum_{i=1}^{n} p_i(t) f(x(t - r_i(t))) = 0,$$

and

$$\dot{x}(t) + \sum_{i=1}^{n} p_i(x(t)) f(x(t - r_i(x(t)))) = 0,$$

where $r_i(\cdot), p_i(\cdot) \in C(\mathbb{R}, \mathbb{R}^+)$.

Chapter 24 - In this paper the authors prove local and global existence results for semilinear differential equations with deviated argument in Banach spaces.

Chapter 25 - Under general conditions on the signed Radon measure μ the authors prove the existence of positive singular solutions for the problem $\Delta u+\mu u^p=0$, $p>1$ on bounded *NTA* domains in $\mathbb{R}^n, n \geq 2$. These results extend the recent ones proved by some authors to more general classes of potentials and domains. A new proof based on a simple fixed point argument is also given.

Chapter 26 - The paper studies the so-called weighted pseudo almost periodic functions recently introduced by the author. Properties of those weighted pseudo almost periodic functions are discussed including a composition result of weighted pseudo almost periodic functions. The obtained results, subsequently, are utilized to study the existence and uniqueness of a weighted pseudo almost periodic solution to some non-autonomous abstract differential equations.

Chapter 27 - The authors present a fixed point theorem for a sum of two convex-valued multi functions acting on a weakly compact, hyper convex subset of a normed space. The theorem is a multivalued version of a result of D. Bugajewski.

All the articles have been previously published in Journal of Evolution Equations, Volume 2 by Nova Science Publishers. It was submitted for appropriate modifications in an effort to encourage wider dissemination of research.

In: Research on Evolution Equation Compendium. Volume 2 ISBN: 978-1-60456-752-6
Editor: Gaston M. N'Guerekata

DOUBLE SCALE CONVERGENCE AND HOMOGENIZATION OF QUASILINEAR PARABOLIC EQUATIONS

Xuming Xie*
Department of Mathematics, Morgan State University
1700 E. Cold Spring Lane, Baltimore, MD 21251

Abstract

We study the homogenization of the nonlinear parabolic equation

$$(u_\varepsilon)_t - \operatorname{div}_x(F(D_x u_\varepsilon, \frac{x}{\varepsilon}, \frac{t}{\varepsilon^2})) = f(t,x)$$

where $F(\xi,y) = F(\xi, y+m, s+k)$ for any integer vectors $k,m \in \mathbf{Z}^n$. We shall use double-scale convergence to derive the homogenized equations and prove a corrector result.

AMS Subject Classification: 35A35
Key Words: Nonlinear parabolic equations; Homogenization; Double-scale convergence; Monotonicity

1 Introduction

The behavior of inhomogeneous material, where the inhomogeneities are on a small scale, is of considerable interest in material sciences, mechanical engineering and in many technological applications, especially those involving composite materials. The theory of homogenization was created to model and predict the behavior of such material, when the inhomogeneities are on such a scale much smaller than the linear dimension of the system. In these situations, continuum models, in which rapid oscillation of the material properties is averaged, are sufficient to describe the behavior of the system. Moreover, they have the advantage of avoiding extensive numerical computations needed when the small scale behavior is taken into account. There have been numerous publications on homogenization theory and its applications in the past decades, we refer to the books by Bensoussan et al [2], Sanchez-Palencia [11] and Jikov et al [7].

For homogenization of nonlinear equations, a standard approach is using energy methods by Tartar and Murat; however, in some cases, it is difficult to work out energy method since the construction of test functions could be tricky. A two scale convergence method

*E-mail address: xxie@morgan.edu

was first introduced by Nguetseng [10], later elaborated by Allaire [1]; the advantage of this method is that the homogenized equations and proof of convergence are obtained in a single step. Recently Caffarelli [3] proposed an alternative approach in which the homogenization be considered as a viscosity process. Following the idea by Caffarelli, Garcia-Azorero et al [5] proved that the uniform limit of the solutions at scale ε to quasilinear parabolic equations is a viscosity solution to the homogenized equation. The use of viscosity solution methods was first considered by Evans [4] for divergence structure PDE.

In this paper, we shall use the method of double-scale convergence to derive the homogerized equation of the quasilinear parabolic equations considered in [5]. Double-scale convergence was introduced in [6] and [8]. It is useful for homogenization of linear parabolic equations when the scale of time variable is different from that of space variable. Since it is a generalization of two scale convergence method, the double convergence method has the same advantage as two scale convergence method and is simpler than viscosity process used in [5]. We obtain that solutions to the microscopic equations double-scale converges to solutions of the same homogenized equations as in [5]; we also obtain a corrector result.

2 Formulation of the Problem

We consider the following nonlinear parabolic equation

$$(u_\varepsilon)_t - \operatorname{div}_x(F(D_x u_\varepsilon, \frac{x}{\varepsilon}, \frac{t}{\varepsilon^2})) = f(t,x) \text{ in } U \tag{2.1}$$

where $F(\xi,y,s)$ is a vector field in $\mathbf{R}^n \times \mathbf{R}^n \times \mathbf{R}$ satisfying the following conditions:
(1) *Periodicity*: $F(\xi,y,s)$ is 1-periodic in y and s, i.e.,

$$F(\xi,y+m,s+k) = F(\xi,y,s)$$

for all $m \in \mathbf{Z}^n, k \in \mathbf{Z}$.
(2) *Monotonicity*:

$$\lambda|\xi-\eta|^2 \leq < F(\xi,y,s) - F(\eta,y,s), \xi-\eta > \leq \Lambda|\xi-\eta|^2 \tag{2.2}$$

for some positive constants λ and Λ which are independent of ε, for all $\xi,\eta \in \mathbf{R}^n$ and for a.e. y and s.
(3) *Growth*: There exists $C > 0$ such that

$$|F(\xi,y,s)| \leq C|\xi| \tag{2.3}$$

where C is positive and independent of ε.
(4) *regularity*: $F(\xi,y,s)$ is continuous in ξ and measurable in y and s.

Remark 2.1. Conditions (1) - (3) are the same as in [5], the regularity condition (4) is the well known Caratheodory condition, which is weaker than the regularity condition required in [5]. In [5], $F(\xi,y,s)$ is required to satisfy Lipschitz condition in ξ for a.e $(y,s) \in \mathbf{R}^n \times \mathbf{R}$.

Lemma 2.2. *For $f(t,x) \in L^\infty([0,T],L^2(U))$, there exists a unique solution $u_\varepsilon \in L^2([0,T],H_0^1(U)) \cap C([0,T],L^2(U)), \partial_t(u_\varepsilon) \in L^2([0,T],H^{-1}(U))$ to equation (2.1) with initial and boundary condition*

$$u_\varepsilon|_{t=0} = 0, \qquad u_\varepsilon|_{\partial U} = 0. \tag{2.4}$$

Furthermore, we have following estimates

$$\|u_\varepsilon\|_{L^2([0,T]\times U)} \leq C, \quad \|\nabla_x u\|_{L^2([0,T]\times U)} \leq C, \tag{2.5}$$

$$\|F(\nabla_x u_\varepsilon, \frac{x}{\varepsilon}, \frac{t}{\varepsilon^2})\|_{L^2([0,T]\times U)} \leq C, \tag{2.6}$$

where C is a positive constant and independent of ε.

Proof. The existence and uniqueness and (2.5) follow from the standard Galerkin Method, see Theorem 30.A of [12] with evolution triple $V = H_0^1(U), H = L^2(U), V^* = H^{-1}(U)$. (2.6) follows from (2.5) and the growth condition (2.3). □

Our goal in this paper is to investigate the behavior of the sequence $u_\varepsilon(t,x)$ of solutions to (2.1) as $\varepsilon \to 0$. We expect that $u_\varepsilon(t,x)$ converges to a function $u(t,x)$ in a double scale sense , which will be explained shortly. We are going to derive the effective equations that the function $u(t,x)$ will satisfy. We are also going to estimate how fast that $u_\varepsilon(t,x)$ will converge to $u(t,x)$, i.e the so called corrector results.

3 Double Scale Convergence and the Main Theorem

To investigate the behavior of the sequence $u_\varepsilon(t,x)$ as $\varepsilon \to 0$, we are going to use the notion of double-scale convergence which is a generalization of two-scale convergence. It was introduced by Holmbom [6] and Goudon and Poupaud [8] to study linear parabolic equations. Let $C_\#^\infty(Y)$ denote the space of those C^∞ functions periodic on Y, where $Y = [0,1] \times [0,1]^n$. Let $dZ = dt\ dx\ ds\ dy$.

Definition 3.1. $\{u^\varepsilon(t,x)\} \subset L^2([0,T] \times \mathbf{U})$ double-scale converges to $u(t,x,s,y) \in L^2([0,T] \times \mathbf{U} \times Y)$ iff for any $\phi(t,x,s,y) \in C^\infty([0,T] \times \mathbf{U}, C_\#^\infty(Y))$,

$$\lim_{\varepsilon \to 0} \int_0^T \int_{\mathbf{U}} u^\varepsilon(t,x) \cdot \phi(t,x,\frac{t}{\varepsilon^2},\frac{x}{\varepsilon}) dx dt = \int_0^T \int_{\mathbf{U}} \int_Y u(t,x,s,y) \cdot \phi(t,x,s,y) dZ.$$

Remark 3.2. The above definition uses C^∞ test functions ϕ; however as pointed out in [1, 6], test function $\phi(t,x,s,y)$ could be any "admissible" test function. Functions in spaces $L^2([0,T] \times \mathbf{U}, C_\#(Y))$, or $L^2_\#(Y, C([0,T] \times \mathbf{U}))$ are admissible.

Recall the space $W_2^1([0,T],H_0^1(U),L^2(U)) = \{u(t,x) : u \in L^2([0,T],H_0^1(U)), \partial_t u \in L^2([0,T],H^{-1}(U))\}$. The following compactness theorem ([6],Theorem 3.1) will be essential:

Theorem 3.3. *(i) If $\{u^{\varepsilon}(t,x)\}$ is a bounded sequence in $L^2([0,T]\times \mathbf{U})$, then there exists $u(t,x,s,y)\in L^2([0,T]\times \mathbf{U}, L^2_{\#}(Y))$ such that a subsequence of $\{u^{\varepsilon}(t,x)\}$ double-scale converges to $u(t,x,s,y)$ in the sense of Definition 3.1.*
(ii) If $\{u^{\varepsilon}(t,x)\}$ is a bounded sequence in $W^1_2([0,T],H^1_0(\mathbf{U}),L^2(\mathbf{U}))$, then there exist $u(t,x)\in L^2([0,T],H^1(\mathbf{U}))$ and $u_1(t,x,s,y)\in L^2([0,T]\times \mathbf{U}, L^2([0,1],H^1_{\#}([0,1]^n)))$ such that a subsequence of $\{u^{\varepsilon}(t,x)\}$ double-scale converges to $u(t,x)$ and a subsequence of $\nabla_x u^{\varepsilon}$ double-scale converges to $\nabla_x u+\nabla_y u_1$ in the sense of Definition 3.1.

Proof. See Theorem 3.1 in [6]. □

Remark 3.4. In [8], double-scale convergence is defined for a sequence of measures; a compact theorem is also obtained (see Proposition 4.4 in [8]).

The lemma below is essential in the proof of the main homogenization theorem.

Lemma 3.5. *Assume that $\{u^{\varepsilon}\}$ is a bounded sequence in $W^1_2([0,T];H^1_0(U),L^2(U))$, and $u(t,x)$ and $u_1(t,x,s,y)$ are as in Theorem 3.3, then*

$$\lim_{\varepsilon\to 0}\int_0^T\int_U \frac{1}{\varepsilon}(u^{\varepsilon}(t,x)-u(t,x))\psi(t,x,\frac{t}{\varepsilon^2},\frac{x}{\varepsilon})dt\,dx=\int_0^T\int_U\int_Y u_1(t,x,s,y)\psi(t,x,s,y)dZ \tag{3.1}$$

for any $\psi(t,x,s,y)=b(t)c(x)p(y)q(s)$, where $b(t)\in C_0^{\infty}([0,T]), c(x)\in C_0^{\infty}(U), p(y)\in L^2_{\#}([0,1]^n), q(s)\in L^2_{\#}([0,1])$.

Proof. See Corollary 3.3 in [6]. □

Now we are ready to prove the following main theorem:

Theorem 3.6. *The solution sequence $u_{\varepsilon}(t,x)$ as in Lemma 2.2 double-scale converges to a function $u(t,x)$ in $L^2([0,T],H^1_0(U))$, and $\nabla_x u_{\varepsilon}$ double-scale converges to $\nabla_x u(t,x)+\nabla_y u_1(t,x,s,y)$ where (u,u_1) is the unique solution in $L^2([0,T],H^1_0(U))\times L^2\left([0,T]\times U; L^2([0,1],H^1_{\#}([0,1]^n))\right)$ of the homogenized problem*

$$u_t-\, div_x\left[\int_Y F(\nabla_x u(t,x)+\nabla_y u_1,y,s)dyds\right]=f \quad in\ U, \tag{3.2}$$

$$(u_1)_s-\, div_y\, F(\nabla_x u(t,x)+\nabla_y u_1,y,s)=0 \quad in\ U\times Y, \tag{3.3}$$

$$u|_{t=0}=0, u|_{\partial U}=0 \tag{3.4}$$

$$u_1(t,x,s,y)\ Y\text{-periodic in } (s,y). \tag{3.5}$$

Proof. The varational formulation of (2.1) is following

$$\int_U u_{\varepsilon}(T,x)\phi(T,x)dx-\int_0^T\int_U u_{\varepsilon}\partial_t\phi\,dx\,dt + \int_0^T\int_U F(\nabla u_{\varepsilon},\frac{x}{\varepsilon},\frac{t}{\varepsilon^2})\cdot\nabla\phi\,dx\,dt=\int_0^T\int_U f\phi\,dx\,dt \tag{3.6}$$

for $\phi(t,x)\in W^1_2([0,T],H^1_0(U),L^2(U))$.

From Lemma 2.2 and Theorem 3.3, up to a subsequence, there are $u(t,x), u_1(t,x,s,y)$ and $F_0(t,x,y,s)$ such that

$$\begin{aligned}
&u_\varepsilon(t,x) \text{ double-scale converges to } u(t,x)\\
&\nabla_x u_\varepsilon(t,x) \text{ double-scale converges to } \nabla_x u(t,x)+\nabla_y u_1(t,x,s,y)\\
&F(\nabla_x u_\varepsilon, \frac{x}{\varepsilon}, \frac{t}{\varepsilon^2}) \text{ double-scale converges to } F_0(t,x,y,s).
\end{aligned} \tag{3.7}$$

Let

$$\phi(t,x) = b(t)\psi_1(x) + \varepsilon b(t)\phi_1(x, \frac{x}{\varepsilon}, \frac{t}{\varepsilon^2}), \tag{3.8}$$

where $\phi_1(x,y,s) = c(x)p(y)q(s), b(t) \in C_0^\infty([0,T]), \psi_1(x), c(x) \in H_0^1(U)), p(y) \in C_\#^\infty([0,1]^n)$, $q(s) \in C_\#^\infty([0,1])$.

Plugging (3.8) into (3.6)

$$\begin{aligned}
-\int_0^T \int_U u_\varepsilon \partial_t [b(t)\psi_1(x) + \varepsilon b(t)\phi_1(x, \frac{x}{\varepsilon}, \frac{t}{\varepsilon^2})] dxdt\\
+\int_0^T \int_U F(\nabla u_\varepsilon, \frac{x}{\varepsilon}, \frac{t}{\varepsilon^2}) \cdot [b(t)\nabla_x \psi_1 + b(t)\nabla_y \phi_1(x, \frac{x}{\varepsilon}, \frac{t}{\varepsilon^2})] dx\, dt\\
= \int_0^T \int_U f[b(t)\psi_1(x) + \varepsilon b(t)\phi_1(x, \frac{x}{\varepsilon}, \frac{t}{\varepsilon^2})] dx\, dt.
\end{aligned} \tag{3.9}$$

Note

$$\begin{aligned}
&\int_0^T \int_U u_\varepsilon \partial_t [\varepsilon b(t)\phi_1(x, \frac{x}{\varepsilon}, \frac{t}{\varepsilon^2})] dxdt\\
&= \int_0^T \int_U \frac{1}{\varepsilon} u_\varepsilon b(t)(\phi_1)_s(x, \frac{x}{\varepsilon}, \frac{t}{\varepsilon^2}) dxdt + \int_0^T \int_U \varepsilon u_\varepsilon b'(t)\phi_1(x, \frac{x}{\varepsilon}, \frac{t}{\varepsilon^2}) dxdt\\
&= \int_0^T \int_U \frac{1}{\varepsilon}(u_\varepsilon - u) b(t)(\phi_1)_s(x, \frac{x}{\varepsilon}, \frac{t}{\varepsilon^2}) dxdt + \int_0^T \int_U \frac{1}{\varepsilon} u(x,t) b(t)(\phi_1)_s(x, \frac{x}{\varepsilon}, \frac{t}{\varepsilon^2}) dxdt\\
&\quad + \int_0^T \int_U \varepsilon u_\varepsilon b'(t)\phi_1(x, \frac{x}{\varepsilon}, \frac{t}{\varepsilon^2}) dxdt\\
&= \int_0^T \int_U \frac{1}{\varepsilon}(u_\varepsilon - u) b(t)(\phi_1)_s(x, \frac{x}{\varepsilon}, \frac{t}{\varepsilon^2}) dxdt + \int_0^T \int_U \varepsilon \partial_t [b(t)u(x,t)]\phi_1(x, \frac{x}{\varepsilon}, \frac{t}{\varepsilon^2}) dxdt\\
&\quad + \int_0^T \int_U \varepsilon u_\varepsilon b'(t)\phi_1(x, \frac{x}{\varepsilon}, \frac{t}{\varepsilon^2}) dxdt.
\end{aligned} \tag{3.10}$$

Taking limit $\varepsilon \to 0$ and using Lemma 3.5, we have

$$\begin{aligned}
\lim_{\varepsilon\to 0} \int_0^T \int_U u_\varepsilon \partial_t [\varepsilon b(t)\phi_1(x, \frac{x}{\varepsilon}, \frac{t}{\varepsilon^2})] dxdt &= \lim_{\varepsilon\to 0} \int_0^T \int_U \frac{1}{\varepsilon}(u_\varepsilon - u) b(t)(\phi_1)_s(x, \frac{x}{\varepsilon}, \frac{t}{\varepsilon^2}) dxdt\\
&= \int_0^T \int_U u_1(x,t,y,s) b(t)(\phi_1)_s(x,y,s) dZ.
\end{aligned} \tag{3.11}$$

Taking double-scale limit in (3.9) and using (3.11), we have

$$-\int_0^T\int_U u\partial_t b\psi_1(x) - \int_0^T\int_U u_1(x,t,y,s)b(t)(\phi_1)_s(x,y,s)dZ$$
$$+\int_0^T\int_U\int_Y F_0(t,x,y,s)\cdot(b(t)\nabla\psi_1 + b(t)\nabla_y\phi_1(t,x,y,s))dZ = \int_0^T\int_U fb(t)\psi_1(t,x)dtdx. \tag{3.12}$$

In above equation, let $\phi_1 = 0$, we obtain:

$$-\int_0^T\int_U u\partial_t b\psi_1(x) + \int_0^T\int_U\int_Y b(t)F_0(t,x,y,s)\cdot\nabla\psi_1 dZ = \int_0^T\int_U fb(t)\psi_1(t,x)dtdx. \tag{3.13}$$

Letting $\psi_1 = 0$ in (3.12), we have

$$-\int_0^T\int_U u_1(x,t,y,s)b(t)(\phi_1)_s(x,y,s)dZ + \int_0^T\int_U\int_Y b(t)F_0(t,x,y,s)\cdot\nabla_y\phi_1(x,y,s)dZ = 0. \tag{3.14}$$

Now the question is to identify $F_0(t,x,y,s)$ in terms of u, u_1 and F. To this end, for $\tau > 0$, $\Phi(t,x,y) \in C_0^\infty(U\times Y\times[0,T])$ and functions $u_1^k(t,x,s,y) \in C_0^\infty(U\times Y\times[0,T])$ such that

$$u_1^k(t,x,s,y) \to u_1(t,x,s,y) \text{ strongly in } L^2([0,T]\times U, L^2([0,1], H^1_\#([0,1]^n))), \tag{3.15}$$

let

$$v_\varepsilon^k(t,x) = \nabla_x[u(t,x) + \varepsilon u_1^k(t,x,\frac{x}{\varepsilon},\frac{t}{\varepsilon^2})] + \tau\Phi(t,x,\frac{x}{\varepsilon}).$$

Then

$$v_\varepsilon^k(t,x) \text{ double-scale converges to } \nabla_x u(t,x) + \nabla_y u_1^k(t,x,y,s) + \tau\Phi(t,x,y) \equiv v^k(t,x,y). \tag{3.16}$$

From monotonicity condition

$$\int_0^T\int_U \left[F(\nabla u_\varepsilon,\frac{x}{\varepsilon},\frac{t}{\varepsilon^2}) - F(v_\varepsilon^k,\frac{x}{\varepsilon},\frac{t}{\varepsilon^2})\right]\cdot(\nabla_x u_\varepsilon - v_\varepsilon^k) \geq 0. \tag{3.17}$$

Using equation (3.6) with $\phi = u_\varepsilon$,

$$\int_0^T\int_U F(\nabla u_\varepsilon,\frac{x}{\varepsilon},\frac{t}{\varepsilon^2})\cdot\nabla u_\varepsilon dxdt = -\frac{1}{2}\int_U u_\varepsilon^2(T)dx + \int_0^T\int_U fu_\varepsilon dx\, dt. \tag{3.18}$$

Combining (3.17) and (3.18),

$$-\frac{1}{2}\int_U u_\varepsilon^2(T,x)\,dx + \int_0^T\int_U fu_\varepsilon\,dxdt - \int_0^T\int_U F(v_\varepsilon^k,\frac{x}{\varepsilon},\frac{t}{\varepsilon^2})\cdot\nabla u_\varepsilon\,dxdt$$
$$-\int_0^T\int_U F(\nabla u_\varepsilon,\frac{x}{\varepsilon},\frac{t}{\varepsilon^2})\cdot v_\varepsilon^k\,dxdt + \int_0^T\int_U F(v_\varepsilon^k,\frac{x}{\varepsilon},\frac{t}{\varepsilon^2})\cdot v_\varepsilon^k\,dxdt \geq 0. \tag{3.19}$$

Since $u_\varepsilon(T,x)$, up to a subsequence, weakly converges to $u(T,x)$ in $L^2(U)$, it follows that

$$\lim_{\varepsilon\to 0}\int_U u_\varepsilon^2(T,x)\,dx \geq \int_U u^2(T,x)\,dx. \tag{3.20}$$

Taking the double scale limit in (3.19) and using (3.20), we have

$$-\frac{1}{2}\int_U u^2(T,x)\,dx+\int_0^T\int_U fu\,dxdt-\int_0^T\int_U\int_Y F(v^k,y,s)\cdot[\nabla_x u+\nabla_y u_1]\,dZ$$
$$-\int_0^T\int_U\int_Y F_0(t,x,y,s)\cdot v^k\,dZ+\int_0^T\int_U\int_Y F(v^k,y,s)\cdot v^k\,dZ\geq 0. \quad (3.21)$$

By growth, regularity conditions and Lebesgue's dominated convergence theorem, and (3.15)

$$\lim_{k\to\infty}\int_0^T\int_U\int_Y F(v^k,y,s)\cdot[\nabla_x u+\nabla_y u_1]dZ$$
$$=\int_0^T\int_U\int_Y F(\nabla_x u+\nabla_y u_1+\tau\Phi,y,s)\cdot[\nabla_x u+\nabla_y u_1]\,dZ, \quad (3.22)$$

$$\lim_{k\to\infty}\int_0^T\int_U\int_Y F(v^k,y,s)\cdot v^k\,dZ$$
$$=\int_0^T\int_U\int_Y F(\nabla u+\nabla_y u_1+\tau\Phi,y,s)\cdot[\nabla u+\nabla_y u_1+s\Phi]\,dZ, \quad (3.23)$$

$$\lim_{k\to\infty}\int_0^T\int_U\int_Y F_0(t,x,y,s)\cdot v^k dZ$$
$$=\int_0^T\int_U\int_Y F_0(t,x,y)\cdot[\nabla u+\nabla_y u_1+\tau\Phi]\,dZ. \quad (3.24)$$

Leting $k\to\infty$ in (3.21) and using (3.22)-(3.24), we have

$$\left[-\frac{1}{2}\int_U u^2(T)dx+\int_0^T\int_U fu\,dxdt-\int_0^T\int_U\int_Y F_0(t,x,y,s)\cdot\nabla_x u\,dZ\right]$$
$$-\int_0^T\int_U\int_Y F_0(t,x,y,s)\cdot\nabla_y u_1\,dZ$$
$$+\tau\int_0^T\int_U\int_Y[F(\nabla u+\nabla_y u_1+\tau\Phi,y,s)-F_0(t,x,y,s)]\cdot\Phi\,dZ\geq 0. \quad (3.25)$$

Replacing ϕ in (3.6) by $u(t,x)$, we have

$$-\int_U u_\varepsilon(T,x)\,u(T,x)dx+\int_0^T\int_U u_\varepsilon\partial_t u\,dxdt+\int_0^T\int_U fu\,dxdt$$
$$-\int_0^T\int_U F(\nabla u_\varepsilon,\frac{x}{\varepsilon},\frac{t}{\varepsilon^2})\cdot\nabla_x u\,dydxdt=0. \quad (3.26)$$

Taking the double-scale limit in the above

$$-\int_U u(T,x)u(T,x)dx+\int_0^T\int_U u\partial_t u\,dxdt+\int_0^T\int_U fudxdt-\int_0^T\int_U\int_Y F_0(t,x,y,s)\cdot\nabla_x udZ=0, \quad (3.27)$$

which is equivalent to

$$-\frac{1}{2}\int_U u^2(T)dx+\int_0^T\int_U fu\,dxdt-\int_0^T\int_U\int_Y F_0(t,x,y,s)\cdot\nabla_x u\,dZ=0. \tag{3.28}$$

Consider a sequence of functions of $b(t)\phi_1(x,y,s)$ that converges strongly to $u_1(t,x,s,y)$ in $L^2([0,T]\times U,L^2([0,1],H^1_\#([0,1]^n)))$. Passing limit in (3.14), we have

$$\int_0^T\int_U\int_Y F_0(t,x,y,s)\cdot\nabla_y u_1 dZ=\int_0^T\int_U\int_{Y^{n+1}} u_1\partial_s u_1 dZ=0. \tag{3.29}$$

From (3.28) and above , (3.25) becomes

$$\int_0^T\int_U\int_Y [F(\nabla u+\nabla_y u_1+\tau\Phi,y,s)-F_0(t,x,y,s)]\cdot\Phi\,dZ\geq 0. \tag{3.30}$$

Letting $\tau\to 0$ we have

$$\int_0^T\int_U\int_Y [F(\nabla u+\nabla_y u_1,y,s)-F_0(t,x,y,s)]\cdot\Phi\,dZ\geq 0, \tag{3.31}$$

which implies

$$F_0(t,x,y,s)=F(\nabla_x u+\nabla_y u_1,y,s). \tag{3.32}$$

Finally, (3.2) follows from (3.13) and (3.32); and (3.3) follows from (3.14) and (3.32). □

Remark 3.7. Our homogenized equation (3.2) is the same as (II.7) in [5], (3.3) is the same as (II.3) in [5].

Theorem 3.8. *Assume that* $u_1(t,x,s,y)\in L^2([0,T]\times U,C^1_\#(Y))$*, then*

$$\lim_{\varepsilon\to 0}\|u_\varepsilon(t,x)-u(t,x)-\varepsilon u_1(x,t,\frac{t}{\varepsilon^2},\frac{x}{\varepsilon})\|_{L^2([0,T],H^1(U))}=0.$$

Proof. We consider

$$v_\varepsilon=\nabla_x\left[u(t,x)+\varepsilon u_1(x,t,\frac{x}{\varepsilon},\frac{t}{\varepsilon^2})\right],$$

which double-scale converges to $\nabla u+\nabla_y u_1$.

By monotonicity condition:

$$\int_0^T\int_U\left[F(\nabla u_\varepsilon,\frac{x}{\varepsilon},\frac{t}{\varepsilon^2})-F(v_\varepsilon,\frac{x}{\varepsilon},\frac{t}{\varepsilon^2})\right]\cdot(\nabla u_\varepsilon-v_\varepsilon)dtdx\geq\lambda\int_0^T\int_U|\nabla u_\varepsilon-v_\varepsilon|^2dxdt. \tag{3.33}$$

As in the proof of Theorem 3.6 , the left hand side of (3.33) goes to zero, which implies that the sequence $\nabla_x\left[u_\varepsilon-u(t,x)-\varepsilon u_1(x,t,\frac{x}{\varepsilon},\frac{t}{\varepsilon^2})\right]$ converges to zero in $\left[L^2([0,T]\times U)\right]^n$. □

References

[1] G. Allaire, *Homogenization and two-scale convergence,* SIAM *J. Math. Anal.* **23.6** (1992) 1482-1518.

[2] A. Bensoussan, J.L. Lions, G. Papanicolau. *Asymptotic Analysis for Periodic Structures.* (1978) Amsterdam, North Holland.

[3] Caffarelli, L. A. A note on nonlinear homogenization. *Comm. Pure. Appl. Math.* LII (1999) 829-838.

[4] L.C. Evans. Periodic homogenization of certain fully nonlinear partial differential equations, *Proc. Roy. Soc. Edinbeugh sect. A*, **120** (1992) 245-265.

[5] J. Garcia-Azorero, C.E. Gutierrez, I. Peral, Homogenization of quasilinear parabolic equations in periodic media, *Comm. Partial Diff. Equa.* vol 28 (2003) 1887-1910.

[6] A. Holmbom, Homogenization of parabolic equations an alternative approach and some corrector-type results. *Applications of Mathematics* **42**(1997) 321-342.

[7] V. Jikov, S. Koslov, O. Oleinik, *Homogenization of Differential Operators and Integral Functions*, (1994) Berlin, Springer.

[8] T. Goudon, F. Poupaud, Homogenization of transport equations: weak mean field approximation. SIAM *J Math. Anal.* **36**(2004) 856-881.

[9] F. Murat, L. Tartar. H-convergence, topics in the mathematical modeling of composite materials, L. Cherkaev, R.V. Kohn ed., *Progress in nonlinear differential equations and their applications*, Birkhauser, Boston, 1998, 21-43.

[10] G, Nguetseng, *A general convergence result for a functional related to the theory of homogenization,* SIAM *J. Math. Anal.* **20** (1989) 608-623.

[11] E. Sanchez-Palencia, Non-homogenous Media and Vibration theory. *Lecture notes in physics* **127** Berlin, Springer.

[12] E. Zeidler, *Nonlinear Functional Analysis and Its Application,* IIB (1990), Berlin, Springer.

[13] X. Xie, *A note on homogenization of quasilinear parabolic equations,* preprint, 2004.

In: Research on Evolution Equation Compendium. Volume 2 ISBN: 978-1-60456-752-6
Editor: Gaston M. N'Guerekata

SECOND GRADE FLUIDS WITH ENHANCED VISCOSITY AS DYNAMICAL SYSTEMS

Ahmed S. Bonfoh*
The Abdus Salam International Centre for Theoretical Physics,
Mathematics Section, Strada Costiera 11, 34014 Trieste, Italy

Abstract

We consider a modified $2d$ second grade incompressible fluid with a higher order viscosity term, and show that this nonlinear evolution system possesses an exponential attractor. This object gives a more precise description of the asymptotic behaviour of dynamical systems than the global attractor.

Keywords: modified second grade fluids, global attractor, exponential attractors.
2000 AMS MS Classifications: 35A05, 35B45, 35B40

1 Introduction

The study of the longtime behaviour of solutions to nonlinear evolution equations arising from mechanics and physics is a capital issue, as it is important, to understand and predict their evolution. Many dissipative systems possess a global attractor, which is a compact set lying in the phase space, and which attracts uniformly the trajectories starting from bounded sets when time goes to infinity. In other words, the global attractor is the unique largest bounded invariant set which embodies all the permanent regimes corresponding to all possible initial data. However, the global attractor may present some major defaults for practical purposes. Indeed, the rate of attraction of the trajectories may be small and very sensitive to perturbations. In order to give a more precise description of the longtime behaviour of such systems, C. Foias *et al.* have introduced a new object called an exponential attractor. This object is a compact and positively invariant set which contains the global attractor, has finite fractal dimension and attracts all the trajectories starting from bounded sets at a uniform exponential rate. Moreover, exponential attractors are more robust than global attractors with respect to approximations and perturbations (see [6]). In this paper, we consider a modified $2d$ fluid of second grade and investigate the existence of exponential attractors.

*E-mail address: abonfoh@ictp.it, Phone: +390402240220, Fax: +39040224163

2 Setting of the Problem

We consider a modified second grade incompressible fluid in two-dimensional space by adding the higher order viscosity term $\mu\Delta^2 u$, that is,

$$\begin{aligned} \frac{\partial}{\partial t}(u-\alpha\Delta u)+\mu\Delta^2 u-\nu\Delta u+\operatorname{curl}(u-\alpha\Delta u)\times u &= f+\nabla p, \\ \operatorname{div} u &= 0, \\ u|_{\partial\Omega} &= 0, \\ u|_{t=0} &= u_0. \end{aligned} \tag{2.1}$$

The function $u=u(x,t)$ is the velocity, $p=p(x,t)$ is the pressure, the density of the fluid is supposed to be $\rho=1$, $f=f(x)$ is the given external body force, and the parameters α, μ and ν are given strictly positive constants (ν is the kinematic viscosity). The case $\mu=0$ corresponds to the usual fluids of second grade, which has been studied in [3], [4], [5] and in the references therein. For both cases $\mu=0$ and $\alpha=0$, (2.1) reduces to the Navier-Stokes equations. Let us remark that a superviscosity term is often added for the numerical solution of the turbulent Navier-Stokes equations (see [7]).

We assume that the fluid fills a simply-connected bounded open set Ω, with smooth and connected boundary $\partial\Omega$ of class $\mathcal{C}^{4,1}$. We denote by $\mathbf{n}$ the unit normal vector to $\partial\Omega$, directed outside Ω.

We denote by $\|.\|$ and $(.,.)$ the usual norm and scalar product in $L^2(\Omega)$ (and also in $L^2(\Omega)^2$). We define the following functional sets: $\mathcal{V}=\{u\in\mathcal{C}_0^\infty(\Omega)^2,\ \operatorname{div} u=0\}$, $H=$ the closure of $\mathcal{V}$ in $L^2(\Omega)^2$, and $V=$ the closure of $\mathcal{V}$ in $H_0^1(\Omega)^2$. V is a Hilbert space with the scalar product $(u,v)_V=(u,v)+\alpha(\nabla u,\nabla v)$. We also define the Hilbert space $W=\{u\in V,\ \operatorname{curl}(u-\alpha\Delta u)\in L^2(\Omega)\}$ endowed with the scalar product $(u,v)_W=(u,v)_V+(\operatorname{curl}(u-\alpha\Delta u),\operatorname{curl}(v-\alpha\Delta v))$. Note that, in two spatial dimension, the operator curl is a scalar. That is, for $u=(u_1,u_2)$, we have $\operatorname{curl} u=\frac{\partial u_2}{\partial x_1}-\frac{\partial u_1}{\partial x_2}$ and $\times u=(-\frac{\partial u_2}{\partial x_1},\frac{\partial u_1}{\partial x_2})$. It has been shown in [3] that $W=\{u\in[H^3(\Omega)\cap H_0^1(\Omega)]^2,\ \operatorname{div} u=0\}$ and also the norms $\|q\|_{H^3(\Omega)^2}$ and $\|\operatorname{curl}(q-\alpha\Delta q)\|$ are equivalent on W.

Throughout this paper, the same letter c (and sometimes c_i, $i=0,1,2,\ldots$) denotes positive constants which may change from line to line.

3 Existence of the Global Attractor

The existence and uniqueness of solutions of problem (2.1) with $\mu=0$ have been investigated, for instance, in [3] and [4]. These results allowed I. Moise *et al.* to consider the semigroup $\{S(t)\}_{t\geq 0}$ of operators $S(t):W\to W$, $u_0\mapsto u(t)$, $u(t)$ being the solution at time t of (2.1) with $\mu=0$. They proved that this semigroup is asymptotically compact (it is not compact), weakly continuous and possesses the global attractor in W via energy equations (cf. [9]). Now, when $\mu>0$, we can show that the semigroup $\{S(t)\}_{t\geq 0}$ generated by (2.1) is uniformly compact, continuous, Lipschitz on bounded sets of $H^4(\Omega)^2$, and possesses the global attractor in W, which is bounded in $H^4(\Omega)^2$ if $f\in H^2(\Omega)^2$ (see, e.g., [1] and [10] for more details about the theory of global attractors).

We state the following results.

Theorem 3.1 *If $u_0 \in W$ and $f \in H^1(\Omega)^2$, then there exists a unique solution u to problem (2.1) satisfying $u \in L^\infty(\mathbb{R}^+;W) \cap L^2(0,T;W \cap H^4(\Omega)^2)$, with $\frac{\partial u}{\partial t} \in L^\infty(\mathbb{R}^+;V)$, for any $T > 0$.*

Theorem 3.2 *We assume that $f \in H^1(\Omega)^2$. Then, the semigroup $\{S(t)\}_{t\geq 0}$ possesses the global attractor $\mathcal{A}$ in W. Furthermore, if $f \in H^2(\Omega)^2$, then the attractor $\mathcal{A}$ is bounded in $H^4(\Omega)^2$.*

The proofs of Theorem 3.1 and 3.2 are based on showing, firstly, that $\operatorname{curl}(u - \alpha\Delta u) \in L^\infty(\mathbb{R}^+;L^2(\Omega)) \cap L^2(0,T;H^1(\Omega))$, and then using the equivalence of norms. We can note that the norms $\|\operatorname{curl}(u-\alpha\Delta u)\|_{H^1(\Omega)}$ and $\|u\|_{H^4(\Omega)^2}$ are also equivalent on $W \cap H^4(\Omega)^2$. Here, a new property is the boundedness of the attractor $\mathcal{A}$ in $H^4(\Omega)^2$. This result comes from the existence of a bounded absorbing set B_2 for $\{S(t)\}_{t\geq 0}$ in $H^4(\Omega)^2$. This is a consequence of showing that, for any u satisfying Theorem 3.1, $\operatorname{curl}(u - \alpha\Delta u)$ belongs to a bounded absorbing set of $H^1(\Omega)$, for all time $t \geq t_2$, for some $t_2 > 0$. Let us give a proof. We set $b = \operatorname{curl}(u - \alpha\Delta u)$ and remark that $\operatorname{curl}(b \times u) = u.\nabla b$, for all $u \in W \cap H^4(\Omega)^2$. If u is the solution of (2.1) satisfying Theorem 3.1, then b is the unique function in $L^\infty(\mathbb{R}^+;L^2(\Omega)) \cap L^2(0,T;H^1(\Omega))$, for all $T > 0$, solution of the following system (see [4]):

$$\begin{aligned} \frac{\partial b}{\partial t} - \frac{\mu}{\alpha}\Delta b + u.\nabla b + (\tfrac{\mu}{\alpha} - \nu)\Delta \operatorname{curl} u &= \operatorname{curl} f, \\ b|_{t=0} &= \operatorname{curl}(u_0 - \alpha\Delta u_0). \end{aligned} \tag{3.1}$$

We consider a special basis $\{w_j\}_{j\geq 1}$ of $H^1(\Omega)$, which is orthogonal in $L^2(\Omega)$, where w_j, $j \in \mathbb{N}^\star$, is the solution of the homogeneous Neumann problem:

$$\begin{aligned} w_j - \frac{\mu}{\alpha}\Delta w_j &= \lambda_j w_j, \\ \frac{\partial w_j}{\partial \mathbf{n}} &= 0, \end{aligned} \tag{3.2}$$

and where $0 < \lambda_1 \leq \lambda_2 \leq \lambda_3 \ldots$, $\lambda_j \to +\infty$ for $j \to +\infty$, $\{\lambda_j\}_{j\geq 1}$ being the family of eigenvalues associated with the family of eigenfunctions $\{w_j\}_{j\geq 1}$. We can note that w_j is also the solution of the following variational problem:

$$(w_j, v) + \frac{\mu}{\alpha}(\nabla w_j, \nabla v) = \lambda_j (w_j, v), \quad \forall v \in H^1(\Omega). \tag{3.3}$$

Since $\partial\Omega$ is regular enough, w_j belongs to $H^2(\Omega)$, and, the norms $\|w - \frac{\mu}{\alpha}\Delta w\|$ and $\|w\|_{H^2(\Omega)}$ are equivalent on $H^2(\Omega)$.

We now implement a Galerkin method with the basis $\{w_j\}_{j\geq 1}$. Noting that $(u.\nabla b, b) = 0$ (see [3]), we formally obtain

$$\begin{aligned} &\frac{1}{2}\frac{d}{dt}(\|b\|^2 + \tfrac{\mu}{\alpha}\|\nabla b\|^2) + \|b - \tfrac{\mu}{\alpha}\Delta b\|^2 + \tfrac{\mu}{\alpha}(\nabla(u.\nabla b), \nabla b) \\ &\quad + (\tfrac{\mu}{\alpha} - \nu)[\tfrac{\mu}{\alpha}(\nabla\Delta \operatorname{curl} u, \nabla b) + (\Delta \operatorname{curl} u, b)] \\ &= \|b\|^2 + \tfrac{\mu}{\alpha}\|\nabla b\|^2 + \tfrac{\mu}{\alpha}(\nabla \operatorname{curl} f, \nabla b) + (\operatorname{curl} f, b). \end{aligned} \tag{3.4}$$

Therefore,

$$\begin{aligned} &\frac{d}{dt}(\|b\|^2 + \tfrac{\mu}{\alpha}\|\nabla b\|^2) + c_1\|b\|^2_{H^2(\Omega)} \leq c_2(\|b\|^2 + \tfrac{\mu}{\alpha}\|\nabla b\|^2) \\ &\quad + \tau\|\nabla u\|^2_{L^\infty(\Omega)}\|b\|^2_{H^2(\Omega)} + c_3\|u\|^2_{H^4(\Omega)^2} + c_4\|f\|^2_{H^2(\Omega)^2}, \end{aligned} \tag{3.5}$$

for any $\tau > 0$. If $u_0 \in W$ and $f \in H^1(\Omega)^2$, then u satisfies Theorem 3.1, and $\|u(t)\|_{L^\infty(\Omega)} \leq c_1\|\nabla u(t)\|_{L^\infty(\Omega)} \leq c_2\|u(t)\|_W \leq c$, $\forall t \geq 0$. Using this latter fact together with a proper choice of τ, equation (3.5) reduces to

$$\begin{aligned} &\tfrac{d}{dt}(\|b\|^2 + \tfrac{\mu}{\alpha}\|\nabla b\|^2) + c_1\|b\|^2_{H^2(\Omega)} \\ &\leq c_2(\|b\|^2 + \tfrac{\mu}{\alpha}\|\nabla b\|^2) + c_3\|f\|^2_{H^2(\Omega)^2}. \end{aligned} \tag{3.6}$$

If $f \in H^2(\Omega)^2$, then we deduce the existence of a bounded absorbing set B_2 for $\{S(t)\}_{t\geq 0}$ in $H^4(\Omega)^2$, by applying the uniform Gronwall lemma to (3.6), and using the equivalence of norms.

4 Existence of Exponential Attractors

Let E be a metric space, X a compact subset of E and consider a continuous semigroup $\{S(t)\}_{t\geq 0}$, on E, mapping X into X.

We recall the definition of an exponential attractor for $\{S(t)\}_{t\geq 0}$.

Definition 4.1 *A compact set $\mathcal{M}$ is called an exponential attractor for $\{S(t)\}_{t\geq 0}$ for the topology of E if:*
- $\mathcal{M}$ contains the global attractor; that is, $\mathcal{A} \subset \mathcal{M}$;
- $\mathcal{M}$ is positively invariant under $S(t)$; that is, $S(t)\mathcal{M} \subset \mathcal{M}$, $\forall t \geq 0$;
- the fractal dimension of $\mathcal{M}$ is finite;
- there exists a constant $c_0 > 0$ such that, for every bounded subset $B \subset X$, there exists a constant $c_1(B) > 0$ such that

$$\mathrm{dist}_E(S(t)B, \mathcal{M}) \leq c_1 e^{-c_0 t}, \quad \forall t \geq 0,$$

where dist_E is the Hausdorff semi-distance with respect to the metric of E:

$$\mathrm{dist}_E(A,B) = \sup_{a\in A}\inf_{b\in B}\|a-b\|_E.$$

Sufficient conditions ensuring the existence of exponential attractors in Hilbert spaces are given in [6]. It depends on a dichotomy principle called the squeezing property.

Definition 4.2 *The semigroup $\{S(t)\}_{t\geq 0}$ verifies the squeezing property on X if, for a real number η belonging to $[0, \frac{1}{4}[$, there exists a projection $P_{N^\star} : E \to E$, with finite rank $N^\star(\eta)$, and a time $t^\star$ such that*

$$\begin{aligned} &\forall\, (\phi,\psi) \in X^2, \\ \text{if} \quad &\|(I - P_{N^\star})(S(t^\star)\phi - S(t^\star)\psi)\|_E \geq \|P_{N^\star}(S(t^\star)\phi - S(t^\star)\psi)\|_E, \\ \text{then} \quad &\|S(t^\star)\phi - S(t^\star)\psi\|_E \leq \eta\|\phi - \psi\|_E. \end{aligned} \tag{4.1}$$

Proposition 4.1 *If $\{S(t)\}_{t\geq 0}$ satisfies the squeezing property on X and if $S(t^\star)$ is Lipschitz on X with Lipschitz constant L, then there exists an exponential attractor $\mathcal{M}$ for $\{S(t)\}_{t\geq 0}$ on X such that the fractal dimension of $\mathcal{M}$ is bounded as follows:*

$$d_F(\mathcal{M}) \leq N^\star \max(1, \tfrac{\ln(16L+1)}{\ln 2}) \; . \tag{4.2}$$

In the order to verify the squeezing property, Babin and Nicolaenko introduced in [2] a method based on a decomposition of the difference of two trajectories. A consequence of their result is stated as follows (see [8]).

Proposition 4.2 *Let E and V be two Hilbert spaces such that the inclusion $V \subset E$ is compact. Let $X \subset E$ be a closed set and $S(t): X \to X$ be a semigroup. Let us furthermore assume that there exists a projection $P_{N^\star}$ with finite rank $N^\star$ such that $\|(I-P_{N^\star})y\|_E \leq c(N^\star)\|y\|_V$, $\forall y \in V$, where $c(N^\star) \to 0$ as $N^\star \to +\infty$. If there exists φ^1, φ^2 satisfying $S(t)\phi - S(t)\psi = \varphi^1(t) + \varphi^2(t)$ and $\|\varphi^1(t)\|_E^2 \leq d(t)\|\phi-\psi\|_E^2$, $\|\varphi^2(t)\|_V^2 \leq h(t)\|\phi-\psi\|_E^2$, $\forall \phi, \psi \in X$, where $d(t)$ is continuous and satisfies $\lim_{t\to+\infty} d(t) = 0$ and $h(t)$ is continuous, then $\{S(t)\}_{t\geq 0}$ enjoys the squeezing property on X for the topology of E.*

We now want to apply this result to show the existence of an exponential attractor on the set $X = \overline{\bigcup_{t\geq t_1} S(t)B_2}^W$, where B_2 is a bounded absorbing set for $\{S(t)\}_{t\geq 0}$ in $H^4(\Omega)^2$ and t_1 is such that $S(t)B_2 \subset B_2$, $\forall t \geq t_1$. The set X is compact and positively invariant by $S(t)$. Let u_1 and u_2 be two solutions of (2.1) satisfying Theorem 3.1. We then write $u = u_1 - u_2$ as sum of two functions $u = u^1 + u^2$, where u^1 and u^2 are the solutions of the following problems:

$$\begin{aligned} \tfrac{\partial}{\partial t}(u^1 - \alpha\Delta u^1) - \tfrac{\mu}{\alpha}\Delta(u^1 - \alpha\Delta u^1) = 0, \\ \operatorname{div} u^1 = 0, \\ u^1|_{\partial\Omega} = 0, \\ u^1|_{t=0} = u_0, \end{aligned} \tag{4.3}$$

and

$$\begin{aligned} \tfrac{\partial}{\partial t}(u^2 - \alpha\Delta u^2) - \tfrac{\mu}{\alpha}\Delta(u^2 - \alpha\Delta u^2) + \operatorname{curl}(u_1 - \alpha\Delta u_1) \times u_1 \\ -\operatorname{curl}(u_2 - \alpha\Delta u_2) \times u_2 + (\tfrac{\mu}{\alpha} - \nu)\Delta u = 0, \\ \operatorname{div} u^2 = 0, \\ u^2|_{\partial\Omega} = 0, \\ u^2|_{t=0} = 0, \end{aligned} \tag{4.4}$$

respectively. Similarly to the decomposition of u, we consider the difference $b = b_1 - b_2$ of the two corresponding solutions of (3.1) and write $b = b^1 + b^2$, where

$$\begin{aligned} \tfrac{\partial b^1}{\partial t} + b^1 - \tfrac{\mu}{\alpha}\Delta b^1 = 0, \\ b^1|_{t=0} = \operatorname{curl}(u_0 - \alpha\Delta u_0), \end{aligned} \tag{4.5}$$

and

$$\begin{aligned} \tfrac{\partial b^2}{\partial t} + b^2 - \tfrac{\mu}{\alpha}\Delta b^2 + u_1.\nabla b_1 - u_2.\nabla b_2 + (\tfrac{\mu}{\alpha} - \nu)\Delta \operatorname{curl} u - b = 0, \\ b^2|_{t=0} = 0. \end{aligned} \tag{4.6}$$

We formally take the L^2-scalar product of (4.5) with b^1, and obtain

$$\tfrac{d}{dt}\|b^1\|^2 + 2\|b^1\|^2 + \tfrac{2\mu}{\alpha}\|\nabla b^1\|^2 = 0. \tag{4.7}$$

Therefore,

$$\tfrac{d}{dt}\|b^1\|^2 + c\|b^1\|^2 \leq 0, \tag{4.8}$$

and

$$\|b^1(t)\| \le c_1 e^{-ct}\|b_0\|,\ \int_0^t \|b^1\|^2_{H^1(\Omega)}\,d\tau \le c\|b_0\|^2,\ \ \forall t \ge 0. \tag{4.9}$$

Finally, we deduce from the equivalence of norms previously noticed that

$$\|u^1(t)\|_W \le c_1 e^{-ct}\|u_0\|_W,\ \int_0^t \|u^1\|^2_{H^4(\Omega)^2} d\tau \le c_2\|u_0\|^2_W,\ \ \forall t \ge 0. \tag{4.10}$$

We now note that $u_1.\nabla b_1 - u_2.\nabla b_2 = u_1.\nabla b + u.\nabla b_2$, and equation (4.6) becomes

$$\begin{gathered}\frac{\partial b^2}{\partial t} + b^2 - \frac{\mu}{\alpha}\Delta b^2 + u_1.\nabla b + u.\nabla b_2 + (\tfrac{\mu}{\alpha} - \nu)\Delta \operatorname{curl} u - b = 0,\\ b^2|_{t=0} = 0.\end{gathered} \tag{4.11}$$

Still working with the basis $\{w_j\}_{j\ge 1}$ considered in Section 3, we obtain, by proceeding as for (3.4),

$$\begin{aligned}&\tfrac{1}{2}\tfrac{d}{dt}(\|b^2\|^2 + \tfrac{\mu}{\alpha}\|\nabla b^2\|^2) + \|b^2 - \tfrac{\mu}{\alpha}\Delta b^2\|^2 + \tfrac{\mu}{\alpha}(\nabla(u_1.\nabla b), \nabla b^2)\\ &\quad + (u_1.\nabla b, b^2) + \tfrac{\mu}{\alpha}(\nabla(u.\nabla b_2), \nabla b^2) + (u.\nabla b_2, b^2)\\ &\quad + (\tfrac{\mu}{\alpha} - \nu)[\tfrac{\mu}{\alpha}(\nabla\Delta \operatorname{curl} u, \nabla b^2) + (\Delta \operatorname{curl} u, b^2)]\\ &\qquad = \tfrac{\mu}{\alpha}(\nabla b, \nabla b^2) + (b, b^2).\end{aligned} \tag{4.12}$$

Therefore,

$$\begin{aligned}&\tfrac{d}{dt}(\|b^2\|^2 + \tfrac{\mu}{\alpha}\|\nabla b^2\|^2) + c_1\|b^2\|^2_{H^2(\Omega)} \le c_2(\|b^2\|^2 + \tfrac{\mu}{\alpha}\|\nabla b^2\|^2)\\ &\quad + \sigma_1(\|u_1\|^2_{L^\infty(\Omega)} + \|\nabla u_1\|^2_{L^\infty(\Omega)})\|b\|^2_{H^2(\Omega)}\\ &\quad + \sigma_2(\|u\|^2_{L^\infty(\Omega)} + \|\nabla u\|^2_{L^\infty(\Omega)})\|b_2\|^2_{H^2(\Omega)}\\ &\quad + \sigma_3\|u\|^2_{H^4(\Omega)^2} + \sigma_4(\|\nabla b\|^2 + \|b\|^2),\end{aligned} \tag{4.13}$$

for any $\sigma_1, \sigma_2, \sigma_3, \sigma_4 > 0$. We note that $\|u\|^2_{H^4(\Omega)^2} \le c\|b\|^2_{H^1(\Omega)}$, and $\|b\|^2_Y \le \|b^1\|^2_Y + \|b^2\|^2_Y$, where $Y = H^1(\Omega)$ or $H^2(\Omega)$. According to the results in Section 3, if $u_{01}, u_{02} \in X$, then $u_1(t), u_2(t) \in X$, $\forall t \ge 0$. We recall that X is a bounded set of $H^4(\Omega)^2$. This is equivalent to the fact that $b_1(t)$ and $b_2(t)$ belong to a bounded set of $H^1(\Omega)$. We also have $\int_0^t \|b_i\|^2_{H^2(\Omega)}\,dt \le c(t)$, $i = 1, 2$, for all $t \ge 0$. Using these boundedness and choosing properly constants σ_1, σ_2 , σ_3 and σ_4, we get

$$\begin{aligned}&\tfrac{d}{dt}(\|b^2\|^2 + \tfrac{\mu}{\alpha}\|\nabla b^2\|^2) + c_1\|b^2\|^2_{H^2(\Omega)}\\ &\le c_2(1 + \|b_2\|^2_{H^2(\Omega)})(\|b^2\|^2 + \tfrac{\mu}{\alpha}\|\nabla b^2\|^2)\\ &\quad + c_3\|b^1\|^2\|b_2\|^2_{H^2(\Omega)} + c_4\|b^1\|^2_{H^2(\Omega)}.\end{aligned} \tag{4.14}$$

Finally, we obtain by applying the Gronwall lemma that

$$\begin{aligned}\|b^2(t)\|^2_{H^1(\Omega)} &\le c_1 e^{c_2\int_0^t(1+\|b_2\|^2_{H^2(\Omega)})d\tau}[\int_0^t \|b^1\|^2\|b_2\|^2_{H^2(\Omega)}\,ds\\ &\quad + \int_0^t \|b^1\|^2_{H^2(\Omega)}\,ds],\ \ \forall t > 0.\end{aligned} \tag{4.15}$$

Noting (4.9) and the equivalence of norms, we get

$$\|u^2(t)\|_{H^4(\Omega)^2} \le c_1 e^{c_2\int_0^t(1+\|b_2\|^2_{H^2(\Omega)})d\tau}\|u_0\|_W,\ \ \forall t > 0. \tag{4.16}$$

Both estimates (4.10) and (4.16) allow us to say that $\{S(t)\}_{t\ge 0}$ enjoys the squeezing property on X. Moreover, $S(t)$ is Lipschitz on X; and we consider P_{N^*} as the projection on the space generated by the N^* first eigenfunctions of the operator $-\Delta$ on $V \cap H^2(\Omega)^2$.

We then have the following result.

Theorem 4.1 *The semigroup $\{S(t)\}_{t\ge 0}$ possesses an exponential attractor $\mathcal{M}$ on X.*

Acknowledgments

I am currently an Abdus Salam International Centre for Theoretical Physics Post-Doctoral Research Fellow, and I would like to thank the Centre for all its support.

References

[1] A.V. Babin and M.I. Vishik, *Attractors of evolution equations,* North-Holland, Amsterdam, London, New York, Tokyo, 1992.

[2] A. Babin and B. Nicolaenko, Exponential attractors of reaction-diffusion systems in an unbounded domain, *J. Dyn. Diff. Equ.* **7** (1995), 567-589.

[3] D. Cioranescu and E.H. Ouazar, Existence and uniqueness for fluids of second grade, *Nonlin. Part. Diff. Equ., Collège de France, Pitman* **109** (1984), 178-197.

[4] D. Cioranescu and V. Girault, Weak and classical solutions of a family of second grade fluids, *Int. J. Non-Linear Mechanics* **32**(2) (1997), 317-335.

[5] J.E. Dunn and R.L. Fosdick, Thermodynamics, stability and boundedness of fluids of complexity two and fluids of second grade, *Arch. Rat. Mech. Anal.* **56**(3) (1974), 191-252.

[6] A. Eden, C. Foias, B. Nicolaenko and R. Temam, *Exponential attractors for dissipative evolution equations,* Masson, Paris, 1994.

[7] J.L. Lions, *Quelques Méthodes de Résolutions des Problèmes aux Limites Non Linéaires*, Dunod, Paris, 1969.

[8] A. Miranville, Exponential attractors for a class of evolution equations by a decomposition method, *C. R. Acad. Sci. Paris*, **328**(1) (1999), 145-150.

[9] I. Moise, R. Rosa and X. Wang, Attractors for non-compact semigroups via energy equations, *Nonlinearity* **11**(7) (1998), 1369-1393.

[10] R. Temam, *Infinite Dimensional Dynamical Systems in Mechanics and Physics,* (2nd Edition) Springer-Verlag, Berlin, Heidelberg, New York, 1997.

In: Research on Evolution Equation Compendium. Volume 2
Editor: Gaston M. N'Guerekata
ISBN: 978-1-60456-752-6

PERIODIC SOLUTIONS OF SOME EVOLUTION EQUATIONS WITH INFINITE DELAY*

Khalil Ezzinbi[1]**†** ***and James H. Liu***[2]**‡**
[1]Faculté des Sciences Semlalia, D épartement de Mathématiques, B.P. 2390, Marrakesh, Morocco
[2]Department of Mathematics and Statistics James Madison University, Harrisonburg, VA 22807, USA

Abstract

In this work, we study the existence of periodic solutions for some partial functional differential equations with infinite delay. We assume that the linear part is not necessarily densely defined and satisfies the Hille-Yosida condition, and the phase space is chosen to be C_g for some decreasing function g from $(-\infty,0]$ to $[1,\infty)$. We also present a related Massera type result, namely the existence of a bounded solution on $\mathbb{R}^+$ implies the existence of a periodic solution.

Keywords and phrases : Periodic solutions, infinite delay, Kuratowski's measure of non-compactness, condensing map.

1 Introduction

The purpose of this work is to study the existence of periodic solutions for the following partial functional differential equation with infinite delay

$$\begin{cases} u'(t) = Au(t) + f(t,u_t), \ t > 0, \\ u(s) = \phi(s), \ s \le 0, \end{cases} \tag{1.1}$$

where A is an unbounded linear operator on a Banach space X. Here, the operator A is not necessarily densely defined on X, and satisfies the Hille-Yosida condition: there exist $M_0 \ge 1$ and $\omega_0 \in \mathbb{R}$ such that

$$(\omega_0, +\infty) \subset \rho(A) \text{ and } |R(\lambda, A)^n| \le \frac{M_0}{(\lambda - \omega_0)^n}, \text{ for } n \in \mathbb{N} \text{ and } \lambda > \omega_0,$$

where $\rho(A)$ is the resolvent set of A and $R(\lambda,A) = (\lambda - A)^{-1}$. The initial data ϕ belongs to some phase space C_g to be defined later.

*This research is supported by a TWAS Research Grant under contract: No 03-030 RG/MATHS/AF/AC
†E-mail address: ezzinbi@ucam.ac.ma
‡E-mail address: liujh@jmu.edu

The function $f : \mathbb{R}^+ \times C_g \to X$ is continuous and T–periodic in t. As usual, the history function $u_t \in C_g$ is defined by

$$u_t(s) = u(t+s),\ s \leq 0.$$

Periodic solutions are derived for Eq.(1.1) by [5] in a general phase space with axioms when A is densely defined; by [6, 7] in C_g when A is time dependent and densely defined; and by [4] in a general phase space with axioms when A is non-densely defined. In this work we use the techniques developed in [4, 5, 6, 7] to present an extension of [4, 5, 6, 7] by studying Eq.(1.1) in C_g when A is non-densely defined.

This work is organized as follows: In Section 2, we recall some results about the existence, uniqueness, and estimations of solutions. In Section 3, we prove that the Poincaré map in condensing, which is then used to derive the existence of periodic solutions by using certain boundedness of solutions. In section 4, we prove a Massera type theorem when f is linear with respect to the second argument, showing that the existence of a bounded solution on $\mathbb{R}^+$ is sufficient for the existence of periodic solutions.

2 Solutions and Their Estimations

First, we make the following assumptions:

($\mathbf{H}_1$) A satisfies the Hille-Yosida condition.

($\mathbf{H}_2$) f is continuous, and Lipschitzian with respect to the second argument.

Lemma 2.1. *[3] Let A_0 be the part of the operator A in $D(A)$, which is defined by*

$$\begin{cases} D(A_0) = & \{x \in D(A) : Ax \in \overline{D(A)}\}, \\ A_0 x = & Ax. \end{cases}$$

Then A_0 generates a C_0-semigroup $(T_0(t))_{t\geq 0}$ on $\overline{D(A)}$.

Now, we define the phase space C_g for Eq.(1.1) as in [6].

Lemma 2.2. *[6] There exists an integer $K_0 > 1$ such that*

$$(\frac{1}{2})^{K_0-1} M_0 < 1,$$

where $M_0 = \sup_{t\in[0,T]} |T_0(t)|$. Next, let $w_0 = \frac{T}{K_0}$, then there exists a function g on $(-\infty, 0]$ such that $g(0) = 1$, $g(-\infty) = \infty$, g is decreasing on $(-\infty, 0]$, and for $d \geq w_0$ one has

$$\sup_{s\leq 0} \frac{g(s)}{g(s-d)} \leq \frac{1}{2}.$$

For the function g given in the above lemma, we define the continuous functions space C_g

$$C_g = \Big\{\phi : \phi \in C((-\infty, 0]; X) \text{ and } \lim_{\theta\to-\infty} \frac{|\phi(\theta)|}{g(\theta)} = 0\Big\}.$$

If we provide C_g with the norm

$$|\phi|_g = \sup_{s\leq 0} \frac{|\phi(s)|}{g(s)}, \quad \phi \in C_g,$$

then C_g is a Banach space.

Next, we recall some results concerning existence and uniqueness of integral solutions for Eq.(1.1).

Definition 2.3. [1] Let $\phi \in C_g$. A function $u : \mathbb{R} \to X$ is called an integral solution of Eq.(1.1) if the following conditions hold:
(i) u is continuous on $[0,\infty)$,
(ii) $u_0 = \phi$,
(iii) $\int_0^t u(s)\,ds \in D(A)$, for $t \geq 0$,
(iv) $u(t) = \phi(0) + A\int_0^t u(s)\,ds + \int_0^t f(s,u_s)ds$, for $t \geq 0$.

Theorem 2.4. *[2] Assume that* $(\mathbf{H}_1)$ *and* $(\mathbf{H}_2)$ *hold. Then, for all* $\phi \in C_g$ *such that* $\phi(0) \in \overline{D(A)}$, *Eq.*(1.1) *has a unique integral solution u. Moreover, let u and v be two integral solutions of Eq.(1.1) on* $(-\infty, L]$, $L > 0$. *Then for* $t \in [0,L]$,

$$|u_t - v_t|_g \leq M_1 |u_0 - v_0|_g e^{k_1 t},$$

where M_1 *and* k_1 *are some positive constants.*

The integral solution of Eq.(1.1) is given by the following variation of constants formula

$$u(t) = \begin{cases} T_0(t)\phi(0) + \lim\limits_{\lambda\to+\infty} \displaystyle\int_0^t T_0(t-s)B_\lambda f(s,u_s)ds\ , \ t \geq 0, \\ \phi(t), \quad t \leq 0, \end{cases}$$

where $B_\lambda = \lambda(\lambda - A)^{-1}$, and $u(t) \in \overline{D(A)}$, $t \geq 0$.

If $\overline{D(A)} = X$, then the integral solutions coincide with the known mild solutions given in [5] and references therein. Later on, we will follow [4, 5, 6] and call “integral solutions” as “solutions”.

As an immediate consequence of the above theorem, we conclude that solutions of Eq.(1.1) are locally bounded in t and ϕ. We can get more information on the estimation of solutions.

Lemma 2.5. *[6] Let u be a continuous function on* $(-\infty, T]$ *such that* $u_0 \in C_g$. *Then for any* $0 \leq h < r \leq T$ *with* $r - h \geq w_0$ *(*w_0 *is from Lemma 2.2), one has*

$$|u_r|_g \leq \max\left(\sup_{s\in[h,r]} |u(s)|,\ \frac{1}{2}|u_h|_g\right).$$

In the sequel, we examine the Kuratowski's measure of non-compactness, which will be used in the next section to study periodic solutions via fixed points of a condensing operator. The Kuratowski's measure of non-compactness (or the α measure) for a bounded set H of a Banach space Y with norm $|\cdot|_Y$ is defined as

$$\alpha(H) = \inf\Big\{d > 0 : H \text{ has a finite cover of diameter } < d\Big\}.$$

The following are some basic properties of the α measure of non-compactness.

Lemma 2.6. *Let B_1 and B_2 be bounded sets of a Banach space Y. Then*
$\alpha(B_1) \leq dia(B_1)$. $(dia(B_1) = \sup\{|x-y|_Y : x,y \in B_1\}.)$
$\alpha(B_1) = 0$ if and only if B_1 is precompact.
$\alpha(\lambda B_1) = |\lambda|\alpha(B_1)$, $\lambda \in \Re$. $(\lambda B_1 = \{\lambda x : x \in B_1\})$
$\alpha(B_1 \cup B_2) = \max\{\alpha(B_1), \alpha(B_2)\}$.
$\alpha(B_1 + B_2) \leq \alpha(B_1) + \alpha(B_2)$. $(B_1 + B_2 = \{x+y : x \in B_1, y \in B_2\})$
$\alpha(B_1) \leq \alpha(B_2)$ if $B_1 \subseteq B_2$.

In the sequel, we assume that

$(\mathbf{H}_3)$ $T_0(t)$ is compact on $\overline{D(A)}$ whenever $t > 0$.

Next, define $C_g^0 = \{\phi \in C_g : \phi(0) \in \overline{D(A)}\}$. For $D \subset C_g^0$ and $u(\phi)$ the unique solution with $u_0(\phi) = \phi$, we define $W_l(D) = \{u_l(\phi) : \phi \in D\}$ and $W_{[h,r]}(D) = \{u_{[h,r]}(\phi) : \phi \in D\}$, where $u_{[h,r]}$ means the restriction of u on $[h,r]$.

Lemma 2.7. *[6] Assume that $(\mathbf{H}_1)$, $(\mathbf{H}_2)$ and $(\mathbf{H}_3)$ are satisfied and let $D \subset C_g^0$. Then for any $0 \leq h < r \leq T$ with $r - h \geq w_0$ (w_0 is from Lemma 2.2), one has*

$$\alpha(W_r(D)) \leq \max\left\{\alpha(W_{[h,r]}(D)),\ \frac{1}{2}\alpha(W_h(D))\right\}.$$

Lemma 2.8. *Let $D \subset C_g^0$ be bounded. Then $\alpha(W_{[h,r]}(D)) = 0$ for any $0 < h < r \leq T$.*

Proof. Let $0 < h < r \leq T$. For $\phi \in D$ and $t \in [h,r]$, one has

$$u(t,\phi) = T_0(t)\phi(0) + \lim_{\lambda \to +\infty} \int_0^t T_0(t-s)B_\lambda f(s, u_s(\cdot,\phi))ds.$$

Define w by

$$w(t,\phi) = \lim_{\lambda \to +\infty} \int_0^t T_0(t-s)B_\lambda f(s, u_s(\cdot,\phi))ds, \text{ for } \phi \in D \text{ for } t \in [h,r].$$

Then

$$\lim_{\lambda \to +\infty} \int_0^t T_0(t-s)B_\lambda f(s,u_s(\cdot,\phi))ds = T_0(\varepsilon) \lim_{\lambda \to +\infty} \int_0^{t-\varepsilon} T_0(t-\varepsilon-s)B_\lambda f(s,u_s(\cdot,\phi))ds$$
$$+ \lim_{\lambda \to +\infty} \int_{t-\varepsilon}^t T_0(t-s)B_\lambda f(s,u_s(\cdot,\phi))ds.$$

Since $T_0(\varepsilon)$ is compact, there exists a compact set W_ε such that

$$\left\{T_0(\varepsilon)\left(\lim_{\lambda \to +\infty} \int_0^{t-\varepsilon} T_0(t-\varepsilon-s)B_\lambda f(s,u_s(\cdot,\phi))ds\right) : \phi \in D\right\} \subseteq W_\varepsilon.$$

From the boundedness of f, there exists a positive constant a such that

$$\left|\lim_{\lambda \to +\infty} \int_{t-\varepsilon}^t T_0(t-s)B_\lambda f(s,u_s(\cdot,\phi))ds\right| \leq a\varepsilon, \text{ uniformly in } \phi \in D.$$

We deduce that the set $\{w(t,\phi) : \phi \in D\}$ is totally bounded and therefore is relatively compact in X.

To prove the equicontinuity, let $0 < h \le t_0 < t \le r$. Then

$$|w(t,\phi) - w(t_0,\phi)| \le \left| \lim_{\lambda \to +\infty} \int_{t_0}^{t} T_0(t-s)B_\lambda f(s,u_s(\cdot,\phi))ds \right| + \left| (T_0(t-t_0) - I) \lim_{\lambda \to +\infty} \int_{0}^{t_0} T_0(t_0-s)B_\lambda f(s,u_s(\cdot,\phi))ds \right|.$$

There exists a positive constant b such that

$$\left| \lim_{\lambda \to +\infty} \int_{t_0}^{t} T_0(t-s)B_\lambda f(s,u_s(\cdot,\phi))ds \right| \le b(t-t_0), \text{ uniformly in } \phi \in D.$$

Moreover, $W_0 = \left\{ \lim_{\lambda \to +\infty} \int_{0}^{t_0} T_0(t_0-s)B_\lambda f(s,u_s(\cdot,\phi))ds : \phi \in D \right\}$ is relatively compact and it's well known that

$$\lim_{h \to 0^+} (T_0(h) - I)u = 0, \text{ uniformly in } u \in W_0.$$

On the other hand,

$$T_0(t)\phi(0) = T_0(t-h)T_0(h)\phi(0). \tag{2.1}$$

Since $T_0(h)$ is compact, using (2.1) we get that the family $\{T_0(\cdot)\phi(0) : \phi \in D\}$ is equicontinuous in $C([h,r];X)$. Consequently, we deduce that

$$\lim_{t-t_0 \to 0^+} |u(t,\phi) - u(t_0,\phi)| = 0, \text{ uniformly in } \phi \in D.$$

It follows that $\left\{ u(\cdot,\phi)_{|[h,r]} : \phi \in D \right\}$ is equicontinuous. By Arzéla-Ascoli's theorem, we deduce that $\left\{ u(\cdot,\phi)_{|[h,r]} : \phi \in D \right\}$ is relatively compact in $C([h,r];X)$. Hence the result is true by using Lemma 2.6.

3 Existence of Periodic Solutions

In this section, we study periodic solutions of Eq.(1.1). For this purpose, we use the Poincaré map P defined by

$$P(\phi) = u_T(\cdot,\phi), \;\; \phi \in C_g^0,$$

where $u(\cdot,\phi)$ denotes the (integral) solution of Eq.(1.1) with $u_0(\cdot,\phi) = \phi$. In the sequel, we will show that P has a fixed point. We note that a fixed point of P gives rise to a periodic solution.

First, we prove that the operator P is condensing.

Theorem 3.1. *Assume that* $(\mathbf{H}_1)$, $(\mathbf{H}_2)$ *and* $(\mathbf{H}_3)$ *hold. Then the Poincaré map P is condensing in C_g^0.*

Proof. Since

$$|P(\phi_1)-P(\phi_2)|_g = |u_T(\phi_1)-u_T(\phi_2)|_g \le M_1 e^{k_1 T}|\phi_1-\phi_2|_g,$$

it implies that P is continuous and takes bounded sets into bounded sets. Let $D \subset C_g^0$ be bounded with $\alpha(D) > 0$. Then, using lemmas 2.7 and 2.8 (w_0 is from Lemma 2.2),

$$\begin{aligned}
\alpha(P(D)) = \alpha(W_T(D)) &\le \max\left\{\alpha(W_{[T-w_0,T]}(D)),\ \frac{1}{2}\alpha(W_{T-w_0}(D))\right\}\\
&= \frac{1}{2}\alpha(W_{T-w_0}(D))\\
&\le \frac{1}{2}\max\left\{\alpha(W_{[T-2w_0,T-w_0]}(D)),\ \frac{1}{2}\alpha(W_{T-2w_0}(D))\right\}\\
&= (\frac{1}{2})^2\alpha(W_{T-2w_0}(D))\\
&\le (\frac{1}{2})^2\max\left\{\alpha(W_{[T-3w_0,T-2w_0]}(D)),\ \frac{1}{2}\alpha(W_{T-3w_0}(D))\right\}\\
&= (\frac{1}{2})^3\alpha(W_{T-3w_0}(D))\\
&\quad\ldots\ldots\\
&\le (\frac{1}{2})^{K_0-1}\max\left\{\alpha(W_{[0,w_0]}(D)),\ \frac{1}{2}\alpha(D)\right\}.
\end{aligned} \tag{3.1}$$

Next, for $\cdot \in [0,w_0]$,

$$W_{[0,w_0]}(D) \subseteq \{T_0(\cdot)\phi(0) : \phi \in D\} + \left\{\lim_{\lambda\to+\infty}\int_0^{\cdot} T_0(\cdot - h)B_\lambda f(h,u_h(\phi))dh : \phi \in D\right\}.$$

And for $t \in [0,w_0]$,

$$|T_0(t)\phi_1(0) - T_0(t)\phi_2(0)| \le M_0\,|\phi_1(0)-\phi_2(0)| \le M_0|\phi_1-\phi_2|_g, \tag{3.2}$$

where $M_0 = \sup_{t\in[0,T]}|T_0(t)|$. Consequently, using an argument similar to that in [6], we see that for $\cdot \in [0,w_0]$, one has

$$\alpha\{T_0(\cdot)\phi(0) : \phi \in D\} \le M_0\alpha(D).$$

Also, we can see that for $\cdot \in [0,w_0]$,

$$\alpha\{\lim_{\lambda\to+\infty}\int_0^{\cdot} T_0(\cdot - h)B_\lambda f(h,u_h(\phi))dh : \phi \in D\} = 0.$$

Therefore we have that

$$\alpha(W_{[0,w_0]}(D)) \le M_0\alpha(D).$$

Thus, we have

$$\begin{aligned}
\alpha(P(D)) &\le (\frac{1}{2})^{K_0-1}\max\left\{M_0\alpha(D),\ \frac{1}{2}\alpha(D)\right\}\\
&\le \left[(\frac{1}{2})^{K_0-1}M_0\right]\alpha(D) < \alpha(D).
\end{aligned} \tag{3.3}$$

This completes the proof.

To derive periodic solutions, we list the following fixed point theorems.

Theorem 3.2 (Sadovskii's fixed point theorem). *[8] Let P be a condensing operator on a Banach space Y, i.e., P is continuous and takes bounded sets into bounded sets, and $\alpha(P(B)) < \alpha(B)$ for every bounded set B of Y with $\alpha(B) > 0$. If $P(H) \subseteq H$ for a convex, closed and bounded set H of Y, then P has a fixed point in H.*

Theorem 3.3. *[7] Suppose $S_0 \subseteq S_1 \subseteq S_2$ are convex bounded subsets of a Banach space Y, S_0 and S_2 are closed, and S_1 is open in S_2, and suppose P is a condensing operator in Y. If $P^j(S_1) \subseteq S_2$, $j \geq 0$, and there is a number $N(S_1)$ such that $P^k(S_1) \subseteq S_0$, $k \geq N(S_1)$, then P has a fixed point.*

Note that the Sadovskii's fixed point theorem requires that the Poincaré operator maps a bounded set into itself. Thus some kind of boundedness of the solutions is required. As for Theorem 3.3, some asymptotic boundedness of the solutions is required. Thus we list the following definitions.

Definition 3.4. [6] The solutions of Eq.(1.1) are said to be *locally strictly bounded* if there exists a constant $B > 0$ such that $|\phi|_g \leq B$ implies that its solution satisfies $\|u(t,\phi)\| \leq B$ for $t \in [0,T]$.

Definition 3.5. [7] The solutions of Eq.(1.1) are said to be *ultimate bounded* if there is a bound $B > 0$, such that for each $B_3 > 0$, there is a $K > 0$, such that $|\phi|_g \leq B_3$ and $t \geq K$ imply that its solution satisfies $\|u(t,\phi)\| < B$.

We now study the relationship between the boundedness and the periodicity of solutions of Eq.(1.1). By using Theorem 3.2 and Definition 3.4, we have

Theorem 3.6. *Let $(\mathbf{H}_1)$, $(\mathbf{H}_2)$ and $(\mathbf{H}_3)$ be satisfied. If the solutions of Eq.(1.1) are locally strictly bounded (or assume that solutions are non-increasing in norm $\|\cdot\|$ on $[0,T]$), then Eq.(1.1) has a T periodic solution.*

Proof. Let $H = \{\phi \in C_g^0 : |\phi|_g \leq B\}$ with B from Definition 3.4. Then H is convex, closed and bounded in C_g^0. Next, for $u(\cdot) = u(\cdot,\phi)$ with $\phi \in H$, the locally strictly boundedness implies that $\|u(t)\| \leq B$ for $t \in [0,T]$. Then we obtain from Lemma 2.5 that

$$\begin{aligned} |P(\phi)|_g = |u_T(\phi)|_g &\leq \max\left\{ \sup_{s\in[0,T]} \|u(s)\|, \ \frac{1}{2}|\phi|_g \right\} \\ &\leq \max\left\{ B, \ \frac{1}{2}B \right\} = B. \end{aligned} \tag{3.4}$$

Thus the result is true by using Theorem 3.2.

By using Theorem 3.3 and Definition 3.5, we have the following result, whose proof is omitted here since it is similar to one in [7].

Theorem 3.7. *Let $(\mathbf{H}_1)$, $(\mathbf{H}_2)$ and $(\mathbf{H}_3)$ be satisfied. If the solutions of Eq.(1.1) are ultimate bounded, then Eq.(1.1) has a T periodic solution.*

4 A Massera Type Theorem

In this section, we study the existence of a periodic solution for the following nonhomogeneous partial functional differential equation

$$\begin{cases} \frac{d}{dt}x(t) = Ax(t) + L(t, x_t) + h(t), \ t > 0, \\ x_0 = \phi, \end{cases} \tag{4.1}$$

where L is a continuous function from $\mathbb{R}^+ \times C_g$ into X, linear with respect to the second argument and T-periodic in t; h is a continuous T-periodic function with values in X. We will obtain a Massera type theorem for Eq. (4.1). More precisely, we have

Theorem 4.1. *Let* $(\mathbf{H}_1)$ *and* $(\mathbf{H}_3)$ *be satisfied. If Eq. (4.1) has a bounded solution* y *(with* $y(0) \in \overline{D(A)}$*) on* $\mathbb{R}^+$ *in the sense that* $\sup_{t \in \mathbb{R}^+} |y(t)| < +\infty$, *then it has a* T*-periodic solution.*

Proof. Let y be a bounded solution of Eq. (4.1) on $\mathbb{R}^+$. From the definition of C_g, we know that $\{y_t : t \geq 0\}$ is also bounded in C_g. Let

$$\mathcal{D} := \overline{co}\{y_{nT} : n \in \mathbb{N}\},$$

where $\overline{co}$ denotes the closure of the convex hull. Then $\mathcal{D}$ is a nonempty bounded closed convex subset of C_g^0.

Let $\psi \in \mathcal{D}$. Then there exists a sequence $(\psi^k)_k \subset co\{y_{nT} : n \in \mathbb{N}\}$ such that

$$\psi^k = \sum_{i=1}^{n_k} \alpha_i^k y_{n_i^k T}, \alpha_i^k \geq 0, \sum_{i=1}^{n_k} \alpha_i^k = 1, \text{ and } \left|\psi - \psi^k\right|_g \to 0 \text{ as } k \to \infty.$$

Then, one has

$$\begin{aligned} x_T(\cdot, \psi^k) &= \sum_{i=1}^{n_k} \alpha_i^k x_T(\cdot, y_{n_i^k T}) \\ &= \sum_{i=1}^{n_k} \alpha_i^k y_{n_i^k T + T}. \end{aligned}$$

Thus, $x_T(\cdot, \psi^k) \in \mathcal{D}$. Since $\left|x_T(\cdot, \psi) - x_T(\cdot, \psi^k)\right|_g \to 0$ as $k \to +\infty$, and $\mathcal{D}$ is closed, we deduce that $P(\psi) = x_T(\cdot, \psi) \in \mathcal{D}$ for all $\psi \in \mathcal{D}$, which implies that $P(\mathcal{D}) \subset \mathcal{D}$. By Theorem 3.2, we conclude that Eq. (4.1) has a T-periodic solution.

References

[1] M. Adimy, H. Bouzahir and K. Ezzinbi, Existence for a class of partial functional differential equations with infinite delay, *Nonlin. Anal.,* **46**(2001), 91-112,

[2] M. Adimy, H. Bouzahir and K. Ezzinbi, Local existence and stability for a class of partial functional differential equations with infinite delay, *Nonlin. Anal.,* **48**(2002), 323-348.

[3] W. Arendt, C.J.K. Batty, M. Hieber and F. Neubrander, Vector Valued Laplace Transforms and Cauchy Problems, *Monographs in Mathematics*. Vol. 96, Birkhauser Verlag, 2001.

[4] R. Benkhalti and H. Bouzahir and K. Ezzinbi, Existence of Periodic solutions for some partial functional differential equations with infinite delay, *J. Math. Anal. Appl.*, **256**(2001), 257-280.

[5] H. Henriquez, Periodic solutions of quasi-linear partial functional differential equations with unbounded delay, *Funkcial. Ekvac.,* **37**(1994), 329-343.

[6] J. Liu, Periodic solutions of infinite delay evolution equations, *J. Math. Anal. Appl.*, **247**(2000), 627-644.

[7] J. Liu, T. Naito and N. Minh, Bounded and Periodic solutions of infinite delay evolution equations, *J. Math. Anal. Appl.*, **286**(2003), 705-712.

[8] B. Sadovskii, On a fixed point principle, *Funct. Anal. Appl.,* **2**(1967), 151-153.

In: Research on Evolution Equation Compendium. Volume 2 ISBN: 978-1-60456-752-6
Editor: Gaston M. N'Guerekata

INTERNAL POLLUTION AND DISCRIMINATING SENTINEL IN POPULATION DYNAMICS PROBLEM

***O. Nakoulima*[1]**and S. Sawadogo*[2†]

[1]Université des Antilles et de la Guyane, Facult é des Sciences Exactes et Naturelles, Département de Math ématiques/Informatiques, Campus de Fouillole, 97159 Pointe à Pitre Cedex, France.

[2]Université de Ouagadougou, UFR/Sciences Exactes et Appliqu ées, Département de Math ématiques/Informatiques, 03 BP 7021 Ouagadougou 03-Burkina Faso

Abstract

The so called Lions sentinel method is applied to some time varying system with two time scale variables and with missing data. As an example, some population dynamics problem with missing data is considered. The two time scale variables represent the running time variable and the age of individuals. This insight seems to be new. Building the object sentinel leads to some null-controllability problem with restricted control variables and two time scale variables.

AMS Subject Classification: 93B05, 92D25, 35B37, 35K05
Key Words: population dynamics, controllability, sentinels.

1 Introduction

1.1 Population Dynamics Problem with Missing Data

The unknown real function y depends on variables t,a,x where $t \in (T_0,T)$ stands for the running time, $a \in (0,A)$ for the age of individuals and $x \in \Omega \subset \mathbb{R}^N$ for the space variable. The number $y(t,a,x)$ is the distribution of $a-year$ old individuals at time t at the point x.

For relevant notation we agree on the following. If some function ϕ depends on several variables $(x_1,x_2,..x_n)$ then we denote by $\phi(x_k)$ or $\phi(x_k,x_l)$the function of the other variables. Possible confusion is avoided in precising the definition domain.

The function y has to satisfy the following two time scale varying equation

$$\begin{cases} \frac{\partial y}{\partial t}+\frac{\partial y}{\partial a}-\Delta y+\mu y=f+\lambda\widehat{f} \text{ in } Q=(T_0,T)\times(0,A)\times\Omega=U\times\Omega \\ y=0 \text{ in } \Sigma=(T_0,T)\times(0,A)\times\partial\Omega=U\times\Omega \\ y(T_0)=y^0+\tau\widehat{y}^0 \text{ in } Q_A=(0,A)\times\Omega \\ y(0)=\int_0^A\beta(a)y(a)\,da \text{ in } Q_T=(T_0,T)\times\Omega \end{cases} \tag{1}$$

*E-mail address: onakouli@univ-ag.fr
†E-mail address: sawasom@yahoo.fr

It is assumed that Ω is open and bounded with c^2 boundary $\Gamma = \partial\Omega$ and $\mu(t,a,x) \geq 0$, $\beta(t,a,x) \geq 0$. The parameters of the problem have the following sense : the bound $T > 0$ is the horizon of the problem (the lower bound of running time scale is T_0 in order to avoid confusion with the lower bound of age time scale which is actually 0), the bound A is the expectation of life, the weight β is the natural fertility rate, the function $\mu = \mu(t,x,a)$ is the natural death rate of $a-$year old individuals at time $t > 0$ and in the position x, the function f corresponds to external flow and $y^0 = y^0(a,x)$ is the initial distribution of individuals.

Convenient assumptions for mesurability and integrability of functions are made. In particular $f \in L^2(Q)$. The data of the state equation (1) are incomplete in the following sense :

(a) $\widehat{f} \in L^2(Q)$ and $\widehat{y}^0 \in L^2(Q_A)$ are unknown functions with

$$\left\|\widehat{f}\right\|_{L^2(Q)} \leq 1, \ \|\widehat{y}^0\|_{L^2(Q_A)} \leq 1$$

(b) $\lambda \in \mathbb{R}$ and $\tau \in \mathbb{R}$ are unknown numbers and small enough.

In the model (1), the term $\tau\widehat{y}^0$ represents missing data where as $\lambda\widehat{f}$ stands for pollution term. The general question we want to address is

$$\left\{\begin{array}{l}\text{for any given extra observation of the system (1),}\\ \text{can we evaluate } \lambda\widehat{f}\text{, without any attempt at computing } \tau\widehat{y}^0?\end{array}\right. \tag{H$_0$}$$

Lions[8] first investigated parabolic patterns. Here we construct sentinels when the supports of the observation function and of the control function are included in two different open subsets of $\mathbb{R}^N$. This point of view has already been proposed by Nakoulima [10].

The sentinels theory relies on three features: the state equation, the observation system and some particular evaluation function : the sentinel itself.

State equation: The equation (1) governs the state y. We assume the following [1] :

$$(H1)\left\{\begin{array}{l}\beta \in L^\infty_+((0,A)\times(T_0,T)\times\Omega),\\ \exists\delta \in (0,A) \text{ s.t. } \beta(a) = 0 \text{ for } a \in (\delta,A);\end{array}\right.$$

$$(H2)\ \mu \in L^\infty_{loc}([0,A);L^\infty((T_0,T)\times\Omega)),\ \mu \geq 0 \text{ a.e. in } Q_A;$$

$$(H3)\left\{\begin{array}{l}\forall t, T_0 < t < T_0 + A,\ \forall x \in \Omega,\ \lim\limits_{a\longrightarrow A} \int_0^t \mu(\iota, a-t+\iota, x)d\iota = +\infty,\\ \forall t, A < t < T,\ \forall x \in \Omega,\ \lim\limits_{a\longrightarrow A} \int_0^a \mu(t-a+\alpha,\alpha,x)d\alpha = +\infty.\end{array}\right.$$

For the biological comments about the model and for the basic existence of the solution to (1) we refer to [2, 7, 11]. For the sake of simplicity, we use indifferently y, $y(t,a,x;\lambda,\tau)$ or $y(\lambda,\tau)$ to denote the unique solution of (1). It is relevant since λ and τ are fixed parameters.

Observation system: the observation is the knowledge, along some time period, of some function y_{obs} which is defined on some strip over some nonempty open subset $O \subset \Omega$, called observatory, as follows

$$y_{obs} = m_0 + \sum_{i=1}^{M} \beta_i m_i. \tag{2}$$

where the functions $m_0, m_1, ..., m_M$ are given in $L^2(U \times O)$ but where the real coefficients β_i are unknown. We assume the coefficients β_i are "small". We refer to the terms $\beta_i m_i$ as the interference terms. We can assume without loss of generality that

$$\text{the functions } m_i \text{ are linearly independent.} \tag{3}$$

Sentinel. Let

$$h_0 \in L^2(U \times O) \tag{4}$$

and let $\omega \subset \Omega$, open and nonempty and set $\mathcal{G} = U \times \omega$. For any control function $w \in L^2(\mathcal{G})$, set

$$S(\lambda,\tau) = \int_U \int_O h_0 y(t,a,x;\lambda,\tau)dtdadx + \int_U \int_\omega wy(t,a,x;\lambda,\tau)dtdadx.$$

Choose now $w \in L^2(\mathcal{G})$ such that the following holds:

(i). S is stationary with respect to the missing terms $\tau\widehat{y}^0$, i.e.

$$\frac{\partial S}{\partial \tau}(0,0) = 0 \quad \forall \widehat{y}^0; \tag{6}$$

(ii). S is stationary with respect to interference terms $\beta_i m_i$,i.e.

$$\int_U \int_O h_0 m_i dtdadx + \int_U \int_\omega w m_i dtdadx = 0, \; 1 \leq i \leq M; \tag{7}$$

and

(iii). the norm $\|w\|_{L^2(U\times\omega)}$ is minimal among control functions in $L^2(\mathcal{G})$ which satisfy the above conditions

$$\|w\|_{L^2(\mathcal{G})} = \text{minimum}. \tag{8}$$

Remark 1.1 *According to (8), the function S if it exists, is unique. We refer to function S as the sentinel.*

Remark 1.2 *If the functions m_i are null functions, no information comes from (7). The sentinel S is defined only by (6) and (8). If $m_i \neq 0$, the sentinel S is defined by (6), (7) and (8) and it is called a discriminating sentinel.*

Remark 1.3 *The support $\operatorname{supp}(m_i)$ of functions m_i is assumed to be included in O. Suppose $\omega \cap O = \emptyset$ then $\int_U \int_\omega w m_i dtdadx = 0$. There are then no contraints on the control w. Therefore, it suffices to choose h_0 such that h_0 is orthogonal to each m_i. So, without loss of generality, it may be assumed that*

$$\omega \subset O. \tag{9}$$

1.2 Equivalence to Null-Controllability with Constraints on the Control

Here it will be shown that the existence of such a control function w dealing (6)-(8) is equivalent to null-controllability property for the system with constrained control.

Consider the function $y_\tau = \frac{\partial y}{\partial \tau}$ where y corresponds to parameter values $\lambda = 0$, $\tau = 0$. The function y_τ is the solution of the problem

$$\begin{cases} \frac{\partial y_\tau}{\partial t} + \frac{\partial y_\tau}{\partial a} - \Delta y_\tau + \mu y_\tau = 0 \text{ in } Q; \\ y_\tau = 0 \text{ on } \Sigma; \\ y_\tau(T_0) = \widehat{y}^0 \text{ in } Q_A; \\ y_\tau(0) = \int_0^A \beta(a) y_\tau(a) da \text{ in } Q_T. \end{cases} \tag{10}$$

Under the assumptions $(H_1)-(H_3)$ the linear problem (10) gets one only solution y_τ such that $y_\tau(t,A,x)=0$. For the details of the proof we refer to [2, 7, 11].

We now consider the stationary condition (6). The stationary condition holds if and only if

$$\int_U \int_O h_0 y_\tau dt da dx + \int_U \int_\omega w y_\tau dt da dx = 0, \ \forall\, \widehat{y}^0, \|\widehat{y}^0\|_{L^2(Q_A)} \leq 1. \tag{11}$$

In order to transform the equation (11), introduce now the classical adjoint state. More precisely, consider the solution $q = q(t,a,x)$ of the linear problem

$$\begin{cases} -\frac{\partial q}{\partial t} - \frac{\partial q}{\partial a} - \Delta q + \mu q = \beta q(0) + h_0 \chi_O + w \chi_\omega \text{ in } Q, \\ q = 0 \quad \text{in } \Sigma, \\ q(T) = 0 \quad \text{in } Q_A, \\ q(A) = 0 \quad \text{in } Q_T, \end{cases} \tag{12}$$

where χ_O and χ_ω are indicator functions for the open sets O and ω respectively. There is one only solution in $L^2(Q)$ as some consequence of the fixed point theorem for contracting mapping [2, 3]. The so called adjoint state q depends on the unknown w and its utility comes from the following process.

First, multiply both members of the differential equation in (12) by y_τ, and integrate by parts over Q

$$\int_U \int_O h_0 y_\tau dt da dx + \int_U \int_\omega w y_\tau dt da dx = \int_0^A \int_\Omega q(0,a,x) \widehat{y}^0 da dx\,, \tag{13}$$

$$\forall \widehat{y}^0 \in L^2(Q_A)\ \|\widehat{y}^0\|_{L^2(Q_A)} \leq 1.$$

Thus, the condition (6) (or (11)) holds if and only if

$$q(0,a,x) = 0 \text{ a.e } (a,x) \in (0,A) \times \Omega. \tag{14}$$

Then, consider the constraints (7). Let $\mathcal{K} \subset L^2(U \times \omega)$ be the subspace generated by the family $\{m_i \chi_\omega\}_{1 \leq i \leq M}$

$$\mathcal{K} = span\{m_i \chi_\omega | 1 \leq i \leq M\}. \tag{15}$$

There is one unique $k_0 \in \mathcal{K}$ such that

$$\int_U \int_O h_0 m_i dt da dx + \int_U \int_\omega k_0 m_i dt da dx = 0, 1 \leq i \leq M. \tag{16}$$

Denoting by $\mathcal{K}^\perp$ the orthogonal subspace of $\mathcal{K}$ in $L^2(U \times \omega)$, the condition (7) holds if and only if

$$w - k_0 = v \in \mathcal{K}^\perp. \tag{17}$$

The above considerations show that finding the control w such that the pair (w,S) satisfies (6)–(8) comes to find the control v such that the pair (v,q) satisfies the following system

$$v \in \mathcal{K}^{\perp},\ q \in L^2(Q) \tag{18}$$

$$\left\{\begin{array}{rcll} -\frac{\partial q}{\partial t} - \frac{\partial q}{\partial a} - \Delta q + \mu q & = & \beta q(0) + h_0\chi_O + k_0\chi_\omega + v\chi_\omega & \text{in} \quad Q, \\ q & = & 0 & \text{on} \quad \Sigma, \\ q(T) & = & 0 & \text{in} \quad Q_A, \\ q(A) & = & 0 \text{ in} & Q_T, \end{array}\right. \tag{19}$$

$$q(T_0) = 0 \text{ in } Q_A, \tag{20}$$

$$\|v\|_{L^2(U\times\omega)} = \text{minimum}. \tag{21}$$

Problem (18)-(21) stands for some null-controllability problem with constrained control variable v.

1.3 Controllability Problem

For the problem (18)-(21), two matters are considered. The first one consists in solving the null-controllability problem, and the second one consists in characterizing the optimal solution (21) by some optimality system. The problem (18)-(21) is solved when $\mathcal{K} = \{0\}$ (i.e. setting without constraints or free constraints) in several issues by various methods [1], [4]. In the present paper both points are considered in the general setting $\mathcal{K} \neq \{0\}$.

2 Null-Controllability with Constraints on the Control

Consider the following general null-controllability problem. Find some pair (v,q) such that

$$v \in \mathcal{K}^{\perp}, q \in L^2(Q); \tag{22}$$

$$\left\{\begin{array}{rcll} -\frac{\partial q}{\partial t} - \frac{\partial q}{\partial a} - \Delta q + \mu q & = & \beta q(0) + h + v\chi_\omega & \text{in} \quad Q, \\ q & = & 0 \text{ in} & \Sigma, \\ q(T) & = & 0 \text{ in} & Q_A, \\ q(A) & = & 0 \text{ in} & Q_T, \end{array}\right. \tag{23}$$

$$q(T_0) = 0 \text{ in } Q_A, \tag{24}$$

$$\|v\|_{L^2(U\times\omega)} = \text{minimum}, \tag{25}$$

where $h \in L^2(Q)$ (for $h = h_0\chi_O + k_0\chi_\omega$ we recognize (18)-(21)).

2.1 Observability Inequality Adapted to Constraints

The observability inequality we are looking for is some consequence of Carleman's inequality. In order to state Carleman's inequality, introduce now some objects and notations. Choose first some auxiliary function $\psi \in C^2(\overline{\Omega})$ which satisfies the following conditions :

$$\psi(x) > 0\ \forall x \in \Omega,\ \psi(x) = 0\ \forall x \in \Gamma,\ |\nabla\psi(x)| \neq 0\ \forall x \in \overline{\Omega} - \omega_0,$$

where ω_0 denotes any open set such that $\overline{\omega}_0 \subset \omega$ (for example ω_0 can be some small enough open ball). Such a function ψ exists according to A. Fursikov and O. Yu. Imanuvilov [6].

For any positive parameter value λ define then the following weight functions

$$\varphi(t,a,x) = \frac{e^{\lambda\psi(x)}}{at(T-t)}, \quad \alpha(t,a,x) = \frac{e^{2\lambda\|\psi\|_\infty} - e^{\lambda\psi(x)}}{at(T-t)}, \tag{26}$$

and adopt the following notations

$$\begin{cases} L &= \frac{\partial}{\partial t} + \frac{\partial}{\partial a} - \Delta + \mu I, \\ L^* &= -\frac{\partial}{\partial t} - \frac{\partial}{\partial a} - \Delta + \mu I, \\ \mathcal{V} &= \left\{\rho \in C^\infty(\overline{Q}),\ \rho = 0 \text{ on } \Sigma\right\}. \end{cases} \tag{27}$$

Now the inequality can be formulated. There exist numbers $\lambda_0 > 1$ and $s_0 > 1$ and there exists some number $C > 0$ which only depends on the open sets Ω and ω, such that for any $\lambda > \lambda_0$, for any $s \geq s_0 = s_0(\mu, T, \lambda)$ and for any $\rho \in \mathcal{V}$ the following holds:

$$\begin{aligned} &\int_Q \frac{e^{-2s\alpha}}{s\varphi}\left(|\rho_t + \rho_a|^2 + |\Delta\rho|^2\right) dtdadx + \int_Q s\lambda^2\varphi e^{-2s\alpha}|\nabla\rho|^2\, dtdadx \\ &\qquad + \int_Q s^3\lambda^4\varphi^3 e^{-2s\alpha}|\rho|^2\, dtdadx \\ &\leq C\left(\int_Q e^{-2s\alpha}|L\rho|^2\, dtdadx + \int_{T_0}^T \int_0^A \int_\omega s^3\lambda^4\varphi^3 e^{-2s\alpha}|\rho|^2\, dtdadx\right). \end{aligned} \tag{28}$$

The above inequality is the global Carleman's inequality for which we refer to [1] and [12].

We are now concerned with our own inequality. Since φ does not vanish, set

$$\theta = \frac{e^{s\alpha}}{\varphi\sqrt{\varphi}} \text{ or } \frac{1}{\theta} = \varphi\sqrt{\varphi}e^{-s\alpha}. \tag{29}$$

Then $\theta \in C^2(Q)$ and $\frac{1}{\theta}$ is bounded. By substitution in (28) the following inequality holds

$$\int_Q \frac{1}{\theta^2}|\rho|^2\, dtdadx \leq C\Big(\int_Q \frac{1}{\theta^2\varphi^3 s^3\lambda^4}|L\rho|^2\, dtdadx + \int_U\int_\omega \frac{1}{\theta^2}|\rho|^2\, dtdadx\Big). \tag{30}$$

As a consequence of the boundeness of $\frac{1}{\theta}$ and $\frac{1}{\varphi^3 s^3\lambda^4}$, some first observability inequality comes

$$\int_Q \frac{1}{\theta^2}|\rho|^2\, dtdadx \leq C\Big(\int_Q |L\rho|^2\, dtdadx + \int_U\int_\omega |\rho|^2\, dtdadx\Big). \tag{31}$$

Another observability inequality follows which is adapted to the constraints setting using the subspace $\mathcal{K}$ (15).

Recall first that

$$\mathcal{K} \text{ is finite dimensional.} \tag{32}$$

Denote by

$$P = \text{the orthogonal projection operator from } L^2(\mathcal{G}) \text{ on } \mathcal{K}. \tag{33}$$

The following lemma is the key for our results

Lemma 2.1 *Assume*

$$\left\{\begin{array}{l} \text{any function } k\in \mathcal{K}\cap L^2(U,H^1(\omega)) \text{ such that} \\ \frac{\partial k}{\partial t}+\frac{\partial k}{\partial a}-\Delta k+\mu k=0 \text{ in } \mathcal{G} \text{ is identically zero in } \mathcal{G}\,. \end{array}\right. \tag{34}$$

Then there exists some positive constant $C>0$ such that :

$$\int_Q \frac{1}{\theta^2}|\rho|^2\,dtdadx \le C\Big(\int_Q |L\rho|^2\,dtdadx+\int_{\mathcal{G}}|\rho-P\rho|^2\,dtdadx\Big)\ \forall \rho\in \mathcal{V}\,. \tag{35}$$

Remark 2.2 *The above assumption has been already set by Lions [8] for the parabolic system.*

Proof 2.1 *Suppose that (35) does not hold. Then*

$$\left\{\begin{array}{l} \forall j\in \mathbf{N}^*, \exists \rho_j\in \mathcal{V}\,, \int_U\int_\Omega \frac{1}{\theta^2}\left|\rho_j\right|^2 dtdadx=1, \\ \int_U\int_\Omega \left|L\rho_j\right|^2 dtdadx\le \frac{1}{j} \text{ and } \int_{\mathcal{G}} \left|(\rho_j-P\rho_j)\chi_\omega\right|^2 dtdadx \le \frac{1}{j}. \end{array}\right. \tag{36}$$

The forthcomming proof consists in extracting some subsequence $(\rho_{j_k})_k$ such that the following contradiction holds

$$\lim_k \int_Q \frac{1}{\theta^2}\left|\rho_{j_k}\right|^2 dtdadx \to 0.$$

Denote by $(h|g)_{L^2(\mathcal{G})}$ the natural scalar product in the Hibert space $L^2(\mathcal{G})$. Let $\{k_1,k_2,...,k_M\}$ be some orthonormal basis of $\mathcal{K}$.

***1)** We show first that for any $i=1,2,...,M$ the numerical sequence $((\rho_j|k_i)_{L^2(\mathcal{G})})_{j\in\mathbb{N}^*}$ is bounded or equivalently that the sequence $\left(\left\|P\rho_j\right\|^2_{L^2(\mathcal{G})}\right)_j$ is bounded*

Start with the norm inequality

$$\left(\int_{\mathcal{G}} \frac{1}{\theta^2}\left|P\rho_j\chi_\omega\right|^2 dtdadx\right)^{\frac{1}{2}} \le \left(\int_{\mathcal{G}} \frac{1}{\theta^2}\left|\rho_j\chi_\omega\right|^2 dtdadx\right)^{\frac{1}{2}} + \left(\int_{\mathcal{G}} \frac{1}{\theta^2}\left|(\rho_j-P\rho_j)\chi_\omega\right|^2 dtdadx\right)^{\frac{1}{2}}. \tag{37}$$

Since $\frac{1}{\theta^2}$ is bounded and by (36) it follows that there is some number γ

$$\forall j\in\mathbb{N}, \int_{\mathcal{G}} \frac{1}{\theta^2}\left|P\rho_j\chi_\omega\right|^2 dtdadx \le \gamma. \tag{38}$$

Since $\mathcal{K}$ is finite dimensional, norms are equivalent. Particularly the mappings

$$k \longmapsto \int_{\mathcal{G}} |k|^2\,dtdadx \text{ and } k\longmapsto \int_{\mathcal{G}} \frac{1}{\theta^2}|k|^2\,dtdadx$$

are equivalent norms on $\mathcal{K}$. There is then some number γ'

$$\forall j\in\mathbb{N}, \int_{\mathcal{G}} \left|P\rho_j\chi_\omega\right|^2 dtdadx \le \gamma'.$$

The relation $((\rho_j - P\rho_j)\chi_\omega \in \mathcal{K}^\perp, \forall j \in \mathbf{N}^*$ *means the following*

$$((\rho_j - P\rho_j)\chi_\omega | k_i)_{L^2(\mathcal{G})} = 0 \; \forall i, 1 \le i \le M, \; \forall j \in \mathbf{N}^*. \tag{39}$$

Thus

$$P\rho_j\chi_\omega = \sum_{i=1}^{M} (P\rho_j\chi_\omega | k_i)_{L^2(\mathcal{G})} k_i = \sum_{i=1}^{M} (\rho_j | k_i)_{L^2(\mathcal{G})} k_i. \tag{40}$$

and from orthonormality

$$\int_{\mathcal{G}} |P\rho_j\chi_\omega|^2 \, dtdadx = \sum_{i=1}^{M} \left| (\rho_j | k_i)_{L^2(\mathcal{G})} \right|^2 = \|P\rho_j\|^2_{L^2(\mathcal{G})} \tag{41}$$

thus

$$\|P\rho_j\|^2_{L^2(\mathcal{G})} \le \gamma. \tag{43}$$

2) *Since* $(P\rho_j\chi_\omega)_{j\in\mathbb{N}}$ *is bounded and*

$$\|(\rho_j - P\rho_j)\chi_\omega\|^2_{L^2(\mathcal{G})} = \int_{\mathcal{G}} |(\rho_j - P\rho_j)\chi_\omega|^2 \, dtdadx \to 0,$$

then the sequence $(\rho_j\chi_\omega)_{j\in\mathbb{N}^*}$ *is bounded. There is some weakly convergent subsequence still denoted by* $(\rho_j\chi_\omega)_{j\in\mathbb{N}^*}$ *such that*

$$\rho_j\chi_\omega \rightharpoonup g \text{ weakly in } L^2(\mathcal{G}). \tag{45}$$

Since subsequences have the same limit as convergent sequence

$$\rho_j\chi_\omega - P\rho_j\chi_\omega \to 0 \text{ strongly in } L^2(\mathcal{G}). \tag{46}$$

Since $\mathcal{K}$ *finite dimensional, there is some possibly sub-subsequence still denoted by* $(\rho_j\chi_\omega)_{j\in\mathbb{N}^*}$ *such that*

$$P\rho_j\chi_\omega \to \zeta \text{ strongly in } L^2(\mathcal{G}). \tag{47}$$

As a consequence of (46) and (47),(45)

$$\rho_j\chi_\omega \longrightarrow g = \zeta \text{ strongly in } L^2(\mathcal{G}).$$

Then (46)

$$P\rho_j\chi_\omega \to g = \zeta \text{ strongly in } L^2(\mathcal{G}).$$

Since P is some compact operator and by (45)

$$P\rho_j\chi_\omega \to Pg \text{ strongly in } L^2(\mathcal{G}).$$

Therefore $Pg = g$ *and so* $g \in \mathcal{K}$.

3) *There is however some better knowledge about the function g. Actually* $g = 0$. *From (36)* $L\rho_j \to 0$ *strongly in* $L^2(Q)$. *Then* $L\rho_j\chi_\omega \longrightarrow 0$ *strongly in* $L^2(\mathcal{G})$.

Thus $L\rho_j\chi_\omega \rightharpoonup 0$ *weakly in* $\mathcal{D}'(\mathcal{G})$ *and so* $Lg = 0$. *Following (34)* $g = 0$ *on* $\mathcal{G}$.

4) The two preceding steps deal the conclusion

$$\rho_j \chi_\omega \to 0 \text{ strongly in } L^2(\mathcal{G}).$$

Since $\rho_j \in \mathcal{V}$, the observability inequality (31) applies as follows:

$$\forall j \in \mathbb{N}, \int_Q \frac{1}{\theta^2} |\rho_j|^2 dtdadx \leq C \left(\int_Q |L\rho_j|^2 dtdadx + \int_{\mathcal{G}} |\rho_j|^2 dtdadx \right). \quad (48)$$

It is clear then that

$$\lim_j \int_Q \frac{1}{\theta^2} |\rho_j|^2 dtdadx = 0.$$

The proof is now completed.

2.2 Existence of Optimal Control Variable

Consider now the following symetric bilinear form

$$\forall \rho \in \mathcal{V}, \forall \widehat{\rho} \in \mathcal{V}, \ a(\rho, \widehat{\rho}) = \int_Q L\rho L\widehat{\rho} dtdadx + \int_{\mathcal{G}} (\rho - P\rho)(\widehat{\rho} - P\widehat{\rho}) dtdadx. \quad (49)$$

According to Lemma 3, this symetric bilinear form is a scalar product on $\mathcal{V}$. Let V be the completion of $\mathcal{V}$ with respect to the related norm :

$$\rho \longmapsto \|\rho\|_V = \sqrt{a(\rho, \rho)}. \quad (50)$$

The closure of $\mathcal{V}$ is the Hilbert space V. Remark that the norm $\|.\|_V$ is related with the right member of the inequality (35). Similarly, the left member of (35) leads to the norm

$$\forall \rho \in \mathcal{V}, |\rho|_\theta = \left(\int_Q \frac{1}{\theta^2} |\rho|^2 dtdadx \right)^{\frac{1}{2}}.$$

The completion of $\mathcal{V}$ is the weighted Hilbert space usually denoted by $L^2_{\frac{1}{\theta}}$. The inequality (35) has the following meaning

$$\forall \rho \in \mathcal{V}, |\rho|_\theta \leq C \|\rho\|_V. \quad (51)$$

This inequality extends to $\rho \in V$. This shows that V is continuously imbedded in $L^2_{\frac{1}{\theta}}$. We can state the following

Lemma 2.3 *Let us consider h such that*

$$h \in L^2(Q) \text{ and } \theta h \in L^2(Q). \quad (52)$$

The linear form defined by

$$\rho \in V \mapsto \int_Q h\rho dtdadx$$

is continuous.

Proof 2.2 *From Cauchy-Schwarz inequality in $L^2(Q)$*

$$\forall \rho \in \mathcal{V}\,, \qquad \left|\int_Q h\rho dt da dx\right| \leq \left(\int_Q \theta^2 \left|h\right|^2 dt da dx\right)^{\frac{1}{2}} \times \left(\int_Q \frac{1}{\theta^2} \left|\rho\right|^2 dt da dx\right)^{\frac{1}{2}} \tag{53}$$

$$\forall \rho \in \mathcal{V}\,, \qquad \left|\int_Q h\rho dt da dx\right| \leq \left\|\theta h\right\|_{L^2} \left|\rho\right|_\theta \leq C \left\|\theta h\right\|_{L^2} \left\|\rho\right\|_V.$$

Since $\overline{\mathcal{V}} = V$ and the function is linear $\mathcal{V}$, the inequality extends to V :

$$\forall \rho \in V, \left|\int_Q h\rho dt da dx\right| \leq C \left\|\rho\right\|_V. \tag{54}$$

that was to be proved.

By the Lax-Milgram theorem [5], for any function h as above, there exists one and only one solution $\rho_\theta \in V$ of the variational equation:

$$a(\rho_\theta, \rho) = \int_Q h\rho dt da dx \ \forall \rho \in V. \tag{55}$$

Set

$$v_\theta = -(\rho_\theta - P\rho_\theta \chi_\omega)\chi\omega \tag{56}$$

then

$$q_\theta = L\rho_\theta. \tag{57}$$

Remark 2.4 *In the statement of the null-controllability problem, there are boundary and initial or end conditions. These conditions concern the values of the control or state functions at the points of the boudary for example. The solutions dealt by means of functionnal analysis are not functions but elements of function spaces which are equivalence classes. As a consequence boundary or initial or end values of the solutions have to be considered in function spaces. Such a question has been adressed by Lions-Magenes. We refer to [9] for deriving the following trace theorems in regular opens sets Ω.*

Assume $q \in L^2(U \times \Omega) \simeq L^2\left(U, L^2\left(\Omega\right)\right)$ and $\Delta q \in H^{-1}(U, L^2(\Omega))$. Then

$$q|_{U\times\Gamma} \in H^{-1}(U, H^{-\frac{1}{2}}(\Gamma)).$$

The meaning of $q_{|\Sigma}$, the trace of q on Σ, is clear.

Assume $q \in L^2(U \times \Omega) \simeq L^2\left([T_0, T] \times [0, A], L^2\left(\Omega\right)\right)$ and $\frac{\partial q}{\partial t} + \frac{\partial q}{\partial a} \in L^2(U, H^{-2}(\Omega))$. Then

$$q \in c\left([0,A], L^2([T_0,T], H^{-2}(\Omega))\right) \cap c\left([T_0,T], L^2([0,A], H^{-2}(\Omega))\right).$$

That means there exists some function $\widetilde{q} : [T_0, T] \times [0, A] \rightarrow L^2([T_0, T], H^{-2}(\Omega))$ standing for $q \in L^2(U \times \Omega)$ which is separately continuous so that the following values in $L^2([T_0, T], H^{-2}(\Omega))$ get sense

$$\forall (t, a) \in U, q\ (t, a) = \widetilde{q}\ (t, a)$$

and

$$\begin{aligned} q(T) &\in L^2([0,A],H^{-2}(\Omega)) \\ q(0) &\in L^2([T_0,T],H^{-2}(\Omega)) \\ q(A) &\in L^2([T_0,T],H^{-2}(\Omega)) \end{aligned}$$

Proposition 2.1 *Keep the notations in the statement of the null-controllability problem. Assume (34) and the function h as in(52). Then there exists some couple* (v,q) *is such that (22)-(24) hold.*

Proof 2.3 *The couple* $(\nu,q)=(\nu_\theta,q_\theta)$ *defined above from* ρ_θ *stands for the solution. Since* $\rho_\theta \in V$ *then* $v_\theta \in L^2(\mathcal{G})$ *and* $q_\theta \in L^2(Q)$. *Since* $P\rho_\theta\chi_\omega \in \mathcal{K}$ *then* $v_\theta = -(\rho_\theta - P\rho_\theta\chi_\omega)\chi\omega \in \mathcal{K}^\perp$. *By direct substitution in the formulas (49), (55) and (57) it follows*

$$\forall \rho \in V, \int_Q q_\theta L\rho dtdadx + \int_{\mathcal{G}} (\rho_\theta - P\rho_\theta)(\rho - P\rho)dtdadx = \int_Q h\rho dtdadx. \quad (58)$$

Because $P\rho\chi_\omega \in \mathcal{K}$, *it reduces to*

$$\forall \rho \in V, \int_Q q_\theta L\rho dtdadx = \int_Q h\rho dtdadx - \int_{\mathcal{G}} (\rho_\theta - P\rho_\theta)\rho dtdadx, \quad (59)$$

i.e.

$$\forall \rho \in V, \int_Q q_\theta L\rho dtdadx = \int_Q h\rho dtdadx + \int_{\mathcal{G}} v_\theta \rho dtdadx. \quad (60)$$

In the duality frame $\mathcal{D}(Q),\mathcal{D}'(Q)$ *(60) means that*

$$L^* q_\theta = h + v_\theta\chi_\omega \text{ in } \mathcal{D}'(Q). \quad (61)$$

Besides $h+v_\theta\chi_\omega \in L^2(Q)$, *then* $L^*q_\theta \in L^2(Q)$.

Since $q_\theta \in L^2(Q)$ *and* $\Delta q_\theta \in H^{-1}(U,L^2(\Omega))$ *and by the above remark* $q_{\theta|_{U\times\Gamma}} \in H^{-1}(U,H^{-\frac{1}{2}}(\Gamma))$. *Similarly, since* $q_\theta \in L^2(Q))$ *and* $\frac{\partial q_\theta}{\partial t} + \frac{\partial q_\theta}{\partial a} \in L^2(U,H^{-2}(\Omega))$ $q_\theta(0) \in L^2([0,A],H^{-2}(\Omega))$, $q_\theta(T) \in L^2([0,A],H^{-2}(\Omega))$; $q_\theta(0) \in L^2([T_0,T],H^{-2}(\Omega))$ *and* $q_\theta(A) \in L^2([T_0,T],H^{-2}(\Omega))$.

Taking into account (61), integrate by parts

$$\begin{aligned} \forall \rho \in \mathcal{V} \int_Q q_\theta L\rho dtdadx &+ \int_U \left\langle q_\theta, \frac{\partial \rho}{\partial \nu} \right\rangle_{H^{-\frac{1}{2}}(\Gamma),H^{\frac{1}{2}}(\Gamma)} dtda \\ &+ \int_0^T [\langle q_\theta(t,0),\rho(t,0)\rangle_{H^{-2}(\Omega),H^2(\Omega)} - \langle q_\theta(t,A),\rho(t,A)\rangle_{H^{-2}(\Omega),H^2(\Omega)}]dt \\ &+ \int_0^A [\langle q_\theta(T_0,a),\rho(T_0,a)\rangle_{H^{-2}(\Omega),H^2(\Omega)} - \langle q_\theta(T,a),\rho(T,a)\rangle_{H^{-2}(\Omega),H^2(\Omega)}]da \\ &= \int_Q (h+v_\theta\chi_\omega)\rho dtdadx. \quad (62) \end{aligned}$$

By (60) since $\mathcal{V} \subset V$, it follows

$$\forall \rho \in \mathcal{V}, \int_U \left\langle q_\theta, \frac{\partial \rho}{\partial \nu} \right\rangle_{H^{-\frac{1}{2}}(\Gamma), H^{\frac{1}{2}}(\Gamma)} dtda$$
$$+ \int_0^T [\langle q_\theta(t,0), \rho(t,0)\rangle_{H^{-2}(\Omega),H^2(\Omega)} - \langle q_\theta(t,A), \rho(t,A)\rangle_{H^{-2}(\Omega),H^2(\Omega)}]dt$$
$$+ \int_0^A [\langle q_\theta(T_0,a), \rho(T_0,a)\rangle_{H^{-2}(\Omega),H^2(\Omega)} - \langle q_\theta(T,a), \rho(T,a)\rangle_{H^{-2}(\Omega),H^2(\Omega)}]da$$
$$= 0. \quad (63)$$

Then, successively, we get $q_\theta = 0$ on Σ, $q_\theta(T_0) = 0$ and $q_\theta(T) = 0$ in Q_A, $q_\theta(0) = 0$ and $q_\theta(A) = 0$ in Q_T. Since $q_\theta(0) = 0$ we have $L^* q_\theta = \beta q_\theta(0) + h + v_\theta \chi_\omega$. Hence the proof is completed.

Remark 2.5 *Adapted observability inequality (35) shows that the choice of the scalar product on $\mathcal{V}$ is not unique. Thus, there exists an infinity of control functions v such that (22)-(24) hold.*

Consider the set of *control variables* v such that (22)-(24) hold. By Proposition 4 this set is nonempty and it is clearly convex and closed in $L^2(\mathcal{G})$. Therefore, there exists a unique *control variable* $\widehat{v}_\theta$ of minimal norm in $L^2(\mathcal{G})$. We state then the following :

Theorem 2.6 *Assume (34). For every h satisfying (52), there exists some control variable v such that (22)-(24) hold. Moreover, we can get a unique control $\widehat{v}_\theta$ such that (25) holds.*

Now, we are concerned with the optimality system for $\widehat{v}_\theta$.

3 Optimality System for the Optimal Control Variable

Let $\widehat{q}_\theta$ be the solution of system (23) submitted to the optimal control $\widehat{v}_\theta$. A classical way to derive the optimality system for the couple $(\widehat{v}_\theta, \widehat{q}_\theta)$ is the penalization method by Lions [8].

Describe now the method. Let $\varepsilon > 0$. Define the functional

$$\begin{aligned} J_\varepsilon(v,q) &= \frac{1}{2}\|v\|^2_{L^2(\mathcal{G})} \\ &+ \frac{1}{2\varepsilon}\left\| -\frac{\partial q}{\partial t} - \frac{\partial q}{\partial a} - \Delta q + \mu q - \beta q(t,0,x) - h - v\chi_\omega \right\|^2_{L^2(Q)} \end{aligned} \quad (64)$$

for any couple (v,q) such that

$$\begin{cases} v \in \mathcal{K}^\perp, \; q \in L^2(Q), \\ -\frac{\partial q}{\partial t} - \frac{\partial q}{\partial a} - \Delta q + \mu q - \beta q(0) \in L^2(Q), \\ q = 0 \text{ on } \Sigma; q(T) = 0 \text{ in } Q_A; q(A) = 0 \text{ in } Q_T; \\ q(T_0) = 0 \text{ in } Q_A. \end{cases} \quad (65)$$

Consider the minimization problem

$$\inf\{J_\varepsilon(v,q) \mid (v,q) \text{ subject to } (65)\}. \quad (66)$$

Proposition 3.1 *Under the assumptions of Theorem 8, the minimization problem has an optimal solution. There exists* $(v_\varepsilon, q_\varepsilon)$ *such that*

$$J_\varepsilon(v_\varepsilon, q_\varepsilon) = \min\{J_\varepsilon(v,q) \mid (v,q) \text{ subject to (65)}\}. \tag{67}$$

Proof 3.1 *Let* (v_n, q_n) *be a minimizing sequence satisfying (65). The sequence* $(J_\varepsilon(v_n, q_n))_n$ *is bounded from above*

$$J_\varepsilon(v_n, q_n) \leq \gamma(\varepsilon) \tag{68}$$

then

$$\begin{cases} \|v_n\|_{L^2(\mathcal{G})} \leq C(\varepsilon) \\ \left\| -\frac{\partial q_n}{\partial t} - \frac{\partial q_n}{\partial a} - \Delta q_n + \mu q_n - \beta q_n(0) - h - v_n \chi_\omega \right\|_{L^2(Q)} \leq \sqrt{\varepsilon} C(\varepsilon). \end{cases} \tag{69}$$

There is some subsequence of $(v_n)_n$, *still denoted by* $(v_n)_n$, *such that*

$$v_n \rightharpoonup v_\varepsilon \text{ weakly in } L^2(\mathcal{G}). \tag{70}$$

As a consequence (65) the (sub)sequence $(q_n)_n$ *is bounded*

$$\|q_n\|_{L^2(Q)} \leq C. \tag{71}$$

There is some subsequence of $(q_n)_n$, *still denoted by* $(q_n)_n$ *such that*

$$q_n \rightharpoonup q_\varepsilon \text{ weakly in } L^2(Q). \tag{72}$$

Then

$$\liminf J_\varepsilon(v_n, q_n) \geq J_\varepsilon(v_\varepsilon, q_\varepsilon). \tag{73}$$

We deduce that $(v_\varepsilon, q_\varepsilon)$ *is a unique optimal control, from the strict convexity of* J_ε.

Now, we study the convergence of $(v_\varepsilon, q_\varepsilon)_\varepsilon$.

Proposition 3.2 *Let* $((v_\varepsilon, q_\varepsilon))_\varepsilon$ *be the family of solutions of (67). Then*

$$\begin{cases} \lim_{\varepsilon\to 0} v_\varepsilon = \widehat{v}_\theta \text{ weakly in } L^2(U\times\omega) \\ \lim_{\varepsilon\to 0} q_\varepsilon \rightharpoonup \widehat{q}_\theta \text{ weakly in } L^2(Q). \end{cases} \tag{74}$$

Proof 3.2 $(\widehat{v}_\theta, \widehat{q}_\theta)$ *satisfies (22)-(24), then from the structure (64) of* $J_\varepsilon(v,q)$, *we have of course*

$$\begin{cases} \|v_\varepsilon\|_{L^2(\mathcal{G})} \leq C \\ \left\| -\frac{\partial q_\varepsilon}{\partial t} - \frac{\partial q_\varepsilon}{\partial a} - \Delta q_\varepsilon + \mu q_\varepsilon - \beta q_\varepsilon(t,0,x) - h - v_\varepsilon \chi_\omega \right\|_{L^2(Q)} \leq C\sqrt{\varepsilon}. \end{cases} \tag{75}$$

Where C's are various constants independent of ε.
q_ε *verifies (65) (75) then*

$$\|q_\varepsilon\|_{L^2(Q)} \leq C. \tag{76}$$

There are a subseqence of $(v_\varepsilon, q_\varepsilon)_\varepsilon$*, again denoted by* $(v_\varepsilon, q_\varepsilon)_\varepsilon$, $v_0 \in L^2(\mathcal{G})$ *and* $q_0 \in L^2(Q)$ *such that*

$$v_\varepsilon \rightharpoonup v_0 \text{ weakly in } L^2(\mathcal{G}), v_0 \in \mathcal{K}^\perp;\ q_\varepsilon \rightharpoonup q_0 \text{ weakly in } L^2(Q). \tag{77}$$

Then $q_\varepsilon \rightharpoonup q_0$ *weakly in* $\mathcal{D}'(Q)$ *and by the weak continuity of the operator* L^* *in* $\mathcal{D}'(Q)$ *it follows* $L^* q_\varepsilon \rightharpoonup L^* q_0$ *weakly in* $\mathcal{D}'(Q)$*. Moreover the traces functions are continuous, then the pair* (v_0, q_0) *satisfies the system*

$$\begin{cases} -\frac{\partial q_0}{\partial t} - \frac{\partial q_0}{\partial a} - \Delta q_0 + \mu q_0 = \beta q_0(0) + h + v_0 \chi_\omega \ in\ Q, \\ q_0 = 0 \quad on\ \Sigma, \\ q_0(T) = 0 \quad in\ Q_A, \\ q_0(A) = 0 \quad in\ Q_T. \end{cases} \tag{78}$$

$$q_0(T_0) = 0 \ \ in\ Q_A. \tag{79}$$

From the following estimate

$$\|v_\varepsilon\|^2_{L^2(\mathcal{G})} \le J_\varepsilon(v_\varepsilon, q_\varepsilon) \tag{80}$$

we get

$$\|v_0\|^2_{L^2(\mathcal{G})} \le \liminf J_\varepsilon(v_\varepsilon, q_\varepsilon). \tag{81}$$

Since the pair $(\widehat{v}_\theta, \widehat{q}_\theta)$ *satisfies (22)-(24)*

$$\liminf J_\varepsilon(v_\varepsilon, q_\varepsilon) \le \|\widehat{v}_\theta\|^2_{L^2(\mathcal{G})}.$$

Thus

$$\|v_0\|_{L^2(\mathcal{G})} \le \|\widehat{v}_\theta\|_{L^2(\mathcal{G})}$$

and then

$$\|v_0\|_{L^2(\mathcal{G})} = \|\widehat{v}_\theta\|_{L^2(\mathcal{G})}.$$

Hence $v_0 = \widehat{v}_\theta$.
Since (78) has a unique solution, it follows $q_0 = \widehat{q}_\theta$.

Let us express the optimality conditions satisfied by $(v_\varepsilon, q_\varepsilon)$, as optimal solution of (67).

Proposition 3.3 *The assumptions are as in Theorem 8. The couple* $(v_\varepsilon, q_\varepsilon)$ *is optimal solution of problem (67) if and only if there is one function* ρ_ε *such that the triplet* $\{v_\varepsilon, q_\varepsilon, \rho_\varepsilon\}$ *satisfies the following so called optimality system*

$$\begin{cases} -\frac{\partial q_\varepsilon}{\partial t} - \frac{\partial q_\varepsilon}{\partial a} - \Delta q_\varepsilon + \mu q_\varepsilon = \beta q_\varepsilon(0) + h + v_\varepsilon \chi_\omega + \varepsilon \rho_\varepsilon \ in\ Q, \\ q_\varepsilon = 0 \ \ in\ \ \Sigma, \\ q_\varepsilon(T) = 0 \quad in\ \ Q_A, \\ q_\varepsilon(A) = 0 \ \ in\ \ Q_T, \end{cases} \tag{82}$$

$$q_\varepsilon(T_0) = 0 \ \ in\ Q_A, \tag{82}$$

$$\begin{cases} \frac{\partial \rho_\varepsilon}{\partial t} + \frac{\partial \rho_\varepsilon}{\partial a} - \Delta \rho_\varepsilon + \mu \rho_\varepsilon = 0 \ in\ Q, \\ \rho_\varepsilon = 0 \ in\ \Sigma, \\ \rho_\varepsilon(0) = \int_0^A \beta(a) \rho_\varepsilon(a) da \ in\ Q_T, \end{cases} \tag{83}$$

$$v_\varepsilon = -(\rho_\varepsilon - P \rho_\varepsilon \chi_\omega) \chi_\omega. \tag{84}$$

Proof 3.3 *Express the Euler-Lagrange optimality conditions which characterize* $(v_\varepsilon, q_\varepsilon)$. *For any* (v, φ) *such that (65) the following holds*

$$\int_{\mathcal{G}} v_\varepsilon v dt da dx + \frac{1}{\varepsilon}\int_Q \left(-\frac{\partial q_\varepsilon}{\partial t} - \frac{\partial q_\varepsilon}{\partial a} - \Delta q_\varepsilon + \mu q_\varepsilon - \beta q_\varepsilon(0) - h - v_\varepsilon \chi_\omega\right) \times \left(-\frac{\partial \varphi}{\partial t} - \frac{\partial \varphi}{\partial a} - \Delta\varphi + \mu\varphi - \beta\varphi(0) - v\chi_\omega\right) dt da dx = 0. \quad (85)$$

Define the adjoint state

$$\rho_\varepsilon = -\frac{1}{\varepsilon}\left(-\frac{\partial q_\varepsilon}{\partial t} - \frac{\partial q_\varepsilon}{\partial a} - \Delta q_\varepsilon + \mu q_\varepsilon - \beta q_\varepsilon(0) - h - v_\varepsilon \chi_\omega\right). \quad (86)$$

Then (82) holds .
For any (v, φ) *such that (65), (85) becomes*

$$\int_{\mathcal{G}} v_\varepsilon v dt da dx + \int_Q \rho_\varepsilon \left(-\frac{\partial \varphi}{\partial t} - \frac{\partial \varphi}{\partial a} - \Delta\varphi + \mu\varphi - \beta\varphi(0) - v\chi_\omega\right) dt da dx = 0. \quad (87)$$

Integrate by parts in (87). As a consequence the couple $(v_\varepsilon, \rho_\varepsilon)$ *is shown to satisfy*

$$\begin{cases} \frac{\partial \rho_\varepsilon}{\partial t} + \frac{\partial \rho_\varepsilon}{\partial a} - \Delta\rho_\varepsilon + \mu\rho_\varepsilon = 0 \text{ in } Q \\ \rho_\varepsilon = 0 \text{ on } \Sigma \\ \rho_\varepsilon(0) = \int_0^A \beta(a)\rho_\varepsilon(a)da \end{cases} \quad (88)$$

and

$$\int_{\mathcal{G}} (v_\varepsilon + \rho_\varepsilon) v dt da dx = 0, \ \forall v \in \mathcal{K}^\perp. \quad (89)$$

Hence $v_\varepsilon + \rho_\varepsilon\chi_\omega \in \mathcal{K}$. *Since* $v_\varepsilon \in \mathcal{K}^\perp$ *then* $v_\varepsilon + \rho_\varepsilon\chi_\omega = P(v_\varepsilon + \rho_\varepsilon\chi_\omega) = P\rho_\varepsilon\chi_\omega$ *and thus*

$$v_\varepsilon = -(\rho_\varepsilon - P\rho_\varepsilon\chi_\omega)\chi_\omega. \quad (90)$$

Hence the assertion follows.

Remark 3.1 *There is no available information concerning* $\rho_\varepsilon(A)$ *in* Q_T, $\rho_\varepsilon(0)$ *in* Q_A, $\rho_\varepsilon(T)$ *in* Q_A.

A priori estimates for approximate adjoint state $(\rho_\varepsilon)_\varepsilon$ are now looked for. This is the essential point.
From (90) and (75)

$$\|(\rho_\varepsilon - P\rho_\varepsilon\chi_\omega)\chi_\omega\|_{L^2(\mathcal{G})} \leq C. \quad (91)$$

Since $L\rho_\varepsilon = 0$, then

$$\|\rho_\varepsilon\|_V \leq C. \quad (92)$$

Therefore there are a subseqence $(\rho_\varepsilon)_\varepsilon$ still denoted by $(\rho_\varepsilon)_\varepsilon$ and $\widehat{\rho}_\theta \in V$ such that

$$\rho_\varepsilon \rightharpoonup \widehat{\rho}_\theta \text{ weakly in } V. \quad (93)$$

Following the lines of the proof of Lemma 3, we conclude with the weak convergence $\rho_\varepsilon \rightharpoonup \widehat{\rho}_\theta$ in $L^2(\mathcal{G})$. Thus

$$P\rho_\varepsilon\chi_\omega \to \chi_0 \text{ strongly in } L^2(\mathcal{G}) \tag{93}$$

so that $\chi_0 \in \mathcal{K}$.

By (91) and (93) there is some $\chi_1 \in \mathcal{K}^\perp$

$$\rho_\varepsilon - P\rho_\varepsilon\chi_\omega \rightharpoonup \chi_1 \text{weakly in } V$$

so that $\widehat{\rho}_\theta = \chi_0 + \chi_1$. Thus $\chi_0 = P\widehat{\rho}_\theta\chi_\omega$ and

$$\rho_\varepsilon\chi_\omega - P\rho_\varepsilon\chi_\omega \rightharpoonup \widehat{\rho}_\theta\chi_\omega - P\widehat{\rho}_\theta\chi_\omega \text{ weakly in } L^2(\mathcal{G}). \tag{94}$$

The above argument leads to the following statement :

Theorem 3.2 *The assumptions are as in Theorem 8. The couple $(\widehat{v}_\theta, \widehat{q}_\theta)$ is the optimal solution of problem (22)-(25) if and only if there is a function $\widehat{\rho}_\theta$ such that the triplet $(\widehat{v}_\theta, \widehat{q}_\theta, \widehat{\rho}_\theta)$ is solution of the following optimality system*

$$\widehat{v}_\theta \in \mathcal{K}^\perp,\ \widehat{q}_\theta \in L^2(Q),\ \widehat{\rho}_\theta \in V \tag{95}$$

$$\begin{cases} -\frac{\partial \widehat{q}_\theta}{\partial t} - \frac{\partial \widehat{q}_\theta}{\partial a} - \Delta\widehat{q}_\theta + \mu\widehat{q}_\theta = \beta\widehat{q}_\theta(0) + h + \widehat{v}_\theta\chi_\omega \ in\ Q, \\ \widehat{q}_\theta = 0 \quad on\ \ \Sigma, \\ \widehat{q}_\theta(T) = 0 \quad in\ \ Q_A, \\ \widehat{q}_\theta(A) = 0 \quad in\ \ Q_T, \end{cases} \tag{96}$$

$$\widehat{q}_\theta(T_0) = 0 \ \ in\ \ Q_A, \tag{97}$$

$$\begin{cases} \frac{\partial \widehat{\rho}_\theta}{\partial t} + \frac{\partial \widehat{\rho}_\theta}{\partial a} - \Delta\widehat{\rho}_\theta + \mu\widehat{\rho}_\theta = 0 \ in\ Q, \\ \widehat{\rho}_\theta = 0 \quad on\ \ \Sigma, \\ \widehat{\rho}_\theta(0) = \int_0^A \beta(a)\widehat{\rho}_\theta(a)da \ in\ Q_T, \end{cases} \tag{98}$$

$$\widehat{v}_\theta = -(\widehat{\rho}_\theta - P\widehat{\rho}_\theta\chi_\omega)\chi_\omega. \tag{99}$$

4 Discriminating Sentinels

It is clear, when $h = h_0\chi_O + k_0\chi_\omega$ that the problem (18)-(21) and the problem (22)-(25) are the same. The results concerning the former apply to the latter. Consider then the results obtained in the previous sections. More precisely, assume that $\mathcal{K}$ is defined as in (15) and h_0, k_0 are such that (52) holds. Assume also (9). Let $(\widehat{v}_\theta, \widehat{q}_\theta, \widehat{\rho}_\theta)$ be defined as in Theorem 13. Then

$$\widehat{v}_\theta = -(\widehat{\rho}_\theta - P\widehat{\rho}_\theta\chi_\omega)\chi_\omega. \tag{100}$$

Definition - *The function S defined by*

$$S(\lambda,\tau) = \int_U\int_O h_0 y(\lambda,\tau)dtdadx + \int_U\int_\omega (k_0 - (\widehat{\rho}_\theta - P\widehat{\rho}_\theta\chi_\omega))y(\lambda,\tau)dtdadx \tag{101}$$

is called discriminating sentinel defined by the open subsets O and ω, and by the function h_0.

Return to the problem (H_0). The above function S allows to lay down the relevant question.

Because of (6) we can write

$$S(\lambda,\tau) = S(0,0) + \lambda \frac{\partial S}{\partial \lambda}(0,0) \text{ for } \lambda \text{ and } \tau \text{ small,} \tag{102}$$

where S is defined by (101). On the other hand, by (2) and (7)

$$S(\lambda,\tau) = \int_U \int_O h_0 m_0 dt da dx + \int_U \int_\omega \widehat{v}_\theta m_0 dt da dx. \tag{103}$$

So that

$$\int_U \int_O h_0 (m_0 - y_0) dt da dx + \int_U \int_\omega \widehat{v}_\theta (m_0 - y_0) dt da dx = \lambda \frac{\partial S}{\partial \lambda}(0,0). \tag{104}$$

We also have

$$\frac{\partial S}{\partial \lambda}(0,0) = \int_U \int_O h_0 y_\lambda dt da dx + \int_U \int_\omega \widehat{v}_\theta y_\lambda dt da dx \tag{105}$$

where y_λ is given by

$$\begin{cases} \frac{\partial y_\lambda}{\partial t} + \frac{\partial y_\lambda}{\partial a} - \Delta y_\lambda + \mu y_\lambda = \widehat{f} \text{ in } Q, \\ y_\lambda = 0 \text{ in } \Sigma, \\ y_\lambda(T_0) = 0 \text{ in } Q_A, \\ y_\lambda(0) = \int_0^A \beta(a) y_\lambda(a) da \text{ in } Q_T. \end{cases} \tag{106}$$

Let $\widehat{q}_\theta(h_0)$ be the solution (uniquely defined by h_0) of (96). Multiplying (96) by y_λ it follows

$$\int_U \int_O h_0 y_\lambda dt da dx + \int_U \int_\omega \widehat{v}_\theta y_\lambda dt da dx = \int_Q \widehat{q}_\theta(h_0) \widehat{f} dt da dx. \tag{107}$$

Therefore (104) becomes

$$\int_U \int_O h_0 (m_0 - y_0) dt da dx + \int_U \int_\omega \widehat{v}_\theta (m_0 - y_0) dt da dx = \lambda \int_Q \widehat{q}_\theta(h_0) \widehat{f} dt da dx. \tag{108}$$

This equation allows to evaluate the right hand side of (108) and therefore it is an available information concerning $\lambda \widehat{f}$.

A perturbation (or a "pollution") $\widehat{f}$ will be stealthy for the sentinel defined by h_0 if

$$\int_Q \widehat{q}_\theta(h_0) \widehat{f} dt da dx = 0. \tag{109}$$

5 Conclusion

As a conclusion we announce two lines in prolongation of our work.

The first one is concerned with a generalization of the idea of sentinel which tackles new observation or control designs. Indeed it has been shown above that sentinel by J. L. Lions can hoock on some two-time scale varying system with both interior observation and control. But several other designs are realic. For example, only one part of the boundary may be observed, nd the control may be constrained and/or restricted to the boundary. Different mixing are possible which could be solved by means of some setting so called Carmeman inequality. This the aim of some forthcomming paper.

The second line in process is concerned with furtivity as it is defined at the end of above section.

Aknowledgments

Authors wish to express theirs gratitudes to Robert Janin for his many helpful suggestions.

References

[1] B.Ainseba, 2002, Exact and approximate controllability of age and space population dynamics structured model, *J. Math.Anal.Appl.*

[2] B.Ainseba and M.Langlais, 2000, On a population dynamics control problem with age dependence and spatial structure, *J. Math. Anal. Appl.* **248**,455–474.

[3] B. Ainseba et M.Langlais, 1996, Sur un problème de contrôle d'une population structurée en âge et en espace.*C.R.Acad.Sci.Paris,* **t.323**,serie I, 269–274.

[4] B.Ainseba and S.Anita, 2001, Local exact controllability of the age-dependent population dynamics with diffusion.*Abstract Appl.Anal.*,**6**, 357-368.

[5] H.Brezis: *Analyse fonctionnelle.Théorie et application.* Masson, Paris 1983.

[6] A.Fursikov, O.Imanuvilov,1996, Controllability of evolution equation.*Lecture Notes Series* **34**,RIM-GARC, Seoul National University.

[7] M.Giovanna and M.Langlais, 1982, Age-dependent population diffusion with external constraint., *J.Math.Biology*, **14**, 7-94.

[8] J.L.Lions: *Sentinelles pour les systèmes distribués à données incomplètes*.Masson, Paris 1992.

[9] J.L. Lions et M. Magenes . (1968) *Problèmes aux limites non homogènes et applications*. Paris, Dunod,Vol. 1 et 2.

[10] O. Nakoulima, 2004, Contrôlabilité à zéro avec contraintes sur le contrôle. *C. R. Acad. Sci. Paris, Ser.I* **339**/6 405-410.

[11] A. Ouédraogo, O. Traoré, 2003, Sur un problème de dynamique des populations, *Imhotep*. Vol 4 n°1.

[12] S.Sawadogo, 2005, Thèse unique.Université de Ouagadougou.

In: Research on Evolution Equation Compendium. Volume 2 ISBN: 978-1-60456-752-6
Editor: Gaston M. N'Guerekata

Center Manifold and Stability in Critical Cases for Some Partial Functional Differential Equations*

Mostafa Adimy[1,†]***, Khalil Ezzinbi***[2‡] ***and Jianhong Wu***[3§]
[1]Université de Pau et des Pays de l'Adour, Laboratoire
de Mathématiques Appliquées CNRS UMR 5142
Avenue de l'université 64000, Pau, France
[2]Université Cadi Ayyad, Faculté des Sciences Semlalia,
Département de Mathématiques, B.P. 2390, Marrakesh, Morocco
[3]York University, Department of Mathematics and Statistics,
Faculty of Pure and Applied Sciences, 4700 Keele Street North York,
Ontario, Canada, M3J 1P3

Abstract

In this work, we prove the existence of a center manifold for some partial functional differential equations, whose linear part is not necessarily densely defined but satisfies the Hille-Yosida condition. The attractiveness of the center manifold is also shown when the unstable space is reduced to zero. We prove that the flow on the center manifold is completely determined by an ordinary differential equation in a finite dimensional space. In some critical cases, when the exponential stability is not possible, we prove that the uniform asymptotic stability of the equilibrium is completely determined by the uniform asymptotic stability of the reduced system on the center manifold.

Key Words: Hille-Yosida operator, integral solution, semigroup, variation of constants formula, center manifold, attractiveness, reduced system, critical case, asymptotic stability, approximation.
2000 Mathematical Subject Classification: 34K17, 34K19, 34K20, 34K30, 34G20, 47D06.

---

*This research is supported by Grant from CNCPRST (Morocco) and CNRS(France) Ref. SPM 17769, by TWAS Grant under contract Ref. 03-030 RG/MATHS/AF/AC, by the Canada Research Chairs Program, by Natural Sciences and Engineering Research Council of Canada, and by Mathematics for Information Technology and Complex Systems.

†E-mail address: mostafa.adimy@univ-pau.fr
‡E-mail address: ezzinbi@ucam.ac.ma: to whom all correspondence should be sent
§E-mail address: wujh@mathstat.yorku.ca

1 Introduction

The aim of this paper is to study the existence of a center manifold and stability in some critical cases for the following partial functional differential equation

$$\begin{cases} \frac{d}{dt}u(t) = Au(t) + L(u_t) + g(u_t), \ t \geq 0 \\ u_0 = \varphi \in C := C([-r,0];E), \end{cases} \tag{1.1}$$

where A is not necessarily densely defined linear operator on a Banach space E and C is the space of continuous functions from $[-r,0]$ to E endowed with the uniform norm topology. For every $t \geq 0$ and for a continuous $u : [-r,+\infty) \longrightarrow E$, the function $u_t \in C$ is defined by

$$u_t(\theta) = u(t+\theta) \ \text{ for } \theta \in [-r,0].$$

L is a bounded linear operator from C into E and g is a Lipschitz continuous function from C to E with $g(0) = 0$.

In this work, we assume that A is a Hille-Yosida operator: there exist $\omega \in \mathbb{R}$ and $M_0 \geq 1$ such that $(\omega,\infty) \subset \rho(A)$ and

$$\left|(\lambda I - A)^{-n}\right| \leq \frac{M_0}{(\lambda-\omega)^n} \ \text{ for } \lambda \geq \omega \text{ and } n \in \mathbb{N},$$

where $\rho(A)$ is the resolvent set of A. In [21], the authors proved the existence, regularity and stability of solutions of (1.1) when A generates a strongly continuous semigroup, which is equivalent by Hille-Yosida Theorem to that A is a Hille-Yosida operator and $\overline{D(A)} = E$. In [3], the authors used the integrated semigroup approach to prove the existence and regularity of solutions of (1.1) when A is only a Hille-Yosida operator. Moreover, it was shown that the phase space of equation (1.1) is given by

$$Y := \left\{\varphi \in C : \varphi(0) \in \overline{D(A)}\right\}.$$

Assume that the function g is differentiable at 0 with $g'(0) = 0$. Then the linearized equation of (1.1) around the equilibrium zero is given by

$$\begin{cases} \frac{d}{dt}v(t) = Av(t) + L(v_t), \ t \geq 0 \\ v_0 = \varphi \in C. \end{cases} \tag{1.2}$$

If all characteristic values (see section 2) of equation (1.2) have negative real part, then the zero equilibrium of (1.1) is uniformly asymptotically stable. However, if there exists at least one characteristic value with a positive real part, then the zero solution of (1.1) is unstable. In the critical case, when exponential stability is not possible and there exists a characteristic value with zero real part, the situation is more complicated since either stability or instability may hold. The subject of the center manifold is to study the stability in this critical case. For differential equations, the center manifold theory has been extensively studied, we refer to [6], [7], [8], [12], [13], [14], [15], [16], [19], [20] and [24]. In [17] and [22], the authors proved the existence of a center manifold when $\overline{D(A)} = E$.

They established the attractiveness of this manifold when the unstable space is reduced to zero. In [11], the authors proved the existence of a center manifold for a given map. Their approach was applied to show the existence of a center manifold for partial functional differential equations in Banach spaces in the case when the linear part generates a compact strongly continuous semigroup. Recently, in [18], the authors studied the existence of invariant manifolds for an evolutionary process in Banach spaces and in particularly for some partial functional differential equations. For more details about the center manifold theory and its applications in the context of partial functional differential equations, we refer to the monograph [27]. Here we consider equation (1.1) when the domain $D(A)$ is not necessarily dense in E. The nondensity occurs, in many situations, from restrictions made on the space where the equations are considered (for example, periodic continuous solutions, H"older continuous functions) or from boundary conditions (the space C^1 with null value on the boundary is not dense in the space of continuous functions). For more details, we refer to [1], [2], [3], [4] and [5].

The organization of this work is as follows: in section 2, we recall some results of integral solutions and the semigroup solution and we describe the variation of constants formula for the associated non-homogeneous problem of (1.2). We also give some results on the spectral analysis of the linear equation (1.2). In section 3, we prove the existence of a global center manifold. In section 4, we prove that this center manifold is exponentially attractive when the unstable space is reduced to zero. In section 5, we prove that the flow on the center manifold is governed by an ordinary differential equation in a finite dimensional space. In section 6, we prove a result on the stability of the equilibrium in the critical case. We also establish a new reduction principal for equation (1.1). In section 7, we study the existence of a local center manifold when g is only defined and C^1-function in a neighborhood of zero. In the last section, we propose a result on the stability when zero is a simple characteristic value and no characteristic value lies on the imaginary axis.

2 Spectral Analysis and Variation of Constants Formula

In the following we assume

($\mathbf{H}_1$) A is a Hille-Yosida operator.

Definition 2.1. A continuous function $u : [-r, +\infty) \to E$ is called an integral solution of equation (1.1) if

i) $\int_0^t u(s)ds \in D(A)$ for $t \geq 0$,

ii) $u(t) = \varphi(0) + A\left(\int_0^t u(s)ds\right) + L\left(\int_0^t u_s ds\right) + \int_0^t g(u_s)ds$ for $t \geq 0$,

iii) $u_0 = \varphi$.

We will call, without causing any confusion, the integral solution the function u_t, for $t \geq 0$.

Let A_0 be the part of the operator A in $\overline{D(A)}$ which is defined by

$$\begin{cases} D(A_0) = \left\{x \in D(A) : Ax \in \overline{D(A)}\right\} \\ A_0 x = Ax \quad \text{for } x \in D(A_0). \end{cases}$$

The following result is well known (see [3]).

Lemma 2.2. *A_0 generates a strongly continuous semigroup $(T_0(t))_{t\geq 0}$ on $\overline{D(A)}$.*

For the existence and uniqueness of an integral solution of (1.1), we need the following condition.

$(\mathbf{H}_2)$ $g : C \longrightarrow E$ is Lipschitz continuous.

The following result can be found in [3].

Proposition 2.3. *Assume that $(\mathbf{H}_1)$ and $(\mathbf{H}_2)$ hold. Then for $\varphi \in Y$, equation (1.1) has a unique global integral solution on $[-r,\infty)$ which is given by the following formula*

$$u(t) = \begin{cases} T_0(t)\varphi(0) + \lim_{\lambda\to\infty}\int_0^t T_0(t-s)B_\lambda(L(u_s)+g(u_s))\,ds, \; t \geq 0 \\ \varphi(t), \; t \in [-r,0], \end{cases} \tag{2.1}$$

where $B_\lambda = \lambda(\lambda I - A)^{-1}$ for $\lambda \geq \omega$.

Assume that g is differentiable at zero with $g'(0) = 0$. Then the linearized equation of (1.1) at zero is given by equation (1.2). Define the operator $U(t)$ on Y by

$$U(t)\varphi = v_t(.,\varphi),$$

where v is the unique integral solution of equation (1.2) corresponding to the initial value φ. Then $(U(t))_{t\geq 0}$ is a strongly continuous semigroup on Y. One has the following linearized principle.

Theorem 2.4. *Assume that $(\mathbf{H}_1)$ and $(\mathbf{H}_2)$ hold. If the zero equilibrium of $(U(t))_{t\geq 0}$ is exponentially stable, in the sense that there exist $N_0 \geq 1$ and $\varepsilon \geq 0$ such that*

$$|U(t)| \leq N_0 e^{-\varepsilon t} \text{ for } t \geq 0,$$

then the zero equilibrium of equation (1.1) is locally exponentially stable, in the sense that there exist $\delta \geq 0$, $\mu \geq 0$ and $k \geq 1$ such that

$$|x_t(.,\varphi)| \leq k e^{-\mu t} |\varphi| \text{ for } \varphi \in Y \text{ with } |\varphi| \leq \delta \text{ and } t \geq 0,$$

where $x_t(.,\varphi)$ is the integral solution of equation (1.1) corresponding to initial value φ. Moreover, if Y can be decomposed as $Y = Y_1 \oplus Y_2$ where Y_i are U-invariant subspaces of Y, Y_1 is a finite-dimensional space and with $\omega_0 = \lim_{h\to\infty} \frac{1}{h}\log|U(h)|Y_2|$ we have

$$\inf\{|\lambda| : \lambda \in \sigma(U(t)|Y_1)\} \geq e^{\omega_0 t} \text{ for } t \geq 0,$$

where $\sigma(U(t)|Y_1)$ is the spectrum of $U(t)|Y_1$, then the zero equilibrium of equation (1.1) is unstable, in the sense that there exist $\varepsilon \geq 0$ and sequences $(\varphi_n)_n$ converging to 0 and $(t_n)_n$ of positive real numbers such that $|x_{t_n}(.,\varphi_n)| \geq \varepsilon$.

The above theorem is a consequence of the following result. For more details on the proof, we refer to [3].

Theorem 2.5. *[9] Let $(V(t))_{t\geq 0}$ be a nonlinear strongly continuous semigroup on a subset Ω of a Banach space Z and assume that $x_0 \in \Omega$ is an equilibrium of $(V(t))_{t\geq 0}$ such that $V(t)$ is differentiable at x_0, with $W(t)$ the derivative at x_0 of $V(t)$ for each $t \geq 0$. Then, $(W(t))_{t\geq 0}$ is a strongly continuous semigroup of bounded linear operators on Z. If the zero equilibrium of $(W(t))_{t\geq 0}$ is exponentially stable, then x_0 is locally exponentially stable equilibrium of $(V(t))_{t\geq 0}$. Moreover, if Z can be decomposed as $Z = Z_1 \oplus Z_2$ where Z_i are W-invariant subspaces of Z, Z_1 is a finite-dimensional space and with $\omega_1 = \lim_{h\to\infty} \frac{1}{h} \log|W(h)|Z_2|$ we have*

$$\inf\{|\lambda| : \lambda \in \sigma(W(t)|Z_1)\} \geq e^{\omega_1 t} \text{ for } t \geq 0,$$

then the equilibrium x_0 is unstable in the sense that there exist $\varepsilon \geq 0$ and sequences $(y_n)_n$ converging to x_0 and $(t_n)_n$ of positive real numbers such that $|V(t_n)y_n - x_0| \geq \varepsilon$.

Some informations of the infinitesimal generator of $(U(t))_{t\geq 0}$ can be found in [5]. For example, we know that

Theorem 2.6. *The infinitesimal generator A_U of $(U(t))_{t\geq 0}$ on Y is given by*

$$\left\{ \begin{array}{l} D(A_U) = \left\{ \begin{array}{l} \varphi \in C^1([-r,0];E) : \varphi(0) \in D(A),\ \varphi'(0) \in \overline{D(A)} \text{ and} \\ \varphi'(0) = A\varphi(0) + L(\varphi) \end{array} \right\} \\ A_U\varphi = \varphi' \quad \text{for } \varphi \in D(A_U). \end{array} \right.$$

We now make the next assumption about the operator A.

$(\mathbf{H}_3)$ The semigroup $(T_0(t))_{t\geq 0}$ is compact on $\overline{D(A)}$ for $t \geq 0$.

Theorem 2.7. *Assume that $(\mathbf{H}_3)$ holds. Then, $U(t)$ is a compact operator on Y for $t \geq r$.*

Proof. Let $t \geq r$ and $\widetilde{D}$ be a bounded subset of Y. We use Ascoli-Arzela's theorem to show that $\left\{U(t)\varphi : \varphi \in \widetilde{D}\right\}$ is relatively compact in Y. Let $\varphi \in \widetilde{D}$, $\theta \in [-r,0]$ and $\varepsilon \geq 0$ such that $t+\theta-\varepsilon \geq 0$. Then

$$(U(t)\varphi)(\theta) = T_0(t+\theta)\varphi(0) + \lim_{\lambda\to+\infty} \int_0^{t+\theta} T_0(t+\theta-s)B_\lambda L(U(s)\varphi)ds.$$

Note that

$$\begin{aligned} &\int_0^{t+\theta} T_0(t+\theta-s)B_\lambda L(U(s)\varphi)ds \\ &= \int_0^{t+\theta-\varepsilon} T_0(t+\theta-s)B_\lambda L(U(s)\varphi)ds + \int_{t+\theta-\varepsilon}^{t+\theta} T_0(t+\theta-s)B_\lambda L(U(s)\varphi)ds \end{aligned}$$

and

$$\lim_{\lambda\to+\infty}\int_0^{t+\theta-\varepsilon} T_0(t+\theta-s)B_\lambda L(U(s)\varphi)ds = T_0(\varepsilon)\lim_{\lambda\to+\infty}\int_0^{t+\theta-\varepsilon} T_0(t+\theta-\varepsilon-s)B_\lambda L(U(s)\varphi)ds.$$

The assumption $(\mathbf{H}_3)$ implies that

$$T_0(\varepsilon)\left\{\lim_{\lambda\to+\infty}\int_0^{t+\theta-\varepsilon} T_0(t+\theta-\varepsilon-s)B_\lambda L(U(s)\varphi)ds : \varphi\in\widetilde{D}\right\}$$

is relatively compact in E. As the semigroup $(U(t))_{t\geq 0}$ is exponentially bounded, then there exists a positive constant b_1 such that

$$\left|\lim_{\lambda\to+\infty}\int_{t+\theta-\varepsilon}^{t+\theta} T_0(t+\theta-s)B_\lambda L(U(s)\varphi)ds\right| \leq b_1\varepsilon \text{ for } \varphi\in\widetilde{D}.$$

Consequently, the set

$$\left\{\lim_{\lambda\to+\infty}\int_{t+\theta-\varepsilon}^{t+\theta} T_0(t+\theta-s)B_\lambda L(U(s)\varphi)ds : \varphi\in\widetilde{D}\right\}$$

is totally bounded in E. We deduce that $\left\{(U(t)\varphi)(\theta) : \varphi\in\widetilde{D}\right\}$ is relatively compact in E, for each $\theta\in[-r,0]$. For the completeness of the proof, we need to show the equicontinuity property. Let θ, $\theta_0\in[-r,0]$ such that $\theta\geq\theta_0$. Then

$$\begin{aligned}(U(t)\varphi)(\theta)-(U(t)\varphi)(\theta_0) &= (T_0(t+\theta)-T_0(t+\theta_0))\varphi(0)\\ &\quad+\lim_{\lambda\to+\infty}\int_0^{t+\theta} T_0(t+\theta-s)B_\lambda L(U(s)\varphi)ds\\ &\quad-\lim_{\lambda\to+\infty}\int_0^{t+\theta_0} T_0(t+\theta_0-s)B_\lambda L(U(s)\varphi)ds.\end{aligned}$$

Furthermore,

$$\begin{aligned}\int_0^{t+\theta} T_0(t+\theta-s)B_\lambda L(U(s)\varphi)ds &= \int_0^{t+\theta_0} T_0(t+\theta-s)B_\lambda L(U(s)\varphi)ds\\ &\quad+\int_{t+\theta_0}^{t+\theta} T_0(t+\theta-s)B_\lambda L(U(s)\varphi)ds.\end{aligned}$$

Consequently,

$$\begin{aligned}|(U(t)\varphi)(\theta)-(U(t)\varphi)(\theta_0)| &\leq |T_0(t+\theta)-T_0(t+\theta_0)|\,|\varphi(0)|\\ &\quad+\left|\lim_{\lambda\to+\infty}\int_0^{t+\theta_0}(T_0(t+\theta-s)-T_0(t+\theta_0-s))B_\lambda L(U(s)\varphi)ds\right|\\ &\quad+\left|\lim_{\lambda\to+\infty}\int_{t+\theta_0}^{t+\theta} T_0(t+\theta-s)B_\lambda L(U(s)\varphi)ds\right|.\end{aligned}$$

Assumption $(\mathbf{H}_3)$ implies that the semigroup $(T_0(t))_{t\geq 0}$ is uniformly continuous for $t \geq 0$. Then

$$\lim_{\theta\to\theta_0} |T_0(t+\theta) - T_0(t+\theta_0)| = 0.$$

The semigroup $(U(t))_{t\geq 0}$ is exponentially bounded. Consequently, there exists a positive constant b_2 such that

$$\left|\lim_{\lambda\to+\infty}\int_{t+\theta_0}^{t+\theta} T_0(t+\theta-s)B_\lambda L(U(s)\varphi)ds\right| \leq b_2(\theta-\theta_0)$$

and

$$\begin{aligned}\lim_{\lambda\to+\infty}&\int_0^{t+\theta_0} (T_0(t+\theta-s) - T_0(t+\theta_0-s))\, B_\lambda L(U(s)\varphi)ds \\ &= (T_0(\theta-\theta_0) - I)\lim_{\lambda\to+\infty}\int_0^{t+\theta_0} T_0(t+\theta_0-s)B_\lambda L(U(s)\varphi)ds.\end{aligned}$$

We have proved that there exists a compact set $\widetilde{K}_0$ in E such that

$$\lim_{\lambda\to+\infty}\int_0^{t+\theta_0} T_0(t+\theta_0-s)B_\lambda L(U(s)\varphi)ds \in \widetilde{K}_0 \text{ for } \varphi\in\widetilde{D}.$$

Using Banach-Steinhaus's theorem, we obtain

$$\lim_{\theta\to\theta_0} (T_0(\theta-\theta_0) - I)x = 0 \text{ uniformly in } x\in\widetilde{K}_0.$$

This implies that

$$\lim_{\theta\to\theta_0^+} (U(t)\varphi)(\theta) - (U(t)\varphi)(\theta_0) = 0 \text{ uniformly in } \varphi\in\widetilde{D}.$$

We can prove in similar way that

$$\lim_{\theta\to\theta_0^-} (U(t)\varphi)(\theta) - (U(t)\varphi)(\theta_0) = 0 \text{ uniformly in } \varphi\in\widetilde{D}.$$

By Ascoli-Arzela's theorem, we conclude that $\overline{\left\{U(t)\varphi : \varphi\in\widetilde{D}\right\}}$ is compact for $t \geq r$.

Now, we consider the spectral properties of the infinitesimal generator A_U. We denote by E, without causing confusion, the complexication of E. For each complex number λ, we define the linear operator $\Delta(\lambda) : D(A) \to E$ by

$$\Delta(\lambda) = \lambda I - A - L(e^{\lambda\cdot}I), \tag{2.2}$$

where $e^{\lambda\cdot}I : E \to C$ is defined by

$$\left(e^{\lambda\cdot}x\right)(\theta) = e^{\lambda\theta}x,\ x\in E \quad\text{and}\quad \theta\in[-r,0].$$

Definition 2.8. We say that λ is a characteristic value of equation (1.2) if there exists $x \in D(A)\backslash\{0\}$ solving the characteristic equation $\Delta(\lambda)x = 0$.

Since the operator $U(t)$ is compact for $t \geq r$, the spectrum $\sigma(A_U)$ of A_U is the point spectrum $\sigma_p(A_U)$. More precisely, we have

Theorem 2.9. *The spectrum* $\sigma(A_U) = \sigma_p(A_U) = \{\lambda \in \mathbb{C} : \ker \Delta(\lambda) \neq \{0\}\}$.

Proof. Let $\lambda \in \sigma_p(A_U)$. Then there exists $\varphi \in D(A_U) \backslash \{0\}$ such that $A_U\varphi = \lambda\varphi$, which is equivalent to

$$\varphi(\theta) = e^{\lambda\theta}\varphi(0), \text{ for } \theta \in [-r, 0] \text{ and } \varphi'(0) = A\varphi(0) - L(\varphi) \text{ with } \varphi(0) \neq 0.$$

Consequently $\Delta(\lambda)\varphi(0) = 0$. Conversely, let $\lambda \in \mathbb{C}$ such that $\ker \Delta(\lambda) \neq \{0\}$. Then there exists $x \in D(A) \backslash \{0\}$ such that $\Delta(\lambda)x = 0$. If we define the function φ by

$$\varphi(\theta) = e^{\lambda\theta}x \text{ for } \theta \in [-r, 0],$$

then $\varphi \in D(A_U)$ and $A_U\varphi = \lambda\varphi$, which implies that $\lambda \in \sigma_p(A_U)$.

The growth bound $\omega_0(U)$ of the semigroup $(U(t))_{t \geq 0}$ is defined by

$$\omega_0(U) = \inf\left\{\kappa \geq 0 : \sup_{t \geq 0} e^{-\kappa t} |U(t)| < \infty\right\}.$$

The spectral bound $s(A_U)$ of A_U is defined by

$$s(A_U) = \sup\{\mathrm{Re}(\lambda) : \lambda \in \sigma_p(A_U)\}.$$

Since $U(t)$ is compact for $t \geq r$, then it is well known that $\omega_0(U) = s(A_U)$. Consequently, the asymptotic behavior of the solutions of the linear equation (1.2) is completely obtained by $s(A_U)$. More precisely, we have the following result.

Corollary 2.10. *Assume that* $(\mathbf{H}_1)$, $(\mathbf{H}_2)$ *and* $(\mathbf{H}_3)$ *hold. Then, the following properties hold,*
i) if $s(A_U) < 0$, *then* $(U(t))_{t \geq 0}$ *is exponentially stable and zero is locally exponentially stable for equation (1.1);*
ii) if $s(A_U) = 0$, *then there exists* $\varphi \in Y$ *such that* $|U(t)\varphi| = |\varphi|$ *for* $t \geq 0$ *and either stability or instability may hold;*
iii) if $s(A_U) \geq 0$, *then there exists* $\varphi \in Y$ *such that* $|U(t)\varphi| \to \infty$ *as* $t \to \infty$ *and zero is unstable for equation (1.1).*

As a consequence of the compactness of the semigroup $U(t)$ for $t \geq r$ and by Theorem 2.11, p.100, in [10], we get the following general spectral decomposition of the phase space Y.

Theorem 2.11. *There exist linear subspaces of* Y *denoted by* Y_-, Y_0 *and* Y_+ *respectively with*

$$Y = Y_- \oplus Y_0 \oplus Y_+$$

such that
i) $A_U(Y_-) \subset Y_-$, $A_U(Y_0) \subset Y_0$, *and* $A_U(Y_+) \subset Y_+$;
ii) Y_0 *and* Y_+ *are finite dimensional;*

iii) $\sigma(A_U|Y_0) = \{\lambda \in \sigma(A_U) : \mathrm{Re}\lambda = 0\}$, $\sigma(A_U|Y_+) = \{\lambda \in \sigma(A_U) : \mathrm{Re}\lambda \geq 0\}$;
iv) $U(t)Y_- \subset Y_-$ *for* $t \geq 0$, $U(t)$ *can be extended for* $t \leq 0$ *when restricted to* $Y_0 \cup Y_+$ *and* $U(t)Y_0 \subset Y_0, U(t)Y_+ \subset Y_+$ *for* $t \in \mathbb{R}$;
v) for any $0 < \gamma < \inf\{|\mathrm{Re}\lambda| : \lambda \in \sigma(A_U)$ *and* $\mathrm{Re}\lambda \neq 0\}$, *there exists* $K \geq 0$ *such that*

$$\begin{aligned} |U(t)P_-\varphi| &\leq Ke^{-\gamma t}\,|P_-\varphi| \ \ \textit{for } t \geq 0, \\ |U(t)P_0\varphi| &\leq Ke^{\frac{\gamma}{3}|t|}\,|P_0\varphi| \ \ \textit{for } t \in \mathbb{R}, \\ |U(t)P_+\varphi| &\leq Ke^{\gamma t}\,|P_+\varphi| \ \ \textit{for } t \leq 0, \end{aligned}$$

where P_-, P_0 *and* P_+ *are projections of* Y *into* Y_-, Y_0 *and* Y_+ *respectively.*
Y_-, Y_0 *and* Y_+ *are called stable, center and unstable subspaces of the semigroup* $(U(t))_{t\geq 0}$.

The following result deals with the variation of constants formula for equation (1.1) which are taken from [5]. Let $\langle X_0 \rangle$ be defined by

$$\langle X_0 \rangle = \{X_0 c : c \in E\},$$

where the function X_0c is defined by

$$(X_0c)(\theta) = \begin{cases} 0 & \text{if } \theta \in [-r, 0), \\ c & \text{if } \theta = 0. \end{cases}$$

We introduce the space $Y \oplus \langle X_0 \rangle$, endowed with the following norm

$$|\varphi + X_0c| = |\varphi| + |c|.$$

The following result is taken from [5].

Theorem 2.12. *The continuous extension* $\widetilde{A_U}$ *of the operator* A_U *defined on* $Y \oplus \langle X_0 \rangle$ *by:*

$$\begin{cases} D(\widetilde{A_U}) = \left\{\varphi \in C^1([-r,0];E) : \varphi(0) \in D(A) \textit{ and } \varphi'(0) \in \overline{D(A)}\right\} \\ \widetilde{A_U}\varphi = \varphi' + X_0(A\varphi(0) + L(\varphi) - \varphi'(0)), \end{cases}$$

is a Hille-Yosida operator on $Y \oplus \langle X_0 \rangle$*: there exist* $\widetilde{\omega} \in \mathbb{R}$ *and* $\widetilde{M_0} \geq 1$ *such that* $(\widetilde{\omega}, \infty) \subset \rho(\widetilde{A_U})$ *and*

$$\left|(\lambda I - \widetilde{A_U})^{-n}\right| \leq \frac{\widetilde{M_0}}{(\lambda - \widetilde{\omega})^n} \ \ \textit{for } \lambda \geq \widetilde{\omega} \textit{ and } n \in \mathbb{N},$$

with $\rho(\widetilde{A_U})$ *the resolvent set of* $\widetilde{A_U}$.
Moreover, the integral solution u *of equation (1.1) is given for* $\varphi \in Y$, *by the following variation of constants formula*

$$u_t = U(t)\varphi + \lim_{\lambda\to\infty} \int_0^t U(t-s)\widetilde{B_\lambda}(X_0 g(u_s))\,ds \ \ \textit{for } t \geq 0, \tag{2.3}$$

where $\widetilde{B_\lambda} = \lambda(\lambda I - \widetilde{A_U})^{-1}$ *for* $\lambda \geq \widetilde{\omega}$.

Remarks. i) Without loss of generality, we assume that $\widetilde{M_0} = 1$. Otherwise, we can renorm the space $Y \oplus \langle X_0 \rangle$ in order to get an equivalent norm for which $\widetilde{M_0} = 1$.
ii) For any locally integrable function $\rho : \mathbb{R} \to E$, one can see that the following limit exists:

$$\lim_{\lambda\to\infty} \int_s^t U(t-\tau)\widetilde{B_\lambda}X_0\rho(\tau)d\tau \ \text{ for } t \geq s.$$

3 Global Existence of the Center Manifold

Theorem 3.1. *Assume that* $(\mathbf{H}_1)$ *and* $(\mathbf{H}_3)$ *hold. Then, there exists* $\varepsilon \geq 0$ *such that if*

$$Lip(g) = \sup_{\varphi_1 \neq \varphi_2} \frac{|g(\varphi_1) - g(\varphi_2)|}{|\varphi_1 - \varphi_2|} < \varepsilon,$$

then, there exists a bounded Lipschitz map $h_g : Y_0 \to Y_- \oplus Y_+$ *such that* $h_g(0) = 0$ *and the Lipschitz manifold*

$$M_g := \{\varphi + h_g(\varphi) : \varphi \in Y_0\}$$

is globally invariant under the flow of equation (1.1) on Y.

Proof. Let $\mathcal{B} = \mathcal{B}(Y_0, Y_- \oplus Y_+)$ denote the Banach space of bounded maps $h : Y_0 \to Y_- \oplus Y_+$ endowed with the uniform norm topology. We define

$$\mathcal{F} = \{h \in \mathcal{B} : h \text{ is Lipschitz, } h(0) = 0 \text{ and } Lip(h) \leq 1\}.$$

Let $h \in \mathcal{F}$ and $\varphi \in Y_0$. Using the strict contraction principle, one can prove the existence of v_t^{φ} solution of the following equation

$$v_t^{\varphi} = U(t)\varphi + \lim_{\lambda \to \infty} \int_0^t U(t-\tau)\left(\widetilde{B_\lambda} X_0 g(v_\tau^{\varphi} + h(v_\tau^{\varphi}))\right)^0 d\tau, \; t \in \mathbb{R}. \tag{3.1}$$

We now introduce the mapping $T_g : \mathcal{F} \to \mathcal{B}$ by

$$\begin{aligned} T_g(h)\varphi = & \lim_{\lambda \to \infty} \int_{-\infty}^0 U(-\tau)\left(\widetilde{B_\lambda} X_0 g(v_\tau^{\varphi} + h(v_\tau^{\varphi}))\right)^- d\tau \\ & + \lim_{\lambda \to \infty} \int_{+\infty}^0 U(-\tau)\left(\widetilde{B_\lambda} X_0 g(v_\tau^{\varphi} + h(v_\tau^{\varphi}))\right)^+ d\tau. \end{aligned}$$

The first step is to prove that T_g maps $\mathcal{F}$ into itself. Let $\varphi_1, \varphi_2 \in Y_0$ and $t \in \mathbb{R}$. Suppose that $Lip(g) < \varepsilon$. Then

$$\left|v_t^{\varphi_1} - v_t^{\varphi_2}\right| \leq K e^{\frac{\gamma}{3}|t|} |\varphi_1 - \varphi_2| + 2K|P_0|\varepsilon \left| \int_0^t e^{\frac{\gamma}{3}|t-\tau|} \left|v_\tau^{\varphi_1} - v_\tau^{\varphi_2}\right| d\tau \right|.$$

By Gronwall's lemma, we get that

$$e^{-\frac{\gamma}{3}|t|} \left|v_t^{\varphi_1} - v_t^{\varphi_2}\right| \leq K|\varphi_1 - \varphi_2| e^{2K|P_0|\varepsilon|t|}$$

and

$$\left|v_t^{\varphi_1} - v_t^{\varphi_2}\right| \leq K|\varphi_1 - \varphi_2| e^{\left[\frac{\gamma}{3} + 2K|P_0|\varepsilon\right]|t|} \text{ for } t \in \mathbb{R}.$$

If we choose ε such that

$$2K|P_0|\varepsilon < \frac{\gamma}{6}, \tag{3.2}$$

then

$$\left|v_t^{\varphi_1} - v_t^{\varphi_2}\right| \leq K|\varphi_1 - \varphi_2| e^{\frac{\gamma}{2}|t|} \text{ for } t \in \mathbb{R}. \tag{3.3}$$

Moreover,

$$|T_g(h)\varphi_1 - T_g(h)\varphi_2| \le \int_{-\infty}^{0} |U(-\tau)P_-| \left|g(v_\tau^{\varphi_1} + h(v_\tau^{\varphi_1})) - g(v_\tau^{\varphi_2} + h(v_\tau^{\varphi_2}))\right| d\tau$$
$$+ \int_{0}^{+\infty} |U(-\tau)P_+| \left|g(v_\tau^{\varphi_1} + h(v_\tau^{\varphi_1})) - g(v_\tau^{\varphi_2} + h(v_\tau^{\varphi_2}))\right| d\tau.$$

Consequently,

$$|T_g(h)\varphi_1 - T_g(h)\varphi_2| \le 2\int_{-\infty}^{0} Ke^{\gamma\tau} |P_-| \varepsilon \left|v_\tau^{\varphi_1} - v_\tau^{\varphi_2}\right| d\tau + 2\int_{0}^{+\infty} Ke^{-\gamma\tau} |P_+| \varepsilon \left|v_\tau^{\varphi_1} - v_\tau^{\varphi_2}\right| d\tau.$$

Using inequality (3.3), we obtain

$$|T_g(h)\varphi_1 - T_g(h)\varphi_2| \le 2K^2 |P_-| \varepsilon |\varphi_1 - \varphi_2| \int_{-\infty}^{0} e^{\frac{\gamma}{2}\tau} d\tau + 2K^2 |P_+| \varepsilon |\varphi_1 - \varphi_2| \int_{0}^{+\infty} e^{-\frac{\gamma}{2}\tau} d\tau.$$

It follows that

$$|T_g(h)\varphi_1 - T_g(h)\varphi_2| \le \frac{4\varepsilon}{\gamma} K^2 (|P_-| + |P_+|) |\varphi_1 - \varphi_2|.$$

If we choose ε such that

$$\frac{4\varepsilon}{\gamma} K^2 (|P_-| + |P_+|) < 1,$$

then T_g maps $\mathcal{F}$ into itself.

The next step is to show that T_g is a strict contraction on $\mathcal{F}$. Let $h_1, h_2 \in \mathcal{F}$. For $\varphi \in Y_0$ and for $i = 1, 2$, let v_t^i denote the solution of the following equation

$$v_t^i = U(t)\varphi + \lim_{\lambda\to\infty} \int_0^t U(t-\tau) \left(\widetilde{B_\lambda} X_0 g(v_\tau^i + h_i(v_\tau^i))\right)^0 d\tau \text{ for } t \in \mathbb{R}.$$

Then,

$$\left|v_t^1 - v_t^2\right| \le \varepsilon K |P_0| \left| \int_0^t e^{\frac{\gamma}{3}|t-\tau|} \left(\left|v_\tau^1 - v_\tau^2\right| + \left|h_1(v_\tau^1) - h_1(v_\tau^2)\right| + \left|h_1(v_\tau^2) - h_2(v_\tau^2)\right| \right) d\tau \right|,$$

and

$$\left|v_t^1 - v_t^2\right| \le 2\varepsilon K |P_0| \left| \int_0^t e^{\frac{\gamma}{3}|t-\tau|} \left|v_\tau^1 - v_\tau^2\right| d\tau \right| + \varepsilon K |P_0| |h_1 - h_2| \left| \int_0^t e^{\frac{\gamma}{3}|t-\tau|} d\tau \right|.$$

By Gronwall's lemma, we obtain that

$$\left|v_t^1 - v_t^2\right| \le \frac{3\varepsilon K}{\gamma} |P_0| |h_1 - h_2| e^{\left[\frac{\gamma}{3} + 2K|P_0|\varepsilon\right]|t|} \quad \text{for } t \in \mathbb{R}.$$

By (3.2), we obtain

$$\left|v_t^1 - v_t^2\right| \le \frac{3\varepsilon K}{\gamma} |P_0| |h_1 - h_2| e^{\frac{\gamma}{2}|t|} \quad \text{for all } t \in \mathbb{R}.$$

For $i = 1,2$, we have

$$T_g(h_i)\varphi = \lim_{\lambda\to\infty}\int_{-\infty}^{0} U(-\tau)\left(\widetilde{B_\lambda}X_0 g(v_\tau^i + h_i(v_\tau^i))\right)^- d\tau + \lim_{\lambda\to\infty}\int_{+\infty}^{0} U(-\tau)\left(\widetilde{B_\lambda}X_0 g(v_\tau^i + h_i(v_\tau^i))\right)^+ d\tau.$$

It follows that

$$|T_g(h_1)\varphi - T_g(h_2)\varphi| \le 2K|P_-|\frac{6\varepsilon^2 K}{\gamma^2}|P_0||h_1-h_2| + \frac{K|P_-|\varepsilon}{\gamma}|h_1-h_2| + 2K|P_+|\frac{6\varepsilon^2 K}{\gamma^2}|P_0||h_1-h_2| + \frac{K|P_+|\varepsilon}{\gamma}|h_1-h_2|.$$

Consequently,

$$|T_g(h_1) - T_g(h_2)| \le (|P_-|+|P_+|)\frac{K\varepsilon}{\gamma}\left[|P_0|\frac{12\varepsilon K}{\gamma}+1\right]|h_1-h_2|.$$

We choose ε such that

$$(|P_-|+|P_+|)\frac{K\varepsilon}{\gamma}\left[|P_0|\frac{12\varepsilon K}{\gamma}+1\right] < 1.$$

Then T_g is a strict contraction on $\mathcal{F}$, and consequently it has a unique fixed point h_g in $\mathcal{F}$: $T_g(h_g) = h_g$.
Finally, we show that

$$M_g := \{\varphi + h_g(\varphi) : \varphi \in Y_0\}$$

is globally invariant under the flow on Y. Let $\varphi \in Y_0$ and v be the solution of equation (3.1). We claim that $t \to v_t^\varphi + h_g(v_t^\varphi)$ is an integral solution of equation (1.1) with initial value $\varphi + h_g(\varphi)$. In fact, we have $T_g(h_g)(v_t^\varphi) = h_g(v_t^\varphi)$, $t \in \mathbb{R}$. Moreover, for $t \in \mathbb{R}$, one has

$$T_g(h)(v_t^\varphi) = \lim_{\lambda\to\infty}\int_{-\infty}^{0} U(-\tau)\left(\widetilde{B_\lambda}X_0 g(v_{t+\tau}^\varphi + h_g(v_{t+\tau}^\varphi))\right)^- d\tau + \lim_{\lambda\to\infty}\int_{+\infty}^{0} U(-\tau)\left(\widetilde{B_\lambda}X_0 g(v_{t+\tau}^\varphi + h_g(v_{t+\tau}^\varphi))\right)^+ d\tau,$$

which implies that

$$h_g(v_t^\varphi) = \lim_{\lambda\to\infty}\int_{-\infty}^{t} U(t-\tau)\left(\widetilde{B_\lambda}X_0 g(v_\tau^\varphi + h_g(v_\tau^\varphi))\right)^- d\tau + \lim_{\lambda\to\infty}\int_{+\infty}^{t} U(t-\tau)\left(\widetilde{B_\lambda}X_0 g(v_\tau^\varphi + h_g(v_\tau^\varphi))\right)^+ d\tau.$$

Then, for $t \in \mathbb{R}$, we have

$$v_t^\varphi + h_g(v_t^\varphi) = U(t)\varphi + \lim_{\lambda\to\infty}\int_{0}^{t} U(t-\tau)\left(\widetilde{B_\lambda}X_0 g(v_\tau^\varphi + h_g(v_\tau^\varphi))\right)^0 d\tau + \lim_{\lambda\to\infty}\int_{-\infty}^{t} U(t-\tau)\left(\widetilde{B_\lambda}X_0 g(v_\tau^\varphi + h_g(v_\tau^\varphi))\right)^- d\tau + \lim_{\lambda\to\infty}\int_{+\infty}^{t} U(t-\tau)\left(\widetilde{B_\lambda}X_0 g(v_\tau^\varphi + h_g(v_\tau^\varphi))\right)^+ d\tau.$$

For any $t \geq a$, we have

$$\lim_{\lambda\to\infty}\int_{-\infty}^{t} U(t-\tau)\left(\widetilde{B_\lambda}X_0 g(v_\tau^\varphi + h_g(v_\tau^\varphi))\right)^{-} d\tau$$
$$= \lim_{\lambda\to\infty}\int_{-\infty}^{a} U(t-\tau)\left(\widetilde{B_\lambda}X_0 g(v_\tau^\varphi + h_g(v_\tau^\varphi))\right)^{-} d\tau$$
$$+ \lim_{\lambda\to\infty}\int_{a}^{t} U(t-\tau)\left(\widetilde{B_\lambda}X_0 g(v_\tau^\varphi + h_g(v_\tau^\varphi))\right)^{-} d\tau,$$

and

$$\lim_{\lambda\to\infty}\int_{-\infty}^{a} U(t-\tau)\left(\widetilde{B_\lambda}X_0 g(v_\tau^\varphi + h_g(v_\tau^\varphi))\right)^{-} d\tau$$
$$= U(t-a)\left(\lim_{\lambda\to\infty}\int_{-\infty}^{a} U(a-\tau)\left(\widetilde{B_\lambda}X_0 g(v_\tau^\varphi + h_g(v_\tau^\varphi))\right)^{-} d\tau\right).$$

By the same argument as above, we obtain

$$\lim_{\lambda\to\infty}\int_{+\infty}^{t} U(t-\tau)\left(\widetilde{B_\lambda}X_0 g(v_\tau^\varphi + h_g(v_\tau^\varphi))\right)^{+} d\tau$$
$$= \lim_{\lambda\to\infty}\int_{+\infty}^{a} U(t-\tau)\left(\widetilde{B_\lambda}X_0 g(v_\tau^\varphi + h_g(v_\tau^\varphi))\right)^{+} d\tau$$
$$+ \lim_{\lambda\to\infty}\int_{a}^{t} U(t-\tau)\left(\widetilde{B_\lambda}X_0 g(v_\tau^\varphi + h_g(v_\tau^\varphi))\right)^{+} d\tau,$$

and

$$\lim_{\lambda\to\infty}\int_{+\infty}^{a} U(t-\tau)\left(\widetilde{B_\lambda}X_0 g(v_\tau^\varphi + h_g(v_\tau^\varphi))\right)^{+} d\tau$$
$$= U(t-a)\left(\lim_{\lambda\to\infty}\int_{+\infty}^{a} U(a-\tau)\left(\widetilde{B_\lambda}X_0 g(v_\tau^\varphi + h_g(v_\tau^\varphi))\right)^{+} d\tau\right).$$

Note that

$$h_g(v_a^\varphi) = \lim_{\lambda\to\infty}\int_{+\infty}^{a} U(a-\tau)\left(\widetilde{B_\lambda}X_0 g(v_\tau^\varphi + h_g(v_\tau^\varphi))\right)^{+} d\tau$$
$$+ \lim_{\lambda\to\infty}\int_{-\infty}^{a} U(a-\tau)\left(\widetilde{B_\lambda}X_0 g(v_\tau^\varphi + h_g(v_\tau^\varphi))\right)^{-} d\tau,$$

and in particular

$$v_t^\varphi = U(t-a)v_a^\varphi + \lim_{\lambda\to\infty}\int_{a}^{t} U(t-\tau)\left(\widetilde{B_\lambda}X_0 g(v_\tau^\varphi + h_g(v_\tau^\varphi))\right)^{0} d\tau.$$

Consequently, for any $t \geq a$, we obtain

$$v_t^\varphi + h_g(v_t^\varphi) = U(t-a)\left(v_a^\varphi + h_g(v_a^\varphi)\right) + \lim_{\lambda\to\infty}\int_{a}^{t} U(t-\tau)\left(\widetilde{B_\lambda}X_0 g(v_\tau^\varphi + h_g(v_\tau^\varphi))\right)^{0} d\tau$$
$$+ \lim_{\lambda\to\infty}\int_{a}^{t} U(t-\tau)\left(\widetilde{B_\lambda}X_0 g(v_\tau^\varphi + h_g(v_\tau^\varphi))\right)^{+} d\tau$$
$$+ \lim_{\lambda\to\infty}\int_{a}^{t} U(t-\tau)\left(\widetilde{B_\lambda}X_0 g(v_\tau^\varphi + h_g(v_\tau^\varphi))\right)^{-} d\tau,$$

which implies that for any $t \geq a$,

$$v_t^{\varphi} + h_g(v_t^{\varphi}) = U(t-a)\left(v_a^{\varphi} + h_g(v_a^{\varphi})\right) + \lim_{\lambda\to\infty}\int_a^t U(t-\tau)\left(\widetilde{B_\lambda}X_0 g(v_\tau^{\varphi} + h_g(v_\tau^{\varphi}))\right)d\tau.$$

Finally, we conclude that $v_t^{\varphi} + h_g(v_t^{\varphi})$ is an integral solution of equation (1.1) on $\mathbb{R}$ with initial value $\varphi + h_g(\varphi)$.

Theorem 3.2. *Let v^{φ} be the solution of equation (3.1) on $\mathbb{R}$. Then, for $t \in \mathbb{R}$*

$$\left|v_t^{\varphi}\right| \leq K|\varphi|e^{\frac{\gamma}{2}|t|} \text{ and } \left|v_t^{\varphi} + h_g(v_t^{\varphi})\right| \leq 2K|\varphi|e^{\frac{\gamma}{2}|t|}.$$

Conversely, if we choose ε such that

$$\frac{2K\varepsilon}{\gamma}\left(|P_-| + |P_+| + 3|P_0|\right) < 1,$$

then for any integral solution u of equation (1.1) on $\mathbb{R}$ with $u_t = O(e^{\frac{\gamma}{2}|t|})$, we have $u_t \in M_g$ for all $t \in \mathbb{R}$.

Proof. Let v^{φ} be the solution of equation (3.1). Then, using the estimate (3.3), we obtain that

$$\left|v_t^{\varphi}\right| \leq K|\varphi|e^{\frac{\gamma}{2}|t|} \text{ for } t \in \mathbb{R}$$

and from the fact that $Lip(h) \leq 1$ and $h_g(0) = 0$, we obtain

$$\left|v_t^{\varphi} + h_g(v_t^{\varphi})\right| \leq 2K|\varphi|e^{\frac{\gamma}{2}|t|} \text{ for } t \in \mathbb{R}.$$

Let u be an integral solution of equation (1.1) such that $u_t = O(e^{\frac{\gamma}{2}|t|})$. Then there exists a positive constant k_0 such that $|u_t| \leq k_0 e^{\frac{\gamma}{2}|t|}$ for all $t \in \mathbb{R}$. Note that

$$u_t = U(t-s)u_s + \lim_{\lambda\to\infty}\int_s^t U(t-\tau)\widetilde{B_\lambda}X_0 g(u_\tau)d\tau \text{ for } t \geq s.$$

On the other hand,

$$u_t^+ = U(t-s)u_s^+ + \lim_{\lambda\to\infty}\int_s^t U(t-\tau)\left(\widetilde{B_\lambda}X_0 g(u_\tau)\right)^+ d\tau \text{ for } s \geq t.$$

Moreover, for $s \geq t$ and $s \geq 0$, we have

$$\left|U(t-s)u_s^+\right| \leq Ke^{\gamma(t-s)}|P_+ u_s| \leq k_0 K|P_+|e^{\gamma(t-s)}e^{\frac{\gamma}{2}|s|} = k_0 K|P_+|e^{\gamma t}e^{-\frac{\gamma}{2}s}.$$

Therefore,

$$\lim_{s\to\infty}\left|U(t-s)u_s^+\right| = 0.$$

It follows that

$$u_t^+ = \lim_{\lambda\to\infty}\int_{+\infty}^t U(t-\tau)\left(\widetilde{B_\lambda}X_0 g(u_\tau)\right)^+ d\tau.$$

Similarly, we can prove that

$$u_t^- = \lim_{\lambda\to\infty}\int_{-\infty}^t U(t-\tau)\left(\widetilde{B_\lambda}X_0 g(u_\tau)\right)^- d\tau.$$

We conclude that

$$u_t = u_t^+ + u_t^- + U(t)u_0^0 + \lim_{\lambda\to\infty}\int_0^t U(t-\tau)\left(\widetilde{B_\lambda X_0 g(u_\tau)}\right)^0 d\tau \quad \text{for } t \in \mathbb{R}.$$

Let $\phi \in Y_0$ such that $\phi = u_0$. By Theorem 3.1, there exists an integral solution w of equation (1.1) on $\mathbb{R}$ with initial value $\phi + h_g(\phi)$ such that $w_t \in M_g$ for all $t \in \mathbb{R}$ and

$$\begin{aligned} w_t = U(t)\phi &+ \lim_{\lambda\to\infty}\int_0^t U(t-\tau)\left(\widetilde{B_\lambda X_0 g(w_\tau)}\right)^0 d\tau \\ &+ \lim_{\lambda\to\infty}\int_{-\infty}^t U(t-\tau)\left(\widetilde{B_\lambda X_0 g(w_\tau)}\right)^+ d\tau \\ &+ \lim_{\lambda\to\infty}\int_{+\infty}^t U(t-\tau)\left(\widetilde{B_\lambda X_0 g(w_\tau)}\right)^- d\tau. \end{aligned}$$

Then, for all $t \in \mathbb{R}$, we have

$$\begin{aligned} |u_t - w_t| \leq &\left|\lim_{\lambda\to\infty}\int_0^t U(t-\tau)\left(\widetilde{B_\lambda X_0}\left(g(u_\tau)-g(w_\tau)\right)\right)^0 d\tau\right| \\ &+ \left|\lim_{\lambda\to\infty}\int_{-\infty}^t U(t-\tau)\left(\widetilde{B_\lambda X_0}\left(g(u_\tau)-g(w_\tau)\right)\right)^+ d\tau\right| \\ &+ \left|\lim_{\lambda\to\infty}\int_{+\infty}^t U(t-\tau)\left(\widetilde{B_\lambda X_0}\left(g(u_\tau)-g(w_\tau)\right)\right)^- d\tau\right|. \end{aligned}$$

This implies that

$$\begin{aligned} |u_t - w_t| \leq K\varepsilon \Bigg(&|P_0| \left|\int_0^t e^{\frac{\gamma}{3}|t-\tau|} |u_\tau - w_\tau| \, d\tau\right| \\ &+ |P_-| \int_{-\infty}^t e^{-\gamma(t-\tau)} |u_\tau - w_\tau| \, d\tau + |P_+| \int_t^\infty e^{\gamma(t-\tau)} |u_\tau - w_\tau| \, d\tau \Bigg). \end{aligned}$$

Let $N(t) = e^{-\frac{\gamma}{2}|t|}|u_t - w_t|$ for all $t \in \mathbb{R}$. Then, $\widetilde{N} = \sup_{t\in\mathbb{R}} N(t) < \infty$. On the other hand, we have

$$N(t) \leq K\varepsilon\widetilde{N}\left(|P_0|\left|\int_0^t e^{-\frac{\gamma}{6}|t-\tau|}d\tau\right| + |P_-|\int_{-\infty}^t e^{-\frac{\gamma}{2}(t-\tau)}d\tau + |P_+|\int_t^\infty e^{\frac{\gamma}{2}(t-\tau)}d\tau\right).$$

Finally we arrive at

$$\widetilde{N} \leq \frac{2K\varepsilon}{\gamma}\left(3|P_0| + |P_-| + |P_+|\right)\widetilde{N}.$$

Consequently, $\widetilde{N} = 0$ and $u_t = w_t$ for $t \in \mathbb{R}$.

4 Attractiveness of the Center Manifold

In this section, we assume that there exists no characteristic value with a positive real part and hence the unstable space Y_+ is reduced to zero. We establish the following result on the attractiveness of the center manifold.

Theorem 4.1. *There exist $\varepsilon \geq 0$, $K_1 \geq 0$ and $\alpha \in \left(\frac{\gamma}{3}, \gamma\right)$ such that if $Lip(g) < \varepsilon$, then any integral solution $u_t(\varphi)$ of equation (1.1) on $\mathbb{R}^+$ satisfies*

$$\left|u_t^-(\varphi) - h_g(u_t^0(\varphi))\right| \leq K_1 e^{-\alpha t} \left|\varphi^- - h_g(\varphi^0)\right| \text{ for } t \geq 0. \tag{4.1}$$

Proof. The proof of this theorem is based on the following technically lemma.

Lemma 4.2. *There exist $\varepsilon \geq 0$, $K_0 \geq 0$ and $\alpha \in \left(\frac{\gamma}{3}, \gamma\right)$ such that if $Lip(g) < \varepsilon$, then there is a continuous bounded mapping $p : \mathbb{R}^+ \times Y_0 \times Y_- \to Y_-$ such that any integral solution $u_t(\varphi)$ of equation (1.1) satisfies*

$$u_t^-(\varphi) = p(t, u_t^0(\varphi), \varphi^-) \text{ for } t \geq 0. \tag{4.2}$$

Moreover,

$$|p(t, \phi_1, \psi_1) - p(t, \phi_2, \psi_2)| \leq K_0 \left(|\phi_1 - \phi_2| + e^{-\alpha t} |\psi_1 - \psi_2|\right) \tag{4.3}$$

for all $\phi_1, \phi_2 \in Y_0, \psi_1, \psi_2 \in Y_-$ and $t \geq 0$.

Idea of the proof of Lemma 4.2. The proof is similar to the one given in [22]. Let $\varphi \in Y_0$ and $\psi \in Y_-$. For $t \geq 0$ and $0 \leq \tau \leq t$, we consider the system

$$q(\tau, t, \varphi, \psi) = U(\tau - t)\varphi - \lim_{\lambda \to \infty} \int_\tau^t U(\tau - s) \left(\widetilde{B_\lambda} X_0 g(q(s, t, \varphi, \psi) + p(s, q(s, t, \varphi, \psi), \psi))\right)^0 ds,$$

and

$$p(t, \varphi, \psi) = U(t)\psi + \lim_{\lambda \to \infty} \int_0^t U(t - s) \left(\widetilde{B_\lambda} X_0 g(q(s, t, \varphi, \psi) + p(s, q(s, t, \varphi, \psi), \psi))\right)^- ds.$$

Using the contraction principle, we can prove the existence of q and p. The expression (4.2) and the estimate (4.3) are obtained in a completely similar fashion to that in [22].

Proof of Theorem 4.1. Let M_g be the center manifold of equation (1.1). Then any integral solution lying in M_g must satisfy (4.2). Let $u_t = u_t(\varphi^- + \varphi^0)$ be an integral solution of equation (1.1) on $\mathbb{R}^+$ with initial value $\varphi^- + \varphi^0$. Let $\tau \geq 0$. Then, $u_\tau^0 + h_g(u_\tau^0) \in M_g$ and the corresponding integral solution exists on $\mathbb{R}$ and lies on M_g. This solution can be considered as an integral solution of equation (1.1) starting from $\psi^- + \psi^0$ at 0. Let $v_t = v_t(\psi^- + \psi^0)$ be the integral solution corresponding to $\psi^- + \psi^0$. Using Lemma 4.2, we conclude that

$$u_\tau^- - h_g(u_\tau^0) = p(\tau, u_\tau^0, \varphi^-) - p(\tau, u_\tau^0, \psi^-),$$

which implies that

$$\left|u_\tau^- - h_g(u_\tau^0)\right| \leq K_0 e^{-\alpha\tau} \left|\varphi^- - \psi^-\right|. \tag{4.4}$$

Since $Lip(h) \leq 1$, we have

$$\left|u_\tau^- - h_g(u_\tau^0)\right| \leq K_0 e^{-\alpha\tau} \left(\left|\varphi^- - h_g(\varphi^0)\right| + \left|\varphi^0 - \psi^0\right|\right).$$

The initial values φ^0 and ψ^0 correspond to the solutions of the following equations for $0 \leq s \leq \tau$

$$v_s = U(s-\tau)v_\tau + \lim_{\lambda\to\infty}\int_\tau^s U(s-\sigma)\left(\widetilde{B_\lambda}X_0 g(v_\sigma + p(\sigma, v_\sigma, \varphi^-))\right)^0 d\sigma,$$
$$v_s^* = U(s-\tau)v_\tau^* + \lim_{\lambda\to\infty}\int_\tau^s U(s-\sigma)\left(\widetilde{B_\lambda}X_0 g(v_\sigma^* + p(\sigma, v_\sigma^*, \psi^-))\right)^0 d\sigma.$$

Note that $v_\tau = v_\tau^*$. It follows, for $0 \leq s \leq \tau$, that

$$|v_s - v_s^*| \leq K(1+K_0)\varepsilon|P_0|\int_s^\tau e^{\frac{\gamma}{3}(\sigma-s)}|v_\sigma - v_\sigma^*|d\sigma + KK_0\varepsilon|P_0|\int_s^\tau e^{\frac{\gamma}{3}(\sigma-s)}e^{-\alpha\sigma}d\sigma\left|\varphi^- - \psi^-\right|.$$

Then by Gronwall's lemma, we deduce that there exists a positive constant ν which depends only on constants γ, K, K_0 and ε such that, for $0 \leq s \leq \tau$, we have

$$|v_s - v_s^*| \leq \nu\left|\varphi^- - \psi^-\right|.$$

If we assume that $Lip(g)$ is small enough such that $\nu < 1$, then

$$\left|\varphi^0 - \psi^0\right| \leq \nu\left|\varphi^- - \psi^-\right| \leq \nu\left(\left|\varphi^- - h_g(\varphi^0)\right| + \left|h_g(\varphi^0) - h_g(\psi^0)\right|\right),$$

which gives that

$$\left|\varphi^0 - \psi^0\right| \leq \frac{\nu}{1-\nu}\left|\varphi^- - h_g(\varphi^0)\right|.$$

We conclude that

$$\left|u_t^- - h_g(u_t^0)\right| \leq \frac{1}{1-\nu}K_0 e^{-\alpha t}\left|\varphi^- - h_g(\varphi^0)\right| \text{ for } t \geq 0.$$

As an immediate consequence, we obtain the following result on the attractiveness of the center manifold.

Corollary 4.3. *Assume that $Lip(g)$ is small enough and the unstable space Y_+ is reduced to zero. Then the center manifold M_g is exponentially attractive.*

We also obtain.

Proposition 4.4. *Assume that $Lip(g)$ is small enough and the unstable space Y_+ is reduced to zero. Let w be an integral solution of equation (1.1) that is bounded on $\mathbb{R}$. Then $w_t \in M_g$ for all $t \in \mathbb{R}$.*

Proof. Let w be a bounded integral solution of equation (1.1). Since, the equation (1.1) is autonomous, then for $\sigma \leq 0$, $w_{t'+\sigma}$ is also an integral solution of equation (1.1) for $t' \geq 0$ with initial value w_σ at 0. It follows by the estimation (4.1) that

$$\left|w_{t'+\sigma}^-(\varphi) - h_g(w_{t'+\sigma}^0(\varphi))\right| \leq K_1 e^{-\alpha t'}\left|w_\sigma^- - h_g(w_\sigma^0)\right| \text{ for } t' \geq 0.$$

Let $t \geq \sigma$. Then

$$\left|w_t^- - h_g(w_t^0)\right| \leq K_1 e^{-\alpha(t-\sigma)}\left|w_\sigma^- - h_g(w_\sigma^0)\right| \text{ for } t \geq \sigma. \tag{4.5}$$

Since w is bounded on $\mathbb{R}$, letting $\sigma \to -\infty$, we obtain that $w_t^- = h_g(w_t^0)$ for all $t \in \mathbb{R}$.

5 Flow on the Center Manifold

In this section, we establish that the flow on the center manifold is governed by an ordinary differential equation in a finite dimensional space. In the sequel, we assume that the function g satisfies the conditions of Theorem 4.1. We also assume that the unstable space Y_+ is reduced to zero. Let d be the dimension of the center space Y_0 and $\Phi = (\phi_1, ..., \phi_d)$ be a basis of Y_0. Then there exists d-elements $\Phi = (\phi_1^*, ..., \phi_d^*)$ in Y^*, the dual space of Y, such that

$$\langle \phi_i^*, \phi_j \rangle := \phi_i^*(\phi_j) = \delta_{ij}, \ 1 \leq i, j \leq d,$$

and $\phi_i^* = 0$ on Y_-. Denote by Ψ the transpose of $(\phi_1^*, ..., \phi_d^*)$. Then the projection operator P_0 is given by

$$P_0\phi = \Phi \langle \Psi, \phi \rangle .$$

Since $\left(U^0(t)\right)_{t\geq 0}$ is a strongly continuous group on the finite dimensional space Y_0, by Theorem 2.15, p. 102 in [10], we get that there exists a $d \times d$ matrix G such that

$$U^0(t)\Phi = \Phi e^{Gt} \text{ for } t \geq 0.$$

Let $n \in \mathbb{N}$, $n \geq n_0 \geq \widetilde{\omega}$ and $i \in \{1, ..., d\}$. We define the function x_{ni}^* on E by

$$x_{ni}^*(y) = \left\langle \phi_i^*, \widetilde{B_n}(X_0 y) \right\rangle .$$

Then x_{ni}^* is a bounded linear operator on E. Let x_n^* be the transpose of $(x_{n1}^*, ..., x_{nd}^*)$, then

$$\langle x_n^*, y \rangle = \left\langle \Psi, \widetilde{B_n}(X_0 y) \right\rangle .$$

Consequently,

$$\sup_{n\geq n_0} |x_n^*| < \infty,$$

which implies that $(x_n^*)_{n\geq n_0}$ is a bounded sequence in $\mathcal{L}(E, \mathbb{R}^d)$. Then, we get the following important result.

Theorem 5.1. *There exists $x^* \in \mathcal{L}(E, \mathbb{R}^d)$ such that $(x_n^*)_{n\geq n_0}$ converges weakly to x^* in the sense that*

$$\langle x_n^*, y \rangle \to \langle x^*, y \rangle \ \text{ as } n \to \infty \text{ for } y \in E.$$

For the proof, we need the following fundamental theorem [25, pp. 776]

Theorem 5.2. *Let X be a separable Banach space and $(z_n^*)_{n\geq 0}$ be a bounded sequence in X^*. Then there exists a subsequence $\left(z_{n_k}^*\right)_{k\geq 0}$ of $(z_n^*)_{n\geq 0}$ which converges weakly in X^* in the sense that there exists $z^* \in X^*$ such that*

$$\left\langle z_{n_k}^*, y \right\rangle \to \langle z^*, y \rangle \text{ as } k \to \infty \text{ for } x \in X.$$

Proof of Theorem 5.1. Let Z_0 be a closed separable subspace of E. Since $(x_n^*)_{n\geq n_0}$ is a bounded sequence, by Theorem 5.2 there is a subsequence $\left(x_{n_k}^*\right)_{k\in\mathbb{N}}$ which converges weakly to some $x_{Z_0}^*$ in Z_0. We claim that all the sequence $(x_n^*)_{n\geq n_0}$ converges weakly to $x_{Z_0}^*$ in Z_0. This can be done by way of contradiction. Namely, suppose that there exists a subsequence $\left(x^*_{n_p}\right)_{p\in\mathbb{N}}$ of $(x^*_n)_{n\geq n_0}$ which converges weakly to some $\widetilde{x}^*_{Z_0}$ with $\widetilde{x}^*_{Z_0}\neq x^*_{Z_0}$. Let $\widetilde{u}_t(.,\sigma,\varphi,f)$ denote the integral solution of the following equation

$$\begin{cases} \dfrac{d}{dt}\widetilde{u}(t) = A\widetilde{u}(t) + L(\widetilde{u}_t) + f(t),\ t \geq \sigma \\ \widetilde{u}_\sigma = \varphi \in C, \end{cases}$$

where f is a continuous function from $\mathbb{R}$ to E. Then by using the variation of constants formula and the spectral decomposition of solutions, we obtain

$$P_0\widetilde{u}_t(.,\sigma,0,f) = \lim_{n\to+\infty}\int_\sigma^t U(t-\xi)\left(\widetilde{B}_nX_0f(\xi)\right)^0 d\xi,$$

and

$$P_0\left(\widetilde{B}_nX_0f(\xi)\right) = \Phi\left\langle\Psi,\widetilde{B}_nX_0f(\xi)\right\rangle = \Phi\left\langle x^*_n, f(\xi)\right\rangle.$$

It follows that

$$\begin{aligned} P_0\widetilde{u}_t(.,\sigma,0,f) &= \lim_{n\to+\infty}\Phi\int_\sigma^t e^{(t-\xi)G}\left\langle\Psi,\widetilde{B}_nX_0f(\xi)\right\rangle d\xi, \\ &= \lim_{n\to+\infty}\Phi\int_\sigma^t e^{(t-\xi)G}\left\langle x^*_n, f(\xi)\right\rangle d\xi. \end{aligned}$$

For a fixed $a\in Z_0$, set $f = a$ to be a constant function. Then

$$\lim_{k\to+\infty}\int_\sigma^t e^{(t-\xi)G}\left\langle x^*_{n_k}, a\right\rangle d\xi = \lim_{p\to+\infty}\int_\sigma^t e^{(t-\xi)G}\left\langle x^*_{n_p}, a\right\rangle d\xi \text{ for } a\in Z_0,$$

which implies that

$$\int_\sigma^t e^{(t-\xi)G}\left\langle x^*_{Z_0}, a\right\rangle d\xi = \int_\sigma^t e^{(t-\xi)G}\left\langle \widetilde{x}^*_{Z_0}, a\right\rangle d\xi \text{ for } a\in Z_0.$$

Consequently $x^*_{Z_0} = \widetilde{x}^*_{Z_0}$, which yields a contradiction.
We now conclude that the whole sequence $(x^*_n)_{n\geq n_0}$ converges weakly to $x^*_{Z_0}$ in Z_0. Let Z_1 be another closed separable subspace of X. By using the same argument as above, we obtain that $(x^*_n)_{n\geq n_0}$ converges weakly to $x^*_{Z_1}$ in Z_1. Since $Z_0\cap Z_1$ is a closed separable subspace of E, we get that $x^*_{Z_1} = x^*_{Z_0}$ in $Z_0\cap Z_1$. For any $y\in E$, we define x^* by

$$\langle x^*, y\rangle = \langle x^*_Z, y\rangle,$$

where Z is an arbitrary given closed separable subspace of E such that $y\in Z$. Then x^* is well defined on E and x^* is a bounded linear operator from E to $\mathbb{R}^d$ such that

$$|x^*| \leq \sup_{n\geq n_0}|x^*_n| < \infty,$$

and $(x^*_n)_{n\geq n_0}$ converges weakly to x^* in E.

As a consequence, we conclude that

Corollary 5.3. *For any continuous function $f : \mathbb{R} \to E$, we have*

$$\lim_{n\to+\infty}\int_{\sigma}^{t} U(t-\xi)\left(\widetilde{B}_n X_0 f(\xi)\right)^0 d\xi = \Phi\int_{\sigma}^{t} e^{(t-\xi)G}\langle x^*, f(\xi)\rangle\, d\xi \text{ for } t,\sigma\in\mathbb{R}.$$

Let $\varphi \in Y_0$ such that $\varphi + h_g(\varphi) \in M_g$. From the properties of the center manifold, we know that the integral solution starting from $\varphi + h(\varphi)$ is given by $v_t^{\varphi} + h_g(v_t^{\varphi})$, where v_t^{φ} is the solution of

$$v_t^{\varphi} = U(t)\varphi + \lim_{\lambda\to\infty}\int_0^t U(t-\tau)\left(\widetilde{B_\lambda} X_0 g(v_\tau^{\varphi} + h_g(v_\tau^{\varphi}))\right)^0 d\tau \text{ for } t\in\mathbf{R}.$$

Let $z(t)$ be the component of v_t^{φ}. Then $\Phi z(t) = v_t^{\varphi}$ for $t \in \mathbf{R}$. By Theorem 5.1 and Corollary 5.3, we have

$$\Phi z(t) = v_t^{\varphi} = \Phi e^{Gt} z(0) + \Phi\int_0^t e^{(t-\tau)G}\left\langle x^*, g(v_\tau^{\varphi} + h_g(v_\tau^{\varphi}))\right\rangle d\tau \text{ for } t\in\mathbf{R}.$$

We conclude that z satisfies

$$z(t) = e^{Gt} z(0) + \lim_{n\to\infty}\int_0^t e^{G(t-\tau)}\left\langle x_n^*, g(v_\tau^{\varphi} + h_g(v_\tau^{\varphi}))\right\rangle d\tau \text{ for } t\in\mathbf{R}.$$

Finally we arrive at the following ordinary differential equation, which determines the flow on the center manifold

$$z'(t) = Gz(t) + \langle x^*, g(\Phi z(t) + h_g(\Phi z(t)))\rangle \text{ for } t\in\mathbf{R}. \tag{5.1}$$

6 Stability in Critical Cases

In this section, we suppose that $Lip(g) < \varepsilon$, where ε is given by Theorem 4.1. Here we study the critical case where the unstable space Y_+ is reduced to zero and the exponential stability is not possible, which implies that there exists at least one characteristic value with a real part equals zero.

Theorem 6.1. *Assume that $Lip(g)$ is small enough. Then there exists a positive constant K_2 such that for each $\varphi \in Y$, there exists $\tilde{z}_0 \in \mathbb{R}^d$ such that*

$$\left|u_t^0 - \Phi\tilde{z}(t)\right| + \left|u_t^- - h_g(\Phi\tilde{z}(t))\right| \le K_2 e^{-\alpha t}\left|\varphi^- - h_g(\varphi^0)\right| \text{ for } t\ge 0, \tag{6.1}$$

where $\tilde{z}$ is the solution of the reduced system (5.1) with initial value $\tilde{z}_0$ and u is the integral solution of equation (1.1) with initial value φ.

Proof. Let $\varphi \in Y$ and u be the corresponding integral solution of equation (1.1). Then

$$u_s^0 = U(s-t)u_t^0 + \lim_{\lambda\to\infty}\int_t^s U(s-\tau)\left(\widetilde{B_\lambda} X_0 g(u_\tau^- + u_\tau^0)\right)^0 d\tau \text{ for } 0\le s\le t.$$

Also let $s \to w_{s,t}$ be the solution of the following equation

$$w_{s,t} = U(s-t)u_t^0 + \lim_{\lambda\to\infty}\int_t^s U(s-\tau)\left(\widetilde{B_\lambda} X_0 g(w_{\tau,t} + h_g(w_{\tau,t}))\right)^0 d\tau \text{ for } 0\le s\le t.$$

Then, for $0 \le s \le t$, we have

$$\left|u_s^0 - w_{s,t}\right| \le 2K\varepsilon \left|P_0\right| \int_s^t e^{\frac{\gamma}{3}(\tau - s)} \left|u_\tau^0 - w_{\tau,t}\right| d\tau + K\varepsilon \left|P_0\right| \int_s^t e^{\frac{\gamma}{3}(\tau - s)} \left|u_\tau^- - h_g(u_\tau^0)\right| d\tau.$$

It follows that

$$e^{\frac{\gamma}{3}s} \left|u_s^0 - w_{s,t}\right| \le 2K\varepsilon \left|P_0\right| \int_s^t e^{\frac{\gamma}{3}\tau} \left|u_\tau^0 - w_{\tau,t}\right| d\tau + K\varepsilon \left|P_0\right| \int_s^t e^{\frac{\gamma}{3}\tau} \left|u_\tau^- - h_g(u_\tau^0)\right| d\tau.$$

Using Gronwall's lemma, we obtain that

$$\left|u_s^0 - w_{s,t}\right| \le 2K\varepsilon \left|P_0\right| \int_s^t e^{\mu(\tau - s)} \left|u_\tau^- - h_g(u_\tau^0)\right| d\tau \text{ for } 0 \le s \le t,$$

where $\mu = 2K\varepsilon \left|P_0\right| + \frac{\gamma}{3}$. By Theorem 4.1, we obtain

$$\left|u_s^0 - w_{s,t}\right| \le \left(2KK_1\varepsilon \left|P_0\right| \int_s^t e^{(\mu - \alpha)(\tau - s)} d\tau\right) \left|u_s^- - h_g(u_s^0)\right| \text{ for } 0 \le s \le t.$$

Let us recall that $\alpha \in \left(\frac{\gamma}{3}, \gamma\right)$. Then we can choose ε small enough such that $\mu - \alpha < 0$. Consequently,

$$\left|u_s^0 - w_{s,t}\right| \le \frac{2KK_1\varepsilon \left|P_0\right|}{\alpha - \mu} \left|u_s^- - h_g(u_s^0)\right| \text{ for } 0 \le s \le t, \tag{6.2}$$

and for $s = 0$, we have

$$\left|u_0^0 - w_{0,t}\right| \le \frac{2KK_1\varepsilon \left|P_0\right|}{\alpha - \mu} \left|u_0^- - h_g(u_0^0)\right| \text{ for } t \ge 0.$$

We deduce that $\{w_{0,t} : t \ge 0\}$ is bounded in Y_0 and then there exists a sequence $t_n \longrightarrow \infty$ as $n \longrightarrow \infty$ and $\phi \in Y_0$ such that

$$w_{0,t_n} \longrightarrow \phi \text{ as } n \longrightarrow \infty.$$

Let $\widetilde{w}(.,\phi)$ be the solution of the following equation

$$\widetilde{w}_t(.,\phi) = U(t)\phi + \lim_{\lambda \to \infty} \int_0^t U(t - \tau) \left(\widetilde{B_\lambda X_0} g(\widetilde{w}_\tau(.,\phi) + h_g(\widetilde{w}_\tau(.,\phi)))\right)^0 d\tau \text{ for } t \ge 0.$$

By the continuous dependence on the initial data, we obtain, for all $s \ge 0$

$$\begin{aligned} \widetilde{w}_s(0,\phi) &= \lim_{n\to\infty} \widetilde{w}_s(0, w_{0,t_n}) = \lim_{n\to\infty} \widetilde{w}_s(0, w_{-t_n}(0, u_{t_n})), \\ &= \lim_{n\to\infty} \widetilde{w}_{s-t_n}(0, u_{t_n}) = \lim_{n\to\infty} w_{s,t_n}. \end{aligned}$$

By (4.4) and (6.2), there exists a positive constant K_2 such that

$$\left|u_s^0 - \widetilde{w}_s(0,\phi)\right| \le K_2 e^{-\alpha s} \left|u_0^- - h_g(u_0^0)\right| \text{ for all } s \ge 0.$$

If we put

$$\Phi\widetilde{z}(t) = \widetilde{w}_t(0,\phi) \text{ for } t \in \mathbb{R},$$

then

$$\widetilde{z}(t) = e^{Gt}\widetilde{z}(0) + \int_0^t e^{(t-\tau)G} \langle x^*, g(\Phi\widetilde{z}(\tau) + h_g(\Phi\widetilde{z}(\tau)))\rangle d\tau \text{ for } t \in \mathbb{R}.$$

Finally, the estimation (6.1) follows from (4.4).

Now, we can state the following result on the stability by using the reduction to the center manifold.

Theorem 6.2. *If the zero solution of equation (5.1) is uniformly asymptotically stable (unstable), then the zero solution of equation (1.1) is uniformly asymptotically stable (unstable).*

Proof. Assume that 0 is uniformly asymptotically stable for equation (5.1). For $\varsigma \geq 0$, let

$$B_{\varsigma} = \left\{\varphi^{-} + \varphi^{0} \in Y_{-} \oplus Y_0 : \left|\varphi^{-}\right| + \left|\varphi^{0}\right| < \varsigma\right\},$$

and $M_g \cap B_\rho$ for some $\rho \geq 0$, be the region of attraction of 0 for equation (5.1). First, we prove that 0 is stable for equation (1.1). Let $\varepsilon \geq 0$. Then there exists $\delta < \rho$ such that $|z(t)| < \varepsilon$ for $t \geq 0$, provided that $|z(0)| < \delta$, where z is a solution of (5.1). As 0 is assumed to be uniformly asymptotically stable for equation (5.1), there exists $t_0 = t_0(\delta)$ such that $|z(t)| < \frac{\delta}{2}$, for $t \geq t_0$. Without loss of generality, we can choose δ and t_0 so that $\max(K_1, K_2)e^{-\alpha t_0} < \frac{1}{2}$.

By the continuous dependence on the initial value for equation (1.1), there exists $\delta_1 < \frac{\delta}{2}$ such that if

$$\varphi^{-} + \varphi^{0} \in V_{\delta_1} := \left\{\psi^{-} + \psi^{0} \in Y_{-} \oplus Y_0 : \left|\psi^{0}\right| < \frac{\delta_1}{2}, \left|\psi^{-} - h_g(\psi^{0})\right| < \frac{\delta_1}{2}\right\},$$

then the corresponding integral solution $u_t = u_t(\varphi^{-} + \varphi^{0})$ of equation (1.1) satisfies

$$u_t \in B_\varepsilon \text{ for } t \in [0, t_0].$$

Moreover,

$$\left|u_{t_0}^{-} - h_g(u_{t_0}^{0})\right| < \frac{\delta_1}{2}.$$

Furthermore, by Theorem 6.1, there exists $z_0 \in \mathbb{R}^d$ such that

$$\left|u_t^0 - \Phi\widetilde{z}(t)\right| \leq K_2 e^{-\alpha t}\left|\varphi^{-} - h_g(\varphi^{0})\right| \text{ for } t \geq 0, \tag{6.3}$$

where $\widetilde{z}$ is a solution of the reduced system (5.1) with initial value $\widetilde{z}_0$ such that $|\widetilde{z}_0| < \delta$. It follows that

$$\left|u_{t_0}^{0}\right| < \delta.$$

Consequently, $u_{t_0} \in B_\varepsilon$ and u_t must be in B_ε for all $t \geq 0$. This completes the proof of the stability.

Now we deal with the local attractiveness of the zero solution. For a given integral solution $u(., \varphi)$ of equation (1.1) which is assumed to be bounded for $t \geq 0$, it is well known that the ω-limit set $\omega(\varphi)$ is nonempty, compact, invariant and connected since the map $\varphi \to u_t(\cdot, \varphi)$ is compact for $t \geq r$.

For the attractiveness of 0, let V_δ be chosen as above and $\varphi \in V_\delta$. Then the integral solution u of equation (1.1) starting from φ lies in B_ε. The ω-limit set $\omega(\varphi)$ of u is nonempty and invariant and must be in $M_g \cap B_\varepsilon$. Since the equilibrium 0 of (1.1) is uniformly asymptotically stable, we deduce by Theorem 11.4, p. 111 [23] and by the LaSalle invariance principle that the only invariant set in $M_g \cap B_\varepsilon$ must be zero. Consequently, the ω-limit set $\omega(\varphi)$ is zero and

$$u_t(.,\varphi) \to 0 \text{ as } t \to 0.$$

Assume now that the zero solution of the reduced system (5.1) is unstable. Then there exist $\sigma_1 \geq 0$, a sequence $(t_n)_n$ of positive real numbers and a sequence $(z_n)_n$ converging to 0 such that $|z(t_n,z_n)| \geq \sigma_1$, where $z(.,z_n)$ is a solution of (5.1). On the other hand $\Phi z(.,z_n) + h_g(\Phi z(.,z_n))$ is an integral solution of equation (1.1) and

$$|\Phi z(t_n,z_n) + h_g(\Phi z(t_n,z_n))| \geq (1 - Lip(h_g))\, |\Phi z(t_n,z_n)|.$$

Moreover, $Lip(h_g)$ can be chosen such that $1 - Lip(h_g) \geq 0$. It follows that

$$|\Phi z(t_n,z_n) + h_g(\Phi z(t_n,z_n))| \geq (1 - Lip(h_g))\, \sigma_2,$$

for some $\sigma_2 \geq 0$. Consequently, the zero solution of equation (1.1) is unstable.

7 Local Existence of the Center Manifold

In this section we prove the existence of the local center manifold when g is only defined in a neighborhood of zero. We assume that

($\mathbf{H}_4$) There exists $\rho_1 \geq 0$ such that $g : B(0,\rho_1) \to E$ is C^1-function, $g(0) = 0$ and $g'(0) = 0$, where $B(0,\rho_1) = \{\varphi \in C : |\varphi| < \rho_1\}$.

For $\rho < \rho_1$, we define the cut-off function $g_\rho : C \to E$ by

$$g_\rho(\varphi) = \begin{cases} g(\varphi) \text{ if } |\varphi| \leq \rho, \\ g(\frac{\rho}{|\varphi|}\varphi) \text{ if } |\varphi| \geq \rho. \end{cases}$$

We consider the following partial functional differential equation

$$\begin{cases} \dfrac{d}{dt}u(t) = Au(t) + L(u_t) + g_\rho(u_t) \text{ for } t \geq 0 \\ u_0 = \varphi \in C. \end{cases} \tag{7.1}$$

Theorem 7.1. *Assume that* ($\mathbf{H}_1$), ($\mathbf{H}_3$) *and* ($\mathbf{H}_4$) *hold. Then there exist* $0 < \rho < \rho_1$ *and Lipschitz continuous mapping* $h_{g_\rho} : Y_0 \to Y_- \oplus Y_+$ *such that* $h_{g_\rho}(0) = 0$ *and the local Lipschitz manifold*

$$M_{g_\rho} = \{\varphi + h_{g_\rho}(\varphi) : \varphi \in Y_0\}$$

is globally invariant under the flow associated to equation (7.1).

Proof. Using the same arguments as in [26], Proposition 3.10, p.95, one can show that g_ρ is Lipschitz continuous with

$$Lip(g_\rho) \leq 2 \sup_{|\varphi|<\rho} |g'(\varphi)|.$$

It follows that $Lip(g_\rho)$ goes to zero when ρ goes to zero. According to Theorem 3.1, we deduce that there exist $\rho \geq 0$ and mapping $h_{g_\rho} : Y_0 \to Y_- \oplus Y_+$ with $h_{g_\rho}(0) = 0$ such that the Lipschitz manifold

$$M_{g_\rho} = \{\varphi + h_{g_\rho}(\varphi) : \varphi \in Y_0\}$$

is globally invariant under the flow of equation (7.1).

Definition 7.2. The local center manifold associated to equation (1.1) is defined by

$$\widetilde{M}_{g_\rho} = \{\varphi + h_{g_\rho}(\varphi) : \varphi \in Y_0\} \cap B(0, \rho).$$

We deduce that the local center manifold contains all bounded integral solutions of equation (1.1) which are bounded by ρ. Moreover, the following interesting results hold.

Theorem 7.3. ***(Attractiveness).*** *Assume that* $(\mathbf{H}_1)$, $(\mathbf{H}_3)$, $(\mathbf{H}_4)$ *hold and the unstable space* Y_+ *is reduced to zero. Then there exist* $0 < \rho < \rho_1$, $K_3 \geq 0$ *and* $\alpha \in \left(\frac{\gamma}{3}, \gamma\right)$ *such that any integral solution* $u_t(\varphi)$ *of equation (1.1) with initial data* $\varphi \in B(0, \rho_1)$, *exists on* $\mathbb{R}^+$ *and satisfies*

$$\left|u_t^-(\varphi) - h_{g_\rho}(u_t^0(\varphi))\right| \leq K_3 e^{-\alpha t} \left|\varphi^- - h_{g_\rho}(\varphi^0)\right| \quad \textit{for } t \geq 0. \tag{7.2}$$

Theorem 7.4. ***(Reduction principle and stability).*** *Assume that* $(\mathbf{H}_1)$, $(\mathbf{H}_3)$, $(\mathbf{H}_4)$ *hold and the unstable space* Y_+ *is reduced to zero. If the zero solution of equation (5.1) is uniformly asymptotically stable (unstable), then the zero solution of equation (1.1) is uniformly asymptotically stable (unstable).*

8 Application

In this section, we assume that

$(\mathbf{H}_5)$ $\sigma(A_U) \cap \{\lambda \in \mathbb{C} : \mathrm{Re}(\lambda) \geq 0\} = \{0\}$ and 0 is a simple eigenvalue of A_U.

In this case, the reduced system (5.1) on the center manifold becomes

$$z'(t) = \upsilon(z(t)),$$

where υ is the scalar function given by

$$\upsilon(z) = \left\langle x^*, g(\Phi z + h_{g_\rho}(\Phi z))\right\rangle \text{ for } z \in \mathbb{R}.$$

Theorem 8.1. *Assume that* $(\mathbf{H}_1)$, $(\mathbf{H}_3)$, $(\mathbf{H}_4)$, $(\mathbf{H}_5)$ *hold and* υ *satisfies*

$$\upsilon(z) = a_m z^m + a_{m+1} z^{m+1} + \dots. \tag{8.1}$$

If m *is odd and* $a_m < 0$, *then the zero solution of equation (1.1) is uniformly asymptotically stable. If* $a_m > 0$ *then the zero solution of equation (1.1) is unstable.*

Proof. The proof is based on Theorem 6.2 and on the following known stability result.

Theorem 8.2. *[6] Consider the scalar differential equation*

$$z'(t) = a_m z^m + a_{m+1} z^{m+1} + \dots. \tag{8.2}$$

If m is odd and $a_m < 0$, then the zero solution of equation (8.2) is uniformly asymptotically stable. If $a_m > 0$, then the zero solution of equation (8.2) is unstable.

Concluding remark. Assumption (8.1) is natural and it is a consequence of the smoothness of the center manifold, which states that if g is a C^k-function, for $k \geq 1$, then h_{g_ρ} is also a C^k-function. Consequently if g is a C^∞-function, then the center manifold h_{g_ρ} is also a C^∞-function. Assumption (8.1) can be obtained by using the approximation of the center manifold h_{g_ρ}. The proof of the smoothness result is omitted here and it can be done in similar way as in [11].

Acknowledgements

A part of this work has been done when the second author was visiting the Abdus Salam International Centre for Theoretical Physics, ICTP, Trieste-Italy. He would like to acknowledge the centre for the support.

References

[1] M. Adimy and K. Ezzinbi, Semi groupes intégrés et équations différentielles à retard en dimension infinie, C. R. *Acad. Sci. Paris*, t. **323**, série I, 481-486, (1996).

[2] M.Adimy and K.Ezzinbi, A class of linear partial neutral functional differential equations with non-dense domain, *Journal of Differential Equations,* **147**, 285-332, (1998).

[3] M. Adimy and K. Ezzinbi, Local existence and linearized stability for partial functional differential equations, *Dynamic Systems and Applications*, Vol. 7, 389-403, (1998).

[4] M. Adimy and K. Ezzinbi, Existence and stability of solutions for a class of partial neutral functional differential equations, *Hiroshima Mathematical Journal*, Vol. 34, No. 3, 251-294, (2004).

[5] M. Adimy, K. Ezzinbi and M. Laklach, Spectral decompostion for some partial neutral functional differential equations, *Canadian Applied Mathematics Quarterly*, Vol. 9, No. 4, 1-34, (2001).

[6] J. Carr, Applications of Center Manifold Theory, *Applied Mathematical Sciences*, Springer-Verlag, Vol. 35, (1981).

[7] S. N. Chow and K. Lu, Invariant manifolds for flows in Banach spaces, *Journal of Differential equations*, **74**, 285-317, (1988).

[8] G. Da Prato and A. Lunardi, Stability, instability and center manifold theorem for fully nonlinear autonomous parabolic equations in Banach spaces, *Archive for Rational Mechanics and Analysis*, Vol. 101, 115-141, (1988).

[9] W. Desch and W. Schappacher, Linearized stability for nonlinear semigroups, in " Differential equations in Banach spaces", (A. Favini and E. Obrecht, Eds), *Lecture Notes in Mathematics*, Springer-Verlag, Vol. 1223, 61-73, (1986).

[10] O. Diekmann, S. A. van Gils, S. M. Verduyn Lunel and H-O Walther, Delay Equations, Functional Complex and Nonlinear Analysis, *Applied Mathematical Sciences*, Springer-Verlag, (1995).

[11] T. Faria, W. Huang and J. Wu, Smoothness of center manifolds for maps and formal adjoints for semilinear FDEs in general Banach spaces, SIAM, *Journal of Mathematical Analysis*, **34**, 173-203, (2002).

[12] J. K. Hale, Critical cases for neutral functional differential equations, *Journal of Differential Equations,* **10**, 59-82, (1971).

[13] J. K. Hale, *Theory of Functional Differential Equations*, Springer-Verlag, (1977).

[14] D. Henry, Geometric Theory of Semilinear Parabolic Equations, *Lecture Notes in Mathematics*, Springer-Verlag, Vol. 840, (1981).

[15] A. Keller, The stable, center-stable, center, center-unstable and unstable manifolds, *Journal of Differential Equations*, **3**, 546-570, (1967).

[16] A. Keller, Stability of the center-stable manifold, *Journal of Mathematical Analysis and Applications,* **18**, 336-344, (1967).

[17] X. Lin, J. W. H. So and J. Wu, Centre manifolds for partial differential equations with delays, *Proceedings of the Royal Society of Edinburgh*, **122**, 237-254, (1992).

[18] N. V. Minh and J. Wu, Invariant manifolds of partial functional differential equations, *Journal of Differential Equations*, **198**, 381-421, (2004).

[19] K. Palmer, On the stability of the center manifold, *Journal of Applied Mathematics and Physics*, (ZAMP), Vol. 38, 273-278, (1987).

[20] S. N. Shimanov, On the stability in the critical case of a zero root for systems with time lag, *Prikl. Mat. Mekh.* **24**, 447-457, (1960).

[21] C. C. Travis and G. F. Webb, Existence and stability for partial functional differential equations, *Transactions of the American Mathematical Society*, **200**, 395-418 (1974).

[22] J. W. H. So, Y. Yang and J. Wu, Center manifolds for functional partial differential equations: Smoothness and attractivity, *Mathematica Japonica*, **48**, 67-81, (1998).

[23] T. Yoshizawa, Stability Theory and the Existence of Periodic Solutions and Almost Periodic Solutions, *Applied Mathematical Sciences*, Springer-Verlag, Vol. 14, (1975).

[24] A. Vanderbauwhede and S. A. Van Gils, Center manifolds and contractions on a scale of Banach spaces, *Journal of Functional Analysis,* **72**, 209-224, (1987).

[25] E. Zeidler, *Nonlinear Functional Analysis and its Applications*, Tome I, Fixed Point theorems, Springer-Verlag, (1993).

[26] G. F. Webb, *Theory of Nonlinear Age-dependent Population Dynamics*, Marcel Dekker, (1985).

[27] J. Wu, Theory and Applications of Partial Functional Differential Equations, *Applied Mathematical Sciences*, Springer-Verlag, Vol. 119, (1996).

In: Research on Evolution Equation Compendium. Volume 2 ISBN: 978-1-60456-752-6
Editor: Gaston M. N'Guerekata

STABILITY RADII OF POSITIVE LINEAR FUNCTIONAL DIFFERENTIAL SYSTEMS IN BANACH SPACES

Pham Huu Anh Ngoc [1]*, ***Nguyen Van Minh*** [2]† ***and Toshiki Naito*** [3]‡
[1]Department of Mathematics, University of Hue,
32 Le Loi Street Hue City, Vietnam
[2]Department of Mathematics, State University
of West Georgia, Carrollton, GA 30118
[3]Department of Mathematics, The University
of Electro-Communications, Chofu, Tokyo 182-8585. Japan

Abstract

In this paper we study stability radii of positive linear functional differential systems in Banach spaces under multi-perturbations and multi-affine perturbations. We prove that for the class of positive systems, complex stability radius, real stability radius and positive stability radius of positive systems under multi-perturbations (or multi-affine perturbations) coincide and they are computed via a simple formula. We illustrated the obtained results by an example.

Key Words: linear functional differential equation in Banach space, positive system, stability radius, multi-perturbation, multi-affine perturbation.
AMS Subject Classification: 34 K20, 93 D09

1 Introduction

Over the last decades there has been an increasing interest in the robust stability of dynamical systems that has many applications in control engineering. In the analysis of robust stability, the notion of stability radius shows to be an appropriate tool. By definition, the stability radius of a given asymptotically stable system $\dot{x}(t) = Ax(t)$ is the *maximal* $\gamma > 0$ for which all the systems of the form

$$\dot{x}(t) = (A + D\Delta E)x(t), \quad \|\Delta\| < \gamma$$

are asymptotically stable. Here, Δ is a unknown disturbance matrix, D and E are given matrices defining the structure of the perturbations.

*E-mail address: phanhngoc@yahoo.com. Corresponding author
†E-mail address: vnguyen@westga.edu
‡E-mail address: naito@e-one.uec.ac.jp

In general, one may consider the disturbance matrix Δ as real or complex. Accordingly, the maximal γ is called complex or real stability radius, respectively. The basic problem in the study of robustness of stability of the system is to characterize and compute these radii in terms of given matrices A, D, E. It is worth mentioning that these two stability radii are in general different. The analysis and computation of the complex stability radius for systems under structured perturbations was done first in [8] and extended later in many subsequent papers (see [9] for a survey up till 1990) while the computation of the real stability radius, being a much more difficult problem, was solved only recently, see e.g. [24].

The situation is much simpler for the class of positive systems. It is shown (see [11], [26]) that if A is a Metzler matrix (i.e. all off-diagonal entries of A are nonnegative) and D, E are nonnegative matrices, then the complex and the real stability radii coincide and can be computed directly by a simple formula. These results have been extended recently to infinite dimensional case (see [1], [2]), for positive continuous time-delay systems (see [27]-[30]) and for discrete time-delay systems (see [13], [19]).

It is worth noticing that the notion of stability radius can be extended to various perturbation types [9]. Among perturbation types, two of the following perturbation types

$$A \to A + \sum_{i=1}^{N} D_i \Delta_i E_i \qquad \text{(Multi-perturbation)} \tag{1}$$

$$A \to A + \sum_{i=1}^{N} \delta_i A_i \qquad \text{(Multi-affine perturbation)} \tag{2}$$

are most well-known in control theory that include the perturbation types studied in the literature. The problem of computing the stability radii of positive linear systems without delays under multi-perturbations was studied in [20]. Recently, in [21], the authors successfully computed the stability radii of positive linear functional differential equations under multi-perturbations and multi-affine perturbations.

Although recently there have been many works dedicated to extending the above-mentioned results to the infinite dimensional case (see e.g. [1], [2], [3], [10], [23], [31]), the problem of computing the stability radii of positive linear functional differential equations in Banach spaces under multi-perturbations and multi-affine perturbations as considered in [21] has not been studied yet. And the main purpose of this paper is to extend the results of [21] to general retarded systems described by linear functional differential equations in Banach spaces of the form

$$\dot{x}(t) = A_0 x(t) + \int_{-h}^{0} d[\eta(\theta)]x(t+\theta),$$

where A_0 is a generator of a compact C_0-semi-group and $\eta(\cdot)$ is a function of bounded variation on $[-h, 0]$ with values in a Banach space. Formulae will be derived for the stability radii of these general systems in terms of the operator A_0 and the function η which define the initial system.

The organization of the paper is as follows. In the next section, we summarize some notations and give preliminary results on C_0- semigroups on Banach spaces and on Banach lattice theory which will be used in the sequel. In Section 3, a lower and upper bound for the complex stability radius of a general functional differential equation under multi-perturbations are given. Then, for the class of positive functional differential equations it will be shown that the complex, real, positive stability radii under multi-perturbations coincide and a simple formula for their computation can be established. In Section 4, we study stability radii of positive functional differential equations under multi-affine perturbations. We also prove that the positive, real, complex stability radii coincide and can be computed by an explicit formula. An example is given to illustrate the obtained results.

2 Preliminaries

Let $(X, \|\cdot\|)$ be a Banach space. An X-valued function $\eta(\cdot) : [\alpha, \beta] \to X$ is said to be of bounded variation if

$$\operatorname{Var}(\eta; \alpha, \beta) := \sup_{P[\alpha,\beta]} \sum_k \|\eta(\theta_k) - \eta(\theta_{k-1})\| < +\infty, \tag{3}$$

where the supremum is taken over the set of all finite partitions of the interval $[\alpha, \beta]$. The set $BV([\alpha, \beta], X)$ of all X-valued functions $\eta(\cdot)$ of bounded variation on $[\alpha, \beta]$ satisfying $\eta(\alpha) = 0$ is a Banach space endowed with the norm $\|\eta\| = \operatorname{Var}(\eta; \alpha, \beta)$. Let Y, U be Banach spaces. Throughout this paper, we denote by $L(Y,U), (\mathcal{L}(Y,U))$ the Banach space of all linear operators (bounded linear operators) from Y to U. Given $\delta(\cdot) \in BV([\alpha, \beta], \mathcal{L}(Y,U))$ then for any continuous functions $\gamma \in C([\alpha, \beta], \mathbb{K})$ and $\phi \in C([\alpha, \beta], Y)$, the integrals

$$\int_\alpha^\beta \gamma(\theta) d[\delta(\theta)] \quad \text{and} \quad \int_\alpha^\beta d[\delta(\theta)]\phi(\theta)$$

exist and are defined respectively as the limits of $S_1(P) := \sum_k^p \gamma(\zeta_k)(\delta(\theta_k) - \delta(\theta_{k-1}))$ and $S_2(P) := \sum_k^p (\delta(\theta_k) - \delta(\theta_{k-1}))\phi(\zeta_k)$ as $d(P) := \max_k |\theta_k - \theta_{k-1}| \to 0$, where $P = \{\theta_1 = \alpha \le \theta_2 \le \cdots \le \theta_n = \beta\}$ is any finite partition of the interval $[\alpha, \beta]$ and $\zeta_k \in [\theta_{k-1}, \theta_k]$ (see e.g. [5]). It is immediate from the definition that

$$\begin{aligned} \| \textstyle\int_\alpha^\beta \gamma(\theta) d[\delta(\theta)] \| &\le \max_{\theta \in [\alpha,\beta]} |\gamma(\theta)| \, \|\delta\|, \\ \| \textstyle\int_\alpha^\beta d[\delta(\theta)]\phi(\theta) \| &\le \max_{\theta \in [\alpha,\beta]} \|\phi(\theta)\| \, \|\delta\|. \end{aligned} \tag{4}$$

Then,

$$L\phi = \int_\alpha^\beta d[\delta(\theta)]\phi(\theta), \quad \phi \in C([-h,0], Y), \tag{5}$$

defines a bounded linear operator from $C([\alpha, \beta], Y)$ to U.

In the subsequent sections the following subspace of $BV([-h,0],X)$ will be used frequently :

$$NBV([-h,0],X) := \{\delta \in BV([-h,0], X); \delta(-h) = 0, \delta \text{ is c.f.l. on } (-h,0)\}. \tag{6}$$

It is clear that $NBV([-h,0],X)$ is closed in $BV([-h,0],X)$ and thus it is a Banach space with the norm $\|\delta\| = \operatorname{Var}(\delta; -h, 0)$. Denote by $C([-h,0],X)$ the Banach space of all continuous functions on $[-h,0]$ with values in X normed by the supremum norm.

Let $(T(t))_{t\geq 0}$ be a strongly continuous semigroup (or shortly, C_0-semigroup) of bounded linear operators on complex Banach space $(X, \|\cdot\|)$. Denote by A the generator of the semigroup $(T(t))_{t\geq 0}$ and $\mathcal{D}(A)$ its domain. That is,

$$\mathcal{D}(A) := \left\{x \in X \,:\, \lim_{t\to 0} \frac{T(t)x - x}{t} \in X\right\}$$

and

$$Ax = \lim_{t\to 0} \frac{T(t)x - x}{t}, \; x \in \mathcal{D}(A).$$

Since A is a closed operator, $\mathcal{D}(A)$ is a Banach space with the graph norm

$$\|x\|_{\mathcal{D}(A)} := \|x\| + \|Ax\|, \qquad x \in \mathcal{D}(A). \tag{7}$$

The resolvent set $\rho(A)$, by definition, consists of all $\lambda \in \mathbb{C}$ for which $(\lambda I_X - A)$ has a bounded linear inverse in X. The complement of $\rho(A)$ in $\mathbb{C}$ is called the spectrum of A and denoted by $\sigma(A)$. With the C_0-semigroup $(T(t))_{t\geq 0}$, we associate the following quantities:

1. The *spectral bound*,
$$s(A) := \sup\{\text{Re}\,\lambda \;\lambda \in \sigma(A)\},$$
where $\sigma(A)$ is spectrum of the linear operator A.

2. The *abscissa of uniform boundedness of the resolvent of A*,
$$s_0(A) := \inf\left\{\omega \in \mathbb{R} \,:\, \{\text{Re}\,\lambda > \omega\} \subset \rho(A) \text{ and } \sup_{Re\lambda>\omega} \|R(\lambda, A)\| < \infty\right\}.$$

3. The *growth bound* $\omega_1(A)$,
$$\omega_1(A) := \inf\{\omega \in \mathbb{R} \,:\, \text{there exists } M > 0 \text{ such that}$$
$$\|T(t)x\| \leq Me^{\omega t}\|x\|_{D(A)} \text{ for all } x \in \mathcal{D}(A) \text{ and } t \geq 0\}.$$

4. The *uniform growth bound* $\omega_0(A)$,
$$\omega_0(A) := \inf\{\omega \in \mathbb{R} \,:\, \text{there exists } M > 0 \text{ such that}$$
$$\|T(t)\| \leq Me^{\omega t} \text{ for all and } t \geq 0\}.$$

It is well-known that

$$-\infty \leq s(A) \leq \omega_1(A) \leq s_0(A) \leq \omega_0(A) \leq +\infty, \tag{8}$$

see, e.g [16], [22].

Next, the C_0-semigroup $T(t)_{t\geq 0}$ is called :

1. Hurwitz stable if $\sigma(A) \subset \mathbb{C}^- := \{\lambda \in \mathbb{C} \,:\, Re\,\lambda < 0\}$,
2. Strictly Hurwitz stable if $s(A) < 0$,

3. Exponentially stable if $\omega_1(A) < 0$,
4. Uniformly exponentially stable if $\omega_0(A) < 0$.

It is well-known that for an eventually norm continuous semigroup, that is,

$$\lim_{t \to t_0} \|T(t) - T(t_0)\| = 0, \quad \text{for some } t_0 \geq 0,$$

$s(A) = \omega_1(A) = s_0(A) = \omega_0(A)$, see e.g. [16]. So, the notations of strictly Hurwitz stability, exponentially stability, uniformly exponentially stability coincide.

To make the presentation self-contained, we give some basic facts on Banach lattices which will be used in the sequel (see, e.g. [15]). Let $X \neq \{0\}$ be a *real* vector space endowed with an order relation $\leq$. Then X is called an *ordered vector space*. Denote the *positive* elements of X by $X_+ := \{x \in X \; : \; 0 \leq x\}$. If furthermore the *lattice property* holds, that is, if $x \vee y := \sup\{x, y\} \in X$, for $x, y \in X$, then X is called a *vector lattice*. It is important to note that X_+ is *generating*, that is,

$$X = X_+ - X_+.$$

Then, the modulus of $x \in X$ is defined by $|x| := x \vee (-x)$. If $\|\cdot\|$ is a norm on the vector lattice X satisfying the *lattice norm property*, that is, if

$$|x| \leq |y| \Rightarrow \|x\| \leq \|y\|, \qquad x, y \in X, \tag{9}$$

then X is called a *normed vector lattice*. If, in addition, $(X, \|\cdot\|)$ is a Banach space then X is called a (real) Banach lattice.

We now extend the notion of Banach lattices to the complex case. For this extension all underlying vector lattices X are assumed to be *relatively uniformly complete,* that is, if for every sequence $(\lambda_n)_{n \in \mathbb{N}}$ in $\mathbb{R}$ satisfying $\sum_{n=1}^{\infty} |\lambda_n| < +\infty$ and for every $x \in X$ and every sequence $(x_n)_{n \in \mathbb{N}}$ in X it holds that

$$0 \leq x_n \leq \lambda_n x \Rightarrow \sup_{n \in \mathbb{N}} \Big(\sum_{i=1}^{n} x_i \Big) \in X.$$

Now let X be a relatively uniformly complete vector lattice. The *complexification* of X is defined by $X_{\mathbb{C}} = X + \imath X$. The modulus of $z = x + \imath y \in X_{\mathbb{C}}$ is defined by

$$|z| = \sup_{0 \leq \phi \leq 2\pi} |(\cos\phi)x + (\sin\phi)y| \in X. \tag{10}$$

A complex vector lattice is defined as the complexification of a relatively uniformly complete vector lattice endowed the modulus (10). If X is normed then

$$\|x\| := \||x|\|, \qquad x \in X_{\mathbb{C}} \tag{11}$$

defines a norm on $X_{\mathbb{C}}$ satisfying the lattice norm property. If X is a Banach lattice then $X_{\mathbb{C}}$ endowed the modulus (10) and the norm (11) is called a complex Banach lattice. Throughout of this paper, for simplicity of presentation, we denote $X, X_{\mathbb{R}}$ instead of $X_{\mathbb{C}}, X$, respectively. Let $E_{\mathbb{R}}, F_{\mathbb{R}}$ be real Banach lattices and $T \in L(E_{\mathbb{R}}, F_{\mathbb{R}})$. Then T is called *positive* and denoted by $T \geq 0$ if $T(E_+) \subset F_+$. By $S \leq T$ we mean $T - S \geq 0$, for $T, S \in L(E_{\mathbb{R}}, F_{\mathbb{R}})$.

An operator $T \in L(E,F)$ is called real if $T(E_{\mathbb{R}}) \subset F_{\mathbb{R}}$. Then a operator $T \in L(E,F)$ is called *positive* ($T \geq 0$) if T is real and $T(E_+) \subset F_+$. We introduce the notations

$$L_{\mathbb{R}}(E,F) := \{T \in L(E,F) \ : \ T \text{ real}\} \ ; \ \mathcal{L}_{\mathbb{R}}(E,F) := \{T \in \mathcal{L}(E,F) \ : \ T \text{ real}\}, \tag{12}$$

$$L_+ := \{T \in L(E,F) \ : \ T \geq 0\} \ ; \quad \mathcal{L}_+(E,F) := \{T \in \mathcal{L}(E,F) \ : \ T \geq 0\}. \tag{13}$$

For $T \in \mathcal{L}_+(E,F)$, we emphasize the simple but important fact

$$\|T\| = \sup_{x \in E_+, \|x\|=1} \|Tx\|. \tag{14}$$

Finally, $\delta \in NBV([\alpha,\beta], \mathcal{L}_{\mathbb{R}}(E,F))$ is called increasing if $\delta(\theta_1) - \delta(\theta_2) \geq 0$ for $\alpha \leq \theta_1 \leq \theta_2 \leq \beta$. In this paper, we always define $\inf \emptyset = +\infty, 0^{-1} = +\infty$.

3 Stability Radii of Linear Abstract Functional Differential Equations under Multi-perturbations

3.1 Abstract Functional Differential Equations

We let X be a complex Banach lattices which makes $C([-h,0],X)$ a Banach lattice as well. We consider a linear retarded system described by the following general functional differential equation

$$\begin{aligned} \dot{x}(t) &= A_0 x(t) + L x_t, \ \ t \geq 0, \ \ x(t) \in X \\ x(\theta) &= \phi_0(\theta), \ \ \theta \in [-h,0]. \end{aligned} \tag{15}$$

where, for each $t \geq 0, x_t \in C := C([-h,0],X)$ is defined by $x_t(\theta) = x(t+\theta), \theta \in [-h,0]$, A_0 is the generator of a C_0-semigroup on X and $L : C([-h,0],X) \to X$ is a bounded linear operator defined by

$$L\phi = \int_{-h}^{0} d[\eta(\theta)]\phi(\theta), \ \ t \geq 0, \ \phi \in C([-h,0],X). \tag{16}$$

Here $\eta(\cdot) \in NBV([-h,0], \mathcal{L}(X))$ is a given $\mathcal{L}(X)$-valued function of bounded variation on $[-h,0]$ such that η vanishes at $-h$ and is c.f.l. on $[-h,0]$. We shall extend the definition of η to $\mathbb{R}$ by setting $\eta(\theta) = \eta(-h) = 0$ for all $\theta \leq -h, \eta(\theta) = \eta(0)$ for all $\theta \geq 0$.

Definition 3.1. *A function $x \in C([-h,0],X)$ is called a solution of the linear retarded system (15)-(16), if*

a) x is right-size differentiable at 0 and continuously differentiable for $t > 0$.

b) $x(t) \in \mathcal{D}(A_0)$ for $t \geq 0$.

c) (15) is satisfied for $t \geq 0$.

To (15)-(16), we associate the following operator A on the Banach space $C([-h,0],X)$. Let A be the differential operator

$$Af = f', \ \mathcal{D}(A) := \{f \in C^1([-h,0],X) \ : \ f(0) \in \mathcal{D}(A_0), \ \ f'(0) = A_0 f(0) + Lf\}. \tag{17}$$

Theorem 3.2. [16] *The operator A defined in (17) is the generator of a C_0-semigroup $(T(t))_{t\geq 0}$ on $C([-h,0],X)$ satisfying the translation property*

$$T(t)f(s) = \begin{cases} f(t+s) \text{ if } t+s \leq 0 \\ T(t+s)f(0) \text{ if } t+s > 0, \end{cases} \tag{18}$$

$f \in C([-h,0],X)$.

Moreover, for $\phi_0 \in \mathcal{D}(A)$ define $x : [-h,\infty) \to X$ by

$$x(t) = \begin{cases} \phi_0(t) \text{ if } -h \leq t \leq 0 \\ T(t)\phi_0(0) \text{ if } t > 0. \end{cases} \tag{19}$$

Then x is the unique solution of the linear retarded system (15)-(16) .

The C_0-semigroup $(T(t))_{t\geq 0}$ in the above theorem is called *solution semigroup* of linear retarded system (15)-(16) and its operator A is called the *solution operator* of the linear retarded system (15)-(16).

Definition 3.3. *The linear retarded system (15)-(16) is called, respectively, Hurwitz stable, strictly Hurwitz stable, exponentially stable, uniformly exponentially stable if its solution semigroup $(T(t))_{t\geq 0}$ is Hurwitz stable, strictly Hurwitz stable, exponentially stable, uniformly exponentially stable.*

Define

$$H(s) := sI_X - A_0 - \int_{-h}^{0} e^{s\theta} d[\eta(\theta)]. \tag{20}$$

$H(s)$ is called the characteristic operator of the linear retarded system (15)-(16). By the retarded resolvent set $\rho(A_0,\eta)$, we understand the set of all $s \in \mathbb{C}$ for which $H(s)$ has a bounded linear inverse in X. The complement of $\rho(A_0,\eta)$ in $\mathbb{C}$ is called the retarded spectrum and denoted by $\sigma(A_0,\eta)$. Then, the following fact can be found in [16], page 224.

$$\sigma(A) = \sigma(A_0,\eta). \tag{21}$$

Assuming that A_0 generates a C_0-compact semigroup $(U(t))_{t\geq 0}$ of bounded linear operators, it was shown in [32, 33, 17, 18] that the initial value problem

$$\begin{aligned} u(t) &= U(t)\phi(0) + \int_0^t U(t-s)L(u_s)ds, \quad t \geq 0, \\ u_0 &= \phi \in \mathcal{C}, \end{aligned}$$

has a unique continuous solution $u(t;\phi)$ for $t \geq -h$, and $\{T(t)\}_{t\geq 0}, T(t)\phi = u_t(\cdot;\phi)$, is a C_0-semigroup of linear (and compact for $t > h$) operators on $\mathcal{C}$, with infinitesimal generator A given as above. Moreover, one has

$$\sigma(A) = \sigma_P(A) = \{s \in \mathbb{C} : H(s)y = 0, \text{ for some } y \in \text{dom}(A_0) \setminus \{0\}\}$$

where $\sigma_p(A)$ denotes the point spectrum of A. In this case, since $(T(t))_{t\geq 0}$ is eventually compact, it is eventually norm continuous. Hence, the above concepts of stability are actually the same if A_0 generates a compact semigroup.

For more information on the theory and applications of abstract functional differential equations we refer the reader to [6, 7, 17, 18, 33].

3.2 Stability Radii

Let X, Y, U_i^0, U_i^1 ($i \in \mathcal{N} := \{i \in \{1,2,...,N\}$) be complex Banach lattices and $D_i^0 \in \mathcal{L}(U_i^0, X), D_i^1 \in \mathcal{L}(U_i^1, X)$ ($i \in \mathcal{N}$), $E \in \mathcal{L}(X,Y)$.

Assume that the retarded system (15)-(16) is Hurwitz stable and A_0, η are subject to multi-perturbations of the type

$$\begin{aligned} A_0 &\rightarrow A_{0\Delta} = A_0 + \textstyle\sum_{i=1}^N D_i^0 \Delta_i E, \qquad \Delta_i \in \mathcal{L}(Y, U_i^0),\ i \in \mathcal{N} \\ \eta &\rightarrow \eta_\delta = \eta + \textstyle\sum_{i=1}^N D_i^1 \delta_i E, \qquad \delta_i \in NBV([-h,0], \mathcal{L}(Y, U_i^1)),\ i \in \mathcal{N} \end{aligned} \tag{22}$$

and thus the perturbed system is described by

$$\begin{aligned} \dot{x}(t) &= (A_0 + \textstyle\sum_{i=1}^N D_i^0 \Delta_i E) x(t) + \int_{-h}^0 d[\eta(\theta) + \textstyle\sum_{i=1}^N D_i^1 \delta_i(\theta) E] x(t+\theta),\ t \geq 0 \\ x(\theta) &= \phi_0(\theta),\ \theta \in [-h, 0]. \end{aligned} \tag{23}$$

Here $D_i^0 \in \mathcal{L}(U_i^0, X), D_i^1 \in \mathcal{L}(U_i^1, X) (i \in \mathcal{N})$ and $E \in \mathcal{L}(X,Y)$ are fixed and describe the *structure* of perturbations, while $\Delta_i \in \mathcal{L}(Y, U_i^0)$ and $\delta_i(\cdot) \in NBV([-h,0], \mathcal{L}(Y, U_i^1)), i \in \mathcal{N}$ are unknown disturbances. We shall measure the size of each perturbation $\tilde{\Delta} := [\Delta, \delta]$ where

$$\Delta := (\Delta_0, ..., \Delta_N),\ \delta := (\delta_0, ..., \delta_N),\ \Delta_i \in \mathcal{L}(Y, U_i^0),\ \delta_i \in NBV([-h,0], \mathcal{L}(Y, U_i^1)),\ i \in \mathcal{N},$$

by the norm

$$\|\tilde{\Delta}\| := \sum_{i=1}^N \|\Delta_i\| + \sum_{i=1}^N \|\delta_i\|, \quad \|\delta_i\| := \mathrm{Var}(\delta_i; -h, 0), i \in \underline{N}. \tag{24}$$

Set

$$\begin{aligned} \mathcal{D}_{\mathbb{C}} &:= \{\tilde{\Delta} = [\Delta, \delta] \ :\ \Delta_i \in \mathcal{L}(Y, U_i^0), \delta_i \in NBV([-h,0], \mathcal{L}(Y, U_i^1)), i \in \underline{N}\}, \\ \mathcal{D}_{\mathbb{R}} &:= \{\tilde{\Delta} = [\Delta, \delta] \ :\ \Delta_i \in \mathcal{L}_{\mathbb{R}}(Y, U_i^0), \delta_i \in NBV([-h,0], \mathcal{L}_{\mathbb{R}}(Y, U_i^1)), i \in \underline{N}\}, \\ \mathcal{D}_{+} &:= \{\tilde{\Delta} = [\Delta, \delta] \ :\ \Delta_i \in \mathcal{L}_{+}(Y, U_i^0), \delta_i \in NBV([-h,0], \mathcal{L}_{\mathbb{R}}(Y, U_i^1)), \\ &\qquad\qquad \delta_i \text{ is increasing for every } i \in \underline{N}\} \end{aligned}$$

Then $\mathcal{D}_{\mathbb{C}}, \mathcal{D}_{\mathbb{R}}, \mathcal{D}_{+}$ is called respectively the *class of complex, real, nonnegative* with respect to the perturbation structure (22). Denote by $A_{\tilde{\Delta}}, ((T_{\tilde{\Delta}}(t))_{t\geq 0}$ the solution operator, solution semigroup of the perturbed system (23), respectively. To study robustness of stability of the retarded system (15)-(16) we introduce the following.

Definition 3.4. *Let the retarded system (15)-(16) be Hurwitz stable. The complex, real, positive stability radius of the system with respect to perturbations of the form (22), measured by the norm (24), is defined respectively by*

$$r_{\mathbb{C}} = \inf\{\|\tilde{\Delta}\| \ :\ \tilde{\Delta} \in \mathcal{D}_{\mathbb{C}},\ \sigma(A_{\tilde{\Delta}}) \not\subset \mathbb{C}^-\}, \tag{25}$$

$$r_{\mathbb{R}} = \inf\{\|\tilde{\Delta}\| \ :\ \tilde{\Delta} \in \mathcal{D}_{\mathbb{R}},\ \sigma(A_{\tilde{\Delta}}) \not\subset \mathbb{C}^-\}, \tag{26}$$

$$r_{+} = \inf\{\|\tilde{\Delta}\| \ :\ \tilde{\Delta} \in \mathcal{D}_{+},\ \sigma(A_{\tilde{\Delta}}) \not\subset \mathbb{C}^-\}. \tag{27}$$

We define the associated transfer functions of the perturbed system (23) by setting

$$G_i^k(s) = EH(s)^{-1}D_i^k \qquad i \in \mathcal{N}, k \in \{0,1\}, \tag{28}$$

where $H(s)^{-1} := \left(sI_X - A_0 - \int_{-h}^{0} e^{\theta s} d[\eta(\theta)]\right)^{-1}$, $Re\, s > s(A)$.
We need the following technical lemmas.

Lemma 3.5. *Let X, Y be Banach spaces and $S \in \mathcal{L}(Y,X), T \in \mathcal{L}(X,Y)$. Then, $I_X - ST$ is invertible if and only if so is $I_Y - TS$.*

Proof. Assume that $I_X - ST$ is invertible. Setting $R := (I_X - ST)^{-1}$, we can check that

$$(I_Y - TS)(I_Y + TRS) = I_Y = (I_Y + TRS)(I_Y - TS).$$

By the same way, we can prove that if $I_Y - TS$ is invertible then so is $I_X - ST$. □

Lemma 3.6. *Assume that the system (15)-(16) is Hurwitz stable and $\sup_{s\in\mathbb{C}, Res\geq 0} \|H(s)^{-1}\| < +\infty$. Then*

(i) *If $\sup_{s\in\mathbb{C}, Res\geq 0} \|G_{i_0}^0(s)\| \neq 0$ for some $i_0 \in \mathcal{N}$ then, for every $\varepsilon > 0$, there exists a complex perturbation $\tilde{\Delta} := [\Delta, \delta] \in \mathcal{D}_{\mathbb{C}}$, such that*

$$\|\tilde{\Delta}\| < \frac{1}{\sup_{s\in\mathbb{C}, Res\geq 0} \|G_{i_0}^0(s)\|} + \varepsilon, \tag{29}$$

and

$$\sigma(A_{\tilde{\Delta}}) \not\subset \mathbb{C}^-. \tag{30}$$

(ii) *If $\sup_{s\in\mathbb{C}, Res\geq 0} \|G_{i_0}^1(s)\| \neq 0$ for some $i_0 \in \mathcal{N}$ then, for every $\varepsilon > 0$, there exists a complex perturbation $\tilde{\Delta} := [\Delta, \delta] \in \mathcal{D}_{\mathbb{C}}$, such that*

$$\|\tilde{\Delta}\| < \frac{1}{\sup_{s\in\mathbb{C}, Res\geq 0} \|G_{i_0}^0(s)\| |e^{-hs}|} + \varepsilon, \tag{31}$$

and

$$\sigma(A_{\tilde{\Delta}}) \not\subset \mathbb{C}^-. \tag{32}$$

(iii) *In particular, if $G_i^0(0) \in \mathcal{L}_+(U_i^0, Y), G_i^1(0) \in \mathcal{L}_+(U_i^1, Y)$ for every $i \in \mathcal{N}$ and $\max\{\max_{i\in\mathcal{N}} \|G_i^0(0)\|, \max_{i\in\mathcal{N}} \|G_i^1(0)\|\} \neq 0$ then, for every $\varepsilon > 0$, there exists a non-negative perturbation $\tilde{\Delta} \in \mathcal{D}_+$ satisfying*

$$\|\tilde{\Delta}\| < \frac{1}{\max\{\max_{i\in\mathcal{N}} \|G_i^0(0)\|, \max_{i\in\mathcal{N}} \|G_i^1(0)\|\}} + \varepsilon. \tag{33}$$

and

$$\sigma(A_{\tilde{\Delta}}) \not\subset \mathbb{C}^-. \tag{34}$$

Proof. (i) It follows from $\sup_{s\in\mathbb{C},Res\geq 0}\|H(s)^{-1}\|<+\infty$ that $\sup_{s\in\mathbb{C},Re\, s\geq 0}\|G^0_{i_0}(s)\|<+\infty$. For $\varepsilon>0$, there exists $s_0\in\mathbb{C}, Re\, s_0\geq 0$ such that

$$\frac{1}{\|G^0_{i_0}(s_0)\|}<\frac{1}{\sup_{s\in\mathbb{C},Re\, s\geq 0}\|G^0_{i_0}(s)\|}+\frac{\varepsilon}{2}$$

Let $u_0\in U^0_{i_0}$ $\|u_0\|=1$ satisfy $\frac{1}{\|G^0_{i_0}(s_0)u_0\|}<\frac{1}{\|G^0_{i_0}(s_0)\|}+\frac{\varepsilon}{2}$. Then, by the Hahn-Banach Theorem, there exists $y^*_0\in(Y)^*, \|y^*_0\|=1$ satisfying

$$y^*_0(G^0_{i_0}(s_0)u_0)=\|G^0_{i_0}(s_0)u_0\|.$$

Define

$$\Delta^0_{i_0}:=\frac{1}{\|G^0_{i_0}(s_0)u_0\|}u_0y^*_0(\cdot)\in\mathcal{L}(Y,U^0_{i_0}).$$

It is easy to see that $\|\Delta^0_{i_0}\|=\frac{1}{\|G^0_{i_0}(s_0)u_0\|}$. Setting $x_0:=H(s_0)^{-1}D^0_{i_0}u_0\in\mathcal{D}(A_0)$, we have $\Delta^0_{i_0}Ex_0=u_0$. Therefore,

$$0\neq x_0=H(s_0)^{-1}D^0_{i_0}\Delta^0_{i_0}Ex_0.$$

This implies

$$((A_0+D^0_{i_0}\Delta^0_{i_0}E)+\int_{-h}^{0}e^{s_0\theta}d[\eta(\theta)])x_0=s_0x_0,\quad x_0\in\mathcal{D}(A_0),x_0\neq 0.$$

We define

$$\tilde{\Delta}:=[\Delta,\delta]=((\Delta_0,...,\Delta_N),(\delta_0,...,\delta_N)),$$

where $\Delta_i=\Delta^0_{i_0}$ if $i=i_0$ otherwise $\Delta_i=0, i\in\mathcal{N}$ and $\delta_i=0$ for all $i\in\mathcal{N}$.
Then, we have

$$\|\tilde{\Delta}\|=\|\Delta^0_{i_0}\|<\frac{1}{\sup_{s\in\mathbb{C},Re\, s\geq 0}\|G^0_{i_0}(s)\|}+\varepsilon\qquad\text{and}\qquad s(A_{\tilde{\Delta}})\geq Re\, s_0\geq 0.$$

(ii) The proof is similar that of (i). For $\varepsilon>0$, there exists $s_1\in\mathbb{C}, Re\, s_1\geq 0$ such that

$$\frac{1}{\|G^1_{i_0}(s_1)\||e^{-s_1h}|}<\frac{1}{\sup_{s\in\mathbb{C},Re\, s\geq 0}\|G^1_{i_0}(s)\||e^{-sh}|}+\frac{\varepsilon}{2}$$

Let $u_1\in U^1_{i_0}, \|u_1\|=1$ satisfy $\frac{1}{\|G^1_{i_0}(s_1)u_1\|}<\frac{1}{\|G^1_{i_0}(s_1)\|}+\frac{\varepsilon}{2}|e^{-hs_1}|$. By the Hahn-Banach Theorem, there exists $y^*_1\in(Y)^*, \|y^*_1\|=1$ satisfying

$$y^*_1(G^1_{i_0}(s_1)u_1)=\|G^1_{i_0}(s_1)u_1\|.$$

Define

$$\Delta^1_{i_0}:=\frac{1}{\|G^1_{i_0}(s_1)u_1\|}u_1y^*_1(\cdot)\in\mathcal{L}(Y,U^1_{i_0}).$$

It is easy to see that $\|\Delta^1_{i_0}\|=\frac{1}{\|G^1_{i_0}(s_1)u_1\|}$. We now consider the following step function

$$\delta_{i_0}(\theta)=\begin{cases}0 \text{ if } \theta=-h,\\ \Delta^1_{i_0}e^{hs_1} \text{ if } \theta\in(-h,0].\end{cases}$$

It is clear that

$$\delta_{i_0} \in NBV([-h,0], \mathcal{L}(Y, U^1_{i_0})), \ \ \|\delta_1\| = \mathrm{Var}(\delta_1; -h, 0) = \|\Delta^1_{i_0}\||e^{hs_1}|, \ \int_{-h}^{0} e^{s_1\theta} d[\delta_1(\theta)] = \Delta^1_{i_0}.$$

Setting $x_1 := H(s_1)^{-1} D^1_{i_0} u_1 \in \mathcal{D}(A_0)$, we have $\int_{-h}^{0} e^{s_1\theta} d[\delta_{i_0}(\theta)]Ex_1 = \Delta^1_{i_0} Ex_1 = u_1$. Therefore,

$$0 \neq x_1 = H(s_1)^{-1} D^1_0 \int_{-h}^{0} e^{s_1\theta} d[\delta_{i_0}(\theta)]Ex_1.$$

This implies

$$(A_0 + \int_{-h}^{0} e^{s_1\theta} d[\eta(\theta) + D^1_{i_0}\delta_{i_0} E])x_1 = s_1 x_1, \quad x_1 \in \mathcal{D}(A_0), x_1 \neq 0.$$

We define $\tilde{\Delta} := [\Delta, \delta] = ((\Delta_0, ..., \Delta_N), (\delta_0, ..., \delta_N))$, where $\delta_i = \delta_{i_0}$ if $i = i_0$ otherwise $\delta_i = 0, i \in \mathcal{N}$ and $\Delta_i = 0$ for every $i \in \mathcal{N}$. Then, we have

$$\|\tilde{\Delta}\| = \|\Delta^1_{i_0}\||e^{s_1 h}| < \frac{1}{\sup_{s\in\mathbb{C}, Re\, s \geq 0} \|G^1_{i_0}(s)\||e^{-sh}|} + \varepsilon \quad \text{and} \quad s(A_{\tilde{\Delta}}) \geq Re\, s_1 \geq 0.$$

(iii) If $G^0_i(0) \in \mathcal{L}_+(U^0_i, Y)$ and $G^1_i(0) \in \mathcal{L}_+(U^1_i, Y)$ for every $i \in \mathcal{N}$ then, by (14), $\|G^0_i(0)\| = \sup_{u\in (U^0_i)_+, \|u\|=1} \|G^0_i(0)u\|$, $\|G^1_i(0)\| = \sup_{u\in (U^1_i)_+, \|u\|=1} \|G^1_i(0)u\|$. Thus we can choose $u_0 \in (U^0_i)_+, u_1 \in (U^1_i)_+$ $\|u_0\| = \|u_1\| = 1$ such that

$$\|G^0_i(0)u_0\| = \|G^0_i(0)\| \ , \ \|G^1_i(0)u_1\| = \|G^1_i(0)\|.$$

Since $G^0_i(0)u_0 \geq 0, G^1_i(0)u_1 \geq 0$, by the Hahn-Banach theorem for positive operators (see e.g. [34] page 249), there exist positive linear forms $y^*_0 \in (Y)^*, y^*_1 \in (Y)^*$ of dual norms $\|y^*_0\| = \|y^*_1\| = 1$ such that $y^*_0 G_i(0)u_0 = \|G_i(0)u_0\|, y^*_1 G_i(0)u_1 = \|G_i(0)u_1\|$. Hence the perturbation $\tilde{\Delta}$ constructed as in (i), (ii) (where $s_0 = s_1 = 0$) is nonnegative. The proof is complete. □

Using Lemma 3.6 we obtain the following estimates for the complex radius $r_{\mathbb{C}}$.

Theorem 3.7. *Let the system (15)-(16) be Hurwitz stable and* $\sup_{s\in\mathbb{C}, Res\geq 0} \|H(s)^{-1}\| < +\infty$. *Assume that* A_0, η *are subjected to multi-perturbations of the form (22). Then, we have*

$$\frac{1}{\max\left\{\max_{i\in\mathcal{N}}\left\{\sup_{s\in\mathbb{C}, Res\geq 0}\|G^0_i(s)\|\right\}, \max_{i\in\mathcal{N}}\left\{\sup_{s\in\mathbb{C}, Res\geq 0}\|G^1_i(s)\|\right\}\right\}} \leq r_{\mathbb{C}}$$

$$\leq \frac{1}{\max\left\{\max_{i\in\mathcal{N}}\left\{\sup_{s\in\mathbb{C}, Res\geq 0}\|G^0_i(s)\|\right\}, \max_{i\in\mathcal{N}}\left\{\sup_{s\in\mathbb{C}, Res\geq 0}\|G^1_i(s)\||e^{-hs}|\right\}\right\}}. \quad (35)$$

Proof. Assume that $r_{\mathbb{C}} < +\infty$. Let $\tilde{\Delta} := [\Delta, \delta]$,

$$\Delta := (\Delta_0, ..., \Delta_N), \ \delta := (\delta_0, ..., \delta_N), \ \Delta_i \in \mathcal{L}(Y, U^0_i), \ \delta_i \in NBV([-h,0], \mathcal{L}(Y, U^1_i)), \ i \in \mathcal{N}$$

satisfy $\sigma(A_{\tilde{\Delta}}) \not\subset \mathbb{C}^-$. By (21), there exists $s_0 \in \mathbb{C}, Res_0 \geq 0$ such that $s_0 \in \sigma\big((A_0 + \sum_{i=1}^N D_i^0 \Delta_i E) + \int_{-h}^0 e^{s_0\theta} d[\eta(\theta) + \sum_{i=1}^N D_i^1 \delta_i(\theta) E]\big)$. Since the system (15)-(16) is Hurwitz stable, $H(s_0) = (s_0 I_X - A_0 - \int_{-h}^0 e^{s_0\theta} d[\eta(\theta)])$ has a bounded linear inverse. Since

$$\left(s_0 I_X - (A_0 + \sum_{i=1}^N D_i^0 \Delta_i E) - \int_{-h}^0 e^{s_0\theta} d[\eta(\theta) + \sum_{i=1}^N D_i^1 \delta_i(\theta) E]\right)$$

$$= \left(I_X - (\sum_{i=1}^N D_i^0 \Delta_i - \sum_{i=1}^N D_i^1 \int_{-h}^0 e^{s_0\theta} d[\delta_i(\theta)]) E H(s_0)^{-1}\right) H(s_0),$$

and since this operator is not invertible, the operator

$$\left(I_X - (\sum_{i=1}^N D_i^0 \Delta_i - \sum_{i=1}^N D_i^1 \int_{-h}^0 e^{s_0\theta} d[\delta_i(\theta)]) E H(s_0)^{-1}\right),$$

is not invertible. Then, it follows from Lemma 3.5 that the operator

$$I_Y - \left(E H(s_0)^{-1} (\sum_{i=1}^N D_i^0 \Delta_i + \sum_{i=1}^N D_i^1 \int_{-h}^0 e^{s_0\theta} d[\delta_i(\theta)]\right)\Bigg)$$

is not invertible. Hence

$$\|E H(s_0)^{-1} \left(\sum_{i=1}^N D_i^0 \Delta_i + \sum_{i=1}^N D_i^1 \int_{-h}^0 e^{s_0\theta} d[\delta_i(\theta)]\right)\| \geq 1.$$

Using the inequalities in (4), we get

$$\max\{\max_{i\in\mathcal{N}} \|G_i^0(s_0)\|, \max_{i\in\mathcal{N}} \|G_i^1(s_0)\|\} (\sum_{i=1}^N \|\Delta_i\| + \sum_{i=1}^N \|\delta_i\|) \geq 1.$$

Hence

$$\max\left\{\max_{i\in\mathcal{N}}\{\sup_{s\in\mathbb{C}, Res\geq 0} \|G_i^0(s)\|\}, \max_{i\in\mathcal{N}}\{\sup_{s\in\mathbb{C}, Res\geq 0} \|G_i^1(s)\|\}\right\} \|\tilde{\Delta}\| \geq 1, \tag{36}$$

or equivalently,

$$\|\tilde{\Delta}\| \geq \frac{1}{\max\{\max_{i\in\mathcal{N}}\{\sup_{s\in\mathbb{C}, Res\geq 0} \|G_i^0(s)\|\}, \max_{i\in\mathcal{N}}\{\sup_{s\in\mathbb{C}, Res\geq 0} \|G_i^1(s)\|\}\}}.$$

By the definition of the complex stability radius $r_{\mathbb{C}}$, we have

$$r_{\mathbb{C}} \geq \frac{1}{\max\{\max_{i\in\mathcal{N}}\{\sup_{s\in\mathbb{C}, Res\geq 0} \|G_i^0(s)\|\}, \max_{i\in\mathcal{N}}\{\sup_{s\in\mathbb{C}, Res\geq 0} \|G_i^1(s)\|\}\}}$$

It remains to prove that

$$r_{\mathbb{C}} \leq \frac{1}{\max\{\max_{i\in\mathcal{N}}\{\sup_{s\in\mathbb{C}, Res\geq 0} \|G_i^0(s)\|\}, \max_{i\in\mathcal{N}}\{\sup_{s\in\mathbb{C}, Res\geq 0} \|G_i^1(s)\| |e^{-hs}|\}\}}.$$

Fix $\varepsilon > 0$. It follows from (36) that

$$\max\{\max_{i\in\mathcal{N}}\{\sup_{s\in\mathbb{C},Res\geq 0}\|G_i^0(s)\|\}, \max_{i\in\mathcal{N}}\{\sup_{s\in\mathbb{C},Res\geq 0}\|G_i^1(s)\|\}\} \neq 0.$$

Taking the Lemma 3.6 (i), (ii) into account, there exists a complex perturbation $\tilde{\Delta} := [\Delta,\delta] \in \mathcal{D}_{\mathbb{C}}$ such that

$$\|\tilde{\Delta}\| < \frac{1}{\max\{\max_{i\in\mathcal{N}}\{\sup_{s\in\mathbb{C},Res\geq 0}\|G_i^0(s)\|\}, \max_{i\in\mathcal{N}}\{\sup_{s\in\mathbb{C},Res\geq 0}\|G_i^1(s)\||e^{-hs}|\}\}} + \varepsilon,$$

and

$$\sigma(A_{\tilde{\Delta}}) \not\subset \mathbb{C}^-.$$

By the definition of the complex stability radius $r_{\mathbb{C}}$,

$$r_{\mathbb{C}} < \frac{1}{\max\{\max_{i\in\mathcal{N}}\{\sup_{s\in\mathbb{C},Res\geq 0}\|G_i^0(s)\|\}, \max_{i\in\mathcal{N}}\{\sup_{s\in\mathbb{C},Res\geq 0}\|G_i^1(s)\||e^{-hs}|\}\}} + \varepsilon.$$

This implies

$$r_{\mathbb{C}} \leq \frac{1}{\max\{\max_{i\in\mathcal{N}}\{\sup_{s\in\mathbb{C},Res\geq 0}\|G_i^0(s)\|\}, \max_{i\in\mathcal{N}}\{\sup_{s\in\mathbb{C},Res\geq 0}\|G_i^1(s)\||e^{-hs}|\}\}}.$$

If $r_{\mathbb{C}} = +\infty$ (for example, E is zero operator) then by Lemma 3.6,

$$\max_{i\in\mathcal{N}}\{\sup_{s\in\mathbb{C},Res\geq 0}\|G_i^0(s)\|\} = \max_{i\in\mathcal{N}}\{\sup_{s\in\mathbb{C},Res\geq 0}\|G_i^1(s)\|\} = 0.$$

Then the equalities in (35) hold. This completes our proof. □

The problem of computation of the real stability radius $r_{\mathbb{R}}$ is much more difficult and has been solved first for the ordinary linear systems of the form $\dot{x}(t) = A_0x(t),\ t \geq 0$, where the system matrix A_0 is subjected to single perturbations (i.e. $N = 1$), see, e.g [24]. Recently, this result has been extended to linear time invariant time delay systems [14]. We note that, by definition, $r_{\mathbb{C}} \leq r_{\mathbb{R}}$, so $r_{\mathbb{C}}$ can be accepted as the lower bound for $r_{\mathbb{R}}$. Unfortunately, as shown in many previous papers (see, e.g. [9]) these two stability radii can be arbitrarily distinct. Therefore, it is an interesting problem to find classes of systems of practical interest for which these two stability radii coincide. Motivated by the results of [11], [12], [28], [29], [30], in this section, we show that for the class of positive functional differential equations, the positive, real and complex stability radius under multi-perturbations coincide and can be computed by a simple formula. We now restrict ourselves to *positive* systems.

Definition 3.8. *System (15)-(16) is called positive if its solution semigroup $(T(t))_{t\geq 0}$ is a positive semigroup, that is, $T(t) \in \mathcal{L}_+(C[-h,0],X)$ for every $t \geq 0$.*

Proposition 3.9. [16] *The system (15)-(16) is positive if A_0 generates a positive semigroup and $L \in \mathcal{L}\big((C[-h,0],X),X\big)$ is a positive operator.*

Recall that, by definition, a linear operator A_0 has a compact resolvent if $\rho(A_0) \neq \emptyset$ and $(sI_X - A_0)^{-1}$ is compact for every $s \in \rho(A_0)$. In order to give characterization of the real stability radius of the positive systems, we need some technical lemmas.

Lemma 3.10. [16] *Suppose that A_0 generates a positive semigroup and $L \in \mathcal{L}\big((C[-h,0], X), X\big)$ is a positive operator. Then*

(i) The "spectral bound function" $\lambda \to s(A_0 + \int_{-h}^{0} e^{\lambda\theta} d[\eta(\theta)])$ is decreasing and continuous from the left on $\mathbb{R}$.

(ii) If, additionally, A_0 has a compact resolvent and there exists a $\lambda_0 \in \mathbb{R}$ with $\sigma(A_0 + \int_{-h}^{0} e^{\lambda_0\theta} d[\eta(\theta)])) \neq \emptyset$, then $\lambda \to s(A_0 + \int_{-h}^{0} e^{\lambda\theta} d[\eta(\theta)])$ is continuous on $\mathbb{R}$ and the spectral bound $s(A)$ of the solution operator of the system (15)-(16) is a unique solution of equation

$$\lambda = s(A_0 + \int_{-h}^{0} e^{\lambda\theta} d[\eta(\theta)])).$$

Moreover

$$s(A_0 + \int_{-h}^{0} e^{\lambda\theta} d[\eta(\theta)]) < \lambda \iff s(A) < \lambda,$$

$$s(A_0 + \int_{-h}^{0} e^{\lambda\theta} d[\eta(\theta)]) > \lambda \iff s(A) > \lambda.$$

In the rest of this paper, we suppose that

Assumption 3.11. *A_0 generates a positive compact C_0-semigroup $(U(t))_{t\geq 0}$ of bounded linear operators on X, $L \in \mathcal{L}\big((C[-h,0],X),X\big)$ is a positive operator and $\sigma(A) \neq \emptyset$.*

Remark 3.12. *As noted in the beginning of this section, if A_0 generates a C_0-compact semigroup then the solution semigroup $(T(t))_{t\geq 0}$ of the system (15)-(16) is eventually compact. Hence, the concepts of stability in Definition 3.3 are actually the same. Furthermore, if Assumption 3.11 is fulfilled then it is easy to see that all conditions of Lemma 3.10(ii) is satisfied.*

Lemma 3.13. *Let $H(s) = sI_X - A_0 - \int_{-h}^{0} e^{s\theta} d[\eta(\theta)]$. Then*

i)

$$H(t_1)^{-1} \geq H(t_2)^{-1} \geq 0 \quad for \;\; t_2 > t_1 > s(A). \tag{37}$$

(ii) Let $s = \alpha + \imath\beta \in \mathbb{C}, \alpha > s(A)$. Then

$$|H(s)^{-1}x| \leq H(\alpha)^{-1}|x|, \; x \in X. \tag{38}$$

Proof. (i) It follows from $L \in \mathcal{L}\big((C[-h,0],X),X\big)$ being a positive operator and the equality

$$(\int_{-h}^{0} e^{t\theta} d[\eta(\theta)])x = \int_{-h}^{0} (e^{t\theta}x) d[\eta(\theta)], \; x \in X,$$

that $\int_{-h}^{0} e^{t\theta} d[\eta(\theta)] \in \mathcal{L}_+(X)$ for every $t \in \mathbb{R}$. Since A_0 is a generator of a positive C_0-semigroup on X and $\int_{-h}^{0} e^{t\theta} d[\eta(\theta)] \in \mathcal{L}_+(X)$, $A_0 + \int_{-h}^{0} e^{t\theta} d[\eta(\theta)]$ also generates a positive C_0-semigroup on X. By Lemma 3.10 (ii), if $t > s(A)$ then $t > s(A_0 + \int_{-h}^{0} e^{t\theta} d[\eta(\theta)])$. From Theorem 1.1 (C-III), page 292 [16], we have

$$(tI_X - A_0 - \int_{-h}^{0} e^{t\theta} d[\eta(\theta)])^{-1} = H(t)^{-1} \geq 0 \qquad \text{for every } \; t > s(A). \tag{39}$$

It is easy to verify that the following resolvent equation holds

$$H(t_1)^{-1} - H(t_2)^{-1} = (t_2 - t_1)H(t_1)^{-1}H(t_2)^{-1} + H(t_1)^{-1}\int_{-h}^{0}(e^{t_1\theta} - e^{t_2\theta})d[\eta(\theta)]H(t_2)^{-1}.$$

It follows from (39) that

$$H(t_1)^{-1} \geq H(t_2)^{-1} \geq 0 \quad for \;\; t_2 > t_1 > s(A).$$

(ii) Let A_0 be the generator of the positive semigroup $(U(t))_{t\geq 0}$ on X. Set $B(s) := \int_{-h}^{0} e^{s\theta}d[\eta(\theta)]$, $s \in \mathbb{C}$. It is well-known that $A_0 + B(s)$ with domain $\mathcal{D}(A_0 + B(s)) = \mathcal{D}(A_0)$ is the generator of a C_0-semigroup $(V_s(t))_{t\geq 0}$ satisfying

$$V_s(t)x = \lim_{n\to\infty} \big(U(\frac{t}{n})e^{\frac{t}{n}B(s)}\big)^n x, \qquad \text{for } t \geq 0,\ x \in X, \tag{40}$$

see e.g [16], page 44. It is easy to see that

$$|(\int_{-h}^{0} e^{s\theta}d[\eta(\theta)])x| \leq (\int_{-h}^{0} e^{s(A)\theta}d[\eta(\theta)])|x|, \quad x \in X,$$

where $s \in \mathbb{C}, Res > s(A)$. Using the above inequality, we could check that $\omega_0(A_0 + \int_{-h}^{0} e^{s\theta}d[\eta(\theta)]) \leq \omega_0(A_0 + \int_{-h}^{0} e^{s(A)\theta}d[\eta(\theta)]), s \in \mathbb{C},\ Res > s(A)$, where $\omega_0(\cdot)$ is *growth bound* of a C_0-semigroup with the generator $(\cdot)$. On the other hand, since A_0 is a generator of a compact C_0-semigroup, so do $(A_0 + \int_{-h}^{0} e^{s\theta}d[\eta(\theta)]), (A_0 + \int_{-h}^{0} e^{s(A)\theta}d[\eta(\theta)])$, see, e.g. [22], page 79. It follows from the *spectral mapping theorem* that

$$\omega_0(A_0 + \int_{-h}^{0} e^{s\theta}d[\eta(\theta)]) = s(A_0 + \int_{-h}^{0} e^{s\theta}d[\eta(\theta)]), \tag{41}$$

$$\omega_0(A_0 + \int_{-h}^{0} e^{s(A)\theta}d[\eta(\theta)]) = s(A_0 + \int_{-h}^{0} e^{s(A)\theta}d[\eta(\theta)]). \tag{42}$$

From (21), $s = (\alpha + \imath\beta) \in \rho(A_0 + \int_{-h}^{0} e^{s\theta}d[\eta(\theta)])$, for every $s \in \mathbb{C}$, Re $s > s(A)$. By Lemma 3.10 (ii) and (42), $s(A) = s(A_0 + \int_{-h}^{0} e^{s(A)\theta}d[\eta(\theta)]) = \omega_0(A_0 + \int_{-h}^{0} e^{s(A)\theta}d[\eta(\theta)])$. Then, we derive that if Re $s = \alpha > s(A)$ then Re $s > \omega_0(A_0 + \int_{-h}^{0} e^{s(A)\theta}d[\eta(\theta)]) \geq \omega_0(A_0 + \int_{-h}^{0} e^{s\theta}d[\eta(\theta)])$. So, we can represent the following

$$H(s)^{-1}x = (sI_X - A_0 - \int_{-h}^{0} e^{s\theta}d[\eta(\theta)])^{-1}x = \int_{0}^{\infty} e^{-st}V_s(t)xdt. \tag{43}$$

From the formulas (40), (43), it is easy to see that

$$|H(s)^{-1}x| \leq \int_{0}^{\infty} e^{-\alpha t}|V_s(t)||x|dt \leq \int_{0}^{\infty} e^{-\alpha t}V_\alpha(t)|x|dt = H(\alpha)^{-1}|x|.$$

□

Remark 3.14. *It follows from the above lemma that if the system (15)-(16) is Hurwitz stable and Assumption 3.11 is fulfilled then* $\sup_{s\in\mathbb{C},Res\geq 0}\|H(s)\| < +\infty$. *In fact, by (37) and (38), we have*

$$|H(s)^{-1}x| \leq H(Re\, s)^{-1}|x| \leq H(0)^{-1}|x|, \qquad x \in X,$$

for every $s \in \mathbb{C}$, $Re\ s \geq 0$. *From the lattice norm property (9) and the definition of the operator norm, we get*

$$\|H(s)^{-1}\| \leq \|H(0)^{-1}\|, \textit{ for every } s \in \mathbb{C}, Res \geq 0.$$

We are now in the position to prove the main result of this section.

Theorem 3.15. *Assume that the linear retarded system (15)-(16) is Hurwitz stable and* A_0, η *are subjected to multi-perturbations of the form (22) where* $D_i^0 \in \mathcal{L}_+(U_i^0, X)$, $D_i^1 \in \mathcal{L}_+(U_i^1, X)$ $(i \in \mathcal{N})$ *and* $E \in \mathcal{L}_+(X,Y)$. *Then, we have*

$$r_{\mathbb{C}} = r_{\mathbb{R}} = r_+ = \frac{1}{\max\{\max_{i\in\mathcal{N}}\|G_i^0(0)\|,\ \max_{i\in\mathcal{N}}\|G_i^1(0)\|\}}. \tag{44}$$

Proof. Suppose that $r_{\mathbb{C}} < +\infty$, as otherwise there is nothing to show. We prove that

$$r_{\mathbb{C}} = \frac{1}{\max\{\max_{i\in\mathcal{N}}\|G_i^0(0)\|,\ \max_{i\in\mathcal{N}}\|G_i^1(0)\|\}}. \tag{45}$$

It follows from (37), (38) and $D_i^0 \in \mathcal{L}_+(U_i^0, X), D_i^1 \in \mathcal{L}_+(U_i^1, X)$ $(i \in \mathcal{N})$ that

$$|G_i^0(\lambda)x| = |(EH(\lambda)^{-1}D_i^0)x| \leq (EH(Re\lambda)^{-1}D_i^0)|x| \leq (EH(0)^{-1}D_i^0)|x|, \quad x \in U_i^0, i \in \mathcal{N},$$

$$|G_i^1(\lambda)y| = |(EH(\lambda)^{-1}D_i^1)y| \leq (EH(Re\lambda)^{-1}D_i^0)|y| \leq (EH(0)^{-1}D_i^1)|y|, \quad y \in U_i^1, i \in \mathcal{N},$$

for every $\lambda \in \mathbb{C}$, $Re\ \lambda \geq 0$. From the lattice norm property (9) and the definition of the operator norm, we have

$$\|G_i^1(\lambda)\| \leq \|G_i^1(0)\| \quad \text{and} \quad \|G_i^1(\lambda)\| \leq \|G_i^1(0)\|,\ i \in \mathcal{N}, \tag{46}$$

for every $\lambda \in \mathbb{C}, Re\ \lambda \geq 0$. Then (45) now follows from (35) and (46).

On the other hand, by $D_i^0 \in \mathcal{L}_+(U_i^0, X), D_i^1 \in \mathcal{L}_+(U_i^1, X)$ $(i \in \mathcal{N})$ and (37), we have

$$G_i^0(0) \in \mathcal{L}_+(U_i^0, Y),\ G_i^1(0) \in \mathcal{L}_+(U_i^1, Y), i \in \mathcal{N}.$$

It follows from Lemma 3.6 and the definition of r_+ that

$$r_+ \leq \frac{1}{\max\{\max_{i\in\mathcal{N}}\|G_i^0(0)\|,\ \max_{i\in\mathcal{N}}\|G_i^1(0)\|\}}. \tag{47}$$

Finally, (44) follows from (45), (47) and the inequalities $r_{\mathbb{C}} \leq r_{\mathbb{R}} \leq r_+$. This completes our proof. □

We illustrate the above result by an example.

Example 3.16. Consider the equation

$$\frac{\partial u(t,x)}{\partial t} = \frac{\partial^2 u(t,x)}{\partial^2 x} - d(x)u(t,x) + b(x)u(t-1,x) \qquad (t \geq 0, x \in [0,1]) \tag{48}$$

with boundary condition

$$\frac{\partial u(t,x)}{\partial x}|_{x=0} = 0 = \frac{\partial u(t,x)}{\partial x}|_{x=1} \qquad (t \geq 0) \tag{49}$$

and initial condition

$$u(s,x) = \psi(s,x), \qquad s \in [-1,0],\ x \in [0,1].$$

Let X be the function space $C[0,1]$ equipped with sup-norm, and let B be defined by

$$Bh = h'',$$

with domain $D(B) := \{h \in C^2[0,1] \ : \ h'(0) = h'(1) = 0\}$. Denote by M_b, M_d the respective multiplication operators for $b,d \in X,\ b,d \geq 0$. Then (48)-(49) takes the abstract time-delay system of the form

$$\begin{aligned} \dot{u}(t) &= Bu(t) - M_d u(t) + M_b u(t-1), \qquad t \geq 0, \\ u_0 &= \psi \in C([-1,0] \times [0,1]). \end{aligned} \tag{50}$$

It is well-known that B generates a positive compact semigroup, see [16], page 230. The same is true for the operator $B - M_d$, see [22], page 79. Moreover, if $\max_{x\in[0,1]}(b(x) - d(x)) < 0$ then the delay system (50) is uniformly exponentially stable, see [16], page 230. This is equivalent to the uniform exponential stability of the solutions of (48)-(49).

We now assume that the functions b,d are subjected to perturbations of the form

$$b(x) \longrightarrow b(x) + \delta_0(x), \qquad d(x) \longrightarrow d(x) + \delta_1(x), \qquad \delta_0,\ \delta_1 \in X.$$

From Theorem 3.15, the solutions of the following perturbed equation

$$\frac{\partial u(t,x)}{\partial t} = \frac{\partial^2 u(t,x)}{\partial^2 x} - (d(x) + \delta_1(x))u(t,x) + (b(x) + \delta_0(x))u(t-1,x) \ (t \geq 0, x \in [0,1]) \tag{51}$$

with boundary condition

$$\frac{\partial u(t,x)}{\partial x}|_{x=0} = 0 = \frac{\partial u(t,x)}{\partial x}|_{x=1} \qquad (t \geq 0) \tag{52}$$

and initial condition

$$u(s,x) = \psi(s,x), \qquad s \in [-1,0],\ x \in [0,1],$$

are uniformly exponentially stable if

$$\max_{x\in[0,1]} |\delta_0(x)| + \max_{x\in[0,1]} |\delta_1(x)| < \frac{1}{\|(B + M_b - M_d)^{-1}\|}.$$

4 Stability Radii of Positive Linear Functional Differential Equations under Multi-affine Perturbations

In this section, we study the robust stability of the positive system (15)-(16) where A_0, η are subjected to multi-affine perturbations of the following form

$$\begin{aligned} A_0 &\rightarrow A_0 + \textstyle\sum_{i=1}^{N_0} \alpha_i A_i \\ \eta &\rightarrow \eta + \textstyle\sum_{j=1}^{N_1} \beta_j \eta_j. \end{aligned} \tag{53}$$

Here $A_i \in \mathcal{L}(X)$, $i \in \underline{N}_0 := \{1,2,...,N_0\}$; $\eta_j \in NBV([-h,0], \mathcal{L}_{\mathbb{R}}(X))$, $j \in \underline{N}_1 := \{1,2,...,N_1\}$ are given and α_i, β_j, $i \in \underline{N}_0, j \in \underline{N}_1$ are unknown scalar parameters. We note that the class of multi-affine perturbations of this type has been considered first in [11], [12] for positive linear systems with no time delays in $\mathbb{R}^n$ and then in [13], [29] for positive time-delay systems in $\mathbb{R}^n$. The problem of computing the stability radius of positive linear functional differential equations in $\mathbb{R}^n$ under multi-affine perturbations has just been done recently in [21]. In this section, we extend these results to general functional differential equations in Banach lattices.

We assume that Assumption 3.11 holds. Denote by $A_{((\alpha_i)_{i\in\underline{N}_0},(\beta_j)_{j\in\underline{N}_1})}$ solution operator of perturbed systems with respect to multi-affine perturbation (53). In view of positivity of the system under consideration, we assume that $A_i \in \mathcal{L}_+(X), i \in \underline{N}_0$ and $\eta_j \in NBV([-h,0], \mathcal{L}_{\mathbb{R}}(X))$ are increasing, for all $j \in \underline{N}_1$. We define the complex and the real stability radii of the linear retarded system (15)-(16) under multi-affine parameter perturbations (53) by setting, for $\mathbb{K} = \mathbb{C}$ and, respectively, $\mathbb{K} = \mathbb{R}$,

$$r^a_{\mathbb{K}} = \inf\{\max(\max_{i\in\underline{N}_0} |\alpha_i|; \max_{j\in\underline{N}_1} |\beta_j|) : \alpha_i, \beta_j \in \mathbb{K}, i \in \underline{N}_0, j \in \underline{N}_1; \sigma\big(A_{((\alpha_i)_{i\in\underline{N}_0},(\beta_j)_{j\in\underline{N}_1})}\big) \not\subset \mathbb{C}^-\}. \tag{54}$$

Similarly, the positive stability radius r^a_+ is obtained by restricting, in the above definition, the disturbances $(\alpha, \beta) := ((\alpha_i)_{i\in\underline{N}_0}, (\beta_j)_{j\in\underline{N}_1})$ to be nonnegative.
We emphasize that by (21), $\sigma\big(A_{((\alpha_i)_{i\in\underline{N}_0},(\beta_j)_{j\in\underline{N}_1})}\big)$ can be replaced by $\sigma(A_0 + \sum_{i=1}^{N_0} \alpha_i A_i, \eta + \sum_{j=1}^{N_1} \beta_j \eta_j)$ in the formula (54). It is clear that

$$r^a_{\mathbb{C}} \leq r^a_{\mathbb{R}} \leq r^a_+. \tag{55}$$

The following theorem shows that the equalities in (55) hold and gives us a computable formula of stability radii $r^a_{\mathbb{K}}, r^a_+, (\mathbb{K} = \mathbb{C}, \mathbb{R})$.

Theorem 4.1. *Suppose the positive linear retarded system (15)-(16) is Hurwitz stable and A_0, η are subjected to multi-affine perturbations of the form (53). If the stability radii of the system are given by (54) then*

$$r^a_{\mathbb{C}} = r^a_{\mathbb{R}} = r^a_+ = \frac{1}{s\Big((-A_0 - \int_{-h}^0 d[\eta(\theta)])^{-1}(\sum_{i=1}^{N_0} A_i + \sum_{j=1}^{N_1} \int_{-h}^0 d[\eta_j(\theta)])\Big)}. \tag{56}$$

Proof. We first show that

$$r^a_+ = \frac{1}{s\Big((-A_0 - \int_{-h}^0 d[\eta(\theta)])^{-1}(\sum_{i=1}^{N_0} A_i + \sum_{j=1}^{N_1} \int_{-h}^0 d[\eta_j(\theta)])\Big)}. \tag{57}$$

Assume that $r_+^a < +\infty$. Let $(\alpha,\beta) = ((\alpha_i)_{i\in\underline{N}_0},(\beta_j)_{j\in\underline{N}_1})$ be an arbitrary nonnegative destabilizing perturbation. That is

$$\sigma(A_0+\sum_{i=1}^{N_0}\alpha_i A_i,\eta+\sum_{j=1}^{N_1}\beta_j\eta_j)\not\subset\mathbb{C}^-.$$

Therefore, there exists $s_0\in\mathbb{C}, Re\ s_0\geq 0$ such that $s_0\in\sigma(A_0+\sum_{i=1}^{N_0}\alpha_i A_i+\int_{-h}^{0}e^{s_0\theta}d[(\eta+\sum_{j=1}^{N_1}\beta_j\eta_j)(\theta)])$. This is equivalent to $1\in\sigma\Big(H(s_0)^{-1}\big(\sum_{i=1}^{N_0}\alpha_i A_i+\sum_{j=1}^{N_1}\beta_j\int_{-h}^{0}e^{s_0\theta}d[\eta_j(\theta)]\big)\Big)$. By Lemma 3.13 (ii), we have

$$|H(s_0)^{-1}(\sum_{i=1}^{N_0}\alpha_i A_i+\sum_{j=1}^{N_1}\beta_j\int_{-h}^{0}e^{s_0\theta}d[\eta_j(\theta)])x|\leq$$

$$H(Re\ s_0)^{-1}|(\sum_{i=1}^{N_0}\alpha_i A_i+\sum_{j=1}^{N_1}\beta_j\int_{-h}^{0}e^{s_0\theta}d[\eta_j(\theta)])x|,\quad \text{for every } x\in X.$$

Since $A_i\in\mathcal{L}_+(X),\ i\in\mathcal{N}_0;\ \eta_j\in NBV([-h,0],\mathcal{L}_{\mathbb{R}}(X)),\ j\in\mathcal{N}_1$ being increasing, using Lemma 3.13 (i), we derive

$$\max(\max_{i\in\underline{N}_0}|\alpha_i|;\max_{j\in\underline{N}_1}|\beta_j|)\Big(H(0)^{-1}(\sum_{i=1}^{N_0}A_i+\sum_{j=1}^{N_1}\int_{-h}^{0}d[\eta_j(\theta)])\Big)|x|\geq$$

$$H(Re\ s_0)^{-1}|(\sum_{i=1}^{N_0}\alpha_i A_i+\sum_{j=1}^{N_1}\beta_j\int_{-h}^{0}e^{s_0\theta}d[\eta_j(\theta)])x|\geq$$

$$|H(s_0)^{-1}(\sum_{i=1}^{N_0}\alpha_i A_i+|\sum_{j=1}^{N_1}\beta_j\int_{-h}^{0}e^{s_0\theta}d[\eta_j(\theta)])x|,$$

for every $x\in X$. Since $\Big(H(0)^{-1}\big(\sum_{i=1}^{N_0}A_i+\sum_{j=1}^{N_1}\int_{-h}^{0}d[\eta_j(\theta)]\big)\Big)$ is a positive bounded linear operator, using the Gelfand's formula for spectral radius $r(\cdot)$ of a bounded linear operator $(\cdot)$ and the lattice norm property (9), we get

$$r\Big(H(0)^{-1}(\sum_{i=1}^{N_0}A_i+\sum_{j=1}^{N_1}\int_{-h}^{0}d[\eta_j(\theta)])\Big)\geq\frac{r\Big(H(s_0)^{-1}\big(\sum_{i=1}^{N_0}\alpha_i A_i+\sum_{j=1}^{N_1}\beta_j\int_{-h}^{0}e^{s_0\theta}d[\eta_j(\theta)]\big)\Big)}{\max(\max_{i\in\underline{N}_0}|\alpha_i|;\max_{j\in\underline{N}_1}|\beta_j|)}.$$

It follows from $1\in\sigma\Big(H(s_0)^{-1}\big(\sum_{i=1}^{N_0}\alpha_i A_i+\sum_{j=1}^{N_1}\beta_j\int_{-h}^{0}e^{s_0\theta}d[\eta_j(\theta)]\big)\Big)$, that

$$r\Big(H(0)^{-1}(\sum_{i=1}^{N_0}A_i+\sum_{j=1}^{N_1}\int_{-h}^{0}d[\eta_j(\theta)])\Big)\geq\frac{1}{\max(\max_{i\in\underline{N}_0}|\alpha_i|;\max_{j\in\underline{N}_1}|\beta_j|)}.\tag{58}$$

On the other hand, by Perron-Frobenius Theorem for positive bounded linear operator (see. e.g [15]),

$$s\Big(H(0)^{-1}(\sum_{i=1}^{N_0}A_i+\sum_{j=1}^{N_1}\int_{-h}^{0}d[\eta_j(\theta)])\Big)=$$

$$r\Big(H(0)^{-1}(\sum_{i=1}^{N_0}A_i+\sum_{j=1}^{N_1}\int_{-h}^{0}d[\eta_j(\theta)])\Big)\in\sigma\Big(H(0)^{-1}(\sum_{i=1}^{N_0}A_i+\sum_{j=1}^{N_1}\int_{-h}^{0}d[\eta_j(\theta)])\Big).$$

This yields, by virtue of (58),

$$\max(\max_{i\in\underline{N}_0}\alpha_i;\ \max_{j\in\underline{N}_1}\beta_j)\geq\frac{1}{s\Big(H(0)^{-1}(\sum_{i=1}^{N_0}A_i+\sum_{j=1}^{N_1}\int_{-h}^{0}d[\eta_j(\theta)])\Big)}.$$

By the definition of the positive stability radius r_+^a,

$$r_+^a\geq\frac{1}{s\Big(H(0)^{-1}(\sum_{i=1}^{N_0}A_i+\sum_{j=1}^{N_1}\int_{-h}^{0}d[\eta_j(\theta)])\Big)}.$$

Setting $\gamma:=s\Big(H(0)^{-1}(\sum_{i=1}^{N_0}A_i+\sum_{j=1}^{N_1}\int_{-h}^{0}d[\eta_j(\theta)])\Big)$. To prove the inverse inequality, using again Perron-Frobenius Theorem for positive bounded linear operator, we have

$$\gamma\in\sigma\Big(H(0)^{-1}(\sum_{i=1}^{N_0}A_i+\sum_{j=1}^{N_1}\int_{-h}^{0}d[\eta_j(\theta)])\Big).$$

This implies

$$0\in\sigma\Big((A_0+\sum_{i=1}^{N_0}\frac{1}{\gamma}A_i),\int_{-h}^{0}d[\eta+\sum_{j=1}^{N_1}\frac{1}{\gamma}\eta_j(\theta)]\Big).$$

It means that nonnegative perturbation (α^*,β^*) defined by $\alpha_i^*=1/\gamma, i\in\underline{N}_0, \beta_j^*=1/\gamma, j\in\underline{N}_1$, is destabilizing. By the definition of r_+^a, we have

$$r_+^a\leq\frac{1}{s\Big(H(0)^{-1}(\sum_{i=1}^{N_0}A_i+\sum_{j=1}^{N_1}\int_{-h}^{0}d[\eta_j(\theta)])\Big)}.$$

Thus, we obtain

$$r_+^a=\frac{1}{s\Big(H(0)^{-1}(\sum_{i=1}^{N_0}A_i+\sum_{j=1}^{N_1}\int_{-h}^{0}d[\eta_j(\theta)])\Big)}.$$

It is important to note that by the above argument, if $r_+^a=+\infty$, then

$$s\Big(H(0)^{-1}(\sum_{i=1}^{N_0}A_i+\sum_{j=1}^{N_1}\int_{-h}^{0}d[\eta_j(\theta)])\Big)=0.$$

So the equality (57) is obvious.

We are now ready to show that $r^a_{\mathbb{C}} = r^a_{\mathbb{R}} = r^a_{+}$. Suppose $r^a_{\mathbb{C}} < +\infty$ and $(\alpha,\beta) = ((\alpha_i)_{i\in\underline{N}_0},(\beta_j)_{j\in\underline{N}_1})$ is an arbitrary complex destabilizing perturbation. By a similar argument as in the beginning, we have

$$\gamma_1 := s\Big(H(0)^{-1}\big(\sum_{i=1}^{N_0}|\alpha_i|A_i + \sum_{j=1}^{N_1}|\beta_j|\int_{-h}^{0} d[\eta_j(\theta)]\big)\Big) \geq 1.$$

Since $\Big(H(0)^{-1}\big(\sum_{i=1}^{N_0}|\alpha_i|A_i + \sum_{j=1}^{N_1}|\beta_j|\int_{-h}^{0} d[\eta_j(\theta)]\big)\Big)$ is a positive bounded operator, it follows from Perron-Frobenius Theorem that

$$\gamma_1 \in \sigma\Big(H(0)^{-1}\big(\sum_{i=1}^{N_0}|\alpha_i|A_i + \sum_{j=1}^{N_1}|\beta_j|\int_{-h}^{0} d[\eta_j(\theta)]\big)\Big).$$

It follows that

$$0 \in \sigma\Big((A_0 + \sum_{i=1}^{N_0}\frac{|\alpha_i|}{\gamma_1}A_i), \int_{-h}^{0} d[\eta + \sum_{j=1}^{N_1}\frac{|\beta_j|}{\gamma_1}\eta_j(\theta)]\Big).$$

It means that nonnegative perturbation $(\alpha^{'},\beta^{'})$ defined by $\alpha_i^{'} = |\alpha_i|/\gamma_1, i\in\underline{N}_0, \beta_j^{'} = |\beta_i|/\gamma_1, j\in\underline{N}_1$, is destabilizing. By the definition of $r^a_{\mathbb{C}}, r^a_{+}$, we conclude

$$r^a_{+} \leq r^a_{\mathbb{C}}.$$

In combining with the inequalities $r^a_{\mathbb{C}} \leq r^a_{\mathbb{R}} \leq r^a_{+}$, this implies $r^a_{\mathbb{C}} = r^a_{\mathbb{R}} = r^a_{+}$. This completes our proof. □

References

[1] A. Fischer, Stability radii of infinite-dimensional positive systems, *Math. Control Signals Systems* **10** (1997) 223-236.

[2] A. Fischer, D. Hinrichsen and N.K. Son, Stability radii of Metzler operators, *Vietnam Journal of Math.* **26** (1998) 147-163.

[3] A. Fischer and J.M.A.M. van Neerven, Robust stability of semi-groups and apllication to stability of delay systems, *Journal of Mathematical Analysis and Applications* **226** (1998) 82-100.

[4] J. Hale, *Theory of Functional Differential Equations*, Acad. Press, New York, 1977.

[5] E. Hille and R. S. Phillips, *Functional Analysis and Semigroups*, Providence, RI: American Mathematical Society, 1957.

[6] Y. Hino, S. Murakami, T. Naito, Nguyen Van Minh, A variation of constants formula for abstract functional differential equations in the phase space, *J. Differential Equations* **179** (2002) 336-355.

[7] Y. Hino, S. Murakami, T. Naito, Functional Differential Equations with Infinite Delay, *Lecture Notes in Math.* **1473**, Springer-Verlag, Berlin-New York, 1991.

[8] D. Hinrichsen and A.J. Pritchard, Stability radius for structured perturbations and the algebraic Riccati equation, *Systems & Control Letters* **8** (1986) 105-113.

[9] D. Hinrichsen and A.J. Pritchard, Real and complex stability radii: a survey, in : D. Hinrichsen and B. Mårtensson, editors, Control of Uncertain Systems, volume 6 of *Progress in System and Control Theory*, Basel. Birkhäuser, 1990, 119-162.

[10] D. Hinrichsen and A.J. Pritchard, Robust stability of linear evolution operators on Banach spaces, SIAM *Journal on Control Optimization* **32** (1994) 1503-1541.

[11] D. Hinrichsen and N.K. Son, μ- analysis and robust stability of positive linear systems, *Appl. Math. and Comp. Sci.* **8** (1998) 253-268.

[12] D. Hinrichsen and N.K. Son, Stability radii of positive discrete-time systems under affine parameter perturbations, *Inter. J. Robust and Nonlinear Control* **8** (1998) 1169-1188.

[13] D. Hinrichsen, N.K. Son and P.H.A. Ngoc, Stability radii of positive higher order difference systems, *Systems & Control Letters* **49** (2003) 377-388.

[14] G. Hu and E. J. Davison, Real stability radii of linear time-invariant time-delay systems, *Systems & Control Letters* **50** (2003) 209-219.

[15] P. Meyer-Nieberg, *Banach Lattices*, Springer-Verlag, Berlin, 1991.

[16] R. Nagel (Ed.), *One-Parameter Semigroups of Positive Operators*, Springer-Verlag, Berlin, 1986.

[17] T. Naito, J.S. Shin, S. Murakami, On solution semigroups of general functional differential equations, *Nonlinear Anal.* **30** (1997) 4565–4576.

[18] T. Naito, J.S. Shin, S. Murakami, The generator of the solution semigroup for the general linear functional differential equation, *Bull. Univ. of Electro-Communications* **11** (1998) 29–38.

[19] P.H.A. Ngoc and N.K. Son, Stability radii of positive linear difference equations under affine parameter perturbations, *Applied Mathematics and Computation* **134** (2003) 577-594.

[20] P.H.A. Ngoc, N.K. Son, Stability radii of linear systems under multi-perturbations, *Numer. Funct. Anal. Optim.* **25** (2004) 221-238.

[21] P.H.A. Ngoc and N.K. Son, *Stability radii of positive linear functional differential equations under multi-perturbations*, To appear in SIAM Journal on Control and Optimization, (2004).

[22] A. Pazy, *Semigroups of Linear Operators and Applications to Partial Differential Equations*, Springer-Verlag, Berlin, 1983.

[23] A.J. Pritchard and S. Townley, Robustness of linear systems, *Journal Differential Equations* **77** (1989) 254-286.

[24] L. Qiu, B. Bernhardsson, A. Rantzer, E.J. Davison, P.M. Young and J.C. Doyle, A formula for computation of the real stability radius, *Automatica* **31** (1995) 879-890.

[25] J. S. Shin, T. Naito, Nguyen Van Minh, On stability of solutions in linear autonomous functional differential equations, *Funkcial. Ekvac.* **43** (2000) 323–337.

[26] N.K. Son and D. Hinrichsen, Robust stability of positive continuous-time systems, *Numer. Funct. Anal. Optim.* **17** (1996) 649-659.

[27] N.K. Son and P.H.A. Ngoc, Complex stability radius of linear time-delay systems, *Vietnam Journal of Mathematics* **26** (1998) 379-383.

[28] N.K. Son, P.H.A. Ngoc, Stability radius of linear delay systems, in *Proceedings of the American Control Conference*, San Diego, California (1999) 815-816.

[29] N.K. Son, P.H.A. Ngoc, Robust stability of positive linear time delay systems under affine parameter perturbations, *Acta Mathematica Vietnamica* **24** (1999) 353-372.

[30] N.K. Son, P.H.A. Ngoc, Robust stability of linear functional differential equations, *Advanced Studies in Contemporary Mathematics* **3**, No.2 (2001) 43-59.

[31] S. Clark, Y. Latushkin, S. Montgomery-Smith, T. Randolph, Stability radius and internal versus external stability in Banach spaces: an evolution semigroup approach, SIAM *J. Control Optim.* **38** (2000) 1757–1793.

[32] C.C. Travis, G.F. Webb, Existence and stability for partial functional differential equations, *Trans. Amer. Math. Soc.* **200** (1974) 394-418.

[33] J. Wu, Theory and Applications of Partial Functional Differential Equations, *Applied Math. Sci.* **119**, Springer, Berlin-New York, 1996.

[34] A.C. Zaanen, *Introduction to Operator Theory in Riesz Spaces*, Springer-Verlag, Berlin, 1997.

In: Research on Evolution Equation Compendium. Volume 2 ISBN: 978-1-60456-752-6
Editor: Gaston M. N'Guerekata

Effects of Nonlocality and Phase Shift Definitions in Generalizing Levinson's Theorem

S.B. Qadri ***and B. Mulligan***
Department of Physics, The Ohio State University Columbus,
Ohio 43210 USA
M.F. Mahmood† ***and J.Y. Al-Khal***
Department of Mathematics, Howard University Washington,
DC 20059, USA

Abstract

This paper considers Levinson's Theorem for nonlocal potentials. The effect of continuum bound states is clarified, and the presence of spurious states is introduced. Differences between definitions of the phase shift are discussed, with special attention given to discontinuities in the phase shift. Examples from nuclear physics emphasize the utility of this theorem and illustrate some of the more delicate aspects of the derivation. A summary is provided which analyzes certain crucial points in the discussion as they relate to previous attempts to obtain a generalized Levinson's theorem.
NUCLEAR REACTIONS Levinson's Theorem, nonlocal potentials continuum bound states, spurious states, Jost functions, Fredholm determinants, definition of the phase shift.

1 Introduction

In this paper, we consider scattering interactions which can be described by a spin-independent, central potential, and focus on the connection between the zero-energy phase shift and the phase shift as the energy of the projectile becomes arbitrarily large. For simplicity, we restrict discussion to the $\ell = 0$ partial wave.

When the scattering potential V is local, a well-known relation exists between the phase shift and the number of bound states of the scattering center. Assuming that V satisfies the restrictions

$$\int_0^\infty r\,|V(r)|\,dr < \infty \quad \text{and} \quad \int_0^\infty r^2\,|V(r)|\,dr < \infty, \tag{1}$$

Levinson [1] showed that phase shift $\delta(E)$ obeys the condition

$$\delta(0) - \delta(\infty) = n\pi, \tag{2}$$

*Naval Reasearch Laboratory, Washington, DC 20375 USA
†Corresponding Author

where n is the number of bound states of V.

This relation, which has come to be known as Levinson's theorem, was first generalized by Swan [2] to include a situation in which the interaction is described by a nonlocal potential. Such a circumstance occurs when the incident particle is identical to or contains particles identical to those of the target, as with collisions of nucleons with light nuclei or electrons with atoms. Under these conditions Swan observed that, in addition to bound states, states excluded due to the Pauli principle also contribute to the zero-energy phase shift. This results in values of n in Eq. (2) not equal to the number of bound states. Specifically, Swan showed that the modified version of Levinson's theorem for the case of potentials with nonlocality which arises from antisymmetrization is

$$\delta(0) - \delta(\infty) = (n+m)\pi \tag{3}$$

where n is (as before) the number of bound states of the target and incident particle and m is the number of states from which the incident particle is excluded by the Pauli principle. Temkin [3] and Rosenberg and Spruch [4] have also investigated Levinson's theorem in the presence of antisymmetrization, and agree with Swan's conclusions.

Subsequently, Jauch [5] extended Levinson's theorem to a large class of potentials for general scattering systems, and Gourdin and Martin [6] discussed the theorem in the context of nonlocal potentials as a class. Jauch's generalization was made under the assumption that a nonlocal potential has no bound states in the continuum. Although this assumption is strictly true for a short range local potential [7], it cannot be made for nonlocal interactions. For example, Martin [8] examined the phase shift due to a nonlocal, separable, central potential and concluded that bound states in the continuum are indeed possible and that one must take such states into account in Levtnson's theorem. In particular, he showed that $\delta(0) - \delta(\infty)$is equal to π times the number of bound states plus the number of continuum bound states.

In this respect, Martin's results are consistent with the earlier conclusions of Swan. In discussing states excluded by the Pauli principle, Swan noted that these states appear as solutions of the scattering integral equation at all scattering energies. Such redundant states are simply continuum bound states which appear in the spectrum of the scattering solution at every positive energy [9], and can be expected in any scattering calculations which take into account antisymmetrization between the incident particle and a target described by a antisyrnmetrized single-particle state. They are a consequence of the fact that any row of a Slater determinant can be added to any other row of the determinant without changing the value of the determinant. Thus, the scattering solution is nonunique because any arbitrary amount of filled states can be added to the scattering state. Additional constraints must be placed on the scattering solution to obtain wavefunction orthogonal to the filled states, as dictated by physical considerations.

Many of the problems associated with the presence of redundant states have been extensively studied and discussed [9-15]. In particular, it has been pointed out that the presence of redundant states is possible only in the case of nonlocal potentials. Also, since the early work, Levinson's theorem has been subjected to repeated examination and extension. For example, Ida [16] has discussed its importance with respect to resolution of nonuniqueness of solutions of the Low equation. Levinson's theorem in the multichannel case has been investigated by a number of authors [17-28].

Martin's result for Levinson's theorem has been replicated by Bertero, et al. [29]. Their derivation was obtained for a large class of nonlocal interactions. On the other hand, Buslaev [30] proved that in spite of the presence of continuum bound states Levinson's theorem holds in its original form for a large class of potentials which is similar to that considered by Bertero, et al. That this apparent discrepancy is due to the definition of the phase shift has been shown by Dreyfus [31]. His discussion follows similar observations by Bolsterli [32] and by Beregi [33], who pointed out that there is an ambiguity in the definition of the phase shift when a continuum bound state is present. Additional work on the generalization of Levinson's theorem to nonlocal potentials also has been carried out by Mills and Reading [34], Horwitz and Marchand [35], Glökle and LeTourneux [36], Wöllenberg [37], and Newton [38]. A review paper on Levinson's theorem was published by Beregi, et al. [39]in 1973.

If for a non-singular local potential the initial conditions imposed on the radial part of the wavefunction $u(r)$ were $u(0) = u'(0) = 0$, the solution would be zero everywhere. In the case of a nonlocal potential, however, it has been shown [40,41] that a nontrivial solution with $u(0) = u'(0) = 0$ can exist. This anomalous type of solution [2,42,43] found with nonlocal potentials is not due to a continuum bound state. A (non-normalizable) state whose wavefunction satisfies the condition $u(0) = u'(0) = 0$ has been named a "spurious state". As is the case with a continuum bound state, a spurious state also results in an ambiguity in the definition of the phase shift at the energy of the spurious state.

In each of the generalizations of Levinson's theorem for nonlocal potentials discussed above, there are important aspects which the authors have not considered. The purpose of the present work is to examine Levinson's theorem in detail and to look into these aspects with the purpose of clarification. All our discussions will be limited to nonlocal potentials which are symmetric, and for which a thorough analysis of solutions of the radial equation has been presented [40]. The extension of the results of Ref. [40] to nonsymmetric nonlocal potentials is rather complicated, and has just been begun [44]. The focus here has been to avoid these complications, and rather to concentrate on fundamental questions of Levinson's theorem as it relates to such basic features of non-locality as the presence of continuum bound states and spurious states.

2 Method of Approach

The procedure which we use here for discussing Levinson's theorem follows the approach of Swan [2]. As pointed out in Eq. (2), for a short range local potential $\delta(0) - \delta(\infty)$ is equal to π times the number of bound states of the potential. Swan noted that in the complex k-plane the Jost function $\mathcal{L}^{+}(k)$for a local potential has a zero on the positive imaginary axis associated with each bound state. Furthermore, for a local potential $\mathcal{L}^{+}(k)$can be shown to be analytic in the upper half plane, and to have no zeros in addition to those associated with the bound states. On this basis, it is possible to use a well-known theorem from the theory of complex variables to form a contour integral which will have as its value π times the number of bounds states.

This theorem concerns a function $f(z)$ which is analytic inside and on a simple closed contour C, except for a finite number of poles inside C, and for which $f(z) \neq 0$ on C.If N

and P are, respectively, the number of zeros and poles inside C , then

$$\frac{1}{2\pi i}\int_C \frac{f(z)'}{f(z)}dz = \ln f|_C = N - P. \tag{4}$$

applying this theorem to the function $\mathcal{L}^+(k)$ using the contour C indicated in Fig. 1 yields

$$\int_C \frac{\mathcal{L}^+(k)'}{\mathcal{L}^+(k)}dz = 2\pi i n. \tag{5}$$

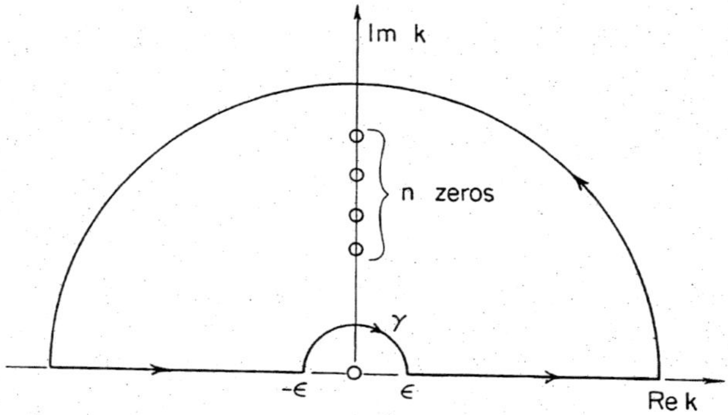

Figure 1. Contour C for integration of in the case of a local potential. Zeros of $\mathcal{L}^+(k)$are indicated by the symbol o.

For a local potential it can also be demonstrated that

$$\mathcal{L}^+(k) = \mid \mathcal{L}^+(k) \mid e^{-i\delta(k)}. \tag{6}$$

Carrying out explicitly the line integral in Eq. (5) gives

$$\int_C \frac{\mathcal{L}^+(k)'}{\mathcal{L}^+(k)}dz = 2i[\delta(0) - \delta(\infty)], \tag{7}$$

where we have used the fact that $\delta(-k) = -\delta(k)$. Combining Eqs. (5) and (7) immediately yields Eq. (2).

In obtaining the result in Eq. (5), the integral around the semicircular contour γ in Fig. 1 has been assumed to go to zero. However, if $\mathcal{L}^+(0) = 0$ the explicit integral in Eq. (7) around the portion γ of the path C will yield $-i\pi$. In that case, the right-hand side of Eq. (7) is replaced by $2i[\delta(0) - \delta(\infty)] - ip$. Then, instead of Eq. (2) we obtain the well-known result

$$\delta(0) - \delta(\infty) = (n + \frac{1}{2})\pi. \tag{8}$$

3 Extension to Nonlocal Potentials

In order to adapt this approach to a discussion of Levinson's theorem for nonlocal potentials it is necessary to investigate in detail the Jost function for a nonlocal potential. Jost and

Pais [45] have shown that, for a local potential, instead of arriving at the properties of the potential by starting with the Jost function $\mathcal{L}^+(k)$, one could equally well have started with the integral equation for the physical solution and obtained the properties of the potential from the Fredholm determinant associated with the kernel of that equation.

As one introduces the concept of nonlocality, this shift of the focus of discussion toward integral equations and the Fredholm determinants related to the kernels of these integral equations becomes very useful. When the potential is nonlocal, the $\ell = 0$ radial equation is of the form

$$\left[\frac{d^2}{dr^2} + k^2\right] u(r) = \int_0^\infty V(r,r')\, u(r')\, dr'. \tag{9}$$

This integro-differential equation can be converted into integral equations by the use of Green's functions which satisfy appropriate boundary conditions. Several integral equations, their solutions, and related quantities are defined in the following paragraph, in analogy with the local potential case.

The regular solution $\varphi(k,r)$ is defined by the boundary conditions

$$\varphi(k,0) = 0; \qquad \phi'(k,0) = 1. \qquad (10).$$

The integral equation for the regular solution $\varphi(k,r)$ is

$$\varphi(k,r) = k^{-1} \sin kr + \int_0^r \int_0^\infty G(k,r,r')\, V(r',s)\, \varphi(k,s)\, ds\, dr', \tag{11}$$

where $G(k,r,r')$ is the Green's function and is equal to $k^{-1} \sin k(r - r')$. The Fredholm determinant associated with the kernel of this integral equation is denoted by $D(k)$. The physical solution $\Psi^+(k,r)$ is defined by the mixed boundary conditions that $\Psi^+(k,r)$ have the asymptotic form

$$\Psi^+ \rightarrow \frac{i}{2}[e^{-ikr} - S^+(k)e^{ikr}] \tag{12}$$

as $r \rightarrow \infty$, and that $\Psi^+(k,r)$ be regular at $r = 0$. $S^+(k)$ is the s-wave scattering matrix element. The physical solution $\Psi^+(k,r)$ and its conjugate $\Psi^-(k,r)$ satisfy the integral equations

$$\Psi^\pm(k,r) = \sin kr + \int_0^\infty \int_0^\infty G^\pm(k,r,r')\, V(r',s)\, \Psi^\pm(k,s)\, ds\, dr', \tag{13}$$

where $G^\pm(k,r) = -k^{-1} e^{\pm ikr_>} \sin kr_<$ with $r_> = \max(r,r')$ and $r_< = \min(r,r')$. The asymptotic form of $\Psi^-(k,r)$ for large r is given by Eq. (12) with S^+ replaced by $S^-(k) = [S^+(k)]^*$ where * represents complex conjugation. The Fredholm determinants associated with the kernels of Eq. (13) are denoted by $D^\pm(k)$. The integral equations for the Jost solutions are

$$f^\pm(k,r) = e^{\pm ikr} - \int_r^\infty \int_0^\infty G(k,r,r')\, V(r',s)\, f^\pm(k,s)\, ds\, dr', \tag{14}$$

The Fredholm determinant associated with the kernel of Eq. (14) is denoted by $\Delta(k)$.

The Fredholm determinants have the following properties: $\Re D^\pm(k), D(k)$, and $\Delta(k)$ are even functions of k, while $\Im D^\pm(k)$ is an odd function of k. Both $D(k)$ and $\Delta(k)$ are real for real k, and $D(k) = \Delta(k)$ for symmetric nonlocal potentials [46].

For a local potential, the reason that one can start either from a Jost function point of view or from a physical integral equation — Fredholm determinant point of view lies in the fact [45] that for a local potential the Jost functions $\mathcal{L}^{\pm}(k)$ and the Fredholm determinants $D^{\pm}(k)$ are identically equal,

$$\mathcal{L}^{\pm}(k) = D^{\pm}(k) \qquad [\text{local potential}]. \tag{15}$$

Also, with the exception of Eq. (13), for a local potential the integral equations become Volterra equations. Thus, the Fredholm determinants $D(k)$ and $\Delta(k)$ become unity in the case of a local potential and can play no explicit role in the description of the scattering process. The Fredholm determinants $D^{\pm}(k)$ are not unity in the case of a local potential, consistent with the fact that they are the only Fredholm determinants needed to establish all of the scattering information.

For a nonlocal potential, the integral equations with which $D(k)$ and $\Delta(k)$ are associated are Fredholm integral equations rather than Volterra integral equations. Thus, the Fredholm determinants $D(k)$ and $\Delta(k)$ are not unity, and it might be anticipated that a complete description of the scattering process would require these determinants in addition to the determinants $D^{\pm}(k)$. Nevertheless, zeros of the Fredholm determinants $D^{\pm}(k)$ and $D(k)$ are not independent. Whenever there is a zero of $D^{\pm}(k)$ for real k there is also associated a zero of $D(k)$ at that same value of k [40].

A zero of $D^{\pm}(k)$ for real $k \neq 0$ is called a continuum bound state [6,8], often referred to by the abbreviation CBS. The existence of a continuum bound state results from the fact that when $D^{\pm}(k) = 0$ for real $k \neq 0$ solutions of the homogeneous integral equations associated with Eq.(13), namely

$$\Psi_h^{\pm}(k,r) = \int_0^\infty\int_0^\infty G^{\pm}(k,r,r')\,V(r',s)\,\Psi_h^{\pm}(k,s)\,ds\,dr', \tag{13}$$

are normalizable. Trivial solutions $\Psi_h^{\pm}(k,r) = 0$ are the only solutions allowed when $D^{\pm}(k) \neq 0$.

A spurious state [40,43] occurs when $D(k)$ is zero and $D^{\pm}(k)$ is nonzero for a real value of k. It is demonstrated in Ref. 40 that at a spurious state the regular solution $\varphi(k,r)$ does not exist, although it is possible to obtain a real solution of Eq. (9) which is regular at the origin.

In the case of a local potential, it is customary [47] to define the Jost functions $\mathcal{L}^{\pm}(k)$ in terms of the Jost solutions at $r = 0$: that is

$$\mathcal{L}^{\pm}(k) = f^{\pm}(k,r)\big|_{r=0}. \tag{17}$$

No ambiguity exists in showing that this definition leads to Eq. (15). Thus Eq. (15) could equally well be used as a definition of $\mathcal{L}^{\pm}(k)$ for a local potential.

For a nonlocal potential, difficulties can arise in using Eq. (17) as the definition of $\mathcal{L}^{\pm}(k)$. The Jost solutions $f^{\pm}(k,r)$ do not exist at a spurious state. At a continuum bound state, $f^{\pm}(k,r)$ may or may not exist. Under circumstances in which there are no difficulties with the existence of $f^{\pm}(k,r)$, that is, at values of k for which $D(k) \neq 0$, for a symmetric nonlocal potential it has been shown [48-51] that $D^{\pm}(k)$ and $\mathcal{L}^{\pm}(k)$ are related by the expression

$$\mathcal{L}^{\pm}(k) = D^{\pm}(k)/D(k). \quad \text{[symmetric nonlocal potential]} \tag{18}$$

However, the derivation of Eq. (18) breaks down when $f^{\pm}(k,r)$ do not exist. Correspondingly, the definition of the Jost functions as given in Eq. (17) will no longer hold if $f^{\pm}(k,r)$ do not exist. This is especially a problem in the case of redundant states, since it would mean a failure of the existence of the Jost functions at any energy. However, it has been shown [52], using the Green's function cancellation method of Krause and Mulligan [53] to construct a potential which produces a redundant state, that under these circumstances Eq. (18) can nevertheless be used as a definition of $\mathcal{L}^{+}(k)$.

In addition, the results in Ref. [52] demonstrate that the Jost functions $\mathcal{L}^{\pm}(k)$ can be directly related to the kernels of integral equations by means of Fredholm determinants without referring to the solutions of the integral equations involved. Consequently, in defining the Jost functions, the question of the existence of Jost solutions at a CBS or in the presence of a redundant state can be avoided. As we shall see, by taking the definition of the Jost functions as the ratio of Fredholm determinants $D^{\pm}(k)$ to $D(k)$ and proceeding with the derivation of Levinson 's theorem, we run into no problems.

There are, nevertheless, other difficulties associated with the derivation of Levinson's theorem in the case of a nonlocal potential. Consider Eqs. (15) and (18). The expression for $\mathcal{L}^{\pm}(k)$ for a local potential is a special case of the expression for $\mathcal{L}^{\pm}(k)$ for a nonlocal potential. Because of the presence of D(k) in Eq.(18), when considering the definition of the phase shift for a nonlocal potential an ambiguity arises. This ambiguity has been discussed in detail in Ref. [41]. The presence of the Fredholm determinant $D(k)$ results in two possible definitions of the phase shift: $\delta_D =$ -phase of $D^{+}(k)$ and $\delta_{\mathcal{L}} =$-phase of $\mathcal{L}^{+}(k)$. The phase shift $\delta_{\mathcal{L}}$ gives a continuous phase at a CBS and discontinuity of $-\pi$ in the phase shift as the energy increases through each spurious state. The phase shift δ_D is continuous at a spurious state, but gives a discontinuity of π with increasing energy to the phase shift at a CBS. Since for a local potential the phase of $\mathcal{L}^{\pm}(k)$ is the same as the phase of $D^{\pm}(k)$, the phase shifts $\delta_{\mathcal{L}}$ and δ_D are identical in that case.

For a nonlocal potential, a derivation of Levinson's theorem can proceed in much the same way as that described in Sec. 2 in the case of a local potential. However, because of the two possible definitions of the phase shift, there will be two possible results for Levinson's theorem. In one case, the function $f(z)$ to be employed in Eq. (4) will be the Jost function $\mathcal{L}^{+}(k)$ defined in Eq. (18); in the other case one would use $D^{+}(k)$.

4 Properties of $D^{+}(k)$, $D(k)$, and $\mathcal{L}^{+}(k)$ for a Nonlocal Potential

For a local potential, it is known that $\mathcal{L}^{+}(k)$ and therefore $D^{+}(k)$ have the property $\mathcal{L}^{+}(k) \equiv D^{+}(k) \neq 0$ for real $k \neq 0$. As we have already noted, in the case of a local potential the Fredholm determinant $D^{+}(k)$ cannot have zeros on the real k-axis. Also, we will see that for a nonlocal potential zeros of $D(k)$ [and thus poles of the Jost function $\mathcal{L}^{+}(k)$] are found on and above the real axis in some cases. For these reasons, in the case of a nonlocal potential the techniques from complex analysis [54] used to obtain the character of the Fredholm determinant $D^{+}(k)$ and the Jost function $\mathcal{L}^{+}(k)$ for a local potential are not sufficient for a general discussion of $D^{+}(k)$, $D(k)$, and $\mathcal{L}^{+}(k)$. However, we have demonstrated [55] that techniques from functional analysis can be used for a wide class of nonlocal potentials to obtain the analytic properties of $D^{+}(k)$, $D(k)$, and $\mathcal{L}^{+}(k)$.

These techniques were developed by Iwasaki and Mulligan [56], who focused on the fact that the kernels of integral Eqs. (11), (13), and (14) are not square integrable. Noting that the usual derivations of the Fredholm alternative [57] do not apply in cases where kernels are not square integrable, they make use of Riesz-Schauder theory [58], which demonstrates that the Fredholm alternative easily can be extended to integral equations for which the operators are compact. The conditions on the nonlocal potential $V(r,r')$ necessary for compactness of the kernels of (11) and (13), are given in Ref. [55] or Ref. [56]. Complete continuity of the kernels of the integral Eqs. (11) and (13) is crucial to establishing the existence of the Fredholm determinants $D^{+}(k)$ and $D(k)$ for a nonlocal potential. It is shown in Ref. [55] that these kernels are compact for k real, from which it follows that the Fredholm determinants $D^{+}(k)$ and $D(k)$ exist for k real. Fredholm determinants can then be defined for complex k by extending these definitions, established on the real axis, to the complex plane. The properties of these determinants, established in Ref. [55], are summarized below.

The character of $D^{+}(k)$ and $D(k)$ in the upper half plane are such that $D^{+}(k)$ when so extended will be analytic in that portion of the k plane, while $D(k)$ may not. Thus with nonlocal potentials, as with local potentials, poles of $D^{+}(k)$ are confined to the bottom half of the complex plane. Zeros of $D^{+}(k)$ in the upper half of the complex k plane must lie either on the real axis or on the positive imaginary axis. In the case of a local potential, $D^{+}(k)$ cannot have a zero for k real [40]; since for a local potential $\mathcal{L}^{+}(k) \neq 0$ for real $k \neq 0$ [47], it follows that for a local potential $D^{+}(k)$ must not be zero for real $k \neq 0$. On the other hand, the modified definition of $\mathcal{L}^{+}(k)$ in the case of a nonlocal potential, given by Eq. (18), allows for a zero of $D^{+}(k)$ on the real k axis as long as such a zero is accompanied by a corresponding zero of $D(k)$, resulting in a bound state in the continuum. Such zeros of $D^{+}(k)$ and $D(k)$ on the real k axis will occur in pairs, symmetric about the imaginary axis.

That such a zero of $D(k)$ is possible on the real k axis has been established by many examples [40]. However, because the boundary conditions associated with the equation for the regular solution are not the same as those associated with the physical solution, in general it would not be expected that $D(k)$ would be zero for the same values of k as is $D^{+}(k)$. Indeed, for a local potential $D(k)$ is unity at all zeros of $D^{+}(k)$. Thus, as mentioned earlier, a zero of $D(k)$ at the same value of k as that for which $D^{+}(k)$ is zero is a special circumstance; usually zeros of $D(k)$ can be expected to occur at values of k for which $D^{+}(k)$ is not zero. However, zeros of $D(k)$ which occur on the real k axis always occur in pairs, symmetric with respect to the origin. Zeros of $D(k)$ off of the real axis occur in sets of four, symmetric with respect to both the real and imaginary k-axes.

Since $D(k)$ is a ratio of two polynomials of the same order, the number of zeros and number of poles in the entire complex plane are equal. We already have pointed out that $D(k)$ will not have poles on the real axis. Its poles will be symmetrically distributed in the upper and lower half planes. If a pair of zeros of D happen to occur on the real axis, then the number of poles in either of the half planes, excluding the real axis, exceeds the number of zeros by 1. If $2m$ zeros of $D(k)$ lie on the real axis (symmetrically located about the imaginary axis) then in either half plane the number of poles exceeds the number of zeros by m, excluding the real axis.

A zero of $D^+(k)$ on the real axis can be of any order. That is, a potential can be constructed which at a given real value k_0 of k produces any specified number of continuum bound states. It has been demonstrated [59] that if the order of the zero of $D^+(k)$ at $k = k_0$ is m, then the order of the zero of $D(k)$ at $k = k_0$ is at most $m+1$ and at least m. Furthermore, a zero of $D(k)$ at $k = k_0$ which occurs for real $k_0 \neq 0$ and for which $D^+(k_0) \neq 0$ must be a simple zero [59]. That is, a zero of $D(k)$ at $k = k_0$ of order m where $m > 1$ must be accompanied by a zero of $D^+(k_0)$ of at least order $m-1$.

As discussed earlier, the Jost function $\mathcal{L}^+(k)$is given by the ratio of $D^+(k)$to $D(k)$. In taking this ratio, poles of $D^+(k)$can be cancelled by poles of $D(k)$ and zeros of $D^+(k)$can be cancelled by zeros of $D(k)$. Thus, as already has been pointed out [52], the Jost function for a nonlocal potential does not usually contain as complete a set of information about the system as the determinants $D^+(k)$and D(k). The zeros of $D(k)$ which do not cancel become poles of $\mathcal{L}^+(k)$and the poles of $D(k)$ which do not cancel become zeros of $\mathcal{L}^+(k)$. Thus, when m continuum bound states occur, there will be $2m$ zeros of $D^+(k)$and $2m$ zeros of $D(k)$ located on the real k axis. These zeros will cancel, and as a result of the m continuum bound states the Jost function $\mathcal{L}^+(k)$will, for $\Im k \geq 0$ have m zeros more than the number of its poles. Since bound states result in zeros of $D^+(k)$on the positive imaginary axis and since these zeros are not cancelled by zeros of $D(k)$, these zeros of $D^+(k)$also will be zeros of $\mathcal{L}^+(k)$in the upper half plane. Thus if a potential is such that it has m continuum bound states and n bound states, the Jost function $\mathcal{L}^+(k)$for that potential will have $(m+n)$ more zeros in the upper half plane than its number of poles in the upper half plane.

For a spurious state, $D(k) = 0$ at the value of k at which the state occurs and $D(k) = 0$ at that same k. As discussed above, when $D^+(k) \neq 0$ only a first order zero of $D(k)$ can occur. Thus at any given value of k there can be only one spurious state. If there are $2s$ such isolated zeros of $D(k)$ on the real axis, then in the upper half plane (and, for that matter, in the lower half plane as well) $D(k)$ will have s more poles than the number of zeros, excluding the real axis. Since $D^+(k)$is analytic in the upper half-plane, if there are no bound states it follows that $\mathcal{L}^+(k)$will have s more zeros than its number of poles (excluding the poles on the real axis) in the upper half of the complex k-plane. Thus, if in addition to s spurious states there are n bound states, in the upper half plane $\mathcal{L}^+(k)$will have $(s+n)$ more zeros than the number of poles.

We are now in a position to discuss Levinson's theorem in terms of $\mathcal{L}^+(k)$or in terms of $D^+(k)$. Before beginning this discussion, however, we investigate the properties of (k) and D (k) at k=0, and associate these properties with solutions of Eq.(9).

5 Solutions at $k = 0$

Newton [60] has pointed out that for a nonsingular, central, local potential $\mathcal{L}^+(k)$can be zero at $k = 0$. Such a zero must be simple, and does not correspond to a normalizable solution of the radial equation. Newton refers to the solution of the radial equation under these circumstances as a half-bound state [61]. In the case of a nonlocal potential, such a state is also possible. It will be characterized by a zero of $D^+(k)$but not a zero of $D(k)$; $\mathcal{L}^+(k)$thus will be zero under this circumstance. It is also possible in the case of a nonlocal potential to have normalizable states at $k = 0$. The Saito potential is an example of a nonlocal potential with a zero-energy bound state, although this case is complicated by the fact that the Saito potential is not symmetric [62]. A nonlocal potential can have more than one of these states, which are characterized by zeros of both $D^+(k)$and $D(k)$, and thus no zero of $\mathcal{L}^+(k)$. Furthermore, if there are q of these states at $k = 0$, then the order of the zeros of $D^+(k)$and $D(k)$ is $2q$. This can easily be seen by realizing that such a state is a continuum bound state which happens to occur at $k = 0$ rather than some positive value of k. A CBS at k_0, where $k_0 > 0$, is characterized by a pair of zeros of $D^+(k)$, one at k_0 and the other at $-k_0$, and a pair of zeros of $D(k)$ at those same positions; zeros of $D^+(k)$and $D(k)$ at $k = 0$ clearly will occur as double zeros.

The Levinson's theorem contour for a half-bound state, since it implies a zero of $\mathcal{L}^+(k)$and $D^+(k)$, is handled for both $\delta_{\mathcal{L}}$ and δ_D in exactly the same manner as for a local potential. Integration of either $\mathcal{L}^+(k)'/\mathcal{L}^+(k)$ or $D^+(k)'/D^+(k)$ clockwise around a semicircle located at the origin yields $-\pi i$, as discussed in Sec. 2.

The treatment of the contour for Levinson's theorem in the case of a zero energy bound state differs depending upon whether one is considering $\delta_{\mathcal{L}}$ or δ_D. For a continuum bound state at zero energy, $\mathcal{L}^+(k)$will not have a zero at $k = 0$, and thus no adjustment of the contour is necessary to accommodate the presence of the state. As discussed above, however, $D^+(k)$will have a double zero at $k = 0$. In order to evaluate the integral of $D^+(k)'/D^+(k)$ around an unambiguous contour in terms of the zero energy phase shift we can slightly displace the double zeros about the origin by η and let $\eta \to 0$ as shown in Fig. 2. The contribution from each of the contours γ in Fig. 2 will be $-\pi i$, resulting in a total contribution as $\eta \to 0$ of $-2\pi i$.

It is also necessary to consider how the presence of the pair of zeros of $D^+(k)$affects the phase shift δ_D at $k = 0$. This question, however, is deferred until the next section.

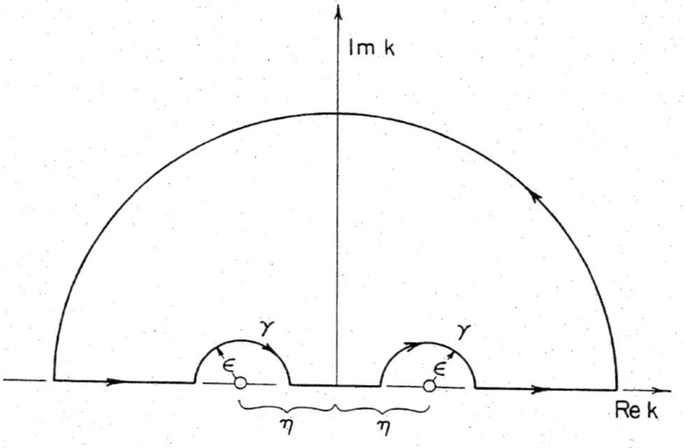

Figure 2. Contour C for integration of $D(k)^+/D(k)$ in the case double zeros of $D^+(k)$ at the origin. These zeros, displaced by η in order to define an unambiguous contour, are indicated by the symbol o.

6 Derivation of the Generalized Levinson's Theorem for the Phase Shift

6.1 Case of n Bound States and One Half-Bound State

For a nonlocal potential with n bound states and one half-bound state, the Jost function $\mathcal{L}^+(k)$will have n zeros on the positive real axis and one zero at the origin, as shown in Fig. 3. There will be no poles on or above the real axis. Thus, making use of Eq. (4) for the contour C of Fig. 3 gives

$$\int_C \frac{\mathcal{L}^+(k)'}{\mathcal{L}^+(k)} dk = 2\pi i n \tag{19}$$

As discussed in the case of a local potential in Sec. 2, the explicit line integral yields

$$\int_C \frac{\mathcal{L}^+(k)'}{\mathcal{L}^+(k)} dk = 2i[\delta_{\mathcal{L}}(0) - \delta_{\mathcal{L}}(\infty)] - \pi i. \tag{20}$$

Thus we obtain the same result as in the case of a local potential, namely

$$\delta_{\mathcal{L}}(0) - \delta_{\mathcal{L}}(\infty) = (n + \frac{1}{2})\pi. \tag{21}$$

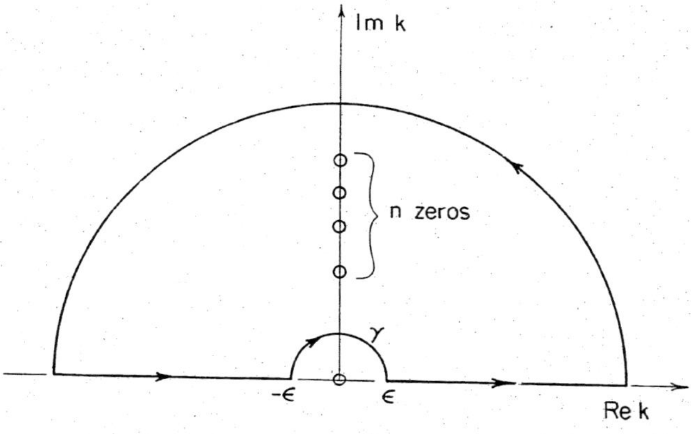

Figure 3. Contour C for the derivation of the Generalized Levinson's Theorem for $\delta_{\mathcal{L}}$ when there are n bound states and a half-bound state at $k = 0$. Zeros of $\mathcal{L}^+(k)$are indicated by the symbol o.

6.2 Case of n Bound States and m Continuum Bound States

As discussed in Sec. 4, the Jost function $\mathcal{L}^+(k)$for a potential with n bound states and m continuum bound states has $n+m$ more zeros than poles in the upper half plane, and has no zeros or poles on the real axis. This set of conditions is indicated in Fig. 4, and applies whether the continuum bound states are at $k = 0$ or $k > 0$.

Calculating the logarithmic derivative of $\mathcal{L}^+(k)$by the two different methods yields

$$\int_C \frac{\mathcal{L}^+(k)'}{\mathcal{L}^+(k)} dk = 2\pi i(n+m) \tag{22}$$

and

$$\int_C \frac{\mathcal{L}^+(k)'}{\mathcal{L}^+(k)} dk = 2i[\delta_{\mathcal{L}}(0) - \delta_{\mathcal{L}}(\infty)]. \tag{23}$$

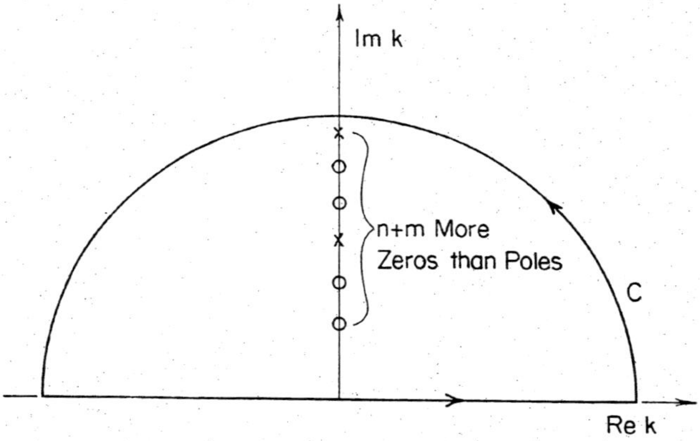

Figure 4. Contour C for the derivation of the Generalized Levinson's Theorem for $\delta_{\mathcal{L}}$ when there are n bound states and m continuum bound states. Zeros and poles of $\mathcal{L}^+(k)$are indicated by o and χ, respectively. At least n zeros of $\mathcal{L}^+(k)$must be on the positive imaginary axis. Additional zeros of $\mathcal{L}^+(k)$and the poles of $\mathcal{L}^+(k)$in the upper half plane may be either on the positive imaginary axis or off the axis in pairs symmetrically located with respect to the imaginary axis. The assertion that there be $n+m$ more zeros than poles in the upper half plane is a condition which takes into account all of the zeros and poles.

Thus, in this case

$$\delta_{\mathcal{L}}(0) - \delta_{\mathcal{L}}(\infty) = (n+m)\pi. \tag{24}$$

If in addition there is a zero of $\mathcal{L}^+(k)$at $k = 0$ we get

$$\delta_{\mathcal{L}}(0) - \delta_{\mathcal{L}}(\infty) = (n+m\frac{1}{2})\pi. \tag{25}$$

6.3 Case of n Bound States and p Spurious States

As discussed in Sec. 4, the Jost function $\mathcal{L}^+(k)$for a potential with n bound states and s spurious states has $n+s$ more zeros than poles above the real axis. In addition, $\mathcal{L}^+(k)$is characterized by $2s$ poles located in pairs on the real axis. This situation is shown in Fig. 5. Using Eq. (4) for the contour C of Fig. 5 we get

$$\int_C \frac{\mathcal{L}^+(k)'}{\mathcal{L}^+(k)} dk = 2\pi i(n+s). \tag{26}$$

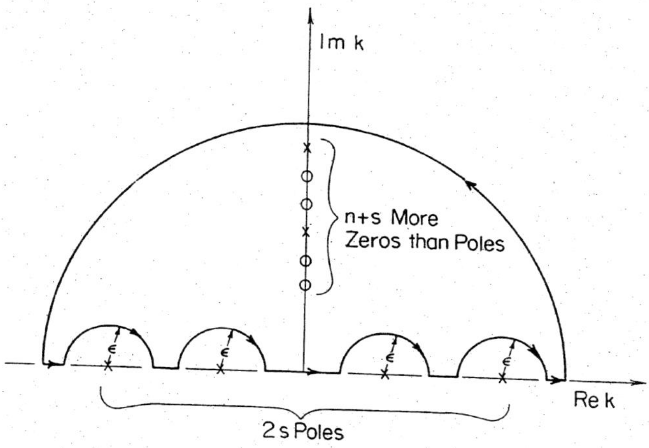

Figure 5. Contour C for the derivation of the Generalized Levinson's Theorem for $\delta_{\mathcal{L}}$ when there are n bound states and s spurious states. Zeros and poles of $\mathcal{L}^+(k)$ are indicated by o and χ respectively. At least n zeros of $\mathcal{L}^+(k)$ must be on the positive imaginary axis and $2s$ poles of $\mathcal{L}^+(k)$ must be on the real axis in pairs symmetric with respect to the imaginary axis. The additional zeros and poles of $\mathcal{L}^+(k)$ in the upper half plane must either be on the positive imaginary axis, or in pairs located symmetrically with respect to that axis. The assertion that there be $n+s$ more zeros than poles in the upper half plane is a condition which takes into account all of the zeros and poles of $\mathcal{L}^+(k)$ above the real axis.

Direct integration of the logarithmic derivative of $\mathcal{L}^+(k)$ around C yields

$$\int_C \frac{\mathcal{L}^+(k)'}{\mathcal{L}^+(k)}\,dk = 2i\left(\delta_{\mathcal{L}}(0)-\delta_{\mathcal{L}}(\infty)+\sum_{i=1}^{s}[\delta_{\mathcal{L}}(k_i+0)-\delta_{\mathcal{L}}(k_i-0)]\right)+2\pi is, \tag{27}$$

where the term $2\pi is$ arises from integrating around the $2s$ semicircles on the real axis, as shown in Fig. 5. When a spurious state occurs at $k = k_i$ there is a phase change of $-\pi$ in going from $k = k_i - \varepsilon$ to $k = k_i + \varepsilon$ in the limit as $\varepsilon \to 0$. This is

$$\delta_{\mathcal{L}}(k_i+0)-\delta_{\mathcal{L}}(k_i-0) = -\pi. \tag{28}$$

Thus Eq. (27) becomes

$$\int_C \frac{\mathcal{L}^+(k)'}{\mathcal{L}^+(k)}\,dk = 2i[\delta_{\mathcal{L}}(0)-\delta_{\mathcal{L}}(\infty)-s\pi]+2\pi is. \tag{29}$$

Taken together, Eqs. (26) and (29) imply

$$\delta_{\mathcal{L}}(0)-\delta_{\mathcal{L}}(\infty) = (n+s)\pi. \tag{30}$$

If, in addition, there is a zero of $\mathcal{L}^+(k)$ at $k = 0$ then we get

$$\delta_{\mathcal{L}}(0)-\delta_{\mathcal{L}}(\infty) = (n+s\frac{1}{2})\pi. \tag{31}$$

6.4 Summary

We have seen that if we use $\delta_{\mathcal{L}}$ as the definition of the phase shift, then Levinson's theorem becomes

$$\delta_{\mathcal{L}}(0)-\delta_{\mathcal{L}}(\infty)=(n+p)\pi, \tag{32}$$

where n is the number of bound states and p is the number of continuum bound states (including those at $k=0$) and/or spurious states. If, also, we have a zero of $\mathcal{L}^{+}(k)$at $k=0$ then Eq. (32) becomes

$$\delta_{\mathcal{L}}(0)-\delta_{\mathcal{L}}(\infty)=(n+p+\frac{1}{2})\pi, \tag{33}$$

7 Derivation of the Generalized Levinson 's Theorem for the Phase Shift

7.1 Case of n Bound States and One Half-Bound State

With respect to bound states and half-bound states, the structure of the poles and zeros of $D^{+}(k)$is the same as that of $\mathcal{L}^{+}(k)$. Thus, as shown in Fig. 6, there are no poles of $D^{+}(k)$on or above the real axis, n zeros of $D^{+}(k)$on the positive imaginary axis corresponding to n bound states, and a simple zero of $D^{+}(k)$at $k=0$ corresponding to a half-bound state. Applying Eq. (4) to the logarithmic derivative of $D^{+}(k)$for the contour C shown in Fig. 6, we get

$$\int_{C}\frac{D^{+}(k)'}{D^{+}(k)}dk=2\pi in. \tag{34}$$

Direct integration yields

$$\int_{C}\frac{D^{+}(k)'}{D^{+}(k)}dk=2i[\delta_{D}(0)-\delta_{D}(\infty)]-\pi i. \tag{35}$$

Therefore

$$\delta_{D}(0)-\delta_{D}(\infty)=(n+\frac{1}{2})\pi. \tag{36}$$

7.2 Case of n Bound States and m Continuum Bound States

As discussed in Sec. 4, $D^{+}(k)$ for a nonlocal potential has no poles on or above the real axis. If the potential has n bound states, then there will be n zeros of $D^{+}(k)$on the positive imaginary axis, as shown in Fig. 7. If, in addition, the potential has m continuum bound states, then there will be $2m$ zeros of$D^{+}(k)$on the real axis, in pairs symmetric about $k=0$. This also is shown in Fig. 7. Using Eq. 4 and integrating around the contour C shown in Fig. 7 yields

$$\int_{C}\frac{D^{+}(k)'}{D^{+}(k)}dk=2\pi in. \tag{37}$$

From the line integral around the contour we get

$$\int_{C}\frac{D^{+}(k)'}{D^{+}(k)}dk=2i\left(\delta_{D}(0)-\delta_{D}(\infty)+\sum_{i=1}^{m}[\delta_{D}(k_i+0)-\delta_{D}(k_i-0)]\right)-2m\pi i, \tag{38}$$

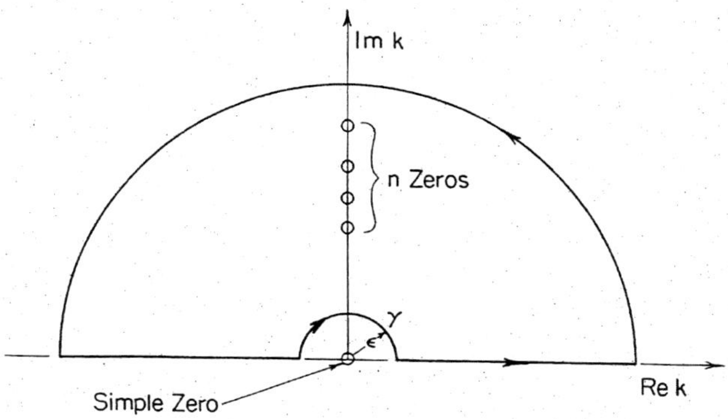

Figure 6. Contour C for the derivation of the Generalized Levinson's Theorem for δ_D when there are n bound states for $k \neq 0$ and a half-bound state at $k = 0$. Zeros of $D^+(k)$ are indicated by o and χ respectively.

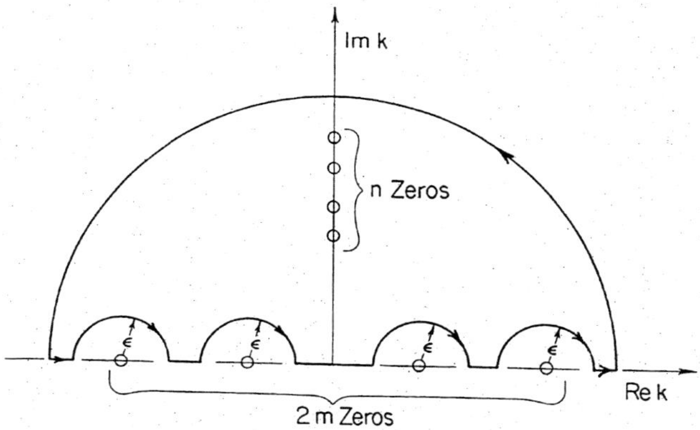

Figure 7. Contour C for the derivation of the Generalized Levinson's Theorem for δ_D when there are n bound states and m continuum bound states. Zeros of $D^+(k)$ are indicated by o.

where the term $-2m\pi i$ results from the integrals around the $2m$ semicircles along the real axis. Since

$$\delta_D(k_i+0)-\delta_D(k_i-0)=\pi. \tag{39}$$

Eq. (39) becomes

$$\int_C \frac{D^+(k)'}{D^+(k)}dk = 2i[\delta_D(0)-\delta_D(\infty)+m]\pi - 2m\pi i, \tag{40}$$

Combining Eqs. (37) and (40) gives

$$\delta_D(0)-\delta_D(\infty)=n\pi. \tag{41}$$

If, in addition, there is a zero of $D^+(k)$of order q at $k=0$, it is necessary to discuss this derivation further. A continuum bound state at $k=0$ requires a double zero of $D^+(k)$. Thus, if q is odd, there will be a half-bound state at $k=0$. Since only one such state is possible according to the result in part A of this section, an odd value of q will add $\frac{1}{2}\pi$ to the right hand side of Eq. (41). There will be no contribution to the right hand side of Eq. (41), however, in the case of a zero-energy continuum bound state. As discussed in Sec. 3 one can move the pair of zeros to give a zero at $k = -\varepsilon$ and one at $k = \varepsilon$; integrating around these two semicircles will yield $-2\pi i$. But, as indicated above, the phase shift is now discontinuous at $k = -\varepsilon$ and one at $k = \varepsilon$, and this discontinuity yields $2\pi i$. This cancels the contribution of the integrals around the semicircles. Thus for n bound states, m continuum bound states (either at $k = 0$ or $k > 0$), and a half-bound state, Levinson's theorem for $\delta_D(k)$ is

$$\delta_D(0) - \delta_D(\infty) = (n + \frac{1}{2})\pi. \tag{42}$$

Actually, there is an ambiguity in the zero energy phase shift δ_D. As the parameters of a nonlocal potential are adjusted, a zero of $D^+(k)$can move from a position on the positive imaginary axis down to $k = 0$. At that point, the bound state becomes a continuum bound state, at zero energy. Whereas the bound state contributed π to the zero energy phase shift δ_D the continuum bound state will contribute nothing. Thus a very small change in the potential parameters can result in a discontinuity of the phase shift δ_D by π at zero energy. Such a discontinuity did not occur in the case of the phase shift δ_L under these circumstances.

7.3 Case of n Bound States and p Spurious States

Corresponding to n bound states there will be n zeros of $D^+(k)$on the positive imaginary axis, as shown in Fig. 8. Since p spurious states correspond to p zeros of $D(k)$, without corresponding zeros of $D^+(k)$on or above the real axis, the logarithmic derivative of $D^+(k)$for the contour C of Fig. 8 will not be affected by the p spurious states. Applying the argument principle then will give

$$\int_C \frac{D^+(k)'}{D^+(k)} dk = 2\pi i n. \tag{43}$$

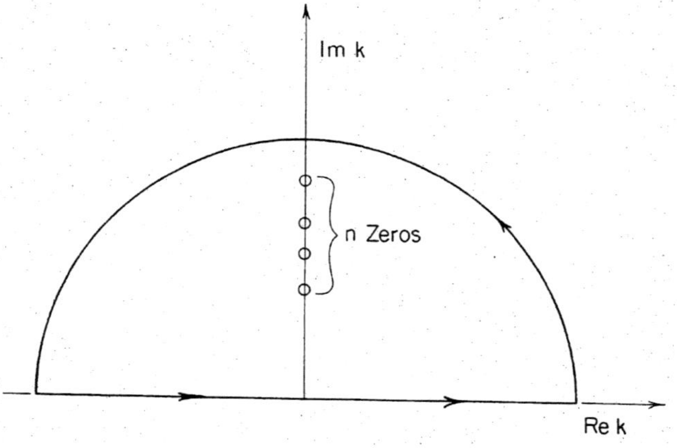

Figure 8. Contour C for the derivation of the Generalized Levinson's Theorem for when there are n bound states and p spurious states. Zeros of $D^+(k)$ are indicated by the symbol o.

Direct integration yields

$$\int_C \frac{D^+(k)'}{D^+(k)} dk = 2i[\delta_D(0) - \delta_D(\infty)]. \tag{44}$$

Therefore

$$\delta_D(0) - \delta_D(\infty) = n\pi. \tag{45}$$

7.4 Summary

We have seen that if we use δ_D as the definition for the phase shift, then Levinson's theorem becomes

$$\delta_D(0) - \delta_D(\infty) = n\pi. \tag{46}$$

where n is the number of bound states. The phase shift does not depend upon the number of continuum bound states and/or spurious states. If, also, we have a first-order zero of $D^+(k)$at $k = 0$, which would be the case if and only if a half-bound state is present, then Eq. (46) becomes

$$\delta_D(0) - \delta_D(\infty) = (n + \frac{1}{2})\pi. \tag{47}$$

8 Examples

In this section we consider two examples of nonlocal potentials and calculate the phase shift $\delta(0) - \delta(\infty)$ using both of the definitions, $D^{\pm}(k) = |D^{\pm}(k)|e^{\mp i\delta_D}$ and $\mathcal{L}^{\pm}(k) = |\mathcal{L}^{\pm}(k)|e^{\mp i\delta_{\mathcal{L}}}$. Both examples will make use of nonlocal potentials of the form

$$V(r, r') = \lambda g(r) g(r'). \tag{48}$$

These two examples have been chosen because the behavior in the complex plane of $D^+(k)$and $D(k)$ for these potentials has been analyzed in a previous paper [55].

8.1 Yamaguchi Potential

In 1954 Yamaguchi [63] introduced a one-term separable nonlocal potential to describe nucleon-nucleon scattering. The Yamaguchi potential is of the form shown in Eq. (48) with

$$g(r) = e^{-\alpha r}. \tag{49}$$

The expressions for $D^+(k)$and $D(k)$ for this potential are given in Refs. [40] and [55]. They can be written in the form

$$D^+(k) = N^+(k)/M^+(k), \tag{50}$$

where

$$N^+(k) = 2\alpha k^2 + i4\alpha^2 k - (\lambda + 2\alpha^3) \tag{51}$$

and

$$M^+(k) = 2\alpha(k + i\alpha)^2 \tag{52}$$

and

$$D(k) = N(k)/M(k), \tag{53}$$

where

$$N(k) = 2\alpha k^2 - (\lambda - 2\alpha^3) \tag{54}$$

and

$$M(k) = 2\alpha(k - i\alpha)(k + i\alpha). \tag{55}$$

From these expressions we see that the roots of $N^+(k)$ are

$$k_{1,2} = -i\alpha \pm \sqrt{\frac{\lambda}{2\alpha}}. \tag{56}$$

No values of λ,α will make $D^+(k)$zero for real values of k; therefore no continuum bound state can be associated with the Yamaguchi form factor. The roots of $N(k)$ are

$$k_{1,2} = \pm\sqrt{\frac{\lambda}{2\alpha} - \alpha^2}. \tag{57}$$

Thus $D(k)$ can be zero for a wide range of values of λ and α. Although the values of λ and α used by Yamaguchi do not generate a spurious state at any energy, if $\lambda > 2\alpha^3$ a spurious state will occur. The zeros of $M^+(k)$ are

$$k_{1,2} = -i\alpha. \tag{58}$$

The zeros of $M(k)$ are

$$k_{1,2} = \pm i\alpha. \tag{59}$$

8.1.1 Phase Shift Calculation Using $\delta_{\mathcal{L}}$

We now use the definition $\mathcal{L}^+(k) = |\mathcal{L}^+(k)|e^{-i\delta_{\mathcal{L}}(k)}$ to obtain the phase shift in two cases.

Case 1. When $\lambda < 2\alpha^3$, there is no spurious state. $\mathcal{L}^+(k)$has one zero in the upper half plane, at $k = i\alpha$, and a pole at $k = i\sqrt{(2\alpha^3 - \lambda)/2\alpha}$.

Using Eq. (4) for $\mathcal{L}^+(k)$to obtain the integral around the contour C shown in Fig. 9 yields

$$\int_C \frac{\mathcal{L}^+(k)'}{\mathcal{L}^+(k)} dk = 0. \tag{60}$$

The line integral around the contour C gives

$$\int_C \frac{\mathcal{L}^+(k)'}{\mathcal{L}^+(k)} dk = \ln \mathcal{L}^+(k)\Big|_0^K + \int_0^K \frac{\mathcal{L}^+(Ke^{i\theta})'}{\mathcal{L}^+(Ke^{i\theta})} Ki\theta e^{i\theta}\, d\theta + \ln \mathcal{L}^+(k)\Big|_{-K}^{0^-}. \tag{61}$$

or

$$\int_C \frac{\mathcal{L}^+(k)'}{\mathcal{L}^+(k)} dk = \ln\left[\frac{\mathcal{L}^+(K)}{\mathcal{L}^+(-K)}\right] + \ln\left[\frac{\mathcal{L}^+(0^-)}{\mathcal{L}^+(0^+)}\right] + \int_0^\pi \frac{\mathcal{L}^+(Ke^{i\theta})'}{\mathcal{L}^+(Ke^{i\theta})} Ki\theta e^{i\theta}\, d\theta. \tag{62}$$

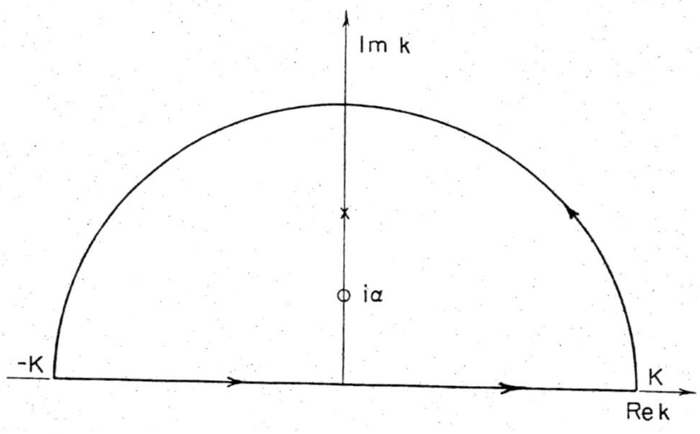

Figure 9. Contour C for integration for Levinson's Theorem for $\delta_{\mathcal{L}}$, showing the poles and zeros of $\mathcal{L}^+(k)$in the upper half of the k plane for the Yamaguchi potential with parameters for which there is no spurious state. [There are no zeros or poles of $\mathcal{L}^+(k)$on the real axis.] Zeros and poles of $\mathcal{L}^+(k)$for this potential are indicated by o and χ, respectively.

The integral on the right hand side of Eq. (62) vanishes as $K \to \infty$. Using the fact that $\delta(-k) = -\delta(k)$, Eq. (62) implies

$$\lim_{K\to\infty} \int_C \frac{\mathcal{L}^+(k)'}{\mathcal{L}^+(k)} dk = 2i[\delta_{\mathcal{L}}(0) - \delta_{\mathcal{L}}(\infty)]. \tag{63}$$

Therefore we get

$$\delta_{\mathcal{L}}(0) - \delta_{\mathcal{L}}(\infty) = 0. \tag{64}$$

Case 2. When $\lambda > 2\alpha^3$, there is a spurious state. The Jost function has one zero at $k = i\alpha$ and two poles, at $k = \pm\sqrt{(\lambda - 2\alpha^3)/2\alpha} = \pm k_1$, on the real axis. Therefore, integrating around the contour C shown in Fig. 10 yields

$$\int_C \frac{\mathcal{L}^+(k)'}{\mathcal{L}^+(k)} dk = 2\pi i. \tag{65}$$

In the limits $\varepsilon \to 0$ and $K \to \infty$, direct integration around the contour shown in Fig. 10 gives

$$\int_C \frac{\mathcal{L}^+(k)'}{\mathcal{L}^+(k)} dk = \ln\left[\frac{\mathcal{L}^+(-k_1-0)}{\mathcal{L}^+(k_1+0)}\right] - \ln\left[\frac{\mathcal{L}^+(-\infty)}{\mathcal{L}^+(+\infty)}\right] + \ln\left[\frac{\mathcal{L}^+(-0)}{\mathcal{L}^+(+0)}\right] - \ln\left[\frac{\mathcal{L}^+(-k_1+0)}{\mathcal{L}^+(k_1-0)}\right] + 2\pi i. \tag{66}$$

or

$$\int_C \frac{\mathcal{L}^+(k)'}{\mathcal{L}^+(k)} dk = 2i\left[\delta_{\mathcal{L}}(0) - \delta_{\mathcal{L}}(\infty) + \delta_{\mathcal{L}}(k_1+0) - \delta_{\mathcal{L}}(k_1-0)\right] + 2\pi i. \tag{67}$$

Because of the spurious state at $k = k_1$, there is a phase change of $-\pi$ in going from $k = k_1 - \varepsilon$ to $k = k_1 + \varepsilon$ That is, in the limit as $\varepsilon \to 0$,

$$\delta_{\mathcal{L}}(k_1+0) - \delta_{\mathcal{L}}(k_1-0) = -\pi. \tag{68}$$

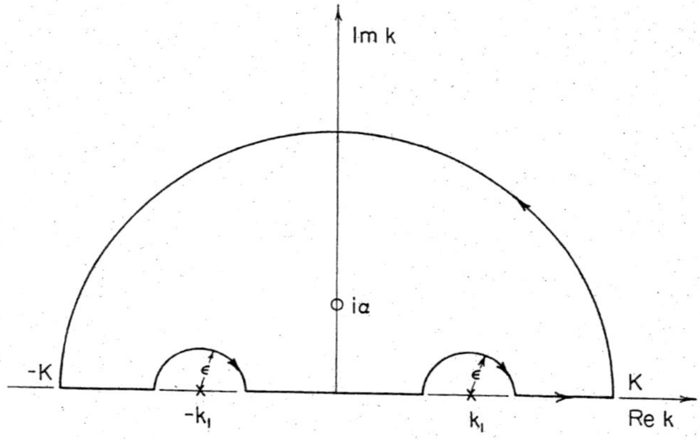

Figure 10. Contour C for integration for Levinson's Theorem for $\delta_{\mathcal{L}}$, showing the poles and zeros of $\mathcal{L}^+(k)$in the upper half of the k plane and on the real axis for the Yamaguehi potential with parameters for which there is a spvrious state at $k = k_1$. Zeros and poles of $\mathcal{L}^+(k)$for this potential are indicated by o and χ, respectively.

Combining Eqs. (65), (67), and (68) yields

$$2\pi i = 2i\,[\delta_{\mathcal{L}}(0) - \delta_{\mathcal{L}}(\infty) - \pi] + 2\pi i, \tag{69}$$

which implies

$$\delta_{\mathcal{L}}(0) - \delta_{\mathcal{L}}(\infty) = \pi. \tag{70}$$

8.1.2 Phase Shift Calculation Using δ_D

Next we use the definition $D^+(k) = |D^+(k)|\,e^{-i\delta_D(k)}$ to get the zero-energy phase shift. In both the case $\lambda > 2\alpha^3$ and $\lambda < 2\alpha^3$ the result will be the same, since for all values of λ and α the zeros of $D^+(k)$ lie in the lower half plane. Using Eq. (4) for $D^+(k)$and integrating around the contour C shown in Fig. 11, we get

$$\int_C \frac{D^+(k)'}{D^+(k)}\,dk = 0. \tag{71}$$

The line integral around the contour yields, in the limit as $K \to \infty$ and $\varepsilon \to 0$,

$$\int_C \frac{D^+(k)'}{D^+(k)}\,dk = \ln D^+(-0) - \ln D^+(-\infty) + \ln D^+(\infty) - \ln D^+(0). \tag{72}$$

Combining Eqs. (7l) and (72) gives

$$\delta_D(0) - \delta_D(\infty) = 0. \tag{73}$$

8.2 Beregi Potential

The second example to be considered here is the Beregi potential. Beregi [33] suggested a one-term separable nonlocal potential of the form given by Eq. (48) with

$$g(r) = e^{-\alpha_1 r} - ae^{-\alpha_2 r}. \tag{74}$$

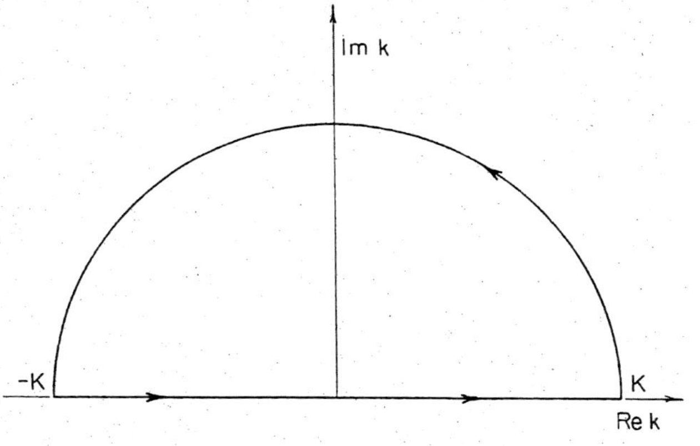

Figure 11. Contour C for integration for Levinson's Theorem for δ_D the Yamaguchi potential. [Note that there are no poles or zeros of $D^+(k)$ in the upper half-plane.]

This potential yields a continuum bound state at 259 .3 MeV and a bound state at -2.225 MeV [41]. The parameters for the Beregi potential are

$$\lambda = -302.73\ fm^{-3},$$

$$\alpha_1 = 2.67\ fm^{-1},$$

$$\alpha_2 = 5.34\ fm^{-1},$$

$$a = 3.0854.$$

The Fredholm determinants for this potential are given in Refs. [41] and [55]. They can be written as

$$D^+(k) = \frac{N^+(k)}{M^+(k)} \tag{75}$$

where

$$\begin{aligned} N^+(k) = {} & k^4[2\alpha_1\alpha_2(\alpha_1+\alpha_2)] \\ & + ik^3(\alpha_1+\alpha_2)[4\alpha_1^2\alpha_2 + 4\alpha_1\alpha_2^2] \\ & + k^2[(\alpha_1+\alpha_2)(-2\alpha_1^3\alpha_2 - 2\alpha_1\alpha_2^3 - \lambda\alpha_2 - \lambda a^2\alpha_1 - 8\alpha_1^2\alpha_2^2 + 4\lambda a\alpha_1\alpha_2] \\ & + ik[(\alpha_1+\alpha_2)(-4\lambda a\alpha_1\alpha_2 - 2\lambda a^2\alpha_1^2 - 4\alpha_1^2\alpha_2^3 - 4\alpha_1^3\alpha_2^2 - 2\lambda\alpha_2^2) \\ & + 8\lambda a\alpha_1^2\alpha_2 + 8\lambda a\alpha_1\alpha_2^2] \\ & + [(\alpha_1+\alpha_2)(2\alpha_1^3\alpha_2^3 + \lambda\alpha_2^3 + \lambda a^2\alpha_1^3) - 4\lambda a\alpha_1^2\alpha_2^2] \end{aligned} \tag{76}$$

and

$$M^+(k) = 2\alpha_1\alpha_2(\alpha_1+\alpha_2)(k+i\alpha_1)^2(k+i\alpha_2)^2 \tag{77}$$

and

$$D(k) = \frac{N(k)}{M(k)} \tag{78}$$

Table 1. Zeros of $N^+(k)$ and $M^+(k)$ for the Beregi Potential (fm^{-1})

Zeros of $N(k)^+$	Zeros of $M^+(k)$
(1.77,0)	(0,-2.67)
(1.77,0)	(0,-2.67)
(0,0.23)	(0,-5.34)
(0,-16.25)	(0,-5.34)

Table 2. Zeros of $N(k)$ and $M(k)$ for the Beregi Potential (fm^{-1})

Zeros of $N(k)$	$Zeros of M(k)$
(1.77,0)	(0,2.67)
(-1.77,0)	(0,-2.67)
(0,11.49)	(0,5.34)
(0,-11.49)	(0,-5.34)

where

$$N(k) = k^4[2\alpha_1\alpha_2(\alpha_1+\alpha_2)] + k^2[(\alpha_1+\alpha_2)(2\alpha_1\alpha_2^3 + 2\alpha_1^3\alpha_2 - \lambda\alpha_2 - \lambda a^2\alpha_1) + 4\alpha_1\alpha_2\lambda a] + [(\alpha_1+\alpha_2)(2\alpha_1^3\alpha_2^3 - \lambda\alpha_2^3 + 2\alpha_1\alpha_2\lambda a(\alpha_1+\alpha_2) - \lambda a^2\alpha_1^3) - 4\alpha_1^2\alpha_2^2\lambda a] \quad (80)$$

and

$$M(k) = 2\alpha_1\alpha_2(\alpha_1+\alpha_2)(\alpha_1^2+k^2)(\alpha_2^2+k^2) \quad (80a)$$

The zeros of $N^+(k)$ and $M^+(k)$ are tabulated in Table I, while the zeros of $N(k)$ and $M(k)$ are tabulated in Table II. The information from these tables is combined in Table III to give the zeros and poles of $\mathcal{L}^+(k)$for this potential.

8.2.1 Phase Shift Calculation Using $\delta_{\mathcal{L}}(k)$

The zero-energy phase shift $\delta_c l$ associated with the Beregi potential can be calculated as discussed in Sec. 6. Using Eq. (4) for $\mathcal{L}^+(k)$and integrating around the contour C of Fig. 12, we have

$$\int_C \frac{\mathcal{L}^+(k)'}{\mathcal{L}^+(k)} dk = 4\pi i. \quad (81)$$

Explicit calculation of the line integral around the contour C yields

$$\int_C \frac{\mathcal{L}^+(k)'}{\mathcal{L}^+(k)} dk = 2i[\delta_{\mathcal{L}}(0) - \delta_{\mathcal{L}}(\infty)]. \quad (82)$$

Table 3. Zeros of $\mathcal{L}^+(k)$ for the Beregi Potential

Zeros	Poles
(0,0.23)	(0,11.49)
(0,2.67)	(0,-11.49)
(0,5.34)	(0,-2.67)
(0,-16.25)	(0,-5.34)

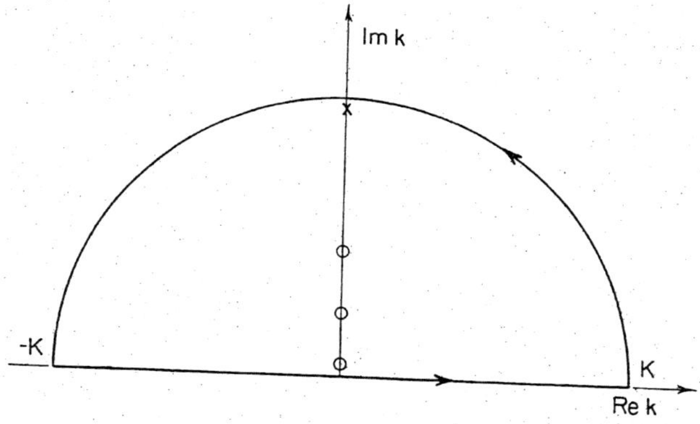

Figure 12. Contour C for integration for Levinson's Theorem for $\delta_{\mathcal{L}}$, showing the poles of $\mathcal{L}^{+}(k)$in the upper half-plane for the Beregi potential [There are no zeros or poles of $\mathcal{L}^{+}(k)$on the real axis.] Zeros and poles are indicated by o and χ, respectively.

This gives

$$\delta_{\mathcal{L}}(0) - \delta_{\mathcal{L}}(\infty) = 2\pi \tag{83}$$

8.2.2 Phase Shift Calculation Using $\delta_{\mathcal{L}}$

The information necessary for the calculation of $\delta_{\mathcal{L}}$ for the Beregi potential is given in Table I. Using the argument principle for the contour C given in Fig. 13, we obtain

$$\int_C \frac{D^+(k)'}{D^+(k)} dk = 2\pi i. \tag{84}$$

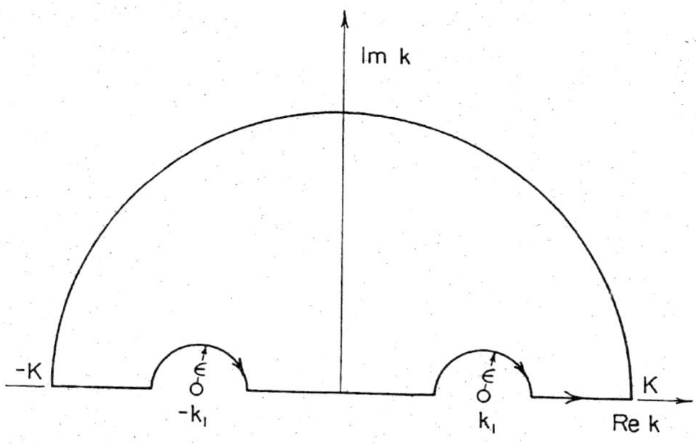

Figure 13. Contour C for integration for Levinson's Theorem for δ_D showing the zeros of $D^+(k)$ on the real axis for the Beregi potential. These zeros are indicated by o.

Taking $k_1 = 1.77 fm^{-1}$, the explicit line integral yields

$$\int_C \frac{D^+(k)'}{D^+(k)} dk = 2i[\delta_D(0) - \delta_D(\infty) + \delta_D(k_1 + 0) - \delta_D(k_1 - 0)] - 2\pi i. \tag{85}$$

The discontinuity in the phase shift at $k = k_1$ is given by

$$\delta_D(k_1 + 0) - \delta_D(k_1 - 0) = \pi. \tag{86}$$

From Eqs. (84), (85), and (86) we get

$$2i[\delta_D(0) - \delta_D(\infty)] = 2\pi i \tag{87}$$

or

$$\delta_D(0) - \delta_D(\infty) = \pi. \tag{88}$$

9 Summary and Conclusion

In this section we compare the proofs of Levinson's theorem given in Secs. 6 and 7 with other proofs in the literature. In so doing, it is important to note that in derivations of Levinson's theorem in the literature some authors have used $\mathcal{L}^+(k)$or $D^+(k)$ and others $\mathcal{L}^-(k)$ or $D^-(k)$. The use of $\mathcal{L}^-(k)$ or $D^-(k)$ instead of $\mathcal{L}^+(k)$or $D^+(k)$ leads to a corresponding change in the contour integral considered from the upper half-plane to the lower half-plane.

As mentioned in the introduction, one of the earliest derivations of the generalized Levinson's theorem for a nonlocal potential was that due to P. Swan [2]. The problem he was considering was that of nonlocality due to antisynimetrization of the incident particle with respect to particles in the target. He found that the natural definition to use was the phase of $\mathcal{L}^-(k)$, that is the phase shift $\delta_{\mathcal{L}}$. [In his paper the notation for $\mathcal{L}^-(k)$ is $f(k)$.] The contour used for his calculation and the pertinent zeros and poles are displayed in Fig. 5 of Ref. [2] and the phase shift calculation is carried out on p. 23 of Ref. [2]. It is interesting to note that although he showed that $\mathcal{L}^-(k)$ could have poles on the real axis, he obtained a zero energy phase shift of zero. Using the approach presented here, for his case we would obtain $\delta_{\mathcal{L}}(0) = \pi$ instead of $\delta_{\mathcal{L}}(0) = 0$. That is, although he recognized the presence of poles on the real axis, Swan did not take into consideration a discontinuity of $-\pi$ in the phase shift as the energy increases through a spurious state.

Also, in his derivation of the generalized Levinson's theorem redundant states occur in his Eq. (54), which reads as follows:

$$\int_{C_1} \frac{f'(k) - \phi_0 \int_0^\infty \phi(r') f'(k,r')\, dr'}{f(k) - \phi_0 \int_0^\infty \phi(r') f(k,r')\, dr'}\, dk = \int_{C_1} \frac{f'(k)}{f(k)}\, dk. \tag{87a}$$

He assumes the existence of the Jost solution $f(k,r)$ under these conditions. As pointed out in Ref. [52], a Jost solution may or may not exist when a continuum bound state or a redundant state occurs. Thus his subsequent derivation is applicable only to those cases in which Jost solutions exist. Jauch's [5] generalized Levinson's theorem was undertaken with the assumption that a nonlocal potential does not have bound states in the continuum. In his generalization of Levinson's theorem, Jauch used the definition of the phase shift as the phase of the S-matrix, that is,

$$S(k) = e^{2i\delta(k)}. \tag{88a}$$

As pointed out in Ref. 41, this definition is not sensitive to the presence of bound states in the continuum or to discontinuities in the phase shift due to poles of $\mathcal{L}^+(k)$on the real axis. Thus Jauch left out the important consideration of the existence of continuum bound states and spurious states in the case of a nonlocal potential.

Shortly after Jauchs paper, Gourdin and Martin [6] noted that a continuum bound state could be associated with a nonlocal potential and that such a state required a modification of Levinson's theorem. Martin [8] showed that for a nonlocal potential with continuum bound states the index in Eq. (2) should be replaced by $\nu+\nu'$), where ν is the number of bound states with energy less than zero and ν' is the number of continuum bound states. He used the definition $\delta_{\mathcal{L}}$ in his derivation, and his result can be expressed in the following form,

$$\delta(0)-\delta(\infty)=(\nu+\nu')\pi. \tag{89}$$

His method does not take into consideration possible zeros of $D(k)$ and thus the possible presence of spurious states.

The derivation of Levinsons theorem presented by Bertero, Talenti, and Viano [29] is based on consideration of the Fredholm determinant $D^+(k)$. Their derivation, which begins on p.635 of Ref. [29], considers the phase shift associated with $D^+(k)$ [denoted in the paper by $F(-k)$] and the contour is closed in the upper half-plane. They consider N zeros in the upper half-plane and $2N'$ zeros on the real axis symmetrically situated about the imaginary axis. There are also q zeros at the origin. In the nomenclature of Bertero et al. a continuum bound state is called a spurious state and states labeled spurious states in our nomenclature are not considered. Figure 1 on p. 637 of Ref. [29] displays the zeros mentioned above. The contour integration in Ref.[29] yields the following equation

$$\int_C \frac{F(k)'}{F(-k)}dk = 2i\left(\eta(0)-eta(\infty)+\sum_{i=1}^{N'}[\eta(k_i+0)-\eta(k_i-0]\right)-2i\pi N'-i\pi q. \tag{90}$$

However, although $F(-k)$ [and thus $\eta(k)$] is not continuous at a continuum bound state, Bertero et al. took the phase shift to be continuous at the continuum bound states (which are in their terminology spurious states), and obtained the following relation

$$\delta(0)-\delta(\infty)=\pi(N+N'+\frac{q}{2}). \tag{91}$$

Had they considered the phase shift to be discontinuous at the continuum bound states they would have obtained the relation (36), but with an ambiguity at zero energy as to whether one should consider zeros of $D^+(k)$ as zero-energy continuum bound states or zero-energy bound states. Also, in addition to not taking into consideration the phase change at continuum bound states, their proof does not refer to the possibility of a spurious state occurring.

Glöckle and LeTourneux [36] studied the generalized Levinson's theorem for composite particles in the framework of the Saito model. They discussed the zeros of $D^+(k)$ [which they designated by $D(k)$] and used the phase of $D^+(k)$ for getting the zero energy phase shift. In their proof, which is given on p. 21 of Ref. [36], they assume m bound states and n continuum bound states. The zeros of $D^+(k)$ corresponding to n continuum bound states lie on the real axis symmetrically about the imaginary axis. They used the integral in Eq. (4)

and obtained the contribution $-2\pi in$ from the $2n$ semicircles around the zeros on the real axis. However in arriving at their Eq. (38), which reads as follows,

$$\int_{-\infty}^{\infty} dk \frac{d}{dk} \ln D(k) = 2i[\delta(0) - \delta(\infty)], \tag{92}$$

they did not take into account the phase change at a continuum bound state. Thus they obtained the relation

$$\delta(0) - \delta(\infty) = (m+n)\pi. \tag{93}$$

Since there is discontinuity of π in the phase of $D(k)$ at each continuum bound state, their result should have been

$$\delta(0) - \delta(\infty) = m\pi. \tag{94}$$

For a large class of nonlocal potentials giving rise to continuum bound states, Buslaev obtained Levinson's theorem in its original form. Buslaev's work is consistent with Bolsterli's results. It is also consistent with our derivation of Levinson's theorem using the definition δ_D when continuum bound states are present. Dreyfus has also studied Levinson's theorem for nonlocal interactions. He pointed out that owing to the two defintions of the scattering phase shift we get two different versions of Levinson's theorem; one of them takes the continuum bound states into account while the other does not. However, neither Buslaev nor Dreyfus take into consideration the possibility of the existence of spurious states. Finally, Newton has considered the point $k = 0$ and proved a generalized Levinson's theorem which includes the possiblity of a bound state or a half-bound state at $k = 0$. His derivation is based on the Fredholm determinant $D^+(k)$. By using the phase of $D^+(k)$ to define a phase shift continuous at a continuum bound state, he obtained a statement of Levinson's theorem which is the same as that which we find here for $\delta_{\mathcal{L}}$. He also presents strong arguments as to why this is the proper statement of the theorem. He does not, however, consider the possibility of spurious states in his derivation.

In studying Levinson's theorem for a nonlocal potential we have taken into account the fact that redundant states, continuum bound states, and spurious states are all possible in the case of a nonlocal potential. This has been done for both of the possible definitions, δ_D and $\delta_{\mathcal{L}}$, of the phase shift for scattering by a nonlocal potential outlined in Ref.[41]. Thus the present work discusses a variety of aspects of Levinson's theorem not previously treated in the case of a nonlocal potential. An important aspect of our ability to extend Levinsons theorem to nonlocal potentials rests upon the results of Ref. [55], in which we introduced functional analytic techniques for determining the character of the Fredholm determinants $D^+(k)$ and $D(k)$ associated with integral equations for scattering solutions of the radial equation.

In discussing the derivation of the generalized Levinson's theorem for both the definitions of the phase shifts we have seen that the generalized Levinson's theorem associated with definition δ_D does not take into account the presence of continuum bound states or spurious states. Also there is ambiguity in δ_D when there are zeros of $D(k)$ at the origin. The definition $\delta_{\mathcal{L}}$ takes into account the presence of continuum bound states as well as spurious states and half-bound states or zero-energy bound states. The definition δ_D fails completely in the presence of redundant states, while the definition $\delta_{\mathcal{L}}$ gives consistent results. All things considered, we find that the phase shift $\delta_{\mathcal{L}}$ seems to be more useful in all the cases considered for the generalization of Levinson's theorem. In this respect we agree with the conclusion of Ref. [41] which prefers the definition δ_L on the basis of different considerations.

References

[1] N. Levinson, *Kgl. Danski Videnskab. Selskab. Mat. Fys.* **25** (1949) No. 9.

[2] P. Swan, *Proc. Roy. Soc. London, Sect.* **A228** (1955) 10.

[3] A. Temkin, *J. Math. Phys.* **2** (1961) 336.

[4] L. Rosenberg and L. Spruch, *Phys. Rev.* **121** (1961) 1720.

[5] J. M. Jauch, *Helv. Phys. Acta* **30** (1957) 143.

[6] M. Gourdin and A. Martin, *Nuovo Cimento* **6** (1957) 757.

[7] Barry Simon, *Comm. Pure and Appl. Math.* **21** (1967) 531.

[8] A. Martin, *Nuovo Cimento* **7** (1958) 607.

[9] I. 0. Krause and B. Mulligan, *Ann. Phys.(N.Y.)* **94** (1975) 31.

[10] E. Feenberg, *Phys. Rev.* **42** (1932) 17.

[11] S. Hochberg, H. S. W. Massey, and L. H. Underhill, *Proc. Phys. Soc. London* **A67** (1954) 957.

[12] S. Saito, *Prog. Theor. Phys.* **40** (1968) 893; 41 (1969) 705.

[13] R. Tamagaki, *Prog. Theo. Phys. Suppl. Extra Number* (1968), p.242

[14] K. Wildermuth and Y. C. Tang, *A Unifical Theory of the Nucleus* (vieweg, Baunschweig, Germany, 1977).

[15] B. Buck, H. Friedrich, and C. Wheatley, *Nucl. Phys.* **A275** (1977) 246.

[16] M. Ida, Prog. *Theor. Phys.* **21** (1959) 625.

[17] R. Haag, *Nuovo Cimento* **5** (1957) 203.

[18] C. Koniski and T. Ogimoto, *Prg. Theo. Phys.* **22** (1959) 807.

[19] J. C. Polkighorne, *Proc. Cambr. Phil. Soc.* **54** (1959) 560.

[20] M. T. Vaughn, R. Aaron, and R. D. Amado, *Phys. Rev.* **124** (1961) 1258.

[21] G. C. Ghirardi, M. Pauli, and A. Rimini, *Ann. Phys. (N.Y.)* **21** (1963) 401.

[22] R. L. Warnock, *Phys. Rev.* **131** (1963) 1320.

[23] J. M. Charap, *Nuovo Cimento* **36** (1965) 419.

[24] M. Kato, *Ann. Phys. (N.Y.)* **31** (1965) 130.

[25] D. Atkinson, K. Dietz, and D. Morgan, *Ann. Phys. (N.Y.)* **37** (1966) 77.

[26] V. G. Kadyskevskii, R. M. Mir-Kasimov., and N. B. Skachkov, *Nuovo Cimento* **55** (1967) 233.

[27] K. L. Nagy, *Acta Phys. Hung*. **24** (1968) 433.

[28] J. B. Hartle and C. E. Jones, *Ann. Phys.(N.Y.)* **38** (1969) 348.

[29] M. Bertero, G. Talenti, and C. A. Viano, *Mud. Phys.* **A113** (1968) 625.

[30] V. S. Buslaev, *Probl. Math. Phys. (Leningrad)* **4** (1969) 43.

[31] T. Dreyfus, thesis, University of Geneva, 1976, unpublished preprint, and *J. Phys. A; Math. Gen.* **9** (1976) L187.

[32] M. Bolsterli, *Phys. Rev.* **182** (1969) 1095.

[33] P. Beregi, *Nucl. Phys.* **A206** (1973) 217.

[34] R. L. Mills and J. F. Reading, *J. Math. Phys.* **10** (1969) 321.

[35] L. Horwitz and J. P. Marchand, *Rocky Mount. J. Math.* **1** (1971) 225.

[36] W. Gltickle and Jean LeTourneux, *Nucl. Phys.* **A269** (1976) 16.

[37] M. Von W 511enberg, *Math. Nachr.* **78** (1977) 223; 78 (1977) 369.

[38] R. C. Newton, *J. Math. Phys.* **18** (1977) 1582.

[39] P. Beregi, B. N. Zakharev and S. A. Niyazgulov, *Soy. J. Particles Nucl.* **4** (1973) 217.

[40] B. Mulligan, L. G. Arnold, B. Bagchi, and T. 0. Krause, *Phys. Rev. C* **13** (1976) 2131.

[41] B. Bagehi, T. 0. Krause, and B. Mulligan, *Phys. Rev. C* **15** (1977) 1623.

[42] M. Coz, L. C. Arnold, and A. D. MacKellar, *Ann. Phys. (N.Y.)* **59** (1970) 219.

[43] L. C. Arnold and A. D. MacKellar, *Phys. Rev. C* **3** (1971) 1095.

[44] B. Bagchi and B. Mulligan, *Phys. Rev. C* **20** (1979) 1973.

[45] R. Jost and A. Pais, *Phys. Rev.* **82** (1951) 840.

[46] L. C. Arnold, B. Bagchi, and B. Mulligan, unpublished.

[47] R. G. Newton, *Scattering Theory of Waves and Particles* (McGraw-Hill, New York 1966).

[48] M. Bertero, G.. Talenti, and C. A. Viano, *Nuovo Cimento* **46** (1966) 337.

[49] C. S. Warke and R. K. Bhaduri, *Nucl. Phys.* **A162** (1971) 289.

[50] Y. Singh and C. S. Warke, *Can. J. Phys.* **49** (1971) 1029.

[51] S. S. Ahmed, *Nuovo Cimento* **23A** (1974) 362.

[52] B. Bagchi, B. Mulligan, and S. B. Qadri, *Phys. Rev. C* **20** (1979) 1251.

[53] T. 0. Krause and B. Mulligan, *J. Math. Phys.* **15** (1974) 770.

[54] H. M. Nussenzveig, *Causality and Dispersion Relations* (Academic Press, New York 1972).

[55] B. Mulligan and S. B. Qadri, *Phys. Rev. C* **24** (1981) 874.

[56] M. Iwasaki and B. Mulligan, *Bull. Am. Phys. Soc.* **22** (1977) 1030; 23 (1978) 629; and to be published.

[57] F. Smithies, *Integral Equations* (Cambridge University Press, Cambridge 1958).

[58] K. Yosida, *Functional Analysis* (Spring-Verlag, Berlin 1974), pp.279-286.

[59] B. Mulligan and A. D. Wolfe, to be published.

[60] R. G. Newton, *J. Math. Phys.* **1** (1960) 319.

[61] R. G. Newton, *J. Math. Phys.* **18** (1977) 1348.

[62] B. Bagchi, B. Mulligan, and S. B. Qadri, *Prog. Theor. Phys.* **60** (1978) 765.

[63] Y. Yamaguchi, *Phys. Rev.* **95** (1954) 1628.

In: Research on Evolution Equation Compendium. Volume 2 ISBN: 978-1-60456-752-6
Editor: Gaston M. N'Guerekata

Asymptotic Behavior of the FitzHugh-Nagumo System

***Weishi Liu*[1]*and *Bixiang Wang*[2]†**
[1]Department of Mathematics, University of Kansas,
Lawrence, KS 66045, USA
[2]Department of Mathematics, New Mexico Institute of Mining and Technology
Socorro, NM 87801, USA

Abstract

For the FitzHugh-Nagumo system defined on $I\!R$, we prove the existence of a compact global attractor in a weighted Sobolev space which contains bounded solutions, in particular, traveling wave and spatially periodic solutions. We also study the behavior of the global attractors as a parameter ϵ goes to zero. Although the limiting system for $\epsilon = 0$ does not possess a bounded attracting set, we show that there exists a constant ϵ_0 such that global attractors for $0 < \epsilon \leq \epsilon_0$ are all contained in a compact subset of the phase space. Furthermore, we construct a compact local attractor for the limiting system and establish the upper semicontinuity of the global attractors of perturbed system and the local attractor of the limiting system.

AMS Subject Classification: Primary 35B40. Secondary 35B41, 35K57, 92C55
Key Words: FitzHugh-Nagumo system, asymptotic behavior, global attractor

1. Introduction

We are interested in the long-time behavior of the FitzHugh-Nagumo system on $I\!R$:

$$\frac{\partial v}{\partial t} = \frac{\partial^2 v}{\partial x^2} + h(v) - w, \tag{1.1}$$

$$\frac{\partial w}{\partial t} = \epsilon(v - \gamma w), \tag{1.2}$$

*E-mail address: wliu@math.ku.edu, supported in part by NSF Grant DMS-0406998 and a General Research Fund of University of Kansas.

†E-mail address: bwang@nmt.edu, supported in part by the Start-up Fund of New Mexico Institute of Mining and Technology.

where, ϵ and γ are positive constants, h is a given nonlinear function (typically, a cubic like function). The FitzHugh-Nagumo system was introduced as a simplification of the Hodgkin-Huxley equation that was derived as a model for the propagation of action potentials in the giant nerve axon of the squid (see, for example, [6, 10, 21]). The nonlinearity in the Hodgkin-Huxley is obtained by assuming a certain form and then fitting the nonlinearity to experimental data by nonlinear least squares or by solving an inverse problem [30], and the cubic like function h in the FitzHugh-Nagumo equation is a simplification that captures some main qualitative behavior of the Hodgkin-Huxley equation.

System (1.1)-(1.2) has been studied extensively along two major directions: the existence and stability of traveling waves, and the long-time behavior of the solutions. The existence of traveling waves for this system was studied in detail through bifurcation analysis in [8], a combination of center manifold reduction and the Lyapunov-Schmidt reduction in [25], and the geometric singular perturbation method in [14, 16]. The stability of the traveling waves was first studied in [13], where the Evan's function method was successfully developed. The method has been extended in [1, 25, 26] and becomes predominant in the investigation of stability problems. The asymptotic behavior of solutions for system (1.1)-(1.2) has been studied in [18, 19, 27]) when the system is defined in a bounded domain. In the case of an unbounded domain, the existence of a global attractor for this system was proved in [23] in the standard $L^2(I\!R^n) \times L^2(I\!R^n)$ space, which is compact in the phase space but does not contain traveling wave solutions. The existence of a global attractor in a locally uniform Sobolev space was showed in [7], which contains traveling wave solutions but is not compact with respect to the norm topology of the phase space. As a consequence, the global attractor in [7] attracts solutions only in a weak topology, but not in the norm topology, of the phase space.

In the present paper, as a necessary step toward the understanding of the global asymptotic behavior, we consider existence of global attractors and study the behavior of these attractors (at the level of invariant sets) as the parameter $\epsilon \to 0$. Note that, if ϵ is set to zero in (1.1)-(1.2), then the component w is conserved, and hence, there cannot be any bounded set that attracts all solutions of the limiting system. However, we will establish, in a suitably weighted space, the existence of a compact global attractor that contains all bounded solutions for all positive but small ϵ and show that the union of all global attractors of the perturbed system (1.1)-(1.2) for positive but small ϵ is actually contained in a compact subset in the phase space. Furthermore, we will construct a compact local attractor for the limiting system and establish the upper semicontinuity of global attractors for the perturbed system at the compact local attractor for the limiting system. Our method is based on the idea of "tail ends" estimates on solutions, by showing that the solutions of system (1.1)-(1.2) are uniformly small when space and time variables are sufficiently large. Actually, along with other things, the uniform smallness of "tail ends" of solutions for large time is not only sufficient, but also necessary for the asymptotic compactness of solution operators of dissipative evolution equations. The idea of "tail ends" estimates on solutions was used in [31] for the Reaction-Diffusion equation and in [23] for the FitzHugh-Nagumo system in the standard $L^2(I\!R^n)$ space, respectively. This paper demonstrates that the method of "tail ends" estimates is effective not only for the standard Sobolev spaces, but also for weighted Sobolev spaces. The existence of global attractors for dissipative equations on unbounded domains can be found in [2, 4, 7, 9, 11, 17, 20, 22, 24] and the references therein.

In order to include bounded solutions, the weight we choose decays with a polynomial order at infinity. A clear drawback of any non-trivial weight is that the weighted norm destroys the transition (in space) invariance and some traveling waves (those connects zero steady-state to other steady-states) approaches zero in a decaying weighted norm as $t \to \infty$. Nonetheless, some dynamics of traveling waves are kept in the weighted norm space and, most importantly, the compactness and upper semicontinuity of perturbed global attractors provide a good platform for further study of asymptotic behavior of the dynamics of the FitzHugh-Nagumo system. For example, an important question is how one defines the semi-flow of the slow dynamics–the limiting system resulting from a time re-scaling by ϵ– on the compact local attractor to capture the essence of asymptotic behavior of the system for $\epsilon > 0$ small. An answer to this question and other related ones will be our further investigation. Another interesting question is what happens if the parameter γ equals zero. Note that $\gamma > 0$ is a dissipative condition for w. While for $\gamma = 0$ the system still defines a continuous semi-group in the weighted space (Theorem 2.1), our approach does require $\gamma > 0$ for existence of compact global attractors. We believe that a certain dissipative condition for w is needed and would also like to examine the situation more carefully in the future.

The paper is organized as follows. In the next section, we describe our main results. In Section 3, the existence and uniqueness of solutions for system (1.1)-(1.2) are proved for initial data in $H^1(I\!R) \times H^1(I\!R)$. We then extend the solution operator to define a continuous dynamical system in a weighted Sobolev space. Section 4 is devoted to uniform estimates of solutions as $t \to \infty$, which are necessary to establish the point dissipativeness and asymptotic compactness of solutions. Particularly, we show that the solutions are uniformly small when space and time variables are large enough. In Section 5, we prove the existence of a global attractor for system (1.1)-(1.2) in a weighted Sobolev space. We first establish the asymptotic compactness of the solutions, and then conclude the existence of a global attractor by a standard result. The behavior of the global attractors as $\epsilon \to 0$ is discussed in the last three sections. We will show that not only global attractors are bounded uniformly in ϵ (in Section 6) but also their union for small $\epsilon > 0$ is contained in a compact subset of the phase space (in Section 7). Finally, in Section 8, we construct a compact local attractor for the limiting system and prove the upper semicontinuity of global attractors as $\epsilon \to 0$.

2. Main results

Consider the following FitzHugh-Nagumo equations:

$$\frac{\partial v}{\partial t} = \frac{\partial^2 v}{\partial x^2} + h(v) - w, \tag{2.1}$$

$$\frac{\partial w}{\partial t} = \epsilon(v - \gamma w), \tag{2.2}$$

for $(t, x) \in (0, \infty) \times I\!R$ with the initial data

$$v(0, x) = v_0(x), \quad w(0, x) = w_0(x), \quad x \in I\!R \tag{2.3}$$

where ϵ and γ are positive constants, h is a smooth function that satisfies, for some positive constants a, α and β and for all $v \in I\!R$,

$$h(0) = 0, \qquad h(v)v \leq -av^2 + \beta, \qquad h'(v) \leq \alpha. \tag{2.4}$$

As a particular example, the cubic function $h = v(1-v)(v-a)$ for $a > 0$ satisfies (2.4) with

$$\beta = \frac{27}{256}(a+1)^4 \quad \text{and} \quad \alpha = \frac{a^2 - a + 1}{3}.$$

To present our results, we introduce the notation used in the paper. Fix $\sigma > 1/2$ and define the weighted Sobolev space $L^2_\sigma(I\!R)$ as

$$L^2_\sigma(I\!R) = \left\{ u: \ u \text{ is measurable, } \int (1+|x|^2)^{-\sigma}|u(x)|^2 dx < \infty \right\}$$

with the norm

$$\|u\|_\sigma = \left(\int (1+|x|^2)^{-\sigma}|u(x)|^2 dx \right)^{1/2}.$$

We also denote $H^1_\sigma(I\!R) \subset L^2_\sigma(I\!R)$ the weighted space with the norm defined by

$$\|u\|^2_{1,\sigma} = \|u\|^2_\sigma + \|\frac{\partial u}{\partial x}\|^2_\sigma.$$

Due to the weight $(1+|x|^2)^{-\sigma}$, a function in $L^2_\sigma(I\!R)$ can have a certain growth as $|x| \to \infty$; in particular, the space $L^2_\sigma(I\!R)$ contains all bounded measurable functions such as traveling wave and spatially periodic solutions. For later purpose, we define a function ϕ_δ:

$$\phi_\delta(x) = (1+|\delta x|^2)^{-\sigma}, \quad x \in I\!R, \tag{2.5}$$

where

$$\delta = \frac{1}{2\sigma}\min\{1, a\} \text{ where } a \text{ is the number in } (2.4). \tag{2.6}$$

Then ϕ_δ satisfies the following inequalities:

$$|\frac{\partial}{\partial x}\phi_\delta(x)| \leq \delta\sigma\phi_\delta(x), \ \forall x \in I\!R, \tag{2.7}$$

and

$$\|u\|^2_\sigma \leq \int \phi_\delta(x)|u(x)|^2 dx \leq \delta^{-2\sigma}\|u\|^2_\sigma. \tag{2.8}$$

Inequality (2.8) shows that the norm $\|\cdot\|_\sigma$ is equivalent to the norm $\left(\int \phi_\delta(x)|u(x)|^2 dx\right)^{1/2}$ for $u \in L^2_\sigma(I\!R)$. As usual, we denote the norm of $L^2(I\!R)$ by $\|\cdot\|$.

We now describe the main results of the paper and delay their proofs to later sections. We emphasize that we will always assume that h satisfies (2.4) and $\sigma > 1/2$.

Our first result shows that one can define a continuous dynamical system for problem (2.1)-(2.3) in the space $L^2_\sigma(I\!R) \times L^2_\sigma(I\!R)$. More precisely, we have

Theorem 2..1. *For $\epsilon > 0$, one can associate problem (2.1)-(2.3) with a continuous semigroup $S(t)_{t\geq 0}$ in $L^2_\sigma(\mathbb{R}) \times L^2_\sigma(\mathbb{R})$ such that, when the initial datum $(v_0, w_0) \in H^1(\mathbb{R}) \times H^1(\mathbb{R})$, $S(t)(v_0, w_0)$ is the unique solution of problem (2.1)-(2.3) in the sense of (3.19)-(3.21).*

Our next result is on the existence of a global attractor of the dynamical system $S(t)_{t\geq 0}$.

Theorem 2..2. *For $\epsilon > 0$, the dynamical system $S(t)_{t\geq 0}$ has a global attractor in $L^2_\sigma(\mathbb{R}) \times L^2_\sigma(\mathbb{R})$, which is a compact invariant set and attracts every bounded set with respect to the norm topology of $L^2_\sigma(\mathbb{R}) \times L^2_\sigma(\mathbb{R})$.*

In the sequel, we denote by $\mathcal{A}_\epsilon$ the global attractor corresponding to a given ϵ. An interesting question is: what happens to the set $\mathcal{A}_\epsilon$ as $\epsilon \to 0$? Naturally, one would like to compare the dynamics for $\epsilon > 0$ with that for $\epsilon = 0$. When $\epsilon = 0$, problem (2.1)-(2.3) reduces to

$$\frac{\partial v}{\partial t} = \frac{\partial^2 v}{\partial x^2} + h(v) - w, \tag{2.9}$$

$$\frac{\partial w}{\partial t} = 0, \tag{2.10}$$

for $(t, x) \in (0, \infty) \times \mathbb{R}$ with the initial data

$$v(0, x) = v_0(x), \quad w(0, x) = w_0(x). \tag{2.11}$$

Given initial data (v_0, w_0), by (2.10), we see $w(t) = w_0$ for all $t \geq 0$. Therefore, problem (2.9)-(2.11) cannot have a bounded set that attracts all solutions in the space $L^2_\sigma(\mathbb{R}) \times L^2_\sigma(\mathbb{R})$. Interestingly enough, we will show that all global attractors $\mathcal{A}_\epsilon$ for small $\epsilon > 0$ are uniformly bounded in $L^2_\sigma(\mathbb{R}) \times L^2_\sigma(\mathbb{R})$ (in fact, in $H^1_\sigma(\mathbb{R}) \times H^1_\sigma(\mathbb{R})$).

Theorem 2..3. *The global attractors $\mathcal{A}_\epsilon$ are uniformly bounded in ϵ for $0 < \epsilon < \min\{1, \frac{a}{2\gamma}\}$ in the space $H^1_\sigma(\mathbb{R}) \times H^1_\sigma(\mathbb{R})$. More precisely, for all $0 < \epsilon < \min\{1, \frac{a}{2\gamma}\}$ and $(v, w) \in \mathcal{A}_\epsilon$,*

$$\|(v, w)\|_\sigma \leq M_1, \tag{2.12}$$

and

$$\|(v, w)\|_{1,\sigma} \leq \sqrt{M_1^2 + M_2^2}, \tag{2.13}$$

where M_1 and M_2 are given by

$$M_1 = \sqrt{\frac{\beta(a^2 + 2a\gamma + 2)}{a^2\gamma} \int \phi_\delta dx}, \tag{2.14}$$

and

$$M_2 = \frac{1}{\gamma}\sqrt{(\gamma^2 + 1)e^{\sigma^2+\sigma+2\alpha}\left(\frac{\beta(\sigma+1)}{\gamma(\sigma^2+\sigma+2\alpha)^2} + \frac{4\beta(a+1)(a\gamma+1)}{3a^2\gamma}\right)\int \phi_\delta dx}, \tag{2.15}$$

where a and β are the constants in (2.4) and δ is the constant given by (2.6).

The next result shows that the union of all global attractors $\mathcal{A}_\epsilon$ is not only bounded, but also compact in $L^2_\sigma(I\!R) \times L^2_\sigma(I\!R)$. More precisely, we have the following conclusion.

Theorem 2..4. *Let* $\epsilon_0 = \min\{1, \frac{a}{2\gamma}\}$. *Then the union* $\bigcup_{0<\epsilon<\epsilon_0} \mathcal{A}_\epsilon$ *of all global attractors is precompact in the space* $L^2_\sigma(I\!R) \times L^2_\sigma(I\!R)$.

Finally, we establish the upper semicontinuity of global attractors at $\epsilon = 0$.

Theorem 2..5. *There exists a local compact attractor* $\mathcal{A}_0$ *for the limiting system* (2.9)-(2.10) *in* $L^2_\sigma(I\!R) \times L^2_\sigma(I\!R)$ *such that*

$$\lim_{\epsilon \to 0} \mathrm{d}_{L^2_\sigma \times L^2_\sigma}(\mathcal{A}_\epsilon, \mathcal{A}_0) = 0,$$

where

$$\mathrm{d}_{L^2_\sigma \times L^2_\sigma}(\mathcal{A}_\epsilon, \mathcal{A}_0) = \sup_{a \in \mathcal{A}_\epsilon} \inf_{b \in \mathcal{A}} \|a - b\|_\sigma.$$

3. A semigroup associated with problem (2.1)-(2.3)

In this section, we prove Theorem 2..1 to establish the dynamical system $S(t)_{t \geq 0}$ associated with problem (2.1)-(2.3). To this end, we first consider the FitzHugh-Nagumo equations defined on a bounded interval $(-L, L)$ and derive some estimates on solutions uniformly in L. We then take the limit of a convergent subsequence of the solutions defined on $(-L, L)$ and show that the limit is a solution of problem (2.1)-(2.3).

We denote by $\Omega_L = (-L, L)$ and assume $(v_0, w_0) \in L^2(I\!R) \times L^2(I\!R)$. Consider the following boundary initial value problem on Ω_L:

$$\frac{\partial v}{\partial t} = \frac{\partial^2 v}{\partial x^2} + h(v) - w, \tag{3.1}$$

$$\frac{\partial w}{\partial t} = \epsilon(v - \gamma w), \tag{3.2}$$

for $(t, x) \in (0, \infty) \times \Omega_L$ with the initial data

$$v(0, x) = v_0(x), \qquad w(0, x) = w_0(x), \quad x \in \Omega_L, \tag{3.3}$$

and the boundary conditions

$$v(t, x) = 0, \quad w(t, x) = 0, \quad (t, x) \in (0, \infty) \times \partial\Omega_L. \tag{3.4}$$

It follows from known results (see, e.g., [18]) that problem (3.1)-(3.4) has a unique solution $(v, w) \in \mathcal{C}([0, \infty), L^2(\Omega_L) \times L^2(\Omega_L))$. In order to obtain a solution of problem (2.1)-(2.3), we next formally derive the uniform estimates in L for solutions of problem (3.1)-(3.4). The estimates hold for smooth functions and will become rigorous by a limiting process. We begin with the estimates in $L^2(\Omega_L) \times L^2(\Omega_L)$.

Lemma 3..1. *Suppose* $(v_0, w_0) \in L^2(I\!R) \times L^2(I\!R)$ *with* $\|(v_0, w_0)\| \leq M$. *Then, for every* $T > 0$, *there is a constant* C *depending only on* ϵ, α, T *and* M *(but independent of* L*), such that, the solution* (v, w) *of problem (3.1)-(3.4) satisfies*

$$\|v(t)\|_{L^2(\Omega_L)} + \|w(t)\|_{L^2(\Omega_L)} \leq C, \quad t \in [0, T], \tag{3.5}$$

$$\int_0^T \|v(t)\|^2_{H^1(\Omega_L)} dt \leq C. \tag{3.6}$$

In the sequel, we denote by C any positive constant which may change values from line to line.

Proof. Multiplying (3.1) and (3.2) by v and w, respectively, and integrating with respect to x over Ω_L, we obtain

$$\frac{1}{2}\frac{d}{dt}\int_{\Omega_L} |v|^2 dx + \int_{\Omega_L} |\partial_x v|^2 dx = \int_{\Omega_L} h(v) v dx - \int_{\Omega_L} vw dx, \tag{3.7}$$

and

$$\frac{1}{2}\frac{d}{dt}\int_{\Omega_L} |w|^2 dx = \epsilon \int_{\Omega_L} vw dx - \epsilon\gamma \int_{\Omega_L} |w|^2 dx. \tag{3.8}$$

It follows from (3.7) and (3.8) that

$$\begin{aligned}\frac{1}{2}\frac{d}{dt}\left(\int_{\Omega_L} |v|^2 dx \right.&+ \left.\int_{\Omega_L} |w|^2 dx\right) + \int_{\Omega_L} |\partial_x v|^2 dx \\ &= \int_{\Omega_L} h(v) v dx + (\epsilon - 1)\int_{\Omega_L} vw dx - \epsilon\gamma \int_{\Omega_L} |w|^2 dx. \end{aligned}\tag{3.9}$$

By (2.4), we have the estimates for the first term on the right-hand side of (3.9):

$$\int_{\Omega_L} h(v) v dx = \int_{\Omega_L} (h(v) - h(0))\, v dx = \int_{\Omega_L} h'(\xi)|v|^2 dx \leq \alpha \int_{\Omega_L} |v|^2 dx. \tag{3.10}$$

Substituting (3.10) into (3.9) and dropping the negative term on the right-hand side of (3.9), we find that there exists a constant C_1 depending on ϵ and α such that, for all $t \geq 0$:

$$\begin{aligned}&\frac{d}{dt}\left(\int_{\Omega_L} |v|^2 dx + \int_{\Omega_L} |w|^2 dx\right) \\ &+2\int_{\Omega_L} |\partial_x v|^2 dx \leq C_1 \left(\int_{\Omega_L} |v|^2 dx + \int_{\Omega_L} |w|^2 dx\right).\end{aligned}\tag{3.11}$$

By Gronwall's Lemma, for all $t \geq 0$, we have

$$\begin{aligned}\int_{\Omega_L} |v(t)|^2 dx + \int_{\Omega_L} |w(t)|^2 dx &\leq e^{C_1 t}\left(\int_{\Omega_L} |v_0|^2 dx + \int_{\Omega_L} |w_0|^2 dx\right) \\ &\leq e^{C_1 t}\left(\int_{I\!R} |v_0|^2 dx + \int_{I\!R} |w_0|^2 dx\right);\end{aligned}$$

in particular, for $t \in [0, T]$, the following holds:

$$\int_{\Omega_L} |v(t)|^2 dx + \int_{\Omega_L} |w(t)|^2 dx \le e^{C_1 T} M^2. \qquad (3.12)$$

Integrating (3.11) with respect to t and using (3.12), we find that there exists a constant C_2 depending on ϵ, α, T and M such that

$$\int_0^T \int_{\Omega_L} |\partial_x v|^2 dx dt \le C_2,$$

which and (3.12) conclude the proof. □

Next, we derive estimates on solutions of problem (3.1)-(3.4) in the space $H^1 \times H^1$.

Lemma 3..2. *Suppose $(v_0, w_0) \in H^1(I\!R) \times H^1(I\!R)$ with $\|(v_0, w_0)\|_{H^1 \times H^1} \le M$. Then, for every $T > 0$, there is a constant C depending only on ϵ, α, T and M (but independent of L), such that, the solution (v, w) of problem (3.1)-(3.4) satisfies*

$$\| (v(t), w(t)) \|_{H^1(\Omega_L) \times H^1(\Omega_L)} \le C, \qquad \|\frac{\partial w}{\partial t}\|_{H^1(\Omega_L)} \le C, \quad t \in [0, T], \qquad (3.13)$$

and

$$\int_0^T \|v(t)\|^2_{H^2(\Omega_L)} dt \le C, \qquad \int_0^T \|\frac{\partial v}{\partial t}\|^2_{L^2(\Omega_L)} dt \le C. \qquad (3.14)$$

Proof. Multiplying (3.1) and (3.2) by $\partial_{xx} v$ and $\partial_{xx} w$, respectively, and integrating with respect to x over Ω_L, one has

$$\frac{1}{2}\frac{d}{dt}\int_{\Omega_L} |\partial_x v|^2 dx + \int_{\Omega_L} |\partial_{xx} v|^2 dx = -\int_{\Omega_L} h(v)\partial_{xx} v dx - \int_{\Omega_L} \partial_x v \, \partial_x w dx,$$

and

$$\frac{1}{2}\frac{d}{dt}\int_{\Omega_L} |\partial_x w|^2 dx = \epsilon \int_{\Omega_L} \partial_x v \, \partial_x w dx - \epsilon\gamma \int_{\Omega_L} |\partial_x w|^2 dx.$$

Therefore,

$$\frac{1}{2}\frac{d}{dt}\left(\int_{\Omega_L} |\partial_x v|^2 dx + \int_{\Omega_L} |\partial_x w|^2 dx\right) + \int_{\Omega_L} |\partial_{xx} v|^2 dx$$
$$= -\int_{\Omega_L} h(v)\partial_{xx} v dx + (\epsilon - 1)\int_{\Omega_L} \partial_x v \, \partial_x w dx - \epsilon\gamma \int_{\Omega_L} |\partial_x w|^2 dx. \qquad (3.15)$$

Using (2.4) to estimate the first term on the right-hand side of (3.15), we obtain

$$-\int_{\Omega_L} h(v)\partial_{xx} v dx \le \alpha \int_{\Omega_L} |\partial_x v|^2 dx. \qquad (3.16)$$

Then, it follows from (3.15) and (3.16) that, for all $t \ge 0$,

$$\frac{d}{dt}\left(\int_{\Omega_L} |\partial_x v|^2 dx + \int_{\Omega_L} |\partial_x w|^2 dx\right) + 2\int_{\Omega_L} |\partial_{xx} v|^2 dx$$
$$\le C_1 \left(\int_{\Omega_L} |\partial_x v|^2 dx + \int_{\Omega_L} |\partial_x w|^2 dx\right), \qquad (3.17)$$

where C_1 is a constant depending on ϵ and α. By Gronwall's Lemma, we find that, for $t \geq 0$,

$$\begin{aligned}\int_{\Omega_L} |\partial_x v(t)|^2 dx + \int_{\Omega_L} |\partial_x w(t)|^2 dx &\leq e^{C_1 t}\left(\int_{\Omega_L} |\partial_x v_0|^2 dx + \int_{\Omega_L} |\partial_x w_0|^2 dx\right) \\ &\leq e^{C_1 t}\left(\int_{I\!R} |\partial_x v_0|^2 dx + \int_{I\!R} |\partial_x w_0|^2 dx\right),\end{aligned}$$

which shows that, for $t \in [0, T]$,

$$\int_{\Omega_L} |\partial_x v(t)|^2 dx + \int_{\Omega_L} |\partial_x w(t)|^2 dx \leq e^{C_1 T} M^2. \tag{3.18}$$

Integrating (3.17) with respect to t and using (3.18), we obtain

$$\int_0^T \int_{\Omega_L} |\partial_{xx} v|^2 dx dt \leq C,$$

which, along with (3.1)-(3.2), (3.18) and Lemma 3..1, implies (3.13) and (3.14). □

We are now in a position to establish the existence of solutions for problem (2.1)-(2.3). Suppose $(v, w) \in \mathcal{C}([0, \infty), L^2(I\!R) \times L^2(I\!R))$. Then we say (v, w) is a solution of problem (2.1)-(2.3) if, $(v(0), w(0)) = (v_0, w_0)$ and, for every $T > 0$,

$$v \in L^2((0,T), H^1(I\!R)), \ \frac{\partial v}{\partial t} \in L^2((0,T), H^{-1}(I\!R)), \ \frac{\partial w}{\partial t} \in L^2((0,T), L^2(I\!R)), \tag{3.19}$$

and (v, w) satisfies the following equalities, for any $\kappa \in C_0^\infty([0, T] \times I\!R)$:

$$\int_0^T \langle \frac{\partial v}{\partial t}, \kappa \rangle_{(H^{-1}, H^1)} dt + \int_0^T \int_{I\!R} \partial_x v \, \partial_x \kappa \, dx dt = \int_0^T \int_{I\!R} (h(v) - w)\kappa \, dx dt, \tag{3.20}$$

$$\int_0^T \int_{I\!R} \frac{\partial w}{\partial t} \kappa \, dx dt = \epsilon \int_0^T \int_{I\!R} (v - \gamma w)\kappa \, dx dt, \tag{3.21}$$

where (H^{-1}, H^1) denotes the duality pairing between $H^{-1}(I\!R)$ and $H^1(I\!R)$.

Lemma 3..3. *Suppose $(v_0, w_0) \in H^1(I\!R) \times H^1(I\!R)$. Then problem (2.1)-(2.3) has a solution (v, w) in the sense of (3.19)-(3.21). In addition,*

$$v \in L^\infty((0,T), H^1(I\!R) \cap L^\infty(I\!R)) \bigcap L^2((0,T), H^2(I\!R)), \quad \frac{\partial v}{\partial t} \in L^2((0,T), L^2(I\!R)),$$

$$w \in L^\infty((0,T), H^1(I\!R) \cap L^\infty(I\!R)), \qquad \frac{\partial w}{\partial t} \in L^\infty((0,T), H^1(I\!R)).$$

Proof. Let θ be a smooth function satisfying $0 \leq \theta(s) \leq 1$ for $s \geq 0$, and

$$\theta(s) = 1 \text{ for } 0 \leq s \leq \frac{1}{2}; \quad \theta(s) = 0 \text{ for } s \geq 1.$$

Then, given $(v_0, w_0) \in H^1(I\!R) \times H^1(I\!R)$, we denote by

$$(v_{0,L}(x), w_{0,L}(x)) = \left(\theta(\frac{|x|^2}{L^2})v_0(x),\ \theta(\frac{|x|^2}{L^2})w_0(x)\right), \qquad x \in I\!R.$$

It is evident that

$$(v_{0,L}, w_{0,L}) \longrightarrow (v_0, w_0) \text{ in } H^1(I\!R) \times H^1(I\!R) \text{ as } L \to \infty.$$

Let (v_L, w_L) be the solution of problem (3.1)-(3.4) defined on Ω_L with the initial datum $(v_{0,L}, w_{0,L})$. Then, we extend (v_L, w_L) to $I\!R$ by defining $(v_L(t,x), w_L(t,x)) = (0,0)$ for $|x| > L$. From Lemma 3..2, for any given $T > 0$, there is a constant C depending only on ϵ, α, T and $\|(v_0, w_0)\|_{H^1 \times H^1}$ (but independent of L) such that

$$\| (v_L(t), w_L(t)) \|_{H^1(I\!R)\times H^1(I\!R)} \leq C, \qquad \|\frac{\partial w_L}{\partial t}\|_{H^1(I\!R)} \leq C, \quad t \in [0,T], \tag{3.22}$$

and

$$\int_0^T \|v_L(t)\|^2_{H^2(I\!R)} dt \leq C, \qquad \int_0^T \|\frac{\partial v_L}{\partial t}\|^2_{L^2(I\!R)} dt \leq C. \tag{3.23}$$

By embedding $H^1(I\!R) \subset L^\infty(I\!R)$ and (3.22), we have

$$\|v_L(t)\|_{L^\infty(I\!R)} + \|w_L(t)\|_{L^\infty(I\!R)} \leq C, \quad t \in [0,T]. \tag{3.24}$$

It follows from (3.22)-(3.24) that there exists a pair of functions (v, w) satisfying

$$v \in L^\infty((0,T), H^1(I\!R) \cap L^\infty(I\!R)) \bigcap L^2((0,T), H^2(I\!R)), \quad \frac{\partial v}{\partial t} \in L^2((0,T), L^2(I\!R)), \tag{3.25}$$

$$w \in L^\infty((0,T), H^1(I\!R) \cap L^\infty(I\!R)), \qquad \frac{\partial w}{\partial t} \in L^\infty((0,T), H^1(I\!R)), \tag{3.26}$$

and a subsequence of (v_L, w_L) (still denoted by (v_L, w_L)) such that, as $L \to \infty$,

$$v_L \longrightarrow v, \quad \frac{\partial v_L}{\partial t} \longrightarrow \frac{\partial v}{\partial t} \quad \text{weakly in } L^2((0,T), L^2(I\!R)),$$

$$w_L \longrightarrow w, \quad \frac{\partial w_L}{\partial t} \longrightarrow \frac{\partial w}{\partial t} \quad \text{star-weakly in } L^\infty((0,T), H^1(I\!R)).$$

Next, we prove (v, w) is a solution of problem (2.1)-(2.3). For that purpose, we take a test function $\kappa \in C_0^\infty([0,T] \times I\!R)$. Then, by (3.1)-(3.2) we have

$$\int_0^T \int_{I\!R} \frac{\partial v_L}{\partial t}\, \kappa\, dxdt + \int_0^T \int_{I\!R} \partial_x v_L\, \partial_x \kappa\, dxdt = \int_0^T \int_{I\!R} (h(v_L) - w_L)\, \kappa\, dxdt, \tag{3.27}$$

$$\int_0^T \int_{I\!R} \frac{\partial w_L}{\partial t}\, \kappa\, dxdt = \epsilon \int_0^T \int_{I\!R} (v_L - \gamma w_L)\, \kappa\, dxdt. \tag{3.28}$$

We now deal with the nonlinear term on the right-hand side of (3.27). Choose a constant K large enough such that supp$(\kappa) \subseteq [0,T] \times \Omega_K$. Then, by a standard compactness argument (see, e.g., [29]), from (3.25), we infer that

$$v_L \longrightarrow v \text{ in } L^2((0,T), L^2(\Omega_K)). \tag{3.29}$$

Therefore, by (3.24)-(3.25) the following holds for the nonlinear term h:

$$\begin{aligned} |\int_0^T \int_{I\!R} h(v_L)\,\kappa\,dxdt &- \int_0^T \int_{I\!R} h(v)\,\kappa\,dxdt| \le \int_0^T \int_{I\!R} |(h(v_L)-h(v))\,\kappa|\,dxdt \\ &\le \int_0^T \int_{\Omega_K} |h'(\xi)(v_L-v)|\,|\kappa|\,dxdt \\ &\le C\int_0^T \int_{\Omega_K} |v_L-v|\,|\kappa|\,dxdt \\ &\le C\int_0^T \int_{\Omega_K} |v_L-v|^2 dxdt \int_0^T \int_{\Omega_K} |\kappa|^2\,dxdt \\ &\longrightarrow 0, \quad \text{as } L\to\infty. \end{aligned} \tag{3.30}$$

Taking the limits of (3.27)-(3.28) as $L \to \infty$, we find that (v, w) satisfies

$$\int_0^T \int_{I\!R} \frac{\partial v}{\partial t}\kappa\,dxdt + \int_0^T \int_{I\!R} \partial_x v\,\partial_x\kappa\,dxdt = \int_0^T \int_{I\!R} (h(v)-w)\kappa\,dxdt, \tag{3.31}$$

$$\int_0^T \int_{I\!R} \frac{\partial w}{\partial t}\kappa\,dxdt = \epsilon\int_0^T \int_{I\!R} (v-\gamma w)\kappa\,dxdt, \tag{3.32}$$

By a standard continuity argument (see, e.g., [28]), from (3.25)-(3.26) we can prove

$$(v,w) \in \mathcal{C}([0,T], H^1(I\!R)\times H^1(I\!R)) \quad \text{and} \quad (v(0),w(0)) = (v_0,w_0). \tag{3.33}$$

Then, by (3.24)-(3.26) and (3.31)-(3.33), Lemma 3..3 follows. □

Note that for any $\sigma > 1/2$ (in fact for any $\sigma \ge 0$), the standard space $L^2(I\!R)$ is contained in the weighted space $L^2_\sigma(I\!R)$. Thus the solutions obtained in Lemma 3..3 are actually functions in $L^2_\sigma(I\!R) \times L^2_\sigma(I\!R)$. Next, we show that the solutions are Lipschitz continuous in the space $L^2_\sigma(I\!R) \times L^2_\sigma(I\!R)$. As an immediate consequence, the solutions in Lemma 3..3 are unique.

Lemma 3..4. *Suppose $(v_0, w_0) \in H^1(I\!R) \times H^1(I\!R)$. Then problem (2.1)-(2.3) possesses a unique solution satisfying (3.19)-(3.21). Furthermore, for every $T > 0$, there exists a constant C depending only on ϵ, α, σ and T such that any two solutions (v_1, w_1) and (v_2, w_2) satisfy the inequality, for $t \in [0, T]$,*

$$\|v_1(t)-v_2(t)\|_\sigma + \|w_1(t)-w_2(t)\|_\sigma \le C\left(\|v_1(0)-v_2(0)\|_\sigma + \|w_1(0)-w_2(0)\|_\sigma\right). \tag{3.34}$$

Proof. Let $V = v_1 - v_2$ and $W = w_1 - w_2$. By Lemma 3..3, both (v_1, w_1) and (v_2, w_2) satisfy equations (2.1) and (2.2) in the distribution sense. Therefore, V and W satisfy

$$\frac{\partial V}{\partial t} = \frac{\partial^2 V}{\partial x^2} + (h(v_1)-h(v_2)) - W, \tag{3.35}$$

$$\frac{\partial W}{\partial t} = \epsilon(V-\gamma W). \tag{3.36}$$

Multiplying (3.35) by $\phi_1 V$ where ϕ_1 is the function given by (2.5) with $\delta = 1$ and integrating the resulting identity with respect to x, we obtain

$$\frac{1}{2}\frac{d}{dt}\int \phi_1 |V|^2 dx \; + \; \int \phi_1 |\partial_x V|^2 = \int (h(v_1) - h(v_2))\,\phi_1 V dx$$
$$- \; \int \partial_x \phi_1 \, (\partial_x V)\, V dx - \int \phi_1 W V dx. \qquad (3.37)$$

The first term on the right-hand side of (3.37) can be estimated from (2.4) as

$$\int (h(v_1) - h(v_2))\,\phi_1 V dx = \int h'(\xi)\phi_1 |V|^2 dx \le \alpha \int \phi_1 |V|^2 dx. \qquad (3.38)$$

Using (2.7), the second term on the right-hand side of (3.37) satisfies

$$|\int \partial_x \phi_1 \,(\partial_x V)\, V dx| \le \sigma \int \phi_1 |\partial_x V|\, |V| dx \le \frac{1}{2}\int \phi_1 |\partial_x V|^2 dx + \frac{1}{2}\sigma^2 \int \phi_1 |V|^2 dx. \qquad (3.39)$$

For the last term on the right-hand side of (3.37), we have

$$|\int \phi_1 W V dx| \le \frac{1}{2}\int \phi_1 |W|^2 dx + \frac{1}{2}\int \phi_1 |V|^2 dx. \qquad (3.40)$$

It follows from (3.37)-(3.40) that there is a constant C_1 depending on α and σ such that

$$\frac{d}{dt}\int \phi_1 |V|^2 dx \le C_1 \left(\int \phi_1 |V|^2 dx + \int \phi_1 |W|^2 dx \right). \qquad (3.41)$$

Similarly, multiplying (3.36) by $\phi_1 W$ and integrating, we find there is a constant C_2 depending on ϵ such that

$$\frac{d}{dt}\int \phi_1 |W|^2 dx \le C_2 \left(\int \phi_1 |V|^2 dx + \int \phi_1 |W|^2 dx \right). \qquad (3.42)$$

Therefore, we have

$$\frac{d}{dt}\left(\int \phi_1 |V|^2 dx + \int \phi_1 |W|^2 \right) \le C \left(\int \phi_1 |V|^2 dx + \int \phi_1 |W|^2 dx \right).$$

Thus, (3.34) follows from Gronwall's inequality. The proof is complete. □

Next, we prove Theorem 2..1 and define a dynamical system for problem (2.1)-(2.3).

Proof of Theorem 2..1. Given $T > 0$, by Lemma 3..4, there is a mapping $\mathcal{G}$ from $H^1(I\!R) \times H^1(I\!R)$ into $\mathcal{C}([0,T], L^2_\sigma(I\!R) \times L^2_\sigma(I\!R))$ such that for each $(v_0, w_0) \in H^1(I\!R) \times H^1(I\!R)$, $\mathcal{G}(v_0, w_0)$ is the unique solution of problem (2.1)-(2.3) in the sense of (3.19)-(3.21). It follows from (3.34) that the mapping $\mathcal{G}$ is continuous from $H^1(I\!R) \times H^1(I\!R) \subseteq L^2_\sigma(I\!R) \times L^2_\sigma(I\!R)$ into $\mathcal{C}([0,T], L^2_\sigma(I\!R) \times L^2_\sigma(I\!R))$. Since $H^1(I\!R) \times H^1(I\!R)$ is a dense subset of $L^2_\sigma(I\!R) \times L^2_\sigma(I\!R)$, $\mathcal{G}$ can be extended uniquely to a mapping $\tilde{\mathcal{G}}$ from $L^2_\sigma(I\!R) \times L^2_\sigma(I\!R)$ into $\mathcal{C}([0,T], L^2_\sigma(I\!R) \times L^2_\sigma(I\!R))$. We then define a semigroup $S(t)_{t\ge 0} : L^2_\sigma(I\!R) \times L^2_\sigma(I\!R) \to L^2_\sigma(I\!R) \times L^2_\sigma(I\!R)$ such that, for every $t \ge 0$ and for $(v_0, w_0) \in L^2_\sigma(I\!R) \times L^2_\sigma(I\!R)$, $S(t)(v_0, w_0) = \tilde{\mathcal{G}}(v_0, w_0)(t)$. It follows from (3.34) that $S(t)_{t\ge 0}$ is a continuous semigroup. □

4. Uniform estimates in time

In this section, we derive *a priori* estimates on solutions of problem (2.1)-(2.3) as $t \to \infty$. We first establish uniform estimates in time in the space $L^2_\sigma(I\!R) \times L^2_\sigma(I\!R)$ and then improve the estimates in $H^1_\sigma(I\!R) \times H^1_\sigma(I\!R)$. Finally, we show that all solutions are uniformly small when space and time variables are sufficiently large.

Lemma 4..1. *Suppose* $\|(v_0, w_0)\|_\sigma \le R$ *and let* $(v(t), w(t)) = S(t)(v_0, w_0)$. *Then,*

$$\|(v(t), w(t))\|_\sigma \le M, \quad \textit{for all } t \ge T, \tag{4.1}$$

and

$$\int_t^{t+d} \|\partial_x v(t)\|^2_\sigma dt \le C(1+d), \quad \textit{for any } \ t \ge T \textit{ and } d > 0, \tag{4.2}$$

where M is a constant depending only on the data $(\epsilon, \gamma, a, \beta, \sigma)$, T depending on the data $(\epsilon, \gamma, a, \beta, \sigma)$ and R, while C depending on the data $(\epsilon, \gamma, a, \beta, \sigma)$, but not on d.

Proof. Multiplying (2.1) and (2.2) by $\epsilon\phi_\delta v$ and $\phi_\delta w$, respectively, and integrating with respect to x, we find

$$\frac{1}{2}\frac{d}{dt}\int_{I\!R} \epsilon\phi_\delta |v|^2 dx + \epsilon\int_{I\!R} v\partial_x v\, \partial_x\phi_\delta dx + \epsilon\int_{I\!R} \phi_\delta|\partial_x v|^2 dx = \epsilon\int_{I\!R}\phi_\delta h(v)v dx - \epsilon\int_{I\!R}\phi_\delta w v dx,$$

and

$$\frac{1}{2}\frac{d}{dt}\int_{I\!R}\phi_\delta|w|^2 dx + \epsilon\gamma\int_{I\!R}\phi_\delta|w|^2 dx = \epsilon\int_{I\!R}\phi_\delta v w dx.$$

Therefore,

$$\begin{aligned}\frac{1}{2}\frac{d}{dt}\left(\int_{I\!R}\epsilon\phi_\delta|v|^2 dx + \int_{I\!R}\phi_\delta|w|^2 dx\right) &+ \epsilon\int_{I\!R} v\partial_x v\, \partial_x\phi_\delta dx + \epsilon\int_{I\!R}\phi_\delta|\partial_x v|^2 dx \\ &+ \epsilon\gamma\int_{I\!R}\phi_\delta|w|^2 dx = \epsilon\int_{I\!R}\phi_\delta h(v) v dx. \end{aligned} \tag{4.3}$$

Since h satisfies (2.4), we have the following estimates for the right-hand side of (4.3):

$$\epsilon\int_{I\!R}\phi_\delta h(v)v dx \le -\epsilon a\int_{I\!R}\phi_\delta|v|^2 dx + \epsilon\beta\int_{I\!R}\phi_\delta dx. \tag{4.4}$$

By (4.3) and (4.4), we get

$$\begin{aligned}\frac{1}{2}\frac{d}{dt}\left(\int_{I\!R}\epsilon\phi_\delta|v|^2 dx + \int_{I\!R}\phi_\delta|w|^2 dx\right) &+ \epsilon\int_{I\!R}\phi_\delta|\partial_x v|^2 dx \\ &+ \epsilon a\int_{I\!R}\phi_\delta|v|^2 dx + \epsilon\gamma\int_{I\!R}\phi_\delta|w|^2 dx \\ &\le \epsilon\beta\int_{I\!R}\phi_\delta dx - \epsilon\int_{I\!R} v\partial_x v\, \partial_x\phi_\delta dx. \end{aligned} \tag{4.5}$$

Next, we estimate the last term on the right-hand side of (4.5). By (2.6) and (2.7) we have

$$\begin{aligned}|\epsilon \int_{\boldsymbol{R}} v\partial_x v\, \partial_x\phi_\delta dx| &\leq \epsilon\delta\sigma \int_{\boldsymbol{R}} |v||\partial_x v|\phi_\delta dx \leq \frac{1}{2}\epsilon\delta\sigma \left(\int_{\boldsymbol{R}} \phi_\delta v^2 dx + \int_{\boldsymbol{R}} \phi_\delta (\partial_x v)^2 dx\right) \\ &\leq \frac{1}{4}\epsilon a \int_{\boldsymbol{R}} \phi_\delta v^2 dx + \frac{1}{4}\epsilon \int_{\boldsymbol{R}} \phi_\delta (\partial_x v)^2 dx. \qquad (4.6)\end{aligned}$$

Note that the last inequality is obtained by (2.6). From (4.5)-(4.6), it follows that

$$\begin{aligned}\frac{1}{2}\frac{d}{dt}\left(\int_{\boldsymbol{R}} \epsilon\phi_\delta |v|^2 dx \right. &+ \left. \int_{\boldsymbol{R}} \phi_\delta |w|^2 dx\right) + \frac{1}{2}\epsilon \int_{\boldsymbol{R}} \phi_\delta |\partial_x v|^2 dx \\ &+ \frac{1}{2}\epsilon a \int_{\boldsymbol{R}} \phi_\delta |v|^2 dx + \epsilon\gamma \int_{\boldsymbol{R}} \phi_\delta |w|^2 dx \leq \epsilon\beta \int_{\boldsymbol{R}} \phi_\delta dx. \qquad (4.7)\end{aligned}$$

Let $C_0 = \min\{a, 2\epsilon\gamma\}$ and $C_1 = 2\epsilon\beta \int_{\boldsymbol{R}} \phi_\delta dx$. Then from (4.7) we get

$$\frac{d}{dt}\left(\int_{\boldsymbol{R}} \epsilon\phi_\delta |v|^2 dx + \int_{\boldsymbol{R}} \phi_\delta |w|^2 dx\right) + C_0 \left(\int_{\boldsymbol{R}} \epsilon\phi_\delta |v|^2 dx + \int_{\boldsymbol{R}} \phi_\delta |w|^2 dx\right) \leq C_1, \quad (4.8)$$

Applying Gronwall's inequality, by $\|(v_0, w_0)\|_\sigma \leq R$ and (2.8), we obtain

$$\begin{aligned}\int_{\boldsymbol{R}} \epsilon\phi_\delta |v(t)|^2 dx + \int_{\boldsymbol{R}} \phi_\delta |w(t)|^2 dx &\leq e^{-C_0 t}\left(\int_{\boldsymbol{R}} \epsilon\phi_\delta |v_0|^2 dx + \int_{\boldsymbol{R}} \phi_\delta |w_0|^2 dx\right) + \frac{C_1}{C_0} \\ &\leq \delta^{-2\sigma} \max\{\epsilon, 1\} e^{-C_0 t} \|(v_0, w_0)\|_\sigma^2 + \frac{C_1}{C_0} \qquad (4.9) \\ &\leq \delta^{-2\sigma} R^2 \max\{\epsilon, 1\} e^{-C_0 t} + \frac{C_1}{C_0} \leq \frac{2C_1}{C_0}, \qquad (4.10)\end{aligned}$$

for $t \geq T = \frac{1}{C_0}(\ln(C_0\delta^{-2\sigma}R^2 \max\{\epsilon, 1\}) - \ln C_1)$. Note that (4.10) implies (4.1). Integrating (4.7) with respect to t, (4.2) follows from (4.10). The proof is complete. □

Notice that if $\epsilon \leq \min\{1, \frac{a}{2\gamma}\}$, then the above constant $C_0 = \min\{a, 2\epsilon\gamma\} = 2\epsilon\gamma$. In this case, for all $t \geq 0$, it follows from (4.9) that

$$\int_{\boldsymbol{R}} \epsilon\phi_\delta |v|^2 dx + \int_{\boldsymbol{R}} \phi_\delta |w|^2 dx \leq \delta^{-2\sigma} e^{-2\epsilon\gamma t} \left(\|v_0\|_\sigma^2 + \|w_0\|_\sigma^2\right) + \frac{\beta}{\gamma} \int_{\boldsymbol{R}} \phi_\delta dx, \quad (4.11)$$

which proves useful to establish uniform bounds of global attractors, independent of small ϵ, in Section 6.

In the sequel, we denote by B the set:

$$B = \left\{(v, w) \in L^2_\sigma(I\!R) \times L^2_\sigma(I\!R) : \ \|(v, w)\|_\sigma \leq M\right\}, \quad (4.12)$$

where M is the constant in (4.1). Then Lemma 4..1 implies the set B is an absorbing set for the dynamical system $S(t)_{t\geq 0}$. Since B itself is bounded, it follows from Lemma 4..1 that there exists a constant T_B depending on the data $(\epsilon, \gamma, a, \beta, \sigma)$ such that

$$S(t)B \subseteq B, \quad \text{for all } t \geq T_B, \quad (4.13)$$

which shows that B is positively invariant for large time, and is useful when we investigate the large-time behavior of solutions.

Next, we estimate the component v of the solution in $H^1_\sigma(I\!R)$.

Lemma 4..2. *Suppose* $\|(v_0, w_0)\|_\sigma \leq R$ *and let* $(v(t), w(t)) = S(t)(v_0, w_0)$. *Then we have*

$$\|\partial_x v(t)\|_\sigma \leq M, \quad \textit{for all } t \geq T, \tag{4.14}$$

where M *is a constant depending only on the data* $(\epsilon, \gamma, a, \beta, \alpha, \sigma)$, T *depending on the data* $(\epsilon, \gamma, a, \beta, \sigma)$ *and* R.

Proof. Multiplying (2.1) by $\phi_1 \partial_{xx} v$ and integrating, we get

$$\int \phi_1 \frac{\partial v}{\partial t}\, \partial_{xx} v dx = \int \phi_1 |\partial_{xx} v|^2 dx + \int \phi_1 h(v) \partial_{xx} v dx - \int \phi_1 w \partial_{xx} v dx. \tag{4.15}$$

Note that

$$\int \phi_1 \frac{\partial v}{\partial t}\, \partial_{xx} v dx = -\frac{1}{2}\frac{d}{dt} \int \phi_1 |\partial_x v|^2 dx - \int \frac{\partial v}{\partial t}\, \partial_x \phi_1\, \partial_x v dx, \tag{4.16}$$

and

$$\int \phi_1 h(v) \partial_{xx} v dx = -\int h(v) \partial_x \phi_1\, \partial_x v dx - \int \phi_1 \partial_x (h(v))\, \partial_x v dx. \tag{4.17}$$

Then, it follows from (4.15)-(4.17) that

$$\begin{aligned}\frac{1}{2}\frac{d}{dt} \int \phi_1 |\partial_x v|^2 dx + \int \phi_1 |\partial_{xx} v|^2 dx = &-\int \left(\frac{\partial v}{\partial t} - h(v)\right) \partial_x \phi_1\, \partial_x v dx \\ &+ \int h' \phi_1 |\partial_x v|^2 dx + \int \phi_1 w\, \partial_{xx} v dx.\end{aligned} \tag{4.18}$$

By (2.1) we have

$$\frac{\partial v}{\partial t} - h(v) = \partial_{xx} v - w. \tag{4.19}$$

Using (2.4) and (4.19), from (4.18) we find

$$\begin{aligned}\frac{1}{2}\frac{d}{dt} \int \phi_1 |\partial_x v|^2 dx + \int \phi_1 |\partial_{xx} v|^2 dx \leq &-\int \partial_{xx} v\, \partial_x \phi_1\, \partial_x v dx \\ &+ \int w \partial_x \phi_1\, \partial_x v dx + \alpha \int \phi_1 |\partial_x v|^2 dx + \int \phi_1 w\, \partial_{xx} v dx.\end{aligned} \tag{4.20}$$

By (2.7) we have

$$\begin{aligned}\left|\int \partial_{xx} v\, \partial_x \phi_1\, \partial_x v dx\right| \leq \sigma \int |\partial_{xx} v|\, |\phi_1|\, |\partial_x v| dx \leq &\frac{1}{2} \int \phi_1 |\partial_{xx} v|^2 dx \\ &+ \frac{1}{2}\sigma^2 \int \phi_1 |\partial_x v|^2 dx.\end{aligned} \tag{4.21}$$

The second and the last term on the right-hand side of (4.20) satisfy

$$\left|\int w \partial_x \phi_1\, \partial_x v dx\right| \leq \sigma \int |\phi_1||w||\partial_x v| dx \leq \frac{1}{2}\sigma \int \phi_1 w^2 dx + \frac{1}{2}\sigma \int \phi_1 |\partial_x v|^2 dx, \tag{4.22}$$

and

$$|\int \phi_1 w\ \partial_{xx} v dx| \le \frac{1}{2}\int \phi_1 |w|^2 dx + \frac{1}{2}\int \phi_1 |\partial_{xx} v|^2 dx. \qquad (4.23)$$

By (4.20)-(4.23) we get

$$\frac{1}{2}\frac{d}{dt}\int \phi_1 |\partial_x v|^2 dx \le \left(\frac{1}{2}\sigma^2 + \frac{1}{2}\sigma + \alpha\right)\int \phi_1 |\partial_x v|^2 dx + \frac{1}{2}(\sigma+1)\int \phi_1 |w|^2 dx,$$

that is,

$$\frac{d}{dt}\|\partial_x v\|_\sigma^2 \le \left(\sigma^2 + \sigma + 2\alpha\right)\|\partial_x v\|_\sigma^2 + (\sigma+1)\|w\|_\sigma^2. \qquad (4.24)$$

Applying the uniform Gronwall's Lemma (see, e.g., [28]), by (4.1)-(4.2) we obtain $\|\partial_x v\|_\sigma^2 \le C$, for all $t \ge T+1$, where T is the constant in (4.1). □

Next, we improve the estimates given by (4.1) for the component w of the solution. Since equation (2.2) has no smoothing effect on w, we cannot obtain an estimate for w in the space $H^1_\sigma(I\!R)$ if the initial datum $w_0 \in L^2_\sigma(I\!R)$. Nevertheless, we can prove w is a sum of two functions: one is regular in the sense it belongs to $H^1_\sigma(I\!R)$ and the other converges to zero uniformly as $t \to \infty$. This decomposition technique has been used by several authors for the FitzHugh-Nagumo equations (see, for example [18]). We split w as $w = w_1 + w_2$ where w_1 is the solution of the initial value problem

$$\frac{\partial w_1}{\partial t} = -\epsilon\gamma w_1, \quad w_1(0) = w_0, \qquad (4.25)$$

and w_2 is the solution of

$$\frac{\partial w_2}{\partial t} = -\epsilon\gamma w_2 + \epsilon v, \quad w_2(0) = 0. \qquad (4.26)$$

It is evident the following energy equation holds for w_1:

$$\|w_1\|_\sigma = e^{-\epsilon\gamma t}\|w_0\|_\sigma, \quad \text{for all } t \ge 0, \qquad (4.27)$$

which implies that w_1 converges to zero uniformly in bounded initial data when $t \to \infty$. Next, we show that w_2 is bounded in the space $H^1_\sigma(I\!R)$ for large time.

Lemma 4..3. *w_2 satisfies*

$$\|\partial_x w_2(t)\|_\sigma \le M, \quad \textit{for all } t \ge T, \qquad (4.28)$$

where M is a constant depending only on the data $(\epsilon, \gamma, a, \beta, \alpha, \sigma)$, T depending on the data $(\epsilon, \gamma, a, \beta, \sigma)$ and R when $\|(v_0, w_0)\|_\sigma \le R$.

Proof. Multiplying (4.26) by $\phi_1 \partial_{xx} w_2$ and then integrating, we get

$$\int \phi_1 \frac{\partial w_2}{\partial t}\ \partial_{xx} w_2 dx = -\epsilon\gamma\int w_2 \phi_1 \partial_{xx} w_2 dx + \epsilon\int v\phi_1 \partial_{xx} w_2 dx. \qquad (4.29)$$

Note that the left-hand side of (4.29) can be rewritten as

$$\int \phi_1 \frac{\partial w_2}{\partial t}\ \partial_{xx} w_2 dx = -\frac{1}{2}\frac{d}{dt}\int \phi_1 |\partial_x w_2|^2 dx - \int \frac{\partial w_2}{\partial t}\ \partial_x \phi_1\ \partial_x w_2 dx. \qquad (4.30)$$

We now deal with the right-hand side of (4.29). For the first term on the right-hand side, we have

$$-\epsilon\gamma \int w_2\phi_1\partial_{xx}w_2 dx = \epsilon\gamma \int \phi_1|\partial_x w_2|^2 dx + \epsilon\gamma \int w_2\partial_x\phi_1\ \partial_x w_2 dx. \tag{4.31}$$

Integrating by parts, we get the identity for the last term on the right-hand side:

$$\epsilon \int v\phi_1\partial_{xx}w_2 dx = -\epsilon \int \phi_1\partial_x v\ \partial_x w_2 dx - \epsilon \int v\partial_x\phi_1\ \partial_x w_2 dx. \tag{4.32}$$

By (4.29)-(4.32) we find

$$\begin{aligned}\frac{1}{2}\frac{d}{dt}\int \phi_1|\partial_x w_2|^2 dx \quad &+ \quad \epsilon\gamma \int \phi_1|\partial_x w_2|^2 dx = \epsilon \int \phi_1\partial_x v\partial_x w_2 dx \\ &- \quad \int \left(\frac{\partial w_2}{\partial t} + \epsilon\gamma w_2 - \epsilon v\right)\partial_x\phi_1\partial_x w_2 dx. \end{aligned}\tag{4.33}$$

By (4.26), the last term on the right-hand side of (4.33) is zero. While, the first term on the right can be estimated as follows.

$$\epsilon \int \phi_1\partial_x v\ \partial_x w_2 dx \le \frac{1}{2}\epsilon\gamma \int \phi_1|\partial_x w_2|^2 dx + \frac{\epsilon}{2\gamma}\int \phi_1|\partial_x v|^2 dx. \tag{4.34}$$

From (4.33) and (4.34) we obtain

$$\frac{d}{dt}\int \phi_1|\partial_x w_2|^2 + \epsilon\gamma \int \phi_1|\partial_x w_2|^2 dx \quad \le \quad \frac{\epsilon}{\gamma}\int \phi_1|\partial_x v|^2 dx \le C, \quad \text{for } t \ge T. \tag{4.35}$$

where the last inequality is obtained by (4.14). Then, (4.28) follows from (4.35) and Gronwall's Lemma. □

In what follows, we derive estimates on solutions for large time and large space variables. Actually, we show that the solutions are uniformly small for large time in the complement of a bounded set, which is crucial for the asymptotic compactness of the dynamical system $S(t)_{t\ge 0}$ associated with problem (2.1)-2.3). The uniform estimates outside a bounded set are stated as follows.

Lemma 4..4. *For every $\eta > 0$, there exists $T(\eta)$ depending only on η, $(\epsilon, \gamma, a, \beta, \sigma)$ and R when $\|(v_0, w_0)\|_\sigma \le R$, such that, for all $t \ge T(\eta)$, the following inequality holds:*

$$\int_{|x|\ge K(\eta)} \left(\phi_1|v|^2 + \phi_1|w|^2\right) dx \le \eta, \tag{4.36}$$

where $K(\eta)$ depends only on η and $(\epsilon, \gamma, a, \beta, \alpha, \sigma)$.

Proof. Let θ be a smooth function satisfying $0 \le \theta(s) \le 1$ for $s \ge 0$, and

$$\theta(s) = 0 \text{ for } 0 \le s \le 1; \quad \theta(s) = 1 \text{ for } s \ge 2.$$

Then, there is a positive constant C such that $|\theta'(s)| \le C$ for all $s \ge 0$. Multiplying (2.1) by $\epsilon\theta(\frac{|x|^2}{k^2})\phi_\delta v$ and then integrating, we find

$$\begin{aligned}\frac{1}{2}\frac{d}{dt}\int \epsilon\theta(\frac{|x|^2}{k^2})\phi_\delta |v|^2 dx &= \epsilon\int \partial_{xx}v\,\theta(\frac{|x|^2}{k^2})\phi_\delta v dx \\ &+ \epsilon\int \theta(\frac{|x|^2}{k^2})\phi_\delta h(v) v dx - \epsilon\int \theta(\frac{|x|^2}{k^2})\phi_\delta w v dx \quad (4.37)\end{aligned}$$

Multiplying (2.2) by $\theta(\frac{|x|^2}{k^2})\phi_\delta w$ and integrating, we get

$$\frac{1}{2}\frac{d}{dt}\int \theta(\frac{|x|^2}{k^2})\phi_\delta |w|^2 dx = \epsilon\int \theta(\frac{|x|^2}{k^2})\phi_\delta v w dx - \epsilon\gamma\int \theta(\frac{|x|^2}{k^2})\phi_\delta |w|^2 dx \quad (4.38)$$

By integration by parts, we have the following identity for the first term on the right-hand side of (4.37):

$$\begin{aligned}\epsilon\int \partial_{xx}v\,\theta(\frac{|x|^2}{k^2})\phi_\delta v dx &= -\epsilon\int \theta(\frac{|x|^2}{k^2})\phi_\delta |\partial_x v|^2 dx - \epsilon\int \theta(\frac{|x|^2}{k^2}) v\,\partial_x v\,\partial_x\phi_\delta dx \\ &- \epsilon\int \phi_\delta v\,\partial_x v\,\partial_x\left(\theta(\frac{|x|^2}{k^2})\right) dx. \quad (4.39)\end{aligned}$$

By (4.37)-(4.39) we obtain

$$\begin{aligned}&\frac{1}{2}\frac{d}{dt}\left(\int \epsilon\theta(\frac{|x|^2}{k^2})\phi_\delta |v|^2 dx + \int \theta(\frac{|x|^2}{k^2})\phi_\delta |w|^2 dx\right) \\ &+\epsilon\int \theta(\frac{|x|^2}{k^2})\phi_\delta |\partial_x v|^2 dx + \epsilon\gamma\int \theta(\frac{|x|^2}{k^2})\phi_\delta |w|^2 dx \\ &= \epsilon\int \theta(\frac{|x|^2}{k^2})\phi_\delta h(v) v dx - \epsilon\int \theta(\frac{|x|^2}{k^2}) v\,\partial_x v\,\partial_x\phi_\delta dx \\ &- \epsilon\int \phi_\delta v\,\partial_x v\,\partial_x\left(\theta(\frac{|x|^2}{k^2})\right) dx. \quad (4.40)\end{aligned}$$

Using (2.4) we have

$$\begin{aligned}&\frac{1}{2}\frac{d}{dt}\left(\int \epsilon\theta(\frac{|x|^2}{k^2})\phi_\delta |v|^2 dx + \int \theta(\frac{|x|^2}{k^2})\phi_\delta |w|^2 dx\right) \\ &+\epsilon\int \theta(\frac{|x|^2}{k^2})\phi_\delta |\partial_x v|^2 dx + \epsilon a\int \theta(\frac{|x|^2}{k^2})\phi_\delta |v|^2 dx + \epsilon\gamma\int \theta(\frac{|x|^2}{k^2})\phi_\delta |w|^2 dx \\ &\le \epsilon\beta\int \theta(\frac{|x|^2}{k^2})\phi_\delta dx - \epsilon\int \theta(\frac{|x|^2}{k^2}) v\,\partial_x v\,\partial_x\phi_\delta dx - \epsilon\int \phi_\delta v\,\partial_x v\,\partial_x\left(\theta(\frac{|x|^2}{k^2})\right) dx. \quad (4.41)\end{aligned}$$

Next, we estimate terms on the right-hand side of (4.41). By (2.7) and (2.6), we have the following bounds for the second term on the right-hand side.

$$
\begin{aligned}
|\epsilon \int \theta(\frac{|x|^2}{k^2}) v\, \partial_x v\, \partial_x \phi_\delta dx| &\leq \delta\sigma\epsilon \int \theta(\frac{|x|^2}{k^2}) \phi_\delta |v||\partial_x v| dx \\
&\leq \frac{1}{4}\epsilon \int \theta(\frac{|x|^2}{k^2}) \phi_\delta |\partial_x v|^2 dx \\
&+ \frac{1}{4}\epsilon a \int \theta(\frac{|x|^2}{k^2}) \phi_\delta |v|^2 dx. \qquad (4.42)
\end{aligned}
$$

By properties of θ, the last term on the right-hand side of (4.41) is bounded by

$$
\begin{aligned}
&|\epsilon \int \phi_\delta v\, \partial_x v\, \partial_x \left(\theta(\frac{|x|^2}{k^2})\right) dx| = |\int \epsilon \phi_\delta v\, \partial_x v\, \theta'(\frac{|x|^2}{k^2})\, \frac{2x}{k^2} dx| \\
&\leq \frac{\sqrt{2}\epsilon C_1}{k} \int_{k \leq |x| \leq \sqrt{2}k} |\phi_\delta||v||\partial_x v| dx \leq \frac{\sqrt{2}\epsilon C_1}{k} \|\partial_x v\|_\sigma \|v\|_\sigma \leq \frac{\epsilon C_2}{k}, \qquad (4.43)
\end{aligned}
$$

for all $t \geq T$, where the last inequality is obtained by Lemmas 4..1 and 4..2, C_2 and T are constants independent of k. By (4.42) and (4.43), it follows from (4.41) that

$$
\begin{aligned}
&\frac{1}{2}\frac{d}{dt}\left(\int \epsilon\theta(\frac{|x|^2}{k^2})\phi_\delta |v|^2 dx + \int \theta(\frac{|x|^2}{k^2})\phi_\delta |w|^2 dx\right) \\
&+\frac{1}{2}\epsilon \int \theta(\frac{|x|^2}{k^2})\phi_\delta |\partial_x v|^2 dx + \frac{1}{2}\epsilon a \int \theta(\frac{|x|^2}{k^2})\phi_\delta |v|^2 dx + \epsilon\gamma \int \theta(\frac{|x|^2}{k^2})\phi_\delta |w|^2 dx \\
&\leq \epsilon\beta \int \theta(\frac{|x|^2}{k^2})\phi_\delta dx + \frac{\epsilon C_2}{k}, \quad \text{for all } t \geq T. \qquad (4.44)
\end{aligned}
$$

Let $C_3 = \min\{a, 2\epsilon\gamma\}$. Then we get from (4.44) that, for all $t \geq T$,

$$
\begin{aligned}
&\frac{d}{dt}\left(\int \epsilon\theta(\frac{|x|^2}{k^2})\phi_\delta |v|^2 dx + \int \theta(\frac{|x|^2}{k^2})\phi_\delta |w|^2 dx\right) \\
&+ C_3 \left(\int \epsilon\theta(\frac{|x|^2}{k^2})\phi_\delta |v|^2 dx + \int \theta(\frac{|x|^2}{k^2})\phi_\delta |w|^2 dx\right) \\
&\leq 2\epsilon\beta \int \theta(\frac{|x|^2}{k^2})\phi_\delta dx + \frac{2\epsilon C_2}{k}. \qquad (4.45)
\end{aligned}
$$

Next, we manipulate the first term on the right-hand side of (4.45), which is bounded by

$$
|2\epsilon\beta \int \theta(\frac{|x|^2}{k^2})\phi_\delta dx| \leq 2\epsilon\beta \int_{|x|\geq k} \theta(\frac{|x|^2}{k^2})\phi_\delta dx \leq 2\epsilon\beta \int_{|x|\geq k} \phi_\delta dx. \qquad (4.46)
$$

Therefore, by (4.45) we find there exists a constant $K(\eta)$ such that for $k \geq K(\eta)$ and $t \geq T$,

$$
\begin{aligned}
&\frac{d}{dt}\left(\int \epsilon\theta(\frac{|x|^2}{k^2})\phi_\delta |v|^2 dx + \int \theta(\frac{|x|^2}{k^2})\phi_\delta |w|^2 dx\right) \\
&+C_3 \left(\int \epsilon\theta(\frac{|x|^2}{k^2})\phi_\delta |v|^2 dx + \int \theta(\frac{|x|^2}{k^2})\phi_\delta |w|^2 dx\right) \leq 2\epsilon\eta. \qquad (4.47)
\end{aligned}
$$

By Gronwall's Lemma, for all $t \geq T$ and $k \geq K(\eta)$, we get

$$\begin{aligned}
&\int_{\mathbb{R}} \epsilon\theta(\frac{|x|^2}{k^2})\phi_\delta|v(t)|^2dx + \int_{\mathbb{R}} \theta(\frac{|x|^2}{k^2})\phi_\delta|w(t)|^2dx \\
&\quad \leq e^{-C_3(t-T)}\left(\int_{\mathbb{R}} \epsilon\theta(\frac{|x|^2}{k^2})\phi_\delta|v(T)|^2dx + \int_{\mathbb{R}} \theta(\frac{|x|^2}{k^2})\phi_\delta|w(T)|^2dx\right) + \frac{2\epsilon\eta}{C_3} \\
&\quad \leq C_4e^{-C_3(t-T)} + \frac{2\epsilon\eta}{C_3}. \qquad (4.48)
\end{aligned}$$

Note the last inequality is obtained by Lemma 4..1. Let $T_1(\eta) = T + 1/C_3 \ln \frac{C_3C_4}{\epsilon\eta}$. Then, by (4.48) we obtain, for $t \geq T_1(\eta)$ and $k \geq K(\eta)$,

$$\int_{\mathbb{R}} \epsilon\theta(\frac{|x|^2}{k^2})\phi_\delta|v(t)|^2dx + \int_{\mathbb{R}} \theta(\frac{|x|^2}{k^2})\phi_\delta|w(t)|^2dx \leq \frac{3\epsilon\eta}{C_3},$$

which implies (4.36) since $\theta(s) = 1$ for $s \geq 2$. The proof is complete. □

As an immediate consequence of Lemma 4..4 and (4.27), we have the following estimates for (v, w_2), which are useful when we establish the asymptotic compactness of the solution in the next section.

Corollary 4..5. *For every $\eta > 0$, there exists $T(\eta)$ depending only on η, $(\epsilon, \gamma, a, \beta, \sigma)$ and R when $\|(v_0, w_0)\|_\sigma \leq R$, such that, for all $t \geq T(\eta)$, the following inequality holds:*

$$\int_{|x|\geq K(\eta)} \left(\phi_1|v|^2 + \phi_1|w_2|^2\right) dx \leq \eta,$$

where $K(\eta)$ depends only on η and $(\epsilon, \gamma, a, \beta, \alpha, \sigma)$.

5. Existence of global attractors

In this section, we show that the dynamical system $S(t)_{t\geq 0}$ has a global attractor in the space $L^2_\sigma(\mathbb{R}) \times L^2_\sigma(\mathbb{R})$. It is known that the global attractor exists if $S(t)_{t\geq 0}$ is asymptotically compact and has a bounded absorbing set. We have already proved that $S(t)_{t\geq 0}$ has a bounded absorbing set, which is given by (4.12). The next lemma establishes the asymptotic compactness of $S(t)_{t\geq 0}$.

Lemma 5..1. *The dynamical system $S(t)_{t\geq 0}$ is asymptotically compact, that is, if (v_n, w_n) is bounded in $L^2_\sigma(\mathbb{R})\times L^2_\sigma(\mathbb{R})$ and $t_n \to \infty$, then $S(t_n)(v_n, w_n)$ is precompact in $L^2_\sigma(\mathbb{R})\times L^2_\sigma(\mathbb{R})$.*

Proof. Assume (v_n, w_n) is bounded in $L^2_\sigma(\mathbb{R}) \times L^2_\sigma(\mathbb{R})$ and $t_n \to \infty$. We want to prove the sequence $\{S(t_n)(v_n, w_n)\}$ has a convergent subsequence in $L^2_\sigma(\mathbb{R}) \times L^2_\sigma(\mathbb{R})$. To this end, we decompose $\{S(t_n)(v_n, w_n)\}$ as a sum of two sequences: one is precompact and the other converges to zero as $n \to \infty$. Given $(v_0, w_0) \in L^2_\sigma(\mathbb{R}) \times L^2_\sigma(\mathbb{R})$, denote by w_1 and w_2 the solutions of problem (4.25) and (4.26), respectively. Then, for $t \geq 0$, we define two maps $S_1(t)$ and $S_2(t)$ from $L^2_\sigma(\mathbb{R})\times L^2_\sigma(\mathbb{R})$ into itself such that $S_1(t)(v_0, w_0) = (0, w_1(t))$

and $S_2(t)(v_0, w_0) = (v(t), w_2(t))$ for $(v_0, w_0) \in L^2_\sigma(I\!R) \times L^2_\sigma(I\!R)$. Thus, $S(t)(v_0, w_0) = S_1(t)(v_0, w_0) + S_2(t)(v_0, w_0)$. Particularly, for the sequence $\{S(t_n)(v_n, w_n)\}$, we have

$$S(t_n)(v_n, w_n) = S_1(t_n)(v_n, w_n) + S_2(t_n)(v_n, w_n). \tag{5.1}$$

Since (v_n, w_n) is bounded in $L^2_\sigma(I\!R) \times L^2_\sigma(I\!R)$, there exists $R > 0$ such that

$$\|(v_n, w_n)\|_\sigma \leq R, \quad \text{for } n \geq 1. \tag{5.2}$$

It then follows from (4.27) that

$$\|S_1(t_n)(v_n, w_n)\|_\sigma \leq e^{-\epsilon\gamma t_n}\|w_n\|_\sigma \leq Re^{-\epsilon\gamma t_n} \to 0, \quad \text{as } n \to \infty. \tag{5.3}$$

Thus, if $S_2(t_n)(v_n, w_n)$ has a convergent subsequence, then so does $S(t_n)(v_n, w_n)$. We now prove the sequence $S_2(t_n)(v_n, w_n)$ is precompact in $L^2_\sigma(I\!R) \times L^2_\sigma(I\!R)$, that is, for any $\eta > 0$, the set $\{S_2(t_n)(v_n, w_n)\}$ has a finite covering of balls of radii less than η. For a positive constant K, let's denote

$$\bar{\Omega}_K = \{x : |x| \leq K\} \quad \text{and} \quad \Omega^c_K = \{x : |x| > K\}.$$

Then by Corollary 4..5, given $\eta > 0$, there exist $K(\eta)$ and $T(\eta)$ such that

$$\|S_2(t)(v_n, w_n)\|_{L^2_\sigma(\Omega^c_{K(\eta)}) \times L^2_\sigma(\Omega^c_{K(\eta)})} \leq \frac{\eta}{4}, \quad \text{for } t \geq T(\eta). \tag{5.4}$$

Since $t_n \to \infty$, there exists $N_1(\eta)$ such that $t_n \geq T(\eta)$ for $n \geq N_1$. Therefore,

$$\|S_2(t_n)(v_n, w_n)\|_{L^2_\sigma(\Omega^c_{K(\eta)}) \times L^2_\sigma(\Omega^c_{K(\eta)})} \leq \frac{\eta}{4}, \quad \text{for } n \geq N_1. \tag{5.5}$$

On the other hand, by Lemmas 4..1-4..3, there exist M_1 and T_1 such that

$$\|S(t)(v_n, w_n)\|_{H^1_\sigma(I\!R) \times H^1_\sigma(I\!R)} \leq M_1 \quad \text{for } t \geq T_1. \tag{5.6}$$

Take N_2 large enough such that $t_n \geq T_1$ for $n \geq N_2$. Then it follows from (5.6) that

$$\|S(t_n)(v_n, w_n)\|_{H^1_\sigma(I\!R) \times H^1_\sigma(I\!R)} \leq M_1 \quad \text{for } n \geq N_2. \tag{5.7}$$

Taking (5.3) into account, by (5.7) we find that there exists M_2 such that

$$\|S_2(t_n)(v_n, w_n)\|_{H^1_\sigma(I\!R) \times H^1_\sigma(I\!R)} \leq M_2 \quad \text{for } n \geq N_2. \tag{5.8}$$

Note (5.8) implies

$$\|S_2(t_n)(v_n, w_n)\|_{H^1_\sigma(\bar{\Omega}_{K(\eta)}) \times H^1_\sigma(\bar{\Omega}_{K(\eta)})} \leq M_2 \quad \text{for } n \geq N_2, \tag{5.9}$$

where $K(\eta)$ is the constant in (5.4). Since $\bar{\Omega}_{K(\eta)}$ is bounded, the norm $\|\cdot\|_{H^1_\sigma(\bar{\Omega}_{K(\eta)})}$ is equivalent to the norm $\|\cdot\|_{H^1(\bar{\Omega}_{K(\eta)})}$. Therefore, by (5.9) we see that there exists M_3 such that

$$\|S_2(t_n)(v_n, w_n)\|_{H^1(\bar{\Omega}_{K(\eta)}) \times H^1_\sigma(\bar{\Omega}_{K(\eta)})} \leq M_3 \quad \text{for } n \geq N_2. \tag{5.10}$$

By the compactness of embedding $H^1(\bar{\Omega}_{K(\eta)}) \subset L^2(\bar{\Omega}_{K(\eta)})$, the sequence $S_2(t_n)(v_n, w_n)$ is precompact in $L^2(\bar{\Omega}_{K(\eta)}) \times L^2(\bar{\Omega}_{K(\eta)})$ and hence in $L^2_\sigma(\bar{\Omega}_{K(\eta)}) \times L^2_\sigma(\bar{\Omega}_{K(\eta)})$. This implies that, for the given $\eta > 0$, $\{S_2(t_n)(v_n, w_n)\}$ has a finite covering in $L^2_\sigma(\bar{\Omega}_{K(\eta)}) \times L^2_\sigma(\bar{\Omega}_{K(\eta)})$ of balls of radii less than $\eta/4$, which along with (5.4) shows that $\{S_2(t_n)(v_n, w_n)\}$ has a finite covering in $L^2_\sigma(I\!R) \times L^2_\sigma(I\!R)$ of balls of radii less than η. In other words, the sequence $\{S_2(t_n)(v_n, w_n)\}$ is precompact in the space $L^2_\sigma(I\!R) \times L^2_\sigma(I\!R)$. Therefore, by (5.1) and (5.3), we know $\{S(t_n)(v_n, w_n)\}$ is precompact in $L^2_\sigma(I\!R) \times L^2_\sigma(I\!R)$. The proof is complete. □

We are now ready to complete Theorem 2..2 which is concerned with the existence of a global attractor.

Proof of Theorem 2..2. By a standard result (see, for example, [3, 5, 12, 15, 28]), we know the dynamical system $S(t)_{t\geq}$ possesses a global attractor if it is asymptotically compact and has a bounded absorbing set. In our case, the asymptotic compactness of $S(t)_{t\geq 0}$ was established by Lemma 5..1 and the bounded absorbing set was given by (4.12). □

6. Uniform bounds of global attractors in ϵ

In this section, we show that the global attractors $\mathcal{A}_\epsilon$ are uniformly bounded as $\epsilon \to 0$. We first derive the uniform bounds in $L^2_\sigma(I\!R) \times L^2_\sigma(I\!R)$.

Lemma 6..1. *For any $\epsilon < \min\{1, \frac{a}{2\gamma}\}$ the solution (v, w) of (2.1)-(2.3) satisfies, for $t \geq 0$,*

$$\|w\|_\sigma^2 \leq \delta^{-2\sigma} e^{-2\epsilon\gamma t} \left(\|v_0\|_\sigma^2 + \|w_0\|_\sigma^2\right) + \frac{\beta}{\gamma} \int \phi_\delta dx, \tag{6.1}$$

and

$$\|v\|_\sigma^2 \leq \frac{a(a-2\epsilon\gamma)+2}{a(a-2\epsilon\gamma)} \delta^{-2\sigma} e^{-2\epsilon\gamma t} \left(\|v_0\|_\sigma^2 + \|w_0\|_\sigma^2\right) + \frac{2\beta(1+a\gamma)}{a^2\gamma} \int \phi_\delta dx, \tag{6.2}$$

where a and β are the constants in (2.4), and δ is the constant given by (2.6). Furthermore, for $t \geq 0$,

$$\begin{aligned} \int_t^{t+1} \|\partial_x v\|_\sigma^2 dt &\leq \frac{2(a+2)(a-2\epsilon\gamma)+4}{3a(a-2\epsilon\gamma)} \delta^{-2\sigma} e^{-2\epsilon\gamma t} \left(\|v_0\|_\sigma^2 + \|w_0\|_\sigma^2\right) \\ &+ \frac{4\beta(a+1)(a\gamma+1)}{3a^2\gamma} \int \phi_\delta dx. \end{aligned} \tag{6.3}$$

Proof. Multiplying (2.1) by $\phi_\delta v$ and integrating with respect to x, we get

$$\frac{1}{2}\frac{d}{dt} \int \phi_\delta |v|^2 dx + \int \phi_\delta |\partial_x v|^2 dx = -\int v\partial_x v\, \partial_x \phi_\delta dx + \int \phi_\delta h(v) v - \int \phi_\delta v w dx. \tag{6.4}$$

By inequality (2.4) for h, we obtain from (6.4) that

$$\begin{aligned} \frac{1}{2}\frac{d}{dt} \int \phi_\delta |v|^2 dx &+ \int \phi_\delta |\partial_x v|^2 dx \leq -\int v\partial_x v\, \partial_x \phi_\delta dx \\ &- a\int \phi_\delta |v|^2 dx + \beta \int \phi_\delta dx - \int \phi_\delta v w dx. \end{aligned} \tag{6.5}$$

We now estimate the terms on the right-hand side of (6.5) as follows. By (2.6) and (2.7) we see the first term on the right-hand side is bounded by

$$\begin{aligned}|\int v\partial_x v\,\partial_x\phi_\delta dx| &\le \delta\sigma\int\phi_\delta|v||\partial_x v|dx \le \frac{1}{2}\delta\sigma\left(\int\phi_\delta|v|^2dx+\int\phi_\delta|\partial_x v|^2dx\right)\\ &\le \frac{1}{4}a\int\phi_\delta|v|^2dx+\frac{1}{4}\int\phi_\delta|\partial_x v|^2dx. \end{aligned}\tag{6.6}$$

For the last term on the right-hand side of (6.5), we have the bounds:

$$|\int\phi_\delta vwdx| \le \frac{a}{4}\int\phi_\delta|v|^2dx+\frac{1}{a}\int\phi_\delta|w|^2dx. \tag{6.7}$$

It follows from (6.5)-(6.7) that

$$\frac{1}{2}\frac{d}{dt}\int\phi_\delta|v|^2dx+\frac{3}{4}\int\phi_\delta|\partial_x v|^2dx+\frac{a}{2}\int\phi_\delta|v|^2dx \le \beta\int\phi_\delta dx+\frac{1}{a}\int\phi_\delta|w|^2dx. \tag{6.8}$$

Since $\epsilon<\min\{1,\frac{a}{2\gamma}\}$, (4.11) implies

$$\int_{\mathbb{R}}\phi_\delta|w|^2dx \le \delta^{-2\sigma}e^{-2\epsilon\gamma t}\left(\|v_0\|_\sigma^2+\|w_0\|_\sigma^2\right)+\frac{\beta}{\gamma}\int\phi_\delta dx, \quad \text{for } t\ge 0. \tag{6.9}$$

By (6.8) and (6.9) we get, for $t\ge 0$,

$$\begin{aligned}&\frac{d}{dt}\int\phi_\delta|v|^2dx+\frac{3}{2}\int\phi_\delta|\partial_x v|^2dx+a\int\phi_\delta|v|^2dx\\ \le\ & 2\beta\left(1+\frac{1}{a\gamma}\right)\int\phi_\delta dx+\frac{2}{a}\delta^{-2\sigma}e^{-2\epsilon\gamma t}\left(\|v_0\|_\sigma^2+\|w_0\|_\sigma^2\right).\end{aligned}\tag{6.10}$$

Dropping the second term on the left side , by Gronwall's inequality we obtain

$$\begin{aligned}\int\phi_\delta|v|^2dx &\le \frac{2\beta}{a}\left(1+\frac{1}{a\gamma}\right)\int\phi_\delta dx+e^{-at}\int\phi_\delta|v_0|^2dx\\ &\quad+\frac{2\delta^{-2\sigma}}{a(a-2\epsilon\gamma)}e^{-2\epsilon\gamma t}\left(\|v_0\|_\sigma^2+\|w_0\|_\sigma^2\right).\end{aligned}\tag{6.11}$$

Since $\epsilon<\frac{a}{2\gamma}$ by assumption, we have

$$e^{-at}\int\phi_\delta|v_0|^2dx \le e^{-2\epsilon\gamma t}\int\phi_\delta|v_0|^2dx \le \delta^{-2\sigma}e^{-2\epsilon\gamma t}\left(\|v_0\|_\sigma^2+\|w_0\|_\sigma^2\right). \tag{6.12}$$

It follows from (6.11) and (6.12) that

$$\int\phi_\delta|v|^2dx \le \frac{2\beta}{a}\left(1+\frac{1}{a\gamma}\right)\int\phi_\delta dx+\left(1+\frac{2}{a(a-2\epsilon\gamma)}\right)\delta^{-2\sigma}e^{-2\epsilon\gamma t}\left(\|v_0\|_\sigma^2+\|w_0\|_\sigma^2\right). \tag{6.13}$$

By (6.9) and (6.13) we get (6.1) and (6.2), respectively. Next, we prove (6.3). Integrating (6.10) from t to $t+1$, we obtain

$$\begin{aligned}\frac{3}{2}\int_t^{t+1}\|\partial_x v\|_\sigma^2dt &\le \int\phi_\delta|v(t)|^2dx+2\beta\left(1+\frac{1}{a\gamma}\right)\int\phi_\delta dx\\ &\quad+\frac{2}{a}\delta^{-2\sigma}\left(\|v_0\|_\sigma^2+\|w_0\|_\sigma^2\right)\int_t^{t+1}e^{-2\epsilon\gamma t}dt.\end{aligned}\tag{6.14}$$

We now deal with the last term on the right-hand side of (6.14). Notice that

$$\int_t^{t+1} e^{-2\epsilon\gamma t}dt \leq e^{-2\epsilon\gamma t} \quad \text{for all } \epsilon > 0 \text{ and } t \geq 0. \tag{6.15}$$

Substituting (6.13) and (6.15) into (6.14), then (6.3) follows. The proof is complete. □

Notice that by adding (6.1) and (6.2) together we get the inequality:

$$\|S(t)(v_0, w_0)\|_\sigma^2 \leq K_1 e^{-2\epsilon\gamma t}\left(\|v_0\|_\sigma^2 + \|w_0\|_\sigma^2\right) + K_2, \quad \text{for all } t \geq 0, \tag{6.16}$$

where K_1 and K_2 are given by

$$K_1 = \frac{2\left(1 + a(a - 2\epsilon\gamma)\right)}{a(a - 2\epsilon\gamma)}\delta^{-2\sigma}, \; K_2 = \frac{\beta(a^2 + 2a\gamma + 2)}{a^2\gamma}\int \phi_\delta dx. \tag{6.17}$$

The uniform boundedness of $\mathcal{A}_\epsilon$ in $L^2_\sigma(I\!R) \times L^2_\sigma(I\!R)$ can now be easily obtained.

Proof of Theorem 2..3. Suppose $(v, w) \in \mathcal{A}_\epsilon$ and $t_n \to \infty$. Then by invariance of $\mathcal{A}_\epsilon$, there exists a sequence of $(v_n, w_n) \in \mathcal{A}_\epsilon$ such that

$$(v, w) = S(t_n)(v_n, w_n) \quad \text{for all } n \geq 1. \tag{6.18}$$

Since $\mathcal{A}_\epsilon \subseteq B$, where B is the bounded absorbing set given by (4.12), we have

$$\|(v_n, w_n)\|_\sigma \leq M, \quad \text{for all } n \geq 1, \tag{6.19}$$

where M depends only on the data $(\epsilon, \gamma, a, \beta, \sigma)$. On the other hand, by applying (6.16) to $S(t_n)(v_n, w_n)$ we get

$$\|S(t_n)(v_n, w_n)\|_\sigma^2 \leq K_1 e^{-2\epsilon\gamma t_n}\|(v_n, w_n)\|_\sigma^2 + K_2, \quad \text{for all } n \geq 1, \tag{6.20}$$

where K_1 and K_2 are constants given by (6.17). It follows from (6.19) and (6.20) that

$$\|S(t_n)(v_n, w_n)\|_\sigma^2 \leq K_1 M^2 e^{-2\epsilon\gamma t_n} + K_2, \quad \text{for all } n \geq 1,$$

which along with (6.18) implies that for any $(v, w) \in \mathcal{A}_\epsilon$, the following inequality holds

$$\|(v, w)\|_\sigma^2 \leq K_1 M^2 e^{-2\epsilon\gamma t_n} + K_2, \quad \text{for all } n \geq 1. \tag{6.21}$$

Taking the limit as $n \to \infty$, we find $\|(v, w)\|_\sigma^2 \leq K_2$ from which (2.12) follows.

To complete the proof, we also need to establish the estimates claimed in (2.13). In what follows, we first derive various uniform bounds in ϵ for the solution in $H^1_\sigma(I\!R) \times H^1_\sigma(I\!R)$ via a sequence of lemmas, and then complete the proof of Theorem 2..3 in the end of this section. □

Lemma 6..2. *For any $\epsilon < \min\{1, \frac{a}{2\gamma}\}$, the solution (v, w) of (2.1)-(2.3) satisfies*

$$\|\partial_x v(t)\|_\sigma^2 \leq K_3 e^{-2\epsilon\gamma t}\left(\|v_0\|_\sigma^2 + \|w_0\|_\sigma^2\right) + K_4, \quad \textit{for } t \geq 1, \tag{6.22}$$

where K_3 and K_4 are given by

$$K_3 = \delta^{-2\sigma} e^{2\epsilon\gamma+\sigma^2+\sigma+2\alpha} \left(\frac{2(a+2)(a-2\epsilon\gamma)+4}{3a(a-2\epsilon\gamma)} + \frac{\sigma+1}{(2\epsilon\gamma+\sigma^2+\sigma+2\alpha)^2} \right), \quad (6.23)$$

and

$$K_4 = e^{\sigma^2+\sigma+2\alpha} \left(\frac{\beta(\sigma+1)}{\gamma(\sigma^2+\sigma+2\alpha)^2} + \frac{4\beta(a+1)(a\gamma+1)}{3a^2\gamma} \right) \int \phi_\delta dx, \quad (6.24)$$

where a and β are the constants in (2.4), and δ is the constant given by (2.6).

Proof. We start with inequality (4.24):

$$\frac{d}{dt} \|\partial_x v\|_\sigma^2 \leq \left(\sigma^2+\sigma+2\alpha\right) \|\partial_x v\|_\sigma^2 + (\sigma+1)\|w\|_\sigma^2, \quad \text{for } t \geq 0. \quad (6.25)$$

Substituting (6.1) into (6.25) we get, for all $t \geq 0$,

$$\frac{d}{dt}\|\partial_x v\|_\sigma^2 - k\|\partial_x v\|_\sigma^2 \leq (\sigma+1)\,\delta^{-2\sigma} e^{-2\epsilon\gamma t} \left(\|v_0\|_\sigma^2 + \|w_0\|_\sigma^2\right) + \frac{\beta(\sigma+1)}{\gamma} \int \phi_\delta dx. \quad (6.26)$$

Here and after, we denote by $\kappa = \sigma^2+\sigma+2\alpha$. Let t and s be positive numbers satisfying $0 \leq t \leq s \leq t+1$. Then multiplying (6.26) by $e^{-\kappa t}$ and integrating with respect to t between s and $t+1$, we obtain

$$\begin{aligned} &\|\partial_x v(t+1)\|_\sigma^2 e^{-\kappa(t+1)} \\ &\leq e^{-\kappa s}\|\partial_x v(s)\|_\sigma^2 + (\sigma+1)\,\delta^{-2\sigma}\left(\|v_0\|_\sigma^2 + \|w_0\|_\sigma^2\right) \int_s^{t+1} e^{-(\kappa+2\epsilon\gamma)t} dt \\ &+ \frac{\beta(\sigma+1)}{\gamma} \int \phi_\delta dx \int_s^{t+1} e^{-\kappa t} dt. \end{aligned} \quad (6.27)$$

The two integrals on the right-hand side of (6.27) have the estimates

$$\int_s^{t+1} e^{-(\kappa+2\epsilon\gamma)t} dt \leq \frac{1}{\kappa+2\epsilon\gamma} e^{-(\kappa+2\epsilon\gamma)s} \text{ and } \int_s^{t+1} e^{-\kappa t} dt \leq \frac{1}{\kappa} e^{-\kappa s}. \quad (6.28)$$

Substituting (6.28) into (6.27), and then multiplying the resulting inequality by $e^{\kappa(t+1)}$ we find

$$\begin{aligned} \|\partial_x v(t+1)\|_\sigma^2 \;\leq\; & e^{\kappa(t+1-s)}\|\partial_x v(s)\|_\sigma^2 + \frac{\beta(\sigma+1)}{\gamma\kappa} e^{\kappa(t+1-s)} \int \phi_\delta dx \\ & + \frac{\sigma+1}{\kappa+2\epsilon\gamma}\delta^{-2\sigma}\left(\|v_0\|_\sigma^2 + \|w_0\|_\sigma^2\right) e^{\kappa(t+1-s)-2\epsilon\gamma s}. \end{aligned} \quad (6.29)$$

Since $0 \leq t \leq s \leq t+1$, integrating (6.29) with respect to s between t and $t+1$ we have

$$\begin{aligned} \|\partial_x v(t+1)\|_\sigma^2 \;\leq\; & \int_t^{t+1} e^{\kappa(t+1-s)}\|\partial_x v(s)\|_\sigma^2 ds + \frac{\beta(\sigma+1)}{\gamma\kappa} \int_t^{t+1} e^{\kappa(t+1-s)} ds \int \phi_\delta dx \\ & + \frac{(\sigma+1)\delta^{-2\sigma}}{\kappa+2\epsilon\gamma}\left(\|v_0\|_\sigma^2 + \|w_0\|_\sigma^2\right) \int_t^{t+1} e^{\kappa(t+1-s)-2\epsilon\gamma s} ds. \end{aligned} \quad (6.30)$$

We now deal with the right-hand side of (6.30). By (6.3), the first term on the right-hand side of (6.30) satisfies

$$\begin{aligned}\int_t^{t+1} e^{\kappa(t+1-s)}\|\partial_x v(s)\|_\sigma^2 ds &\le e^\kappa \int_t^{t+1} \|\partial_x v(s)\|_\sigma^2 ds \\ &\le \frac{2(a+2)(a-2\epsilon\gamma)+4}{3a(a-2\epsilon\gamma)}\delta^{-2\sigma}e^{-2\epsilon\gamma t+\kappa}\left(\|v_0\|_\sigma^2+\|w_0\|_\sigma^2\right) \\ &\quad + \frac{4\beta(a+1)(a\gamma+1)}{3a^2\gamma}e^\kappa\int\phi_\delta dx. \qquad (6.31)\end{aligned}$$

The integrals in the last two terms of the right-hand side of (6.30) can be estimated

$$\int_t^{t+1} e^{\kappa(t+1-s)-2\epsilon\gamma s}ds \le \frac{1}{\kappa+2\epsilon\gamma}e^{-2\epsilon\gamma t+\kappa} \quad \text{and} \quad \int_t^{t+1} e^{\kappa(t+1-s)}ds \le \frac{1}{\kappa}e^\kappa. \qquad (6.32)$$

It follows from (6.30)-(6.32) that, for all $t \ge 0$,

$$\begin{aligned}\|\partial_x v(t+1)\|_\sigma^2 &\le \left(\frac{2(a+2)(a-2\epsilon\gamma)+4}{3a(a-2\epsilon\gamma)}+\frac{\sigma+1}{(\kappa+2\epsilon\gamma)^2}\right)\delta^{-2\sigma}e^{-2\epsilon\gamma t+\kappa}\left(\|v_0\|_\sigma^2+\|w_0\|_\sigma^2\right) \\ &\quad + \left(\frac{\beta(\sigma+1)}{\gamma\kappa^2}+\frac{4\beta(a+1)(a\gamma+1)}{3a^2\gamma}\right)e^\kappa\int\phi_\delta dx,\end{aligned}$$

which along with $\kappa = \sigma^2+\sigma+2\alpha$ implies (6.22). □

As an immediate consequence of Lemma 6..2, we have the following estimates for global attractors.

Lemma 6..3. *Let $0 < \epsilon < \min\{1, \frac{a}{2\gamma}\}$. If $(v, w) \in \mathcal{A}_\epsilon$, then*

$$\|\partial_x v\|_\sigma^2 \le e^{\sigma^2+\sigma+2\alpha}\left(\frac{\beta(\sigma+1)}{\gamma(\sigma^2+\sigma+2\alpha)^2}+\frac{4\beta(a+1)(a\gamma+1)}{3a^2\gamma}\right)\int\phi_\delta dx, \qquad (6.33)$$

where a, β and α are the constants in (2.4) and δ is the constant given by (2.6).

Next, we establish the uniform bounds in ϵ for the solution w_2 of problem (4.26) in $H_\sigma^1(I\!R)$.

Lemma 6..4. *Let $0 < \epsilon < \min\{1, \frac{a}{2\gamma}\}$. If $(v_0, w_0) \in \mathcal{A}_\epsilon$, then the solution w_2 of problem (4.26) satisfies:*

$$\|\partial_x w_2\|_\sigma^2 \le \frac{e^{\sigma^2+\sigma+2\alpha}}{\gamma^2}\left(\frac{\beta(\sigma+1)}{\gamma(\sigma^2+\sigma+2\alpha)^2}+\frac{4\beta(a+1)(a\gamma+1)}{3a^2\gamma}\right)\int\phi_\delta dx, \qquad (6.34)$$

where a, β and α are the constants in (2.4) and δ is the constant given by (2.6).

Proof. We begin with inequality (4.35):

$$\frac{d}{dt}\int\phi_1|\partial_x w_2|^2+\epsilon\gamma\int\phi_1|\partial_x w_2|^2 dx \le \frac{\epsilon}{\gamma}\int\phi_1|\partial_x v|^2 dx, \quad \text{for } t \ge 0. \qquad (6.35)$$

Since $(v_0, w_0) \in \mathcal{A}_\epsilon$, we know for all $t \geq 0$, $(v(t), w(t)) = S(t)(v_0, w_0) \in \mathcal{A}_\epsilon$. Then it follows from Lemma 6..3 that

$$\int \phi_1 |\partial_x v(t)|^2 dx \leq C, \quad \text{for all } t \geq 0, \tag{6.36}$$

where C is given by

$$C = e^{\sigma^2+\sigma+2\alpha} \left(\frac{\beta(\sigma+1)}{\gamma(\sigma^2+\sigma+2\alpha)^2} + \frac{4\beta(a+1)(a\gamma+1)}{3a^2\gamma} \right) \int \phi_\delta dx, \tag{6.37}$$

By (6.35) and (6.36) we get

$$\frac{d}{dt} \int \phi_1 |\partial_x w_2|^2 + \epsilon\gamma \int \phi_1 |\partial_x w_2|^2 dx \quad \leq \quad \frac{\epsilon C}{\gamma}, \quad \text{for } t \geq 0. \tag{6.38}$$

Applying Gronwall's Lemma to (6.38), we obtain

$$\int \phi_1 |\partial_x w_2(t)|^2 dx \leq \frac{C}{\gamma^2}, \quad \text{for all } t \geq 0. \tag{6.39}$$

Then (6.34) follows from (6.39) and (6.37). □

As an immediate consequence of Lemmas 6..3 and 6..4, we have the following bounds in the space $H^1_\sigma(I\!\!R) \times H^1_\sigma(I\!\!R)$ for the dynamical system $S(t)_{t\geq 0}$.

Corollary 6..5. *If $(v_0, w_0) \in \mathcal{A}_\epsilon$ for $0 < \epsilon < \min\{1, \frac{a}{2\gamma}\}$, then the solution (v, w) of problem (2.1)-(2.3) and the solution w_2 of problem (4.26) satisfy:*

$$\|\partial_x v(t)\|^2_\sigma + \|\partial_x w_2(t)\|^2_\sigma \leq K_5, \quad \textit{for all } t \geq 0, \tag{6.40}$$

where K_5 is given by

$$K_5 = \frac{(\gamma^2+1)e^{\sigma^2+\sigma+2\alpha}}{\gamma^2} \left(\frac{\beta(\sigma+1)}{\gamma(\sigma^2+\sigma+2\alpha)^2} + \frac{4\beta(a+1)(a\gamma+1)}{3a^2\gamma} \right) \int \phi_\delta dx, \tag{6.41}$$

where a, β and α are the constants in (2.4) and δ is the constant given by (2.6).

We are now ready to complete the proof of Theorem 2..3.

Proof of Theorem 2..3 (Continued). By invariance of $\mathcal{A}_\epsilon$, there exist $t_n \to \infty$ and $(v_n, w_n) \in \mathcal{A}_\epsilon$ such that

$$(v, w) = S(t_n)(v_n, w_n), \quad \text{for all } n \geq 1. \tag{6.42}$$

Decompose $S(t)$ as in (5.1):

$$S(t)(v_0, w_0) = S_1(t)(v_0, w_0) + S_2(t)(v_0, w_0), \quad \text{for } (v_0, w_0) \in L^2_\sigma(I\!\!R) \times L^2_\sigma(I\!\!R), \tag{6.43}$$

where $S_1(t)(v_0, w_0) = (0, w_1(t))$, $S_2(t)(v_0, w_0) = (v(t), w_2(t))$, w_1 and w_2 are solutions of problem (4.25) and problem (4.26), respectively. It follows from (6.42) and (6.43) that

$$(v, w) = S_1(t_n)(v_n, w_n) + S_2(t_n)(v_n, w_n), \quad \text{for all } n \geq 1. \tag{6.44}$$

Since $(v_n, w_n) \in \mathcal{A}_\epsilon$, we find

$$\|(v_n, w_n)\|_\sigma \leq C, \quad \text{for all } n \geq 1, \tag{6.45}$$

where C is the constant on the right-hand side of (2.12). By (4.27) and (6.45) we get

$$\|S_1(t_n)(v_n, w_n)\|_\sigma \leq e^{-\epsilon\gamma t_n}\|w_n\|_\sigma \leq Ce^{-\epsilon\gamma t_n} \to 0, \quad \text{as } n \to \infty. \tag{6.46}$$

On the other hand,

$$\|S_2(t_n)(v_n, w_n)\|_{H^1_\sigma} \leq \sqrt{K_6^2 + K_7^2}, \tag{6.47}$$

where K_6 and K_7 are given by

$$K_6 = \sqrt{\frac{\beta(a^2 + 2a\gamma + 2)}{a^2\gamma}\int \phi_\delta dx}, \tag{6.48}$$

and

$$K_7 = \frac{1}{\gamma}\sqrt{(\gamma^2+1)e^{\sigma^2+\sigma+2\alpha}\left(\frac{\beta(\sigma+1)}{\gamma(\sigma^2+\sigma+2\alpha)^2} + \frac{4\beta(a+1)(a\gamma+1)}{3a^2\gamma}\right)\int \phi_\delta dx}. \tag{6.49}$$

By (6.47) we see the sequence $\{S_2(t_n)(v_n, w_n)\}$ is bounded in $H^1_\sigma(I\!R) \times H^1_\sigma(I\!R)$. Therefore there exists $(\tilde{v}, \tilde{w}) \in H^1_\sigma(I\!R) \times H^1_\sigma(I\!R)$ such that

$$S_2(t_n)(v_n, w_n) \to (\tilde{v}, \tilde{w}) \text{ weakly in } H^1_\sigma(I\!R) \times H^1_\sigma(I\!R). \tag{6.50}$$

Furthermore, by (6.47) and (6.50) we have

$$\|(\tilde{v}, \tilde{w})\|_{H^1_\sigma \times H^1_\sigma} \leq \liminf_{n\to\infty} \|S_2(t_n)(v_n, w_n)\|_{H^1_\sigma \times H^1_\sigma} \leq \sqrt{K_6^2 + K_7^2}. \tag{6.51}$$

For any $(f, g) \in L^2_\sigma(I\!R) \times L^2_\sigma(I\!R)$, it follows from (6.44) that

$$\langle (v, w), (f, g)\rangle = \langle S_1(t_n)(v_n, w_n), (f, g)\rangle + \langle S_2(t_n)(v_n, w_n), (f, g)\rangle, \tag{6.52}$$

where $\langle \cdot, \cdot \rangle$ denotes the inner product of $L^2_\sigma(I\!R) \times L^2_\sigma(I\!R)$. Taking the weak limit of (6.52) in $L^2_\sigma(I\!R) \times L^2_\sigma(I\!R)$ as $n \to \infty$, by (6.46) and (6.50) we get

$$\langle (v, w), (f, g)\rangle = \langle (\tilde{v}, \tilde{w}), (f, g)\rangle, \quad \text{for all } (f, g) \in L^2_\sigma(I\!R) \times L^2_\sigma(I\!R). \tag{6.53}$$

By (6.53) we find

$$(v, w) = (\tilde{v}, \tilde{w}) \in H^1_\sigma(I\!R) \times H^1_\sigma(I\!R). \tag{6.54}$$

It follows from (6.51) and (6.54) that $\|(v, w)\|_{H^1_\sigma \times H^1_\sigma} \leq \sqrt{K_6^2 + K_7^2}$, which, together with (6.48) and (6.49), concludes the proof. □

7. Compactness of the union of global attractors

In this section, we show that all global attractors are contained in a compact subset of $L^2_\sigma(I\!R) \times L^2_\sigma(I\!R)$. The idea is to prove that the functions in global attractors are uniformly small in ϵ when the space variable is sufficiently large.

Lemma 7..1. *Suppose $0 < \epsilon < \min\{1, \frac{a}{2\gamma}\}$. Then for every $\eta > 0$, there exist a constant $K(\eta)$ depending only on η and the data $(\gamma, a, \beta, \alpha, \sigma)$, but independent of ϵ, such that for all $(v, w) \in \mathcal{A}_\epsilon$, the following inequality holds:*

$$\int_{|x|\geq K(\eta)} (\phi_1|v|^2dx + \phi_1|w|^2dx) \leq \eta. \tag{7.1}$$

Proof. Let $t_n \to \infty$. Since $(v, w) \in \mathcal{A}_\epsilon$, there exists a sequence of $(v_n, w_n) \in \mathcal{A}_\epsilon$ such that

$$(v, w) = S(t_n)(v_n, w_n), \quad \text{for } n \geq 1. \tag{7.2}$$

Set $(v_n(t), w_n(t)) = S(t)(v_n, w_n)$. Then Theorem 2..3 implies that there exists a constant C independent of ϵ such that for all $n \geq 1$ and $t \geq 0$:

$$\|(v_n(t), w_n(t))\|_{H^1_\sigma \times H^1_\sigma} \leq C. \tag{7.3}$$

Proceeding as the proof of Lemma 4..4 but using (7.3) instead of Lemmas 4..1 and 4..2, we find that, given $\eta > 0$, there exists a constant $K_1(\eta)$ depending only on η and the data $(\gamma, a, \beta, \alpha, \sigma)$ such that for $k \geq K_1(\eta)$ and $t \geq 0$:

$$\frac{d}{dt}\left(\int \epsilon\theta(\frac{|x|^2}{k^2})\phi_\delta|v_n(t)|^2dx + \int \theta(\frac{|x|^2}{k^2})\phi_\delta|w_n(t)|^2dx\right)$$
$$+2\epsilon\gamma\left(\int \epsilon\theta(\frac{|x|^2}{k^2})\phi_\delta|v_n(t)|^2dx + \int \theta(\frac{|x|^2}{k^2})\phi_\delta|w_n(t)|^2dx\right) \leq 2\epsilon\eta, \tag{7.4}$$

which is the analogue of (4.47) with $C_3 = \min\{a, 2\epsilon\gamma\} = 2\epsilon\gamma$ since $\epsilon < \frac{a}{2\gamma}$. By Gronwall's Lemma, we can verify from (7.4) that, for $t \geq 0$ and $k \geq \sqrt{2}K_1(\eta)$:

$$\int_{|x|\geq k} \phi_\delta|w_n(t)|^2dx \leq C_1e^{-2\epsilon\gamma t} + \frac{\eta}{\gamma}. \tag{7.5}$$

Next, we derive the estimates similar to (7.5) for $v_n(t)$. For that purpose, multiplying (2.1) by $\theta(\frac{|x|^2}{k^2})\phi_\delta v_n(t)$ we get

$$\frac{1}{2}\frac{d}{dt}\int \theta(\frac{|x|^2}{k^2})\phi_\delta|v_n(t)|^2dx = \int \partial_{xx}v_n(t)\,\theta(\frac{|x|^2}{k^2})\phi_\delta v_n(t)dx$$
$$+\int \theta(\frac{|x|^2}{k^2})\phi_\delta h(v_n(t))v_n(t)dx - \int \theta(\frac{|x|^2}{k^2})\phi_\delta w_n(t)v_n(t)dx. \tag{7.6}$$

By Young's inequality and (7.5), we obtain the following bound for the last term on the right-hand side of (7.6).

$$\begin{aligned}|\int \theta(\frac{|x|^2}{k^2})\phi_\delta w_n(t)v_n(t)dx| &\leq \frac{a}{4}\int \theta(\frac{|x|^2}{k^2})\phi_\delta|v_n(t)|^2dx + \frac{1}{a}\int \theta(\frac{|x|^2}{k^2})\phi_\delta|w_n(t)|^2dx \\ &\leq \frac{a}{4}\int \theta(\frac{|x|^2}{k^2})\phi_\delta|v_n(t)|^2dx + \frac{C_1}{a}e^{-2\epsilon\gamma t} + \frac{\eta}{a\gamma},\end{aligned} \tag{7.7}$$

for all $t \geq 0$ and $k \geq \sqrt{2}K_1(\eta)$, where C_1 is independent of ϵ and the last inequality is obtained by (7.5). Dealing with the first two terms on the right-hand side of (7.6) as the proof of Lemma 4..4 but using (7.3) instead of Lemmas 4..1 and 4..2, after detailed computations, by (7.7) we find that there exists $K_2(\eta)$ such that for $t \geq 0$ and $k \geq K_2(\eta)$:

$$\frac{d}{dt}\int \theta(\frac{|x|^2}{k^2})\phi_\delta|v_n(t)|^2dx + a\int \theta(\frac{|x|^2}{k^2})\phi_\delta|v_n(t)|^2dx \leq \left(1+\frac{2}{a\gamma}\right)\eta + \frac{2C_1}{a}e^{-2\epsilon\gamma t}.$$

By Gronwall's Lemma, for all $t \geq 0$ and $k \geq K_2(\eta)$, we get:

$$\int_{\mathbb{R}} \theta(\frac{|x|^2}{k^2})\phi_\delta|v_n(t)|^2dx \leq \left(1+\frac{2}{a\gamma}\right)\frac{\eta}{a} + \frac{2C_1}{a(a-2\epsilon\gamma)}e^{-2\epsilon\gamma t},$$

which implies for $t \geq 0$ and $k \geq \sqrt{2}K_2(\eta)$:

$$\int_{|x|\geq k} \phi_\delta|v_n(t)|^2dx \leq \left(1+\frac{2}{a\gamma}\right)\frac{\eta}{a} + \frac{2C_1}{a(a-2\epsilon\gamma)}e^{-2\epsilon\gamma t}. \tag{7.8}$$

Let $K_3(\eta) = \sqrt{2}\ \max\{K_1(\eta), K_2(\eta)\}$. Then it follows from (7.2), (7.5) and (7.8) that for all $n \geq 1$ and $k \geq K_3(\eta)$:

$$\int_{|x|\geq k} \phi_\delta|w|^2dx = \int_{|x|\geq k} \phi_\delta|w_n(t_n)|^2dx \leq C_1e^{-2\epsilon\gamma t_n} + \frac{\eta}{\gamma}, \tag{7.9}$$

and

$$\int_{|x|\geq k} \phi_\delta|v|^2dx = \int_{|x|\geq k} \phi_\delta|v_n(t_n)|^2dx \leq \left(1+\frac{2}{a\gamma}\right)\frac{\eta}{a} + \frac{2C_1}{a(a-2\epsilon\gamma)}e^{-2\epsilon\gamma t_n}. \tag{7.10}$$

Taking the limits of (7.9) and (7.10) as $n \to \infty$, then (7.1) follows. □

We now establish the compactness of the union of all global attractors.

Proof of Theorem 2..4. Given $\eta > 0$, we want to show that the set $\bigcup_{0<\epsilon<\epsilon_0} \mathcal{A}_\epsilon$ has a finite covering of balls of radii less than η. For a positive constant K, denote by

$$\bar{\Omega}_K = \{x : |x| \leq K\} \quad \text{and} \quad \Omega_K^c = \{x : |x| > K\}.$$

Lemma 7..1 implies that there exists a constant $K(\eta)$ depending only on η and the data $(\gamma, a, \beta, \alpha, \sigma)$ such that for all $(v, w) \in \bigcup_{0<\epsilon<\epsilon_0} \mathcal{A}_\epsilon$:

$$\|(v, w)\|_{L^2_\sigma(\Omega^c_{K(\eta)})\times L^2_\sigma(\Omega^c_{K(\eta)})} \leq \frac{\eta}{4}. \tag{7.11}$$

On the other hand, by Theorem 2..3, the set $\bigcup_{0<\epsilon<\epsilon_0} \mathcal{A}_\epsilon$ is bounded in $H^1_\sigma(\mathbb{R}) \times H^1_\sigma(\mathbb{R})$, and hence bounded in $H^1_\sigma(\bar{\Omega}_{K(\eta)}) \times H^1_\sigma(\bar{\Omega}_{K(\eta)})$. Since the embedding $H^1_\sigma(\bar{\Omega}_{K(\eta)}) \times H^1_\sigma(\bar{\Omega}_{K(\eta)}) \subset L^2_\sigma(\bar{\Omega}_{K(\eta)}) \times L^2_\sigma(\bar{\Omega}_{K(\eta)})$ is compact, we find that, for the given η, the set $\bigcup_{0<\epsilon<\epsilon_0} \mathcal{A}_\epsilon$ has a finite covering of balls of radii less than $\eta/4$ in $L^2_\sigma(\bar{\Omega}_{K(\eta)}) \times L^2_\sigma(\bar{\Omega}_{K(\eta)})$, which and (7.11) imply that the set $\bigcup_{0<\epsilon<\epsilon_0} \mathcal{A}_\epsilon$ has a finite covering of balls of radii less than η in $L^2_\sigma(\mathbb{R}) \times L^2_\sigma(\mathbb{R})$, as desired. □

8. Upper semicontinuity of global attractors

The upper semicontinuity of global attractors as $\epsilon \to 0$ is established in this section. Note that the limiting system (2.9)-(2.11) has no global attractors in $L^2_\sigma(I\!R) \times L^2_\sigma(I\!R)$. But we will construct a local attractor $\mathcal{A}_0$ for the limiting system such that $\mathcal{A}_\epsilon$ is upper semicontinuous to $\mathcal{A}_0$ as $\epsilon \to 0$.

In order to define a local attractor for the limiting system, we denote by $\mathcal{B}$ the limiting set of $\bigcup_{0<\epsilon<\epsilon_0} \mathcal{A}_\epsilon$, that is,

$$\mathcal{B} = \{(v,w) \in L^2_\sigma \times L^2_\sigma : \text{there are } (v_{\epsilon_n}, w_{\epsilon_n}) \in \mathcal{A}_{\epsilon_n} \text{ with } \epsilon_n \to 0$$

$$\text{such that } (v_{\epsilon_n}, w_{\epsilon_n}) \to (v,w)\}.$$

Let $\mathcal{W}$ be the projection of $\mathcal{B}$ to the w-space:

$$\mathcal{W} = \{w \in L^2_\sigma(I\!R) : \text{there is } v \in L^2_\sigma(I\!R) \text{ such that } (v,w) \in \mathcal{B}\}.$$

Then it follows from Theorem 2..4 that $\mathcal{B}$ is compact in $L^2_\sigma(I\!R) \times L^2_\sigma(I\!R)$, and therefore $\mathcal{W}$ is compact in $L^2_\sigma(I\!R)$. The following lemma states that the limiting system has a local attractor in $L^2_\sigma(I\!R) \times \mathcal{W}$.

Lemma 8..1. *The limiting system* (2.9)-(2.10) *has a local attractor $\mathcal{A}_0$ which is a compact invariant set and attracts every bounded subset of $L^2_\sigma(I\!R) \times \mathcal{W}$ with respect to the norm topology. Further, if we denote by $\mathcal{A}_w$ the global attractor of equation* (2.9) *for a fixed $w \in \mathcal{W}$, then $\mathcal{A}_0$ can be characterized by*

$$\mathcal{A}_0 = \{(v,w) \in L^2_\sigma(I\!R) \times L^2_\sigma(I\!R) : w \in \mathcal{W} \text{ and } v \in \mathcal{A}_w\}.$$

Proof. The proof is similar but simpler than Theorem 2..2, and the details are omitted. □

The relationships between the solutions of system (2.1)-(2.2) and system (2.9)-(2.10) are necessary for proving the upper semicontinuity of global attractors. In this respect, we have

Lemma 8..2. *Suppose $0 < \epsilon < \min\{1, \frac{\alpha}{2\gamma}\}$. If (v_ϵ, w_ϵ) is the solution of system* (2.1)-(2.2) *with the initial datum $(v_1, w_1) \in L^2(I\!R) \times L^2(I\!R)$, and (v,w) is the solution of system* (2.9)-(2.10) *with the initial datum $(v_2, w_2) \in L^2(I\!R) \times L^2(I\!R)$, then we have, for $t \geq 0$,*

$$\|v_\epsilon(t) - v(t)\|^2_\sigma + \|w_\epsilon(t) - w(t)\|^2_\sigma$$

$$\leq e^{Ct}\left(\|v_1 - v_2\|^2_\sigma + \|w_1 - w_2\|^2_\sigma\right) + \epsilon \int_0^t e^{C(t-s)} \|v(s) - \gamma w(s)\|^2_\sigma ds,$$

where C is a constant depending on α and σ, but independent of ϵ.

Proof. Let $V = v_\epsilon - v$ and $W = w_\epsilon - w$. Then it follows from (2.1)-(2.2) and (2.9)-(2.10) that

$$\frac{dV}{dt} - \frac{\partial^2 V}{\partial x^2} = h(v_\epsilon) - h(v) - W, \tag{8.1}$$

$$\frac{dW}{dt} = \epsilon(V - \gamma W) + \epsilon(v - \gamma w). \tag{8.2}$$

Similar to the proof of Lemma 3..4, by (8.1)-(8.2) we can verify that V and W satisfy

$$\frac{d}{dt}\left(\|V\|_\sigma^2 + \|W\|_\sigma^2\right) \le C\left(\|V\|_\sigma^2 + \|W\|_\sigma^2\right) + \epsilon\|v - \gamma w\|_\sigma^2. \tag{8.3}$$

Then Lemma 8..2 follows from Gronwall's lemma and inequality (8.3). □

To distinguish the semigroups associated with the perturbed and the unperturbed system, we denote by $S^\epsilon(t)_{t\ge 0}$ the semigroup for problem (2.1)-(2.3) with $\epsilon > 0$, and $S(t)_{t\ge 0}$ for $\epsilon = 0$. By a continuity argument, it follows from Lemma 8.2 that if $\epsilon_n \to 0$ and $(v_{\epsilon_n}, w_{\epsilon_n}) \to (v, w)$ in $L^2_\sigma(I\!R) \times L^2_\sigma(I\!R)$, then

$$S^{\epsilon_n}(t)(v_{\epsilon_n}, w_{\epsilon_n}) \to S(t)(v, w) \text{ in } L^2_\sigma(I\!R) \times L^2_\sigma(I\!R). \tag{8.4}$$

The following lemma is crucial for establishing the upper semicontinuity of attractors.

Lemma 8..3. *Let $(v_{\epsilon_n}, w_{\epsilon_n}) \in \mathcal{A}_{\epsilon_n}$ and $\epsilon_n \to 0$. Then there exists $(v_0, w_0) \in \mathcal{A}_0$ and a subsequence $(v_{\epsilon_{n_k}}, w_{\epsilon_{n_k}})$ of $(v_{\epsilon_n}, w_{\epsilon_n})$ such that $(v_{\epsilon_{n_k}}, w_{\epsilon_{n_k}}) \to (v_0, w_0)$ in $L^2_\sigma(I\!R) \times L^2_\sigma(I\!R)$.*

Proof. It follows from Theorem 2..4 that the set $\bigcup_{n=1}^{\infty} \mathcal{A}_{\epsilon_n}$ is precompact. Therefore, there exist $(v_0, w_0) \in L^2_\sigma(I\!R) \times L^2_\sigma(I\!R)$ and a subsequence of $(v_{\epsilon_n}, w_{\epsilon_n})$ (still denoted by $(v_{\epsilon_n}, w_{\epsilon_n})$) such that

$$(v_{\epsilon_n}, w_{\epsilon_n}) \to (v_0, w_0) \text{ in } L^2_\sigma(I\!R) \times L^2_\sigma(I\!R). \tag{8.5}$$

Take a sequence $\{t_m\}_{m=1}^\infty$ with $t_m \to \infty$. By the invariance of $\mathcal{A}_{\epsilon_n}$ we know

$$S^{\epsilon_n}(-t_m)(v_{\epsilon_n}, w_{\epsilon_n}) \in \mathcal{A}_{\epsilon_n} \text{ for all } n, m \ge 1.$$

Therefore, for $m = 1$, it follows from the precompactness of $\bigcup_{n=1}^{\infty} \mathcal{A}_{\epsilon_n}$ that there exist (v_1, w_1) and a subsequence $(v_{\epsilon_{n_1}}, w_{\epsilon_{n_1}})$ of $(v_{\epsilon_n}, w_{\epsilon_n})$ such that

$$S^{\epsilon_{n_1}}(-t_1)(v_{\epsilon_{n_1}}, w_{\epsilon_{n_1}}) \to (v_1, w_1) \text{ in } L^2_\sigma(I\!R) \times L^2_\sigma(I\!R) \quad as\ n_1 \to \infty.$$

Similarly, for $m = 2$, there exist (v_2, w_2) and a subsequence $(v_{\epsilon_{n_2}}, w_{\epsilon_{n_2}})$ of $(v_{\epsilon_{n_1}}, w_{\epsilon_{n_1}})$ such that

$$S^{\epsilon_{n_2}}(-t_2)(v_{\epsilon_{n_2}}, w_{\epsilon_{n_2}}) \to (v_2, w_2) \text{ in } L^2_\sigma(I\!R) \times L^2_\sigma(I\!R) \quad as\ n_2 \to \infty.$$

Repeating this procedure for all $m \ge 1$, it follows from a standard diagonal argument that there exists a subsequence $(v_{\epsilon_{n_k}}, w_{\epsilon_{n_k}})$ of $(v_{\epsilon_n}, w_{\epsilon_n})$ such that, for all $m \ge 1$,

$$S^{\epsilon_{n_k}}(-t_m)(v_{\epsilon_{n_k}}, w_{\epsilon_{n_k}}) \to (v_m, w_m) \text{ in } L^2_\sigma(I\!R) \times L^2_\sigma(I\!R) \quad as\ k \to \infty. \tag{8.6}$$

By (8.4) and (8.6) we find that, as $k \to \infty$:

$$(v_{\epsilon_{n_k}}, w_{\epsilon_{n_k}}) = S^{\epsilon_{n_k}}(t_m)\left(S^{\epsilon_{n_k}}(-t_m)(v_{\epsilon_{n_k}}, w_{\epsilon_{n_k}})\right) \to S(t_m)(v_m, w_m). \tag{8.7}$$

It follows from (8.5) and (8.7) that, for all $m \ge 1$, $(v_0, w_0) = S(t_m)(v_m, w_m)$, which, along with $t_m \to \infty$, implies $(v_0, w_0) \in \mathcal{A}_0$. The proof is complete. □

We are now in a position to prove that the global attractors $\mathcal{A}_\epsilon$ are upper semicontinuous at $\epsilon = 0$.

Proof of Theorem 2..5. We argue by contradiction. If Theorem 2..5 is not true, then there exist $\eta > 0$ and a sequence $(v_{\epsilon_n}, w_{\epsilon_n}) \in \mathcal{A}_{\epsilon_n}$ with $\epsilon_n \to 0$ such that

$$\mathrm{d}_{L^2_\sigma \times L^2_\sigma}((v_{\epsilon_n}, w_{\epsilon_n}), \mathcal{A}_0) \geq \eta. \tag{8.8}$$

On the other hand, it follows from Lemma 8..3 that there exists a subsequence $(v_{\epsilon_{n_k}}, w_{\epsilon_{n_k}})$ of $(v_{\epsilon_n}, w_{\epsilon_n})$ such that

$$\mathrm{d}_{L^2_\sigma \times L^2_\sigma}\Big((v_{\epsilon_{n_k}}, w_{\epsilon_{n_k}}), \mathcal{A}_0\Big) = 0,$$

a contradiction, which concludes the proof. □

For simplicity, we only discuss the one-dimensional FitzHugh-Nagumo system in the paper. It is evident that all estimates obtained in Sections 4 and 6 are also valid even the space dimensions are higher than one, as long as the nonlinear function h satisfies (2.4). This means that if there is a dynamical system associated to problem (2.1)-(2.3), and condition (2.4) is satisfied, then Theorems 2..2-2..5 still hold regardless of the space dimensions.

References

[1] J. Alexander, R. Gardner, and C. Jones. A topological invariant arising in the stability analysis of traveling waves. *J. Reine Angew. Math.* **410** (1990), 167-212.

[2] F. Antoci and M. Prizzi. Attractors and global averaging of non-autonomous Reaction-Diffusion equations in $I\!R^N$. *Topological Methods in Nonlinear Analysis*, **20** (2002), 229-259.

[3] A.V. Babin and M.I. Vishik. *Attractors of Evolution Equations*. North Holland, Amsterdam, 1991.

[4] A.V. Babin and M.I. Vishik. Attractors of partial differential evolution equations in an unbounded domain. *Proceedings Royal Society Edinburgh* **116A** (1990), 221-243.

[5] J.M. Ball. Continuity properties and global attractors of generalized semiflows and the Navier-Stokes equations. *J. Nonlinear Sci.* **7** (1997), 475-502.

[6] J. Bell. Some threshold results for models of myelinated nerves. *Mathematical Biosciences* **54** (1981), 181-190.

[7] A.N. Carvalho and T. Dlotko. Partially dissipative systems in locally uniform spaces. *Cadernos de Matematica* **02** (2001), 291-307.

[8] B. Deng. The existence of infinitely many traveling front and back waves in the FitzHugh-Nagumo equations. *SIAM J. Math. Anal.* **22**(1991), 1631-1650.

[9] E. Feireisl, Ph. Laurencot, and F. Simondon. Global attractors for degenerate parabolic equations on unbounded domains. *J. Differential Equations* **129** (1996), 239-261.

[10] R. FitzHugh. Impulses and physiological states in theoretical models of nerve membrane. *Biophys. J.* **1** (1961), 445-466.

[11] O. Goubet and R. Rosa. Asymptotic smoothing and the global attractor of a weakly damped KdV equation on the real line. *J. Differential Equations* **185** (2002), 25-53.

[12] J.K. Hale. *Asymptotic Behaviour of Dissipative Systems.* Providence R.I., 1988.

[13] C. Jones. Stability of the traveling wave solution of the FitzHugh-Nagumo system. *Trans. Amer. Math. Soc.* **286** (1984), 431-469.

[14] C. Jones, N. Kopell, and R. Langer. Construction of the FitzHugh-Nagumo pulse using differential forms. Patterns and dynamics in reactive media (Minneapolis, MN, 1989), 101-115, *IMA Vol. Math. Appl.* **37**, Springer, New York, 1991.

[15] O. Ladyzhenskaya, *Attractors for semigroups and evolution equations* . Cambridge University Press, Cambridge, 1991.

[16] W. Liu and E. Van Vleck. Turning points and traveling waves in FitzHugh-Nagumo equations. *J. Differential Equations* (to appear).

[17] K. Lu and B. Wang. Global attractors for the Klein-Gordon-Schrödinger equation in unbounded domains. *J. Differential Equations* **170** (2001), 281–316.

[18] M. Marion. Finite-dimensional attractors associated with partly dissipative reaction-diffusion systems. *SIAM J. Math. Anal.* **20** (1989), 816-844.

[19] M. Marion. Inertial manifolds associated to partly dissipative reaction-diffusion systems. *J. Math. Anal. Appl.* **143** (1989), 295-326.

[20] I. Moise, R. Rosa, and X. Wang. Attractors for non-compact semigroups via energy equations. *Nonlinearity* **11** (1998), 1369-1393.

[21] J. Nagumo, S. Arimoto, and S. Yosimzawa, An active pulse transmission line simulating nerve axon. *Proc. J. R. E.* **50** (1964), 2061-2070.

[22] M. Prizzi. A remark on reaction-diffusion equations in unbounded domains. *Discrete Contin. Dynam. Syst.* **9** (2003), 281-286.

[23] A. Rodriguez-Bernal and B. Wang. Attractors for partly dissipative reaction diffusion systems in $\mathbb{R}^n$. *J. Math. Anal. Appl.* **252** (2000), 790-803.

[24] R. Rosa. The global attractor for the 2D Navier-Stokes flow on some unbounded domains. *Nonl. Anal. TMA* **32** (1998), 71-85.

[25] B. Sandstede. Stability of multiple-pulse solutions. *Trans. Amer. Math. Soc.* **350** (1998), 429–472.

[26] B. Sandstede. Stability of N-fronts bifurcating from a twisted heteroclinic loop and an application to the FitzHugh-Nagumo equation. *SIAM J. Math. Anal.* **29** (1998), 183–207.

[27] Z. Shao. Existence of inertial manifolds for partly dissipative reaction diffusion systems in higher space dimensions. *J. Differential Equations* **144** (1998), 1-43.

[28] R. Temam. *Infinite Dimensional Dynamical Systems in Mechanics and Physics* (2nd Ed.). Springer-Verlag, New York, 1997.

[29] R. Temam, *Navier-Stokes Equations, Theory and Numerical Analysis* (3rd Ed.). North-Holland, Amsterdam, 1984.

[30] G. J. Wang and J. Beaumont. Parameter estimation of the Hodgkin-Huxley gating model: an inversion procedure. *SIAM J. Appld. Math.* **64** (2004), 1249-1267.

[31] B. Wang. Attractors for reaction diffusion equations in unbounded domains, *Physica D* **128** (1999), 41-52.

In: Research on Evolution Equation Compendium. Volume 2 ISBN: 978-1-60456-752-6
Editor: Gaston M. N'Guerekata

A PARTIAL DIFFERENTIAL EQUATION WITH NONAUTONOMOUS PAST DELAY IN L^1 -PHASE SPACE

S. Boulite,[1*] ***G. Fragnelli,***[2†] ***M. Halloumi***[1‡] ***and L. Maniar***[1§]
[1]University Cadi Ayyad, Faculty of Sciences Semlalia,
Department of Mathematics,
B.P. 2390, 40000 Marrakesh, Morocco.
[2] Dipartimento di Ingegneria dell'Informazione,
Università degli Studi di Siena, via Roma 56, 53100 Siena, Italy

Abstract

In this paper, we consider partial differential equations with nonautonomous past delay in L^1 -phase space . Here, we consider general delay operators given by Stieltjes integrals. Using the Miyadera-Voigt perturbation result, we show the wellposedness of these equations. We study also the robustness of some asymptotic properties, as asymptotic stability, almost periodicity and almost automorphy, under the nonautonomous past delay effect. As an application, we give a dynamical population equation.

AMS Subject Classification: 34G10, 35B15, 35B20, 35B40, 47D06, 47N60
Key Words: Partial differential equations with delay, Miyadera-Voigt perturbation, semigroup, asymptotic stability, almost periodicity, almost automorphy, population equation

*E-mail address: sboulite@ucam.ca.ma
†E-mail address:fragnelli@dii.unisi.it
‡E-mail address: Halloumi@menara.ma
§E-mail address: maniar@ucam.ac.ma

1. Introduction

After their introduction by R. Nagel and S. Brendle [5], partial differential equations with nonautonomous past delay

$$\begin{cases} u'(t) = Bu(t) + \Phi(\tilde{u}_t), & t \geq 0, \\ u(0) = x, \quad u_0 = g, \end{cases} \tag{1.1}$$

have attracted many interest. Here the operator $(B, D(B))$ is a generator of a strongly continuous semigroup $(S(t))_{t\geq 0}$ in a Banach space X, the delay $r \leq +\infty$, and $\Phi : D(\Phi) \subset L^p([-r, 0], X) \longrightarrow X$, $p \geq 1$, is an unbounded linear operator. Moreover, the modified history function $\tilde{u}_t$ is defined by

$$\tilde{u}_t(s) := \begin{cases} U(s, 0)u(t+s) & \text{if } t+s > 0, \\ U(s, s+t)g(t+s) & \text{if } t+s \leq 0, \end{cases}$$

where $(U(t, s))_{-r\leq t\leq s\leq 0}$ is an exponentially bounded backward evolution family on X. In the case $p > 1$, this equation was studied in [14, 16] with general delay operators Φ following the study done in [1, 2] for autonomous past delay (i.e., $U(\cdot, \cdot) = Id$). More precisely, the authors have considered the delay operator

$$\Phi(f) := \int_{-r}^{0} d\eta(\theta) f(\theta), \tag{1.2}$$

where $\eta : [-r, 0] \longrightarrow \mathcal{L}(X)$ is a function of bounded variation. In particular, in these papers, they have studied the wellposedness of (1.1), by showing that the delay operator Φ is a small Miyadera-Voigt perturbation.

More recently, in [15, 16, 22], the authors have considered the critical case $p = 1$, but with the particular unbounded delay operator

$$\Phi(f) := \int_{-r}^{0} \alpha(s) f(s) ds, \quad \alpha \in L^1([-r, 0]). \tag{1.3}$$

They have showed that these operators are also small Miyadera-Voigt perturbations; thus, by the Miyadera-Voigt perturbation Theorem [17, 24], the wellposedness of (1.1) follows.

In this paper, we consider the case $p = 1$ and the general linear delay operator Φ given in (1.2). In particular, in Section 3, by using the same idea as in [25], we show that this operator Φ is still a small Miyadera-Voigt perturbation, and this yields the wellposedness of the delay equation (1.1).

Moreover, in Section 4, we study the persistence of some asymptotic properties, as the asymptotic stability, the almost periodicity, the almost automorphy and the ergodicity, under the nonautonomous past delay effect. More exactly, we give conditions under which the trajectories of the semigroup $(S(t))_{t\geq 0}$ and the modified history functions $t \longmapsto \tilde{u}_t$ (and then the mild solution $t \longmapsto u(t) \in X$) have the same asymptotic properties. These results have been obtained for autonomous past delay by Casarino-Maniar-Piazzera [6, 7, 8] in the finite delay case. Here, assuming the exponential stability of the backward evolution family $U(\cdot, \cdot)$ (which cannot happen in the autonomous past delay case, since $U(\cdot, \cdot) = Id$), we are able to show this also in the infinite delay case. These asymptotic properties have been extensively studied, see, e.g., [4, 10, 12, 19, 20]

Finally, in Section 5, we illustrate our abstract results by the dynamical population equation

$$\begin{cases} u'(t,x) &= \Delta_N u(t,x) - du(t,x) + \int_0^r b(a)v(t,a,x)da - b_1 v(t,r,x),\ t \geq 0, x \in \Omega, \\ v'(t,a,x) &= -\frac{\partial}{\partial a}v(t,a,x) + \Delta_D v(t,a,x) - dv(t,a,x) - b(a)v(t,a,x),\ t \geq 0, \\ & \quad x \in \Omega, 0 \leq a \leq r, \\ v(t,0,x) &= f(x)u(t,x),\ t \geq 0, x \in \Omega, \end{cases} \tag{1.4}$$

where $u(t,x)$ is the density of the population at time t and position $x \in \Omega$, and $v(t,a,x)$ is the density of the subpopulation of pregnant individuals with time of gestation a, that at time t is in position x. This population equation with $b_1 = 0$ was studied in [15], by transforming it in a partial differential equation with nonautonomous past delay (1.1), where the delay operator Φ is of the form (1.3). Here, the term $-b_1 v(t,r,x)$ represents a foreign action on the new born individuals which were born at the time of gestation r. For more explanation of the biological signification of this equation and precise data, see Section 5 and also [15]. We show how to write (1.4) as a partial differential equation with nonautonomous past delay, with a delay operator Φ of the form (1.2) and then we study its wellposedness and its asymptotic behavior.

2. Preliminaries

Recall some definitions and basic results needed for the elaboration of this work.

Definition 2..1. A family $(U(t,s))_{t \leq s, t,s \in I}$ of bounded linear operators on a Banach space X is called an (exponentially bounded backward) evolution family if

(i) $U(t,\tau)U(\tau,s) = U(t,s), U(t,t) = Id$ for all $t \leq \tau \leq s, t, \tau, s \in I$,

(ii) the mapping $(t,s) \longmapsto U(t,s)$ is strongly continuous,

(iii) $\|U(t,s)\| \leq Me^{w(s-t)}$ for some $M \geq 1, w \in \mathbb{R}$ and all $t \leq s, t, s \in I$.

Here $I = [-r, 0]$ or $I = (-\infty, 0]$.

We will use evolution semigroup techniques for which we refer to [9, 21]. To that purpose, we extend $(U(t,s))_{t \leq s,t,s \in I}$ to an evolution family $(\tilde{U}(t,s))_{t \leq s}$ on $\mathbb{R}$.

Definition 2..2. (1) The evolution family $(U(t,s))_{t \leq s,t,s \in I}$ on X is extended to an evolution family $(\tilde{U}(t,s))_{t \leq s}$ by setting

$$\tilde{U}(t,s) := \begin{cases} U(-r,-r), & t \leq s < -r, \ (\text{if } r < \infty) \\ U(-r,s), & t < -r, s \in I, \ (\text{if } r < \infty) \\ U(t,s), & t \leq s, t, s \in I, \\ U(t,0), & t \leq 0 \leq s, \\ U(0,0), & 0 \leq t \leq s. \end{cases}$$

(2) On the space $\tilde{E} := L^p(\mathbb{R}, X)$, we define the corresponding evolution semigroup $(\tilde{T}(t))_{t\geq 0}$ by $(\tilde{T}(t)\tilde{f})(s) := \tilde{U}(s, s+t)\tilde{f}(s+t)$, $\tilde{f} \in \tilde{E}$, $s \in \mathbb{R}, t \geq 0$.

It is easy to prove that the semigroup $(\tilde{T}(t))_{t\geq 0}$ is strongly continuous on $\tilde{E}$. We denote its generator by $(\tilde{G}, D(\tilde{G}))$, where $D(\tilde{G})$ is a dense subset of $C_0(\mathbb{R}, X)$.

Since $(\tilde{G}, D(\tilde{G}))$ is a local operator see [23, Theorem 2.4], we can restrict it to the space $E := L^p(I, X)$ by the following definition.

Definition 2..3. Take

$$D(G) := \{\tilde{f}_{|I} : \tilde{f} \in D(\tilde{G})\}$$

and define $Gf := (\tilde{G}\tilde{f})_{|I}$ for $f = \tilde{f}_{|I} \in D(G)$.

This operator G is not a generator on E. However, if one identifies E with the subspace $\{f \in \tilde{E} : f(s) = 0,\ \forall s \notin I\}$, then E remains invariant under $(\tilde{T}(t))_{t\geq 0}$. As a consequence we obtain the following lemma.

Lemma 2..1. The semigroup $(T_0(t))_{t\geq 0}$ induced by $(\tilde{T}(t))_{t\geq 0}$ on E is

$$(T_0(t)f)(s) = \begin{cases} 0, & s+t > 0, \\ U(s, s+t)f(s+t), & s+t \leq 0, \end{cases} \tag{2.5}$$

and its generator is given by $G_0 = G$, $D(G_0) = \{f \in D(G) : f(0) = 0\}$.

We recall also some needed asymptotic properties for functions. Let J be an interval of $\mathbb{R}$. Let $BC(J, X)$ be the Banach space of all bounded continuous functions from J to X, endowed with the uniform norm, and $BUC(J, X)$ the Banach space of all bounded uniformly continuous functions from J to X. A function $g \in BC(\mathbb{R}, X)$ is said to be almost periodic if for every $\epsilon > 0$ there exists $l(\epsilon) > 0$ such that for every $a \in \mathbb{R}$ the interval $[a, a + l(\epsilon)]$ contains some τ for which the inequality

$$\|g(t+\tau) - g(t)\| \leq \epsilon$$

holds for all $t \in \mathbb{R}$. We note that $g \in AP(\mathbb{R}, X)$.

A function $f \in BC(\mathbb{R}_+, X)$ is called asymptotically almost periodic if there exist a unique $g \in AP(\mathbb{R}, X)$ and a unique $h \in C_0(\mathbb{R}_+, X)$ such that $f = h + g_{|\mathbb{R}_+}$.

A function $f \in BUC(\mathbb{R}_+, X)$ is said to be (uniformly) ergodic if the limit

$$\lim_{\alpha\to 0} \alpha \int_0^{+\infty} e^{-\alpha t} f(\cdot + t)dt$$

exists in $BUC(\mathbb{R}_+, X)$, and totally uniformly ergodic if the function $e^{i\theta}f(\cdot)$ is uniformly ergodic for all $\theta \in \mathbb{R}$.

A function $f \in BC(\mathbb{R}, X)$ is called almost automorphic ($f \in AA(\mathbb{R}, X)$) if for every sequence $(\sigma_n)_{n\in N}$ in $\mathbb{R}$ there exists a subsequence $(s_n)_{n\in N} \subset (\sigma_n)_{n\in\mathbb{N}}$ such that

$$\lim_{n,m\to+\infty} f(t + s_n - s_m) = f(t) \quad \text{for each } t \in \mathbb{R}.$$

This is equivalent to

$$g(t) := \lim_{n\to+\infty} f(t + s_n) \quad \text{and} \quad f(t) = \lim_{n\to+\infty} g(t - s_n)$$

are well defined for each $t \in \mathbb{R}$. The function g is measurable, but not necessarily continuous. Clearly, if the convergence in the definition above is uniform in $t \in \mathbb{R}$, then $f \in AP(\mathbb{R}, X)$. That is $AP(\mathbb{R}, X) \subset AA(\mathbb{R}, X)$.

A function $f \in BC(\mathbb{R}_+, X)$ is called asymptotically almost automorphic if there exist a unique $g \in AP(\mathbb{R}, X)$ and a unique $h \in \mathcal{C}_0(\mathbb{R}_+, X)$ such that $f = h + g_{|\mathbb{R}_+}$.

3. Wellposedness of partial differential equations with nonautonomous past delay

Consider the abstract equation with nonautonomous past delay

$$\begin{cases} u'(t) = Bu(t) + \Phi(\tilde{u}_t), & t \geq 0, \\ u(0) = x, \quad u_0 = g \in L^1([-r, 0], X), \end{cases} \tag{3.6}$$

where $(B, D(B))$ generates a C_0-semigroup $(S(t))_{t\geq 0}$ on a Banach space X, $r \leq +\infty$ and the delay operator $\Phi : D(\Phi) \subset L^1([-r, 0], X) \longrightarrow X$ is linear. The modified history function $\tilde{u}_t$ is defined as

$$\tilde{u}_t(s) := \begin{cases} U(s, 0)u(t+s) & \text{if}\, t+s > 0, \\ U(s, s+t)g(t+s) & \text{if}\, t+s \leq 0, \end{cases}$$

where $\mathcal{U} := (U(t, s))_{-r\leq t\leq s\leq 0}$ is an exponentially bounded backward evolution family on X (with constants ω, M).

It is shown in [1, 14, 16] that the wellposedness of (3.6) is equivalent to show that the operator matrix

$$\mathcal{A} := \begin{pmatrix} B & \Phi \\ 0 & G \end{pmatrix}$$

on the domain

$$D(\mathcal{A}) := \left\{ \begin{pmatrix} x \\ f \end{pmatrix} \in D(B) \times D(G) : f(0) = x \right\}$$

generates a strongly continuous semigroup in the product Banach space $\mathcal{E}_1 = X \times L^1([-r, 0], X)$. Hence, we concentrate now on the last fact. To obtain this claim, we use Miyadera-Voigt perturbation result. For this, we write $\mathcal{A}$ in the form

$$\mathcal{A} = \mathcal{A}_0 + \mathcal{B} := \begin{pmatrix} B & 0 \\ 0 & G \end{pmatrix} + \begin{pmatrix} 0 & \Phi \\ 0 & 0 \end{pmatrix}$$

with domain $D(\mathcal{A}_0) = D(\mathcal{A})$ and $\mathcal{B} : X \times D(\Phi) \longrightarrow \mathcal{E}_1$.

In [14], see also [22] for finite delay, it was shown that the operator $(\mathcal{A}_0, D(\mathcal{A}_0))$ generates a strongly continuous semigroup $(\mathcal{T}_0(t))_{t\geq 0}$ on $\mathcal{E}_1$ given by

$$\mathcal{T}_0(t) = \begin{pmatrix} S(t) & 0 \\ S_t & T_0(t) \end{pmatrix}, \quad t \geq 0,$$

where $S_t : X \longrightarrow L^1([-r,0],X)$ is given by

$$(S_t x)(\sigma) = \begin{cases} U(\sigma,0)S(t+\sigma)x & \text{for } \sigma+t>0, \\ 0 & \text{for } \sigma+t\leq 0, \end{cases}$$

and $(T_0(t))_{t\geq 0}$ is given by (2.5).

By using the Miyadera-Voigt perturbation Theorem [24, Theorem 2.2], we have the following main result.

Theorem 3..1. Assume that Φ is bounded linear from $\mathcal{C}([-r,0],X)\cap L^1([-r,0],X)$ to X satisfying the following condition: there exist $t_0>0$, $q_1 \in]0,1[$ and $c_1>0$ such that

$$(M) \qquad \int_0^{t_0} \|\Phi(S_t x + T_0(t)f)\| dt \leq q_1\|x\| + c_1\|f\|$$

for all $\binom{x}{f} \in D(\mathcal{A})$. Then, the matrix operator $\mathcal{A}$ generates a strongly continuous semigroup $(\mathcal{T}(t))_{t\geq 0}$ on $\mathcal{E}_1$ satisfying the variation of constants formula

$$\mathcal{T}(t)\binom{x}{f} = \mathcal{T}_0(t)\binom{x}{f} + \int_0^t \mathcal{T}_0(t-s)\mathcal{B}\mathcal{T}(s)\binom{x}{f}\, ds \tag{3.7}$$

for all $\binom{x}{f} \in D(\mathcal{A})$ and $t \geq 0$.

Proof. Let $c > c_1$, and the new norm $\|\binom{x}{f}\|_c = \|x\| + c\|f\|_{L^1}$ on $\mathcal{E}_1$, which is equivalent to the original norm $\|\cdot\|_{\mathcal{E}_1}$. Hence, by (M)

$$\begin{aligned}\int_0^{t_0} \|\mathcal{B}\mathcal{T}_0(t)\binom{x}{f}\|_c &= \int_0^{t_0} \|\Phi(S_t x + T_0(t)f)\| dt \\ &\leq q_1\|x\| + c_1\|f\|_{L^1} \\ &\leq \max\left(q_1, \frac{c_1}{c}\right) \|\binom{x}{f}\|_c.\end{aligned}$$

Setting $q = \max\left(q_1, \frac{c_1}{c}\right)$, we have $q<1$ and

$$\int_0^{t_0} \|\mathcal{B}\mathcal{T}_0(t)\binom{x}{f}\|_c \leq q\|\binom{x}{f}\|_c. \tag{3.8}$$

Since $\mathcal{A}_0$ is a generator of a strongly continuous semigroup on $(\mathcal{E}_1, \|\cdot\|_{\mathcal{E}_1})$, and being $\|\cdot\|_{\mathcal{E}_1}$ equivalent to $\|\cdot\|_c$, one has that $\mathcal{A}_0$ is a generator on $(\mathcal{E}_1, \|\cdot\|_c)$. By assumptions, $\mathcal{B} \in \mathcal{L}(D(\mathcal{A}_0), (\mathcal{E}_1, \|\cdot\|_c))$ and, by (3.8), $\mathcal{B}$ is a small Miyadera-Voigt perturbation of the semigroup $\mathcal{T}_0$. Hence, the Miyadera-Voigt perturbation Theorem implies that also $\mathcal{A} = \mathcal{A}_0 + \mathcal{B}$ generates a strongly continuous semigroup on $\mathcal{E}_1$ and satisfies (3.7).

Observe that, for $r=+\infty$, $\mathcal{C}([-\infty,0],X)$ denotes the space $\mathcal{C}_0((-\infty,0],X)$(i.e. the space of continuous functions vanishing at $-\infty$).

Remark 3..1. In the references [1, 2, 15, 16, 22], the above result was shown under a different Miyadera-Voigt condition $(M)'$: there is $t_0>0$, $q \in]0,1[$ such that

$$\int_0^{t_0} \|\Phi(S_t x + T_0(t)f)\| dt \leq q\|\binom{x}{f}\|_{\mathcal{E}_1}.$$

The Miyadera-Voigt condition (M) has appeared for the first time in [25] in the autonomous past delay case.

In the following lemma we give a class of operators Φ which satisfies the Miyadera-Voigt assumption (M).

Lemma 3..2. Let $r \leq \infty$ and $\eta : [-r,0] \longrightarrow \mathcal{L}(X)$ be of bounded variation such that $|\eta|([-r,0]) < \infty$, where $|\eta|$ is the positive Borel measure of $[-r,0]$ defined by the total variation on η. Let $\Phi : \mathcal{C}([-r,0],X) \cap L^1([-r,0],X) \longrightarrow X$ the operator given by

$$\Phi(f) := \int_{-r}^{0} d\eta(\theta) f(\theta). \tag{3.9}$$

Then, Φ fulfills the condition (M).

Proof. Let $0 < t_0 < r$. We have

$$\begin{aligned}
\int_0^{t_0} \|\Phi(S_t x + T_0(t)f)\| dt &= \int_0^{t_0} \| \int_{-r}^{0} d\eta(\theta)(S_t x + T_0(t)f)(\theta)\| dt \\
&\leq \int_0^{t_0} \| \int_{-t}^{0} d\eta(\theta) U(\theta,0) S(\theta+t)x\| dt + \int_0^{t_0} \| \int_{-r}^{-t} d\eta(\theta) U(\theta,\theta+t) f(\theta+t)\| dt \\
&\leq \int_0^{t_0} \int_{-t}^{0} d|\eta|(\theta) \|U(\theta,0)\| \|S(\theta+t)x\| dt + \int_0^{t_0} \int_{-r}^{-t} d|\eta|(\theta) \|U(\theta,\theta+t)\| \|f(\theta+t)\| dt.
\end{aligned}$$

By Fubini's theorem and a change of variables, we obtain

$$\begin{aligned}
\int_0^{t_0} \|\Phi(S_t x + T_0(t)f)\| dt \leq & M e^{|\omega| t_0} \int_0^{t_0} \int_{-t}^{0} d|\eta|(\theta) \|S(\theta+t)x\| dt \\
& + M e^{|\omega| t_0} \int_{-r}^{0} \int_{\theta}^{0} d|\eta|(\theta) \|f(t)\|\, dt \\
& \leq M_\omega (\sup_{0 \leq t \leq t_0} \|S(t)\| t_0 |\eta|([-r,0]) \|x\| + |\eta|([-r,0]) \|f\|_{L_1}) \\
& \leq M_w |\eta|([-r,0]) (t_0 \sup_{0 \leq t \leq t_0} \|S(t)\| \|x\| + \|f\|_{L_1}) \\
& \leq q_1 \|x\| + c_1 \|f\|_{L_1},
\end{aligned}$$

with $M_w := M e^{|w| t_0}$, $q_1 := \sup_{0 \leq t \leq t_0} \|S(t)\| M_w |\eta|([-r,0]) t_0$ and $c_1 := M_w |\eta|([-r,0])$. Choosing t_0 small so that $q_1 < 1$, the condition (M) is then satisfied. □

Corollary 3..1. *Let Φ given as in Lemma 3..2. Then, $(\mathcal{A}, D(\mathcal{A}))$ generates a strongly continuous semigroup $(\mathcal{T}(t))_{t \geq 0}$ on $\mathcal{E}_1$ satisfying the variation of constants formula*

$$\mathcal{T}(t) \binom{x}{f} = \mathcal{T}_0(t) \binom{x}{f} + \int_0^t \mathcal{T}_0(t-s) \mathcal{B} \mathcal{T}(s) \binom{x}{f}\, ds \tag{3.10}$$

for all $\binom{x}{f} \in D(\mathcal{A})$ and $t \geq 0$. Moreover, there is a unique mild solution u of the equation (3.6) *and it is given by*

$$u(t) = \begin{cases} \pi_1 \mathcal{T}(t) \binom{x}{f}, & t \geq 0, \\ f(t), & -r \leq t \leq 0. \end{cases}$$

4. Asymptotic behaviour of partial differential equations with nonautonomous past delay

We study the persistence of some asymptotic properties of the semigroup $(S(t))_{t\geq 0}$ under the modified delay term. More precisely, we give conditions under which the trajectories of the semigroups $S(\cdot)$ and $\mathcal{T}(\cdot)$ have the same asymptotic behavior. First, we show that this holds for $S(\cdot)$ and $\mathcal{T}_0(\cdot)$, and by perturbation techniques the claim will be obtained.

In the following theorem, we show first that $S(\cdot)$ and $(\mathcal{T}_0(t))_{t\geq 0}$ have some common asymptotic properties. The finite delay case can be shown similarly as in [7]; for the asymptotic almost automorphy see [6]. We give here the proof for the infinite delay case.

Theorem 4..1. Assume that the evolution family $(U(t,s))_{t\leq s\leq 0}$ is exponentially stable, and let $x \in X$. If the map $t \longmapsto S(t)x$ is

(i) in $\mathcal{C}_0(\mathbb{R}_+, X)$, the space of functions vanishing at infinity, or

(ii) asymptotically almost periodic, or

(iii) asymptotically almost automorphic, or

(iv) uniformly ergodic, or

(v) totally uniformly ergodic.

then $t \mapsto \mathcal{T}_0(t)\left(\begin{smallmatrix} x \\ f \end{smallmatrix}\right)$ has the same property for every $f \in L^1(\mathbb{R}_-, X)$.

Proof. Let $f \in L^1(\mathbb{R}_-, X)$. Since

$$\mathcal{T}_0(t)\left(\begin{smallmatrix} x \\ f \end{smallmatrix}\right) = \left(\begin{smallmatrix} S(t)x \\ S_t x + T_0(t) f \end{smallmatrix}\right),$$

we have only to verify that the map $t \mapsto S_t x + T_0(t)f$ has the same property as $S(\cdot)x$. To show (i), we should show that $t \mapsto S_t x + T_0(t)f \in \mathcal{C}_0(\mathbb{R}_+, L^1(\mathbb{R}_-, X))$, i.e.,

$$\int_{-\infty}^{0} \|S_t x(\theta) + T_0(t)f(\theta)\| d\theta \underset{t\to\infty}{\longrightarrow} 0.$$

We have

$$\begin{aligned}
\int_{-\infty}^{0} \|S_t x(\theta) + T_0(t)f(\theta)\| d\theta &\leq \int_{-\infty}^{0} \|S_t x(\theta)\| d\theta + \int_{-\infty}^{0} \|T_0(t)f(\theta)\| d\theta \\
\leq \int_{-\infty}^{0} 1_{t+\theta\geq 0} \|U(\theta,0)S(t+\theta)x\| d\theta &+ \int_{-\infty}^{0} 1_{t+\theta\leq 0} \|U(\theta,t+\theta)f(t+\theta)\| d\theta \\
\leq \int_{-\infty}^{0} 1_{\theta\geq -t} \|U(\theta,0)S(t+\theta)x\| d\theta &+ \int_{-\infty}^{-t} \|U(\theta,t+\theta)f(t+\theta)\| d\theta.
\end{aligned}$$

Since $S(\cdot)x$ is vanishing at $+\infty$, there exists $M' > 0$ such that $\sup_{t\geq 0} \|S(t)x\| \leq M'$, and then

$$1_{\theta\geq -t} \|U(\theta,0)S(t+\theta)x\| \leq M'Me^{w\theta} := g(\theta).$$

The function g is integrable on $\mathbb{R}_-$ and

$$1_{\theta\geq -t}\|U(\theta,0)S(t+\theta)x\| \underset{t\longrightarrow+\infty}{\longrightarrow} 0 \qquad \text{a.e. } \theta\in\mathbb{R}_-.$$

Hence, the Lebesgue dominated convergence Theorem implies that

$$\int_{-t}^{0}\|U(\theta,0)S(t+\theta)x\|d\theta \underset{t\longrightarrow+\infty}{\longrightarrow} 0 .$$

We have also

$$\begin{aligned}\|T_0(t)f\|_{L^1} = \int_{-\infty}^{-t}\|U(\theta,t+\theta)f(t+\theta)\|d\theta &= \int_{-\infty}^{0}\|U(\theta-t,\theta)f(\theta)\|d\theta\\ &\leq Me^{-tw}\|f\|_{L^1} \underset{t\longrightarrow+\infty}{\longrightarrow} 0 .\end{aligned}$$

Assume now (ii). It is clear that $T_0(\cdot)f$ is asymptotically almost periodic. Now it is sufficient to show that $\mathbb{R}_+\ni t\longmapsto S_tx$ is asymptotically almost periodic. From definition there exist a unique $g\in AP(\mathbb{R},X)$ and a unique $h\in\mathcal{C}_0(\mathbb{R}_+,X)$ such that

$$S(t)x = h(t)+g(t), \quad t\geq 0. \tag{4.11}$$

Hence,

$$\begin{aligned}S_tx(\theta) &= \begin{cases} U(\theta,0)S(t+\theta)x, & \theta+t>0\\ 0, & \theta+t\leq 0\end{cases}\\ &= \begin{cases} U(\theta,0)g(t+\theta)+U(\theta,0)h(t+\theta), & \theta+t>0\\ 0, & \theta+t\leq 0.\end{cases}\end{aligned}$$

Since $h\in\mathcal{C}_0(\mathbb{R}_+,X)$ and $U(\theta,0)$ is exponentially stable, then, as in (i),

$$h_1(t)(\theta) := 1_{\theta\geq -t}U(\theta,0)h(t+\theta)\in\mathcal{C}_0(\mathbb{R}_+,L^1(\mathbb{R}_-,X)).$$

Set $g_1(t)(\theta) := U(\theta,0)g(t+\theta)$ and $h_2(t)(\theta) := h_1(t)(\theta) - 1_{t+\theta\leq 0}U(\theta,0)g(t+\theta)$ for all $t\in\mathbb{R}$ and $\theta\in\mathbb{R}_-$. Then,

$$\begin{aligned}(g_1(t)+h_2(t))(\theta) &= \begin{cases} g_1(t)(\theta)+h_1(t)(\theta) & \text{for } \theta+t>0,\\ g_1(t)(\theta)-g_1(t)(\theta)=0 & \text{for } \theta+t\leq 0\end{cases}\\ &= S_tx(\theta).\end{aligned} \tag{4.12}$$

Since $h_1\in\mathcal{C}_0(\mathbb{R}_+,L^1(\mathbb{R}_-,X))$ and

$$\int_{-\infty}^{0}1_{t+\theta\leq 0}\|U(\theta,0)g(t+\theta)\|d\theta \leq \|g\|_\infty\int_{-\infty}^{-t}\|U(\theta,0)\|d\theta \underset{t\longrightarrow+\infty}{\longrightarrow} 0,$$

it follows that $h_2\in\mathcal{C}_0(\mathbb{R}_+,L^1(\mathbb{R}_-,X))$. Since g is almost periodic, then for all $\epsilon>0$, there exists $l(\epsilon)>0$ such that for all $a\in\mathbb{R}$, there exists $\tau\in[a,a+l(\epsilon)]$ such that

$$\|g(t+\theta+\tau)-g(t+\theta)\|\leq\epsilon \quad \text{for all} \quad t\in\mathbb{R},\theta\in I\!R_-,$$

and moreover

$$\begin{aligned}\|g_1(t+\tau)-g_1(t)\|_{L^1} &= \int_{-\infty}^{0} \|U(\theta,0)g(t+\theta+\tau)-U(\theta,0)g(t+\theta)\|d\theta \\ &\le \int_{-\infty}^{0} \|U(\theta,0)\|\|g(t+\theta+\tau)-g(t+\theta)\|d\theta \\ &\le \epsilon\int_{-\infty}^{0} \|U(\theta,0)\|d\theta = \epsilon\|U(\cdot,0)\|_{L^1}.\end{aligned}$$

This means that g_1 is almost periodic, and finally $t \longmapsto S_t x$ is asymptotically almost periodic. For the asymptotic almost automorphy, assume that $S(t)x = h(t) + g(t), t \ge 0$, where $g \in AA(\mathbb{R}, X)$ and $h \in \mathcal{C}_0(\mathbb{R}_+, X)$. From (4.12), $S_t x = g_1(t) + h_2(t), t \ge 0$, with $h_2 \in \mathcal{C}_0(\mathbb{R}_+, X)$. To conclude we have to show that g_1 is almost automorphic on $I\!R$. For this, consider a sequence $(\sigma_n)_{n\in N}$ in $I\!R$. Since $g \in AA(I\!R, X)$, there is a subsequence $(s_n)_{n\in N}$ such that

$$\lim_{n,m\longrightarrow+\infty} g(t+\theta+s_n-s_m) = g(t+\theta)$$

for each $t \in I\!R, \theta \in I\!R_-$. Therefore, as for the almost periodicity, one can see that $g_1 \in AA(I\!R, L^1(I\!R_-, X))$. As $(T_0(t))_{t\ge0}$ is exponentially stable then it is asymptotically almost automorphic.

For (iv), we have that the limit $F(\cdot) = \lim_{\alpha\to0} \alpha \int_0^{+\infty} e^{-\alpha t} S(\cdot + t)x dt$ exists in $BUC(\mathbb{R}_+, X)$ for each $x \in X$, and then $F(\cdot) \in BUC(\mathbb{R}_+, X)$. Put

$$F_t(\theta) = \begin{cases} U(\theta,0)F(t+\theta) & \text{for } \theta+t>0 \\ U(\theta,0)F(0) & \text{for } \theta+t\le0. \end{cases}$$

Show first that $t \mapsto F_t \in BUC(\mathbb{R}_+, L^1(\mathbb{R}_-, X))$. For $h > 0$, we have

$$\begin{aligned}&\int_{-\infty}^{0} \|F_{s+h}(\theta)-F_s(\theta)\|d\theta \\ &\le \int_{-\infty}^{0} \|1_{s+h+\theta\ge0}U(\theta,0)F(s+h+\theta)-1_{s+\theta\ge0}U(\theta,0)F(s+\theta)\|d\theta \\ &+ \int_{-\infty}^{0} \|1_{s+h+\theta\le0}U(\theta,0)F(0)-1_{s+\theta\le0}U(\theta,0)F(0)\|d\theta \\ &\le \int_{-s-h}^{-s} \|U(\theta,0)F(s+h+\theta)\|d\theta + \int_{-s}^{0} \|U(\theta,0)\|\|F(s+h+\theta)-F(s+\theta)\|d\theta \\ &+ \int_{-s-h}^{-s} \|U(\theta,0)F(0)\|d\theta \\ &\le 2h\|U(\cdot,0)\|_\infty\|F\|_\infty + \int_{-s}^{0} \|U(\theta,0)\|\|F(s+h+\theta)-F(s+\theta)\|d\theta.\end{aligned}$$

Since $F(\cdot) \in BUC(\mathbb{R}_+, X)$ then, for $\epsilon > 0$ there exists $h_\epsilon > 0$ such that $\|F(s+h+\theta) - F(s+\theta)\| \le \epsilon$ for all $s+\theta \ge 0$ and $0 < h < h_\epsilon$. Hence,

$$\|F_{s+h}(\cdot) - F_s(\cdot)\|_\infty \le h\|U(\cdot,0)\|_\infty\|F\|_\infty + \epsilon\|U(\cdot,0)\|_{L^1}.$$

The same for $h < 0$. To achieve (iv) we show that the limit $\lim_{\alpha\to 0} \alpha \int_0^{+\infty} e^{-\alpha t} S_{\cdot+t} x dt = F_{\cdot}$ exists in $BUC(\mathbb{R}_+, L^1(\mathbb{R}_-, X))$. Let $s \geq 0$.

$$\int_{-\infty}^{0} \|\alpha \int_0^{+\infty} e^{-\alpha t} S_{s+t} x dt(\theta) - F_s(\theta)\| d\theta$$
$$= \int_{-\infty}^{0} \|\alpha \int_0^{+\infty} e^{-\alpha t} S_{s+t} x(\theta) dt - F_s(\theta)\| d\theta$$
$$= \int_{-\infty}^{0} \|\alpha \int_0^{+\infty} 1_{s+t+\theta\geq 0} e^{-\alpha t} U(\theta,0) S(s+t+\theta) x dt - 1_{s+\theta>0} U(\theta,0) F(s+\theta)$$
$$- 1_{s+\theta\leq 0} U(\theta,0) F(0)\| d\theta$$
$$\leq \int_{-s}^{0} \|U(\theta,0)\| \|\alpha \int_0^{+\infty} e^{-\alpha t} S(s+t+\theta) x dt - F(s+\theta)\| d\theta$$
$$+ \int_{-\infty}^{-s} \|U(\theta,0)\| \|\alpha \int_{-s-\theta}^{+\infty} e^{-\alpha t} S(s+t+\theta) x dt - F(0)\| d\theta$$
$$\leq \sup_{s\geq 0} \|\alpha \int_0^{+\infty} e^{-\alpha t} S(s+t) x dt - F(s)\| \|U(\cdot,0)\|_{L^1}$$
$$+ \int_{-\infty}^{-s} e^{\alpha(s+\theta)} \|U(\theta,0)\| \|\alpha \int_0^{+\infty} e^{-\alpha t} S(t) x dt - F(0)\| d\theta$$
$$+ \int_{-\infty}^{-s} \|U(\theta,0)\| \|F(0) - e^{\alpha(s+\theta)} F(0)\| d\theta.$$

Hence

$$\int_{-\infty}^{0} \|\alpha \int_0^{+\infty} e^{-\alpha t} S_{s+t} x dt(\theta) - F_s(\theta)\| d\theta$$
$$\leq 2 \sup_{s\geq 0} \|\alpha \int_0^{+\infty} e^{-\alpha t} S(s+t) x dt - F(s)\| \|U(\cdot,0)\|_{L^1}$$
$$+ M\|F(0)\| \int_{-\infty}^{0} e^{\omega\sigma}(1 - e^{\alpha\sigma}) d\sigma,$$

and this ends the proof of (iv). The same proof yields the assertion (v). □

Using [3, Lemma 7.2] and the variation of constants formula (3.7), we obtain the asymptotic properties of the perturbed semigroup $\mathcal{T}(\cdot)$ and then the ones of the trajectories of the solutions to equation (3.6).

Theorem 4..2. Assume that the evolution family $(U(t,s))_{t\leq s\leq 0}$ is exponentially stable, and that Φ is bounded linear from $\mathcal{C}([-r,0],X) \cap L^1([-r,0],X)$ to X. Assume also that $S(\cdot)x$ satisfies one of the asymptotic properties (i)-(v) for all $x \in X$, and there are $0 \leq q_1 < 1$ and $c_1 > 0$ such that

$$\int_0^{\infty} \|\Phi(S_t x + T_0(t) f)\| dt \leq q_1 \|x\| + c_1 \|f\| \tag{4.13}$$

for all $\binom{x}{f} \in D(\mathcal{A}_0)$. Then, $\mathcal{T}(\cdot)\binom{x}{f}$ has the same asymptotic properties for all $\binom{x}{f} \in \mathcal{E}_1$.

Proof. As in the proof of Theorem 3..1, one can find $0 \le q < 1$ and $c > 0$ such that

$$\int_0^\infty \|\mathcal{B}\mathcal{T}_0(t)\left(\begin{smallmatrix} x \\ f \end{smallmatrix}\right)\|_c \le q\|\left(\begin{smallmatrix} x \\ f \end{smallmatrix}\right)\|_c \tag{4.14}$$

for all $\left(\begin{smallmatrix} x \\ f \end{smallmatrix}\right) \in D(\mathcal{A})$, where $\|\cdot\|_c$ is a norm equivalent to $\|\cdot\|_{\mathcal{E}_1}$. Fix now $\left(\begin{smallmatrix} x \\ f \end{smallmatrix}\right) \in D(\mathcal{A})$. From Theorem 4..1, the trajectory $\mathcal{T}_0(\cdot)\left(\begin{smallmatrix} x \\ f \end{smallmatrix}\right)$ has the same asymptotic properties of $S(\cdot)x$ with respect to the new norm $\|\cdot\|_c$. From Theorem 3..1, we have

$$\mathcal{T}(t)\left(\begin{smallmatrix} x \\ f \end{smallmatrix}\right) = \mathcal{T}_0(t)\left(\begin{smallmatrix} x \\ f \end{smallmatrix}\right) + \int_0^t \mathcal{T}_0(t-s)\mathcal{B}\mathcal{T}(s)\left(\begin{smallmatrix} x \\ f \end{smallmatrix}\right) ds, \quad t \ge 0. \tag{4.15}$$

Hence, it is sufficient to show that the function

$$I\!R_+ \ni t \longmapsto f(t) := \int_{-\infty}^t \mathcal{T}_0(t-s)g(s)ds,$$

satisfies the same properties, where g is defined by

$$g(s) := \begin{cases} \mathcal{B}\mathcal{T}(s)\left(\begin{smallmatrix} x \\ f \end{smallmatrix}\right), & s \ge 0, \\ 0, & s \le 0, \end{cases}$$

For assertions (i)-(ii)-(iv)-(v), by [3, Lemma 7.2] it suffices to show $g \in L^1(I\!R, (\mathcal{E}_1, \|\cdot\|_C))$, and this can be deduced from (4.14) and (3.7). As the semigroup $S(\cdot)$ is bounded, the assertion (iii) follows from [6, Proposition 2.3] and (ii). Finally, all the assertions follow for all $\left(\begin{smallmatrix} x \\ f \end{smallmatrix}\right) \in \mathcal{E}_1$ by density of $D(\mathcal{A})$.

The asymptotic behavior of the solutions of (3.6) can be now obtained.

Corollary 4..1. *Assume that $S(\cdot)x$ satisfies one of the asymptotic properties (i)-(v) for all $x \in X$ and that Φ is a bounded linear from $\mathcal{C}([-r, 0], X) \cap L^1([-r, 0], X)$ to X satisfying (4.13). Then, $t \longmapsto \tilde{u}_t$ and $t \longmapsto u(t)$ have the same asymptotic properties as $S(\cdot)x$.*

5. Application

In this section, we use the abstract results of the previous ones to study the wellposedness and asymptotic behavior of the dynamical population equation

$$\begin{cases} u'(t,x) &= \Delta_N u(t,x) - du(t,x) + \int_0^r b(a)v(t,a,x)da - b_1 v(t,r,x),\ t \ge 0, x \in \Omega, \\ v'(t,a,x) &= -\frac{\partial}{\partial a}v(t,a,x) + \Delta_D v(t,a,x) - dv(t,a,x) - b(a)v(t,a,x),\ t \ge 0, \\ & \quad x \in \Omega, 0 \le a \le r, \\ v(t,0,x) &= f(x)u(t,x),\ t \ge 0, x \in \Omega, \end{cases} \tag{5.16}$$

where Δ_D and Δ_N denote the Laplace operators whit Dirichlet and Neumann conditions, respectively. Here t, r are positive, $a \in [0, r]$, $d, b_1 > 0$ and $b \in L^1([0, r], I\!R_+)$. The space variable x is supposed to vary in $\Omega \subset \mathbb{R}^n$ where Ω is open, connected and bounded with smooth boundary. The condition

$$\frac{\partial}{\partial n}u(t,x) = 0 \quad in \quad \partial\Omega, \tag{5.17}$$

where n denotes, as usual, the outward normal, states that the population cannot cross the boundary.
The condition

$$v(t,a,x)=0 \quad if \quad x\in\partial\Omega,\ t\geq 0,\ a\in[0,r] \tag{5.18}$$

says that no pregnant individual reaches the borderline. Moreover the condition $f(x)=0$ for $x\in\partial\Omega$ is required. The initial conditions

$$u(0,x)=u_0(x), u(\tau,x)=g(\tau,x),\ \tau\in(-r,0), \tag{5.19}$$

$$v(0,a,x)=v_0(a,x),\ v_0(a,x)=f(x)g(-a,x),\ a\in[0,r], \tag{5.20}$$

where v_0, g and u_0 are given functions.

5.1. The population equation as an abstract equation with nonautonomous past delay

In this subsection, we show in details how to transform the population equation (5.16)-(5.20) in a partial differential equation with nonautonomous past delay, and then study its wellposedness and asymptotic behavior. To this purpose, we start by solving the second equation along the characteristic lines in the plane (t,a), namely in the strip $[0,+\infty)\times[0,r]$. Set

$$V(s,x)=v(t_0+s,a_0+s,x),$$

where $a_0\in[0,r], t_0\geq 0$ are fixed, while x varies in Ω and s in $[0,+\infty)$. Rewriting the second equation in (5.16) with $V(s,x)$, one obtains

$$\begin{aligned}&\frac{\partial}{\partial s}V(s,x)=\frac{\partial}{\partial t}v(t_0+s,a_0+s,x)+\frac{\partial}{\partial a}v(t_0+s,a_0+s,x)\\&=\Delta_D v(t_0+s,a_0+s,x)-dv(t_0+s,a_0+s,x)-b(a_0+s)v(t_0+s,a_0+s,x)\\&=\Delta_D V(s,x)-dV(s,x)-b(a_0+s)V(s,x).\end{aligned}$$

To solve the above equation we follow the abstract approach choosing $X=L^1(\Omega)$ as a Banach space and denoting $D(\Delta_D)$ the domain of Δ_D on X. For $t<a$, putting $t_0=0$, we obtain

$$\begin{cases}V'(s,x)=\Delta_D V(s,x)-dV(s,x)-b(a_0+s)V(s,x)\\V(0,x)=v(0,a_0,x)=v_0(a_0,x),\end{cases}$$

which has the unique solution

$$V(s,\cdot)=e^{-\int_0^s b(a_0+\sigma)d\sigma}e^{s(\Delta_D-d)}V(0,\cdot),$$

where d denotes the multiplication operator and $e^{t(\Delta_D-d)}$ the strongly continuous semigroup generated by the linear operator Δ_D-d on X with the appropriate domain. Hence

$$\begin{aligned}v(s,a_0+s,\cdot)&=e^{\int_0^s b(a_0+\sigma)d\sigma}e^{s(\Delta_D-d)}v(0,a_0,\cdot)\\&=e^{\int_{a_0}^{a_0+s} b(\sigma)d\sigma}e^{s(\Delta_D-d)}v_0(a_0,\cdot).\end{aligned}$$

Setting $t = s$ and $a_0 + s = a$, this becomes

$$v(t,a,\cdot) = e^{\int_{a-t}^{a} b(\sigma)d\sigma} e^{s(\Delta_D - d)} v_0(a-t,\cdot).$$

For $t > a$, putting $a_0 = 0$, we obtain

$$\begin{cases} V'(s,x) = \Delta_D V(s,x) - dV(s,x) - b(s)V(s,x) \\ V(0,x) = v(t_0,0,x) = f(x)u(t_0,x). \end{cases}$$

Solving this equation we obtain

$$\begin{aligned} v(t_0+s,s,\cdot) = V(s,\cdot) &= e^{\int_0^s b(\sigma)d\sigma} e^{s(\Delta_D-d)} V(0,\cdot) \\ &= e^{\int_0^s b(\sigma)d\sigma} e^{s(\Delta_D-d)} v(t_0,0,\cdot) \\ &= e^{\int_0^s b(\sigma)d\sigma} e^{s(\Delta_D-d)} f(\cdot)u(t_0,\cdot). \end{aligned}$$

Setting $t_0 + s = t$, this becomes

$$v(t,a,\cdot) = e^{\int_0^a b(\sigma)d\sigma} e^{a(\Delta_D-d)} f(\cdot)u(t-a,\cdot).$$

Substituting $v(t,a,x)$ in the first equation of (5.16), one has, for $t > r$,

$$\begin{aligned} u'(t,x) &= (\Delta_N - d)u(t,x) + \int_0^r b(a)v(t,a,x)da - b_1 v(t,r,x) \\ &= (\Delta_N - d)u(t,x) + \int_0^r b(a)[e^{\int_0^a b(\sigma)d\sigma} e^{a(\Delta_D-d)} f(x)u(t-a,x)]da \\ &- b_1 e^{\int_0^r b(\sigma)d\sigma} e^{r(\Delta_D-d)} f(x)u(t-r,x) \\ &= (\Delta_N - d)u(t,x) + \int_{-r}^0 b(-a)[e^{-\int_0^{-a} b(\sigma)d\sigma} e^{-a(\Delta_D-d)} f(x)u(t+a,x)]da \\ &- b_1 e^{\int_0^r b(\sigma)d\sigma} e^{r(\Delta_D-d)} f(x)u(t-r,x). \end{aligned}$$

For $t < r$, we have

$$\begin{aligned} u'(t,x) &= (\Delta_N - d)u(t,x) + \int_0^r b(a)v(t,a,x)da - b_1 v(t,r,x) \\ &= (\Delta_N - d)u(t,x) + \int_0^t b(a)v(t,a,x)da + \int_t^r b(a)v(t,a,x)da - b_1 v(t,r,x) \\ &= (\Delta_N - d)u(t,x) + \int_{-t}^0 b(-a)[e^{-\int_0^{-a} b(\sigma)d\sigma} e^{-a(\Delta_D-d)} f(x)u(t+a,x)]da \\ &+ \int_{-r}^{-t} b(-a)[e^{-\int_{a-t}^{-a} b(\sigma)d\sigma} e^{t(\Delta_D-d)} v_0(-a-t,x)]da - b_1 e^{-\int_{r-t}^{r} b(\sigma)d\sigma} e^{t(\Delta_D-d)} v_0(r-t,x). \end{aligned}$$

We put

$$U(t,s) = e^{-\int_{-s}^{-t} b(\sigma)d\sigma} e^{(s-t)(\Delta_D-d)}, \quad t \le s \le 0,$$

and

$$\Phi(\varphi) = \int_{-r}^0 b(-s)f(\cdot)\varphi(s)ds - b_1 f(\cdot)\varphi(-r), \quad \varphi \in C([-r,0],X).$$

It is easy to verify that $(U(t,s))_{t\le s\le 0}$ is an evolution family.

The modified history function $\tilde{u}_t$ is

$$\tilde{u}_t(s,\cdot) := \begin{cases} U(s,0)u(t+s,\cdot) & \text{if } t+s>0 \\ U(s,s+t)g(t+s,\cdot) & \text{if } t+s\le 0. \end{cases}$$

Hence,

$$\Phi(\tilde{u}_t(\cdot)) = \int_{-r}^{0} b(-s)f(\cdot)\tilde{u}_t(s,\cdot)ds - b_1 f(\cdot)\tilde{u}_t(-r,\cdot).$$

Since for $t>r$, $\tilde{u}_t(-r,\cdot) = U(-r,0)u(t-r,\cdot)$ and $s\ge -r, t+s\ge t-r\ge 0 \Longrightarrow \tilde{u}_t(s,\cdot) = U(s,0)u(t+s,\cdot)$, through

$$\Phi(\tilde{u}_t) = \int_{-r}^{0} b(-s)f(\cdot)U(s,0)u(t+s,\cdot)ds - b_1 f(\cdot)U(-r,0)u(t-r,\cdot).$$

If we set $B=\Delta_N - d$, then

$$\begin{aligned} Bu(t,\cdot)+\Phi(\tilde{u}_t) &= (\Delta_N - d)u(t,\cdot) + \int_{-r}^{0} b(-a)f(\cdot)U(a,0)u(a+s,\cdot)da \\ &\quad - b_1 f(\cdot)U(-r,0)u(t-r,\cdot) = u'(t,\cdot). \end{aligned}$$

Thus, by setting $u(t) := u(t,\cdot)$, we obtain

$$u'(t) = Bu(t) + \Phi(\tilde{u}_t).$$

For $t\le r$, $\tilde{u}_t(-r,\cdot) = U(-r,t-r)g(t+s,\cdot)$ and

$$\begin{aligned} \Phi(\tilde{u}_t) &= \int_{-r}^{0} b(-s)f(\cdot)\tilde{u}_t(s,\cdot)ds - b_1 f(\cdot)U(-r,t-r)g(t-r,\cdot) \\ &= \int_{r}^{-t} b(-s)f(\cdot)U(s,s+t)g(t+s,\cdot)ds \\ &+ \int_{t}^{0} b(-s)f(\cdot)U(s,0)u(t+s,\cdot)ds - b_1 f(r)U(-r,t-r)g(t-r,\cdot). \end{aligned}$$

Summarizing, the population equation is written as the abstract equation with nonautonomous past

$$\begin{cases} u'(t) = Bu(t) + \Phi(\tilde{u}_t), & t\ge 0, \\ u(0) = x, \quad u_0 = g, \end{cases} \tag{5.21}$$

where the operator $B = \Delta_N - d$ generates an exponentially stable C_0-semigroup $S := (S(t))_{t\ge 0}$ on $X = L^1(\Omega)$.

5.2. Wellposedness and asymptotic behavior of the population equation

We have transformed the population equation (5.16)- (5.20) into the abstract equation with nonautonomous past

$$\begin{cases} u'(t) = Bu(t) + \Phi(\tilde{u}_t), & t\ge 0, \\ u(0) = x, \quad u_0 = g, \end{cases} \tag{5.22}$$

where the operator $B = \Delta_N - d$ generates an exponentially stable C_0-semigroup $S := (S(t))_{t\geq 0}$ on $X = L^1(\Omega)$ and the backward evolution family $U(t,s) = e^{-\int_{-s}^{-t} b(\sigma)d\sigma} e^{(s-t)(\Delta_D - d)}$, $t \leq s \leq 0$, is exponentially stable. Moreover, the operator Φ, given by

$$\Phi(\varphi) = \int_{-r}^{0} b(-s)f(\cdot)\varphi(s)ds - b_1 f(\cdot)\varphi(-r),$$

can be written as a Stieltjes integral

$$\Phi(\varphi) = \int_{-r}^{0} d\eta(s)\varphi(s),$$

where $\eta : [-r, 0] \longrightarrow \mathcal{L}(X)$ is the function with a bounded variation on $[-r, 0]$ given by

$$\eta(s) = -b_1 f(\cdot)1_{[-r,0]} + \int_{0}^{s} b(-\tau)f(\cdot)d\tau.$$

Here, we have considered a delay operator Φ with one discrete delay, but one can consider several discrete delays and the delay operator

$$\Phi(\varphi) = \int_{-r}^{0} b(-s)f(\cdot)\varphi(s)ds - \sum_{i=1}^{n} b_i f(\cdot)\varphi(-r_i),$$

with $0 \leq r_1 < r_2 < ... < r_n = r$. In this case η is the bounded variation function

$$\eta(s) = -\sum_{i=1}^{n} b_i f(\cdot)1_{[-r_i,0]} + \int_{0}^{s} b(-\tau)f(\cdot)d\tau.$$

Hence, by Lemma 3..2, these operators Φ satisfy the Miyadera-Voigt condition (M). Therefore, by Corollary 3..1, there is a unique mild solution of equation (5.22). Moreover, we have the following result.

Theorem 5..1. There is a unique mild solution $u : I\!R_+ \longrightarrow L^1(\Omega)$ of the population equation (5.16)-(5.20). Moreover, if $|\eta|([-r,0]) < d$ then $\|u(t)\|_{L^1(\Omega)} \longrightarrow 0, \quad t \to +\infty$.

Proof. One has

$$\begin{aligned}
&\int_0^{+\infty} \|\Phi(S_t x + T_0(t)f)\|dt = \int_0^{+\infty} \|\int_{-r}^{0} d\eta(\theta)(S_t x + T_0(t)f)(\theta)\|dt \\
&\leq \int_0^{+\infty} \|\int_{-t}^{0} d\eta(\theta)U(\theta,0)S(\theta+t)x\|dt + \int_0^{r} \|\int_{-r}^{-t} d\eta(\theta)U(\theta,\theta+t)f(\theta+t)\|dt \\
&\leq \int_0^{+\infty}\int_{-t}^{0} d|\eta|(\theta)\|U(\theta,0)\|\|S(\theta+t)x\|dt + \int_0^{r}\int_{-r}^{-t} d|\eta|(\theta)\|U(\theta,\theta+t)\|\|f(\theta+t)\|dt \\
&\leq \frac{1}{d}|\eta|([-r,0])\|x\| + |\eta|([-r,0])\|f\|.
\end{aligned}$$

Hence, by assumption $q_1 = \dfrac{1}{d}|\eta|([-r,0]) < 1$, and since S is exponentially stable, by Theorem 4..2, the result follows.

□

References

[1] A. Bátkai and S. Piazzera, Semigroups and linear partial differential equations with delay, *J. Math. Anal. Appl.* **264** (2001), 1-20.

[2] A. Bátkai and S. Piazzera, *Semigroups for Delay Equations on L^P-phase Spaces*, book manuscript, Research Notes in Mathematics **10**, A. K. Peters, Ltd., Wellesley, MA, (2005).

[3] C. J. K. Batty and R. Chill, Bounded convolutions and solutions of inhomogeneous Cauchy problems, *Forum Math.* **11** (1999), 253–277.

[4] S. Boulite, L. Maniar and G. M. N'Guérékata, Almost automorphic solutions for hyperbolic semilinear evolution equations, *Semigroup Forum* 71 (2005), 231–240.

[5] S. Brendle and R. Nagel, Partial functional differential equations with nonautonomous past, *Disc. cont. Dyn. Syst.* **8** (2002), 1-24.

[6] V. Casarino, Almost automorphic groups and semigroups. *Rend. Accad. Naz. Sci. XL Mem. Mat. Appl.* **24** (2000), 219–235.

[7] V. Casarino and S. Piazzera, On the stability of asymptotic properties of perturbed C_0-semigroups, *Forum Math.* **13** (2001), 91-107.

[8] V. Casarino, L. Maniar and S. Piazzera, *The asymptotic behaviour of perturbed evolution families, Differential Integral Equations* **15** (2002), 567–586

[9] C. Chicone and Y. Latushkin, *Evolution Semigroups in Dynamical Systems and Differential Equations, Mathematical Surveys and Monographs* **70**, American Mathematical Society, Rhode Island, (1999).

[10] C. Corduneanu, *Almost Periodic Functions*, 2nd Edition, Chelsea-New York, 1989.

[11] K.J. Engel and R. Nagel, *One-parameter Semigroups for Linear Evolution Equations* , *Graduate Texts in Mathematics* **194**, Springer-Verlag (2000).

[12] A. M. Fink, Almost Periodic Differential Equations, *Lecture Notes in Mathematics* 377, Springer-Verlag, New York, 1974.

[13] G. Fragnelli, Classical solutions for PDEs with nonautonomous past in L^p- spaces, *Bulletin of the Belgian Mathematical Society* **11** (2004), 133–148.

[14] G. Fragnelli and G. Nickel, Partial functional differential equations with nonautonomous past in L^p-phase spaces, *Diff. Int. Equ.* **16** (2003), 327–348.

[15] G. Fragnelli and L. Tonetto, A population equation with diffusion, *J. Math. Anal. Appl.* **289** (2004), 90–99.

[16] G. Fragnelli, An age dependent population equation with diffusion and delayed birth process, *Int. J. Math. Math. Sci.* **20** (2005), 3273–3289.

[17] I. Miyadera, On perturbation theory for semi-groups of operators, *Tôhoku Math. J.* **18** (1966), 299–310.

[18] R. Nagel and G. Nickel, Well-posedness for nonautonomous abstract Cauchy problems, Progr. *Nonlinear Differential Equations Appl.* **50**, Birkhäuser, (2002), 279–293.

[19] G. M. N'Guérékata, *Almost Automorphic Functions and Almost Periodic Functions in Abstract Spaces*, Kluwer Academic / Plenum Publishers, New York, 2001.

[20] G. M. N'Guérékata, *Topics in Almost Automorphy*, Springer-Verlag, New York, 2005.

[21] G. Nickel, Evolution semigroups for nonautonomous Cauchy problems, *Abstr. Appl. Anal.* **2** (1997), 73–95.

[22] G. Nickel and A. Rhandi, Positivity and Stability of Delay Equations with Nonautonomous Past, *Math. Nachr.* **278** (2005), 1–13.

[23] F. Räbiger, A. Rhandi, R. Schnaubelt and J. Voigt, Non-autonomous Miyadera perturbations, *Diff. Int. Eq.* **13** (2000), 341–368.

[24] J. Voigt, On the perturbation theory for strongly continuous semigroups, *Math. Ann.* **229** (1977), 163–171.

[25] M. Stein, H. Vogt, J. Voigt, The modulus semigroup for linear delay equations III. *J. Funct. Anal.* **220** (2005), 388–400.

In: Research on Evolution Equation Compendium. Volume 2 ISBN: 978-1-60456-752-6
Editor: Gaston M. N'Guerekata

SOLVING THE HYPERBOLIC PROBLEM OBTAINED BY TRANSMUTATION OPERATOR

Hikmet Koyunbakan*
Firat University, Department of Mathematics, 23119
Elazig, TURKEY

Abstract

In this paper, we solve the problem of constructing the kernels of transmutation operator and give the generalized transmutation operator for a singular problem.

Key Words: Inverse Problem, Hyperbolic Equation

AMS Subject Classification: 31A25, 35L20.

1. Introduction

The standard problem on spectral theory is to solve (1.1) for y with appropriate boundary conditions, for a given potential $q(x)$. In the unusual application all that can be observed is the asymptotic behaviour of y for large $|x|$, which is taken for the whole spectrum comprising at the scattering data for $q(x)$.

Inverse eigenvalue problems, usually for equations of the form

$$-y'' + q(x)y = \lambda y \tag{1.1}$$

with a variety of boundary conditions, have an extensive literature [3], [4], [5], [6], [14]. Unless $q(x)$ is constrained, one needs the spectra from two sets of boundary conditions, or one spectrum and a sequence of norming constants to uniquely determine $q(x)$. In the past, important works on these problems were done by Gelfand and Levitan [4], Levinson [8] and Levitan [9]. These authors show that the inverse problem is solved by

$$q(x) = 2\frac{d}{dx}H(x, x),$$

—

*E-mail address: hkoyunbakan@gmail.com

where $H(x,t)$ is nucleus function of transmutation operator.

One topic of inverse problem for spectral analysis is transmutation operator [1], [2], [7], [15]. In [5], [11] Gilbert examined some problems for boundary value problems by using this method, in [7], Hryniv and Mykytyuk, constructed transformation operators for Srurm Liouville operators with singular potentials, in [1], Boumenir and Tuan proved the existence of a transmutation operator between two weighted Sturm-Liouville operator and in [15] Volk solved Hyperbolic equation having singularity type $\frac{m^2-\frac{1}{4}}{x^2}$ at zero by using transmutation operator.

In this paper, we obtain the problem which is singularity type $\left(\frac{k}{x}+\frac{m^2-\frac{1}{4}}{x^2}\right)$ and solve it by using the Riemann method. Before giving the main results, we will mention some well-known trues.

We consider the confluent hypergeometric equation

$$x\frac{d^2W}{dx^2}+[c-x]\frac{dW}{dx}-aW=0 \tag{1.2}$$

in which a and c are parameters. In this equation, by considering transmutation $w = exp(-\frac{x}{2})x^{m+\frac{1}{2}}W$, we get

$$-\frac{d^2w}{dx^2}+\left[\frac{1}{4}-\frac{k}{x}+\frac{m^2-\frac{1}{4}}{x^2}\right]w=0 \tag{1.2$'$}$$

where $k=\frac{c}{2}-a$, $m=\frac{c}{2}-\frac{1}{2}$. Standard solution of $(1.2')$ is

$$M(k,m,x)=exp(-\frac{x}{2})x^{m+\frac{1}{2}}F(m-k+\frac{1}{2},2m+1,x), \tag{1.3}$$

where F is solution of (1.2) and sometimes known as *Kummer's* function. [13, p:255-261]

Let E be a topological linear space and L_1 and L_2 be two linear, but not necessarily continuous, operators from E to E, where E_1 and E_2 are closed subspaces in E.

Definition 1.1 A linear invertible operator, X, defined in the whole space E and acting from E_1 to E_2 is called a *transmutation operator* for a pair of operators L_1 and L_2 if it satisfies the following two conditions [9, p:4]

i) The operator X and its inverse X^{-1} are continuous in space E.

ii) The operator equation is of the form $L_1X=XL_2$ or $L_1=XL_2X^{-1}$.

We consider the problem

$$-\frac{d^2w}{dx^2}+\left[\frac{1}{4}-\frac{k}{x}+\frac{m^2-\frac{1}{4}}{x^2}\right]w=0, \qquad 0<x<1 \tag{1.4}$$

$$w(0)=0. \tag{1.5}$$

Equation (1.4) is a Whittaker equation and its solution is Whittaker function $M(k,m,x)$. [13] Furthermore, considering

$$-\frac{d^2w}{dx^2}+\left[\frac{1}{4}-\frac{k}{x}+\frac{m^2-\frac{1}{4}}{x^2}+q(x)\right]w=0, \qquad 0<x<1 \tag{1.6}$$

$$w(0) = 0, \tag{1.7}$$

where $q(x)$ is a real-valued integrable function on $(0,1)$. This function is called *potential function* in spectral theory.

2. Existence of the Transmutation Operator

Theorem 2.1 Let take $M(k,m,x)$ and $\varphi(x)$ as solutions of (1.4),(1.5) and (1.6), (1.7) respectively. When the transmutation operator $X = X_{L_1,L_2}$ can be realized as

$$X\left[M(k,m,x)\right] = \varphi(x) = M(k,m,x) + \int_0^x H(x,t)\, M(k,m,t)dt, \tag{2.1}$$

the kernel $H(x,t)$ is solution of the problem

$$\frac{\partial^2 H(x,t)}{\partial x^2} - \left[-\frac{k}{x} + \frac{m^2 - \frac{1}{4}}{x^2} + q(x)\right] H(x,t) = \frac{\partial^2 H(x,t)}{\partial t^2} - \left[-\frac{k}{t} + \frac{m^2 - \frac{1}{4}}{t^2}\right] H(x,t) \tag{2.2}$$

$$H(x,0) = 0, \tag{2.3}$$

$$2\frac{dH(x,x)}{dx} = q(x). \tag{2.4}$$

Conversely, if function $H(x,t)$ is a solution of equation (2.2) and satisfies conditions (2.3), (2.4) then the operator X_{L_1,L_2} defined by formula (2.1) is a transmutation operator for the pair of L_1 and L_2.

Proof: We define two operators

$$L_1 = -\frac{d^2}{dx^2} + \left[-\frac{k}{x} + \frac{m^2 - \frac{1}{4}}{x^2} + q(x)\right],$$

$$L_2 = -\frac{d^2}{dx^2} + \left[-\frac{k}{x} + \frac{m^2 - \frac{1}{4}}{x^2}\right].$$

Hence, we have

$$L_1X\left[M(k,m,x)\right]=L_1\varphi=-\varphi''+\left[-\frac{k}{x}+\frac{m^2-\frac{1}{4}}{x^2}+q(x)\right]\varphi$$

$$\begin{aligned}
=\; & -M''(k,m,x)-\frac{dH\left(x,x\right)}{dx}M(k,m,x)\\
& -M'(k,m,x)H\left(x,x\right)-\left.\frac{\partial H\left(x,t\right)}{\partial x}M(k,m,t)\right|_{t=x}\\
& -\int_0^x\frac{\partial^2 H\left(x,t\right)}{\partial x^2}M(k,m,t)dt+\left[-\frac{k}{x}+\frac{m^2-\frac{1}{4}}{x^2}+q(x)\right]M(k,m,x)\\
& +\left[-\frac{k}{x}+\frac{m^2-\frac{1}{4}}{x^2}+q(x)\right]\int_0^x H\left(x,t\right)M(k,m,t)dt. \qquad (2.5)
\end{aligned}$$

On the other hand,

$$\begin{aligned}
XL_2\left[M(k,m,x)\right] \;=\; & -M''(k,m,x)+\left[-\frac{k}{x}+\frac{m^2-\frac{1}{4}}{x^2}\right]M(k,m,x)\\
& -\int_0^x H\left(x,t\right)M''(k,m,t)dt\\
& +\int_0^x H\left(x,t\right)\left[-\frac{k}{t}+\frac{m^2-\frac{1}{4}}{t^2}\right]M(k,m,t)dt. \qquad (2.6)
\end{aligned}$$

For third term, integrating by parts

$$\begin{aligned}
\int_0^x H\left(x,t\right)M''(k,m,t)dt \;=\; & H\left(x,t\right)M'(k,m,t)\Big|_{t=0}^{t=x}-\left.\frac{\partial H\left(x,t\right)}{\partial t}M(k,m,t)\right|_{t=x}\\
& +\int_0^x\frac{\partial^2 H\left(x,t\right)}{\partial t^2}M(k,m,t)dt.
\end{aligned}$$

Inserting this value in the (2.6), we obtain

$$\begin{aligned}
& XL_2\left[M(k,m,x)\right]=-M''(k,m,x)+\left[-\frac{k}{x}+\frac{m^2-\frac{1}{4}}{x^2}\right]M(k,m,x)\\
& -H\left(x,t\right)M'(k,m,t)\Big|_{t=0}^{t=x}+\left.\frac{\partial H\left(x,t\right)}{\partial t}M(k,m,t)\right|_{t=x}- \qquad (2.6')\\
& \int_0^x\frac{\partial^2 H\left(x,t\right)}{\partial t^2}M(k,m,t)dt+\int_0^x H\left(x,t\right)\left[-\frac{k}{t}+\frac{m^2-\frac{1}{4}}{t^2}\right]M(k,m,t)dt.
\end{aligned}$$

The operator X is a transmutation operator and satisfies Definition 1.1 (i). Hence, from (2.5) and (2.6′) we get the problem

$$\frac{\partial^2 H(x,t)}{\partial x^2} - \left[-\frac{k}{x} + \frac{m^2 - \frac{1}{4}}{x^2} + q(x)\right] H(x,t) = \frac{\partial^2 H(x,t)}{\partial t^2} - \left[-\frac{k}{t} + \frac{m^2 - \frac{1}{4}}{t^2}\right] H(x,t), \quad (2.7)$$

$$H(x, 0) = 0, \quad (2.8)$$

$$2\frac{dH(x,x)}{dx} = q(x). \quad (2.9)$$

Thus, if an operator $X = X_{L_1,L_2}$ can be realized in form (2.1), then its kernel, i.e., $H(x,t)$ is solution of problems (2.2)-(2.4). Conversely, if a function $H(x,t)$ is solution of problems (2.2)-(2.4), then operator $X = X_{L_1,L_2}$ constructed from (2.1) is a transmutation operator for the pair of L_1 and L_2. To verify this, it suffices to repeat the above computations in reverse order. This completes the proof.

3. Riemann Method

In this section, we solve the problem (2.7)-(2.9) by using Riemann method. Equation (2.7) has a singularity at $x = t = 0$. Thus, under the following substitution

$$z = \frac{(x+t)^2}{4}, \qquad s = \frac{(x-t)^2}{4}, \qquad H(x,t) = (z-s)^{\frac{1}{4}} U(z,s), \quad (3.1)$$

the equation (2.7) is equivalent to

$$\frac{\partial^2 U}{\partial z \partial s} - \frac{1}{4(z-s)}\frac{\partial U}{\partial z} + \frac{1}{4(z-s)}\frac{\partial U}{\partial s} - \frac{1}{4\sqrt{zs}} qU + R(z,s) U = 0, \quad (3.2)$$

where $R(z,s)$ is a function depending on variables z and s.

Equation (3.2) is a Euler-Poisson-Darboux differential equation. For $t = 0$ in (3.1), because of $z = s$, equation (3.2) has a singularity on the characteristic line $z = s$. For $\varepsilon > 0$, we consider characteristic $z = s + \varepsilon$. (*Figure 3.1*) Hence, the problem (2.7)-(2.9) transforms to

$$\frac{\partial^2 U}{\partial z \partial s} - \frac{1}{4(z-s)}\frac{\partial U}{\partial z} + \frac{1}{4(z-s)}\frac{\partial U}{\partial s} - \frac{1}{4\sqrt{zs}} qU + R(z,s) U = 0, \quad (3.3)$$

$$\frac{dU}{dz} + \frac{U}{2z^{\frac{3}{4}}} = \frac{q}{4z^{\frac{1}{2}}}, \quad (3.4)$$

$$U(z, z-\varepsilon) = 0. \quad (3.5)$$

In this section, we shall describe an approach which has considerable affiliation with the Green's function and which is restricted in its application to hyperbolic equation. Now,

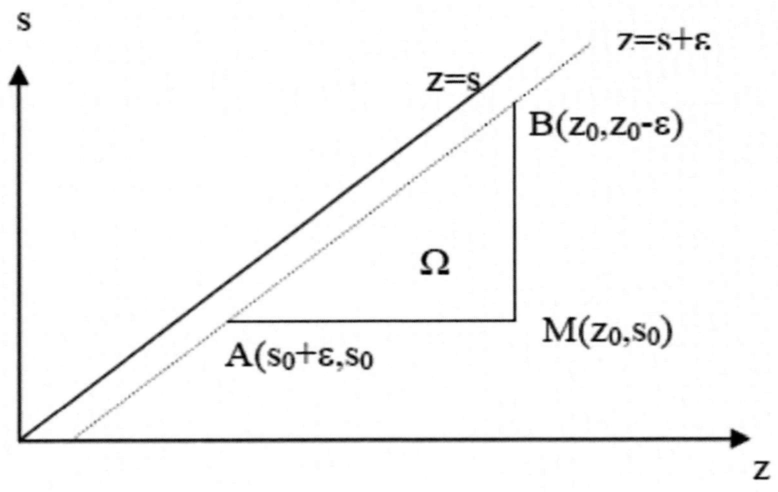

Figure 3.1: Singularity of (3.3) on plane (z,s).

we shall find out solution of problem (3.3)-(3.5) by Riemann method. We consider two operators

$$L\left[U\right]=\frac{\partial^{2}U}{\partial z\partial s}-\frac{1}{4\left(z-s\right)}\frac{\partial U}{\partial z}+\frac{1}{4\left(z-s\right)}\frac{\partial U}{\partial s}+R\left(z,s\right)U,$$

$$L^{*}\left[V\right]=\frac{\partial^{2}V}{\partial z\partial s}+\frac{1}{4\left(z-s\right)}\frac{\partial V}{\partial z}-\frac{1}{4\left(z-s\right)}\frac{\partial V}{\partial s}-\frac{1}{2\left(z-s\right)^{2}}V+R\left(z,s\right)V,$$

where L^* is the adjoint operator of L and V is a solution of conjugate equation. If the first equation multiply by V and the second equation multiply by U, after subtracting and integrating on Ω, it yields

$$\begin{aligned}&\iint_{\Omega} VL\left[U\right]-UL^{*}\left[V\right]dzds\\&=\iint_{\Omega}\frac{1}{2}\frac{\partial}{\partial z}\left(V\frac{\partial U}{\partial s}-U\frac{\partial V}{\partial s}-\frac{1}{2\left(z-s\right)}UV\right)dzds\end{aligned}$$

$$+\iint_{\Omega}\frac{1}{2}\frac{\partial}{\partial s}\left(V\frac{\partial U}{\partial z}-U\frac{\partial V}{\partial z}+\frac{1}{2\left(z-s\right)}UV\right)dzds.$$

Using the Green theorem, we yield

$$\iint_{\Omega} VL[U] - UL^*[V]\, dzds = \int_{AMBA} -\frac{1}{2}\left[V\frac{\partial U}{\partial z} - U\frac{\partial V}{\partial z} - \frac{1}{2(z-s)}UV\right] dz$$
$$+\frac{1}{2}\left[V\frac{\partial U}{\partial s} - U\frac{\partial V}{\partial s} - \frac{1}{2(z-s)}UV\right] ds. \qquad (3.6)$$

As shown in the *Figure 3.1*, since s and z are fixed on AM and MB respectively, it follows that

$$\int_{AM} \frac{1}{2}\left[V\frac{\partial U}{\partial s} - U\frac{\partial V}{\partial s} - \frac{1}{2(z-s)}UV\right] ds = 0$$

and

$$\int_{MB} -\frac{1}{2}\left[V\frac{\partial U}{\partial z} - U\frac{\partial V}{\partial z} + \frac{1}{2(z-s)}UV\right] dz = 0.$$

Furthermore, we have

$$\int_{AM} -\frac{1}{2}\left[V\frac{\partial U}{\partial z} - U\frac{\partial V}{\partial z} + \frac{1}{2(z-s)}UV\right] dz = -\frac{1}{2}UV(M) + \frac{1}{2}UV(A)$$
$$+\int_{AM} U\left[\frac{\partial V}{\partial z} - \frac{V}{4(z-s)}\right] dz$$

and

$$\int_{MB} \frac{1}{2}\left[V\frac{\partial U}{\partial s} - U\frac{\partial V}{\partial s} + \frac{1}{2(z-s)}UV\right] ds = \frac{1}{2}UV(B) - \frac{1}{2}UV(M)$$
$$-\int_{MB} U\left[\frac{\partial V}{\partial s} + \frac{V}{4(z-s)}\right] ds.$$

Inserting these values in (3.6), it is obtained that

$$\iint_{\Omega} \{VL[U] - UL^*[V]\}\, dzds = \int_{BA} -\frac{1}{2}\left[V\frac{\partial U}{\partial z} - U\frac{\partial V}{\partial z} - \frac{UV}{2(z-s)}\right] dz$$
$$+\frac{1}{2}\left[V\frac{\partial U}{\partial s} - U\frac{\partial V}{\partial s} - \frac{UV}{2(z-s)}\right] ds - \frac{1}{2}UV(M)$$
$$+\frac{1}{2}UV(A) + \int_{AM} U\left[\frac{\partial V}{\partial z} - \frac{V}{4(z-s_0)}\right] dz + \frac{1}{2}UV(B)$$
$$-\frac{1}{2}UV(M) - \int_{MB} U\left[\frac{\partial V}{\partial s} + \frac{V}{4(z_0-s)}\right] ds. \qquad (3.7)$$

On the right hand, the values of integrals on AM and MB are not known and on these lines, M is not vanished, then we have to investigate solution of V the following problem.

$$L^*[V]=0, \tag{3.8}$$

$$\frac{\partial V}{\partial z}-\frac{V}{4(z-s_0)}=0 \quad \text{on line } AM, \tag{3.9}$$

$$\frac{\partial V}{\partial s}+\frac{V}{4(z_0-s)}=0 \quad \text{on line } MB, \tag{3.10}$$

$$V=1 \quad \text{at } M \text{ point.} \tag{3.11}$$

Solution of this problem is called Riemann-Green function and it is in the form [12]

$$V(z,s,z_0,s_0)=\frac{(s-z)^{\frac{1}{2}}}{[(s_0-z)(s-z_0)]^{\frac{1}{4}}}F\left(\frac{1}{4},\frac{1}{4},1,\sigma\right),$$

where

$$\sigma=\frac{(s-s_0)(z-z_0)}{(s-z_0)(z-s_0)}$$

and F is a Hypergeometric function such that

$$F\left(\frac{1}{4},\frac{1}{4},1,\sigma\right)=1+\frac{\frac{1}{4}\cdot\frac{1}{4}}{1!\cdot 1}\sigma+\ldots.$$

Hence, we have

$$\begin{aligned} U(M)= &-\frac{UV(A)+UV(B)}{2}+\iint_{\Omega} VL[U]\,dzds \\ &+\int_{BA}\frac{1}{2}\left[V\frac{\partial U}{\partial z}-U\frac{\partial V}{\partial z}-\frac{UV}{2(z-s)}\right]dz \\ &-\frac{1}{2}\left[V\frac{\partial U}{\partial s}-U\frac{\partial V}{\partial s}-\frac{UV}{2(z-s)}\right]ds. \end{aligned} \tag{3.12}$$

Because of being $z=s+\varepsilon$ on BA line,

$$dz=ds \text{ and } \frac{1}{2}\left[V\frac{\partial U}{\partial z}-U\frac{\partial V}{\partial z}-\frac{UV}{2(z-s)}\right]=\frac{1}{2}\left[V\frac{\partial U}{\partial s}-U\frac{\partial V}{\partial s}-\frac{UV}{2(z-s)}\right].$$

On the other hand $L[U]=\frac{1}{4\sqrt{zs}}qU$. From this point of view and (3.12) we obtain

$$U(M)=-\frac{UV(A)+UV(B)}{2}+\int_{s_0}^{z-\varepsilon}\int_{s_0+\varepsilon}^{z_0}\frac{1}{4\sqrt{zs}}qUV\,dzds. \tag{3.13}$$

Last equation is a Volterra- type integral equation. Therefore, its solution is unique and the solution can be proved by successive approximations .

Now, we generalize transmutation operator for the problem (1.6),(1.7). Taking $q_1(x), q_2(x), \ldots, q_n(x)$ as real-valued integrable functions on $(0,1)$, we show that the solution of the following problem can be established by using transmutation operator

$$\frac{d^2w}{dx^2} - \left[\frac{1}{4} - \frac{k}{x} + \frac{m^2 - \frac{1}{4}}{x^2} + \sum_{k=0}^{n} q_k(x)\right] w = 0, \qquad 0 < x < 1 \tag{3.14}$$

$$w(0) = 0.$$

If the solution of the problem

$$\frac{d^2w}{dx^2} - \left[\frac{1}{4} - \frac{k}{x} + \frac{m^2 - \frac{1}{4}}{x^2} + q_1(x)\right] w = 0,$$

$$w(0) = 0$$

is $\varphi_1(x)$, then, we obtain

$$\varphi_1(x) = M(k, m, x) + \int_0^x H_1(x,t)\, M(k, m, x) dt.$$

In the identical manner, when the solution of the problem

$$\frac{d^2w}{dx^2} - \left[\frac{1}{4} - \frac{k}{x} + \frac{m^2 - \frac{1}{4}}{x^2} + q_1(x) + q_2(x)\right] w = 0,$$

$$w(0) = 0$$

is $\varphi_2(x)$, then, we have

$$\varphi_2(x) = X\left[\varphi_1(x,\lambda)\right] = \varphi_1(x,\lambda) + \int_0^x H_2(x,t)\, \varphi_1(t)\, dt.$$

$$= M(k, m, x) + \int_0^x \left[H_2(x,t) + H_1(x,t) + \int_t^x H_1(x,s)\, H_2(s,t)\, ds\right] M(k, m, t) dt.$$

or

$$\varphi_2(x) = M(k, m, x) + \int_0^x K_1(x,t)\, M(k, m, t) dt,$$

where

$$K_1(x,t) = \left[H_2(x,t) + H_1(x,t) + \int_t^x H_1(x,s)\, H_2(s,t)\, ds \right].$$

If this process is continued for $q_1(x), q_2(x), ...q_n(x)$, the solution of the problem

$$\frac{d^2w}{dx^2} - \left[\frac{1}{4} - \frac{k}{x} + \frac{m^2 - \frac{1}{4}}{x^2} + \sum_{k=0}^{n} q_k(x)\right] w = 0, \qquad 0 < x < 1$$

$$w(0) = 0$$

is

$$\varphi_n(x) = X\left[\varphi_{n-1}(x)\right] = M(k,m,x) + \int_0^x K_{n-1}(x,t)\, M(k,m,t)dt. \tag{3.15}$$

In this case, the kernel $K_{n-1}(x,t)$ is solution of the problem

$$\frac{\partial^2 K_{n-1}(x,t)}{\partial x^2} + \left[-\frac{k}{x} + \frac{m^2 - \frac{1}{4}}{x^2} + \sum_{k=0}^{n} q_k(x)\right] K_{n-1}(x,t)$$

$$= \frac{\partial^2 K_{n-1}(x,t)}{\partial t^2} + \left[-\frac{k}{t} + \frac{m^2 - \frac{1}{4}}{t^2}\right] K_{n-1}(x,t), \tag{3.16}$$

$$K_{n-1}(x,0) = 0, \tag{3.17}$$

$$K_{n-1}(x,x) = \frac{1}{2}\int_0^x \sum_{k=0}^{n} q_k(s)ds. \tag{3.18}$$

References

[1] Boimenir, A. and Tuan Vu Kim, Existence and construction of the transmutation operator, *Journal of Math. Phys.*, 45, 7(2004) 2833-2843

[2] Carrol, R. W., *Transmutation theory and applications*, North-Holland, 1985

[3] Gasymov, M.G., On the determination of the Sturm-Liouville equation with peculiarity in zero on two spectrum. *Dan SSSR*, vol.16, 2(1965), 271-276.

[4] Gelfand, I.M. and Levitan, B.M., On the determination of a differential equation from its spectral function, *Amer. Math. Soc.Trans.*, ser. 2,1(1955), 253-304.

[5] Gilbert, Robert P. and Wood, David H. A transmutation approach to underwater sound propagation. *Wave Motion* 8 (1986), no. 5, 383–397.

[6] Hochstadt H., The Inverse Sturm-Liouville Problem, *Commun. on Pure and Appl. Math.*1973, vol.XXVI, 715-729.

[7] Hryniv, R. O. and Mykytyuk Ya. V., Transformation operators for Sturm-Liouville operators with singular potentials, *Math. Phys. Anal. Geom.*, 7(2004), 119-149

[8] Levinson, N., The inverse Sturm-Liouville problem, *Mat.Tidsskr.B.*, 1949, pp. 25-30.

[9] Levitan, B. M., *Inverse Sturm-Liouville problems*, VNU science press, Netherland, 1987.

[10] Levitan, B. M. and Sargsjan, I. S, Introduction to spectral theory: Self-Adjoint ordinary differential operators. *Am. Math. Soc Trans*. 39, 1975.

[11] Lin, Zhongyan; Gilbert, R. P. Numerical algorithm based on transmutation for solving inverse wave equation. *Math. Comput. Modelling* 39 (2004), 13, 1467–1476.

[12] Mackie, A.G., *Boundary value problems*, Oliver and Boyd Ltd., Edinburgh and London, 1965.

[13] Olver, F.W.J., *Asymptotics and Special Functions*, Academic Press, *New York and London*, 1974.

[14] Panakhov E. S.and Koyunbakan H., Inverse problem for singular Sturm-Liouville operator, *Proceeding of IMM of NAS of Azerbaijan*, vol. XVIII(2003), p.113-126.

[15] Volk, V. Y., On Inverse formulas for a differential equation with a singularity at $x = 0$, *Usp. Math. Nauk.* 8 (56), 1953, 141-151.

In: Research on Evolution Equation Compendium. Volume 2 ISBN: 978-1-60456-752-6
Editor: Gaston M. N'Guerekata

ELLIPTIC OPERATORS WITH VARIABLE COEFFICIENTS GENERATING FRACTIONAL RESOLVENT FAMILIES

Miao Li,[1*] ***Fu-Bo Li***[1†] ***and Quan Zheng***[2‡]
[1]Department of Mathematics, Sichuan University,
Chengdu, Sichuan 610064, P.R. China
[2]Department of Mathematics, Central China (Huazhong)
Normal University Wuhan 430079, P. R. China
and Department of Mathematics Huazhnog University of Science and Technology
Wuhan 430074, P. R. China

Abstract

In this paper, we show that under very general conditions, elliptic operator with zero boundary condition can generate fractional resolvent families.

Key Words: elliptic operator, fractional resolvent family, Volterra equations AMS Subject Classification: Primary 45D05, 34G10; Secondary 47D06, 47D09.

Introduction

The theory for the abstract Volterra equations has been developed rapidly due to its applications to many problems in mathematical physics, such as viscoelasticity and heat conduction in materials; see [10] and the references therein.

A Volterra-type equation on a complex Banach space X takes the form

$$u(t) = f(t) + \int_0^t a(t-s)Au(s)\,ds, \quad t \geq 0 \tag{0.1}$$

*E-mail address: limiao1973@hotmail.com; The first and second author were supported by the NSF of China (Grant No. 10501032), and the third by TRAPOYT and the NSF of China (Grant No. 10371046).
†E-mail address: lifubomath@yahoo.com.cn
‡E-mail address: hqzheng@hust.edu.cn

where $a \in L^1_{loc}(\mathbb{R}^+)$ a scalar kernel, A is a closed linear operator on X with dense domain $D(A) \subset X$, and $f : \mathbb{R}^+ \to X$ is continuous. In 1980, Da Prato and Iannelli [3] first introduced the notion of resolvent families, which now plays a central role in the theory of Volterra equations [6, 8, 10]. A strongly continuous family of bounded linear operators on X, $\{R(t)\}_{t\geq 0}$, is called a *resolvent family* for (0.1) if $R(t)$ commutes with A and satisfies the *resolvent equation*

$$R(t)x = x + \int_0^t a(t-s)R(s)Ax\,ds, \quad t \geq 0,\ x \in D(A).$$

If the solution of (0.1) exists, it can be represented by the variation of parameters formula

$$u(t) = \frac{d}{dt}\int_0^t R(t-s)f(s)\,ds.$$

It is well known that when A generates a C_0-semigroup $\{S_1(t)\}_{t\geq 0}$, then the abstract Cauchy problem $u'(t) = Au(t)(t \geq 0)$, $u(0) = x$ has a unique solution with initial value at $x \in D(A)$, and this is saying,

$$S_1(t)x = x + \int_0^t S_1(s)Ax\,ds, \quad x \in D(A),$$

which means that A generates a resolvent family with kernel $a_1(t) \equiv 1$; and similarly, when A generates a cosine function, then the second order equation $u''(t) = Au(t)(t \geq 0)$, $u(0) = x, u'(0) = 0$ has unique solutions which form a resolvent family

$$S_2(t)x = x + \int_0^t (t-s)S_2(s)Ax\,ds, \quad x \in D(A)$$

with kernel $a_2(t) = t$. This motivates us to consider the resolvent family for kernel $a(t) = t^{\alpha-1}$. This class of resolvent families was studied recently by Bajlekova [2], and we call it fractional (or α-times) resolvent family in this paper. When $1 < \alpha < 2$, such resolvent families interpolate between C_0-semigroups and cosine functions. For $A = \Delta$ on $\mathbb{R}$, the Volterra equation corresponding to such resolvent family just interpolates the heat equation and the wave equation (see [5, 7]).

A number of papers are devoted to studying the generation of analytic semigroups by elliptic operators (see, e.g. [9, 11, 12, 13]). Since the class of fractional resolvent families is a generalization of semigroups and cosine functions, it is therefore natural to consider the generation of fractional resolvent families by elliptic operators.

In this paper we consider elliptic operators on $L^2(\Omega)$, where Ω is a bounded domain in $\mathbb{R}^n$. The Gårding's inequality and the numerical range method will be used. We show that under very general conditions, elliptic operators with zero boundary condition can generate α-times resolvent family for some $\alpha \in (1, 2)$. And in some cases, such α can be specified.

1 Preliminaries on fractional resolvent families

In this section we will introduce the notion of fractional resolvent families and quote some of their basic properties as lemmas, for the proofs we refer to [2]. Throughout this section, the operator A is densely defined and $\alpha > 0$.

Definition 1.1. A family $\{S_\alpha(t)\}_{t\geq 0} \subset \mathbf{B}(X)$ is called an *α-times resolvent family for A* if the following conditions are satisfied:

(a) $S_\alpha(t)$ is strongly continuous for $t \geq 0$ and $S_\alpha(0) = I$;

(b) $S_\alpha(t)D(A) \subset D(A)$ and $AS_\alpha(t)x = S_\alpha(t)Ax$ for $x \in D(A)$ and $t \geq 0$;

(c) for $x \in D(A)$, $S_\alpha(t)x$ satisfies

$$S_\alpha(t)x = x + \int_0^t \frac{(t-s)^{\alpha-1}}{\Gamma(\alpha)} S_\alpha(s)Ax\, ds, \quad t \geq 0. \tag{1.1}$$

Definition 1.2. (a) An α-times resolvent family $\{S_\alpha(t)\}_{t\geq 0} \subset \mathbf{B}(X)$ is called *exponentially bounded* if there are constants $M \geq 1$ and $\omega \geq 0$ such that

$$\|S_\alpha(t)\| \leq Me^{\omega t}, \quad t \geq 0. \tag{1.2}$$

If there is an α-times resolvent family $S_\alpha(t)$ for A and satisfying (1.2), we write $(A, S_\alpha(\cdot)) \in \mathcal{C}^\alpha(M,\omega)$ or $A \in \mathcal{C}^\alpha(M,\omega)$. Also, set $\mathcal{C}^\alpha(\omega) = \cup_{M\geq 0}\mathcal{C}^\alpha(M,\omega)$ and $\mathcal{C}^\alpha = \cup_{\omega\geq 0}\mathcal{C}^\alpha(\omega)$.

(b) An α-times resolvent family $\{S_\alpha(t)\}_{t\geq 0}$ is called *analytic* if $S_\alpha(t)$ admits an analytic extension to a sector $\Sigma_{\theta_0} := \{z \in \mathbb{C} | |\arg z| < \theta_0\}$ for some $\theta_0 \in (0, \pi/2]$. $A \in \mathcal{A}^\alpha(\theta_0, \omega_0)$ means that there is an α-times resolvent family $S_\alpha(t)$ for A which can be extended analytically to Σ_{θ_0} and for each $\theta < \theta_0$ and $\omega > \omega_0$ there is $M = M(\theta,\omega)$ such that

$$\|S_\alpha(z)\| \leq Me^{\omega \mathrm{Re} z}, \quad z \in \Sigma_\theta.$$

In addition, denote $\mathcal{A}^\alpha(\theta) = \cup_{\omega\geq 0}\mathcal{A}^\alpha(\theta,\omega)$ and $\mathcal{A}^\alpha = \cup_{0<\theta\leq\pi/2}\mathcal{A}^\alpha(\theta)$.

It is clear that the resolvent family $S_1(t)$ is a C_0-semigroup and $S_2(t)$ is a cosine function. Moreover, we have

Lemma 1.3. *Assume that $A \in \mathcal{C}^\alpha$ for some $\alpha > 2$, then $A \in \mathbf{B}(X)$.*

The following is the generation theorem for α-times resolvent families.

Lemma 1.4. *Let $\alpha \in (0,2]$, the following are equivalent:*

(a) *$A \in \mathcal{C}^\alpha(M,\omega)$;*

(b) *$(\omega^\alpha, \infty) \subset \rho(A)$ and*

$$\|\frac{d^n}{d\lambda^n}(\lambda^{\alpha-1}R(\lambda^\alpha, A))\| \leq \frac{Mn!}{(\lambda-\omega)^{n+1}}, \quad \lambda > \omega, n \in \mathbb{N}\cup\{0\};$$

(c) *$(\omega^\alpha, \infty) \subset \rho(A)$ and there exists a strongly continuous family $\{S(t)\}_{t\geq 0} \subset \mathbf{B}(X)$ satisfying $\|S(t)\| \leq Me^{\omega t}$ for $t \geq 0$ such that*

$$\lambda^{\alpha-1}R(\lambda^\alpha, A)x = \int_0^\infty e^{-\lambda t}S(t)x\, dt, \quad \lambda > \omega, x \in X.$$

And the generation and perturbation for analytic α-times resolvent family are as follows

Lemma 1.5. *Let $\alpha \in (0, 2)$. Then A generates a bounded analytic α-times resolvent family of angle θ_0 if and only if $\rho(A) \supset \Sigma_{\alpha(\pi/2+\theta_0)}$ and for all $0 < \theta < \theta_0$*

$$\|R(\lambda, A)\| \le \frac{C}{|\lambda|}, \quad \lambda \in \Sigma_{\alpha(\pi/2+\theta)}.$$

Lemma 1.6. *Let $\alpha \in (0, 2)$ and $A \in \mathcal{A}^\alpha(\theta, \omega)$. Let B be a closed operator satisfying $D(B) \supset D(A)$ and*

$$\|Bx\| \le a\|Ax\| + b\|x\|, \quad x \in D(A).$$

There exists $\delta > 0$ such that if $0 \le a \le \delta$ then $A + B \in \mathcal{A}^\alpha$.

The following *subordination principle* is very important in the theory of α-times resolvent families.

Lemma 1.7. *Let $0 < \alpha < \beta \le 2$, $\gamma = \alpha/\beta$, $\omega \ge 0$. If $A \in \mathcal{C}^\beta(\omega)$ then $A \in \mathcal{C}^\alpha(\omega^{1/\gamma})$ and the following representation holds:*

$$S_\alpha(t)x = \int_0^\infty \varphi_{t,\gamma}(s) S_\beta(s)x\, ds, \quad t > 0,$$

where $\varphi_{t,\gamma}(s) := t^{-\gamma}\Phi_\gamma(st^{-\gamma})$ and $\Phi_\gamma(z)$ is the Wright function defined by

$$\Phi_\gamma(z) := \sum_{n=0}^{\infty} \frac{(-z)^n}{n!\Gamma(-\gamma n + 2 - \gamma)} = \frac{1}{2\pi i} \int_\Gamma \mu^{\gamma-1} \exp(\mu - z\mu^\gamma)\, d\mu,$$

where $0 < \gamma < 1$ and Γ is a contour which starts and ends at $-\infty$ and encircles the origin counterclockwise. Moreover, $S_\alpha(t)$ admits an analytic extension to the sector $\Sigma_{\min\{\theta(\gamma),\pi\}}$, where $\theta(\gamma) = (1/\gamma - 1)\pi/2$.

Remark 1.8. When $\alpha = 1$ and $\beta = 2$, it is in fact the Weierstrass formula relating C_0-semigroup S_1 with the cosine function S_2

$$S_1(t) = \frac{1}{\sqrt{\pi t}} \int_0^\infty e^{-s^2/4t} S_2(s)\, ds, \quad t > 0.$$

2 Elliptic differential operators on L^2

Throughout this section, we suppose that Ω is a bounded domain in $\mathbb{R}^n$ with boundary $\partial\Omega$ of class C^{2m}. We will use the following notations. If $\mu = (\mu_1, \cdots, \mu_n)$ is any n-tuple of non-negative integers, we set $|\mu| = \mu_1 + \cdots \mu_n$. Let $x = (x_1, \cdots, x_n)$ be a vector in $\mathbb{R}^n$, we set

$$x^\mu = x_1^{\mu_1} \cdots x_n^{\mu_n}.$$

Similarly,

$$D^\mu = D_1^{\mu_1} \cdots D_n^{\mu_n},$$

where

$$D_k = -i\frac{\partial}{\partial x_k}, \quad k = 1, \cdots, n.$$

Let Ω be a domain in $\mathbb{R}^n$, $u(x)$ be a complex valued function defined in Ω. Various norm will be used in the sequel:

$$\|u\|_2 := (\int_\Omega |u(x)|^2\, dx)^{1/2},$$

$$\|u\|_{m,2} := \sum_{|\mu|\le m} \|D^\mu u\|_2,$$

and we will denote

$$(u,v)_m := \sum_{|\mu|\le m} \int_\Omega D^\mu u \overline{D^\mu v}\, dx.$$

Consider the differential operator of order $2m$,

$$A(x,D) = \sum_{|\mu|\le 2m} a_\mu(x) D^\mu, \tag{2.1}$$

where the coefficients a_μ satisfies

$(A1)$ for $|\mu| = 2m$, $a_\mu \in C^m(\overline{\Omega})$; for $|\mu| \le 2m$, $a_\mu(x)$ have derivatives of order $|\mu| - m$ and uniformly bounded;

$(A2)$ the operator $A(x,D)$ is *strongly elliptic*, that is, there exists a constant $C > 0$ such that

$$\mathrm{Re}\Big((-1)^m \sum_{|\mu|=2m} a_\mu(x)\xi^\mu\Big) \ge C|\xi|^{2m}, \quad \forall x \in \overline{\Omega},\ \xi \in \mathbb{R}^n. \tag{2.2}$$

Definition 2.1. For the strongly elliptic operator $A(x,D)$, we define the operator A by

$$Au := A(x,D)u, \quad \text{for } u \in D(A) := H^{2m}(\Omega) \cap H_0^m(\Omega).$$

Lemma 2.2. *(Gårding's inequality) If $A(x,D)$ satisfies $(A1)$ and $(A2)$, then there exist constants $c_0 > 0$ and $\lambda_0 \ge 0$ such that for $u \in H^{2m}(\Omega) \cap H_0^m(\Omega)$*

$$\mathrm{Re}(Au,u)_0 \ge c_0\|u\|_{m,2}^2 - \lambda_0\|u\|_{0,2}^2.$$

Lemma 2.3 ([9]). *Let $A(x,D)$ be strongly elliptic of order $2m$. Then for every λ satisfying $\lambda \ge \lambda_0$ and every $f \in L^2(\Omega)$, there exists a unique $u \in H^{2m}(\Omega) \cap H_0^m(\Omega)$ such that $A(x,D)u + \lambda u = f$.*

If B is a linear operator in a Banach space X its numerical range $S(B)$ is the set

$$S(B) = \{\langle Bx, x^*\rangle : x \in D(B), \langle x, x^*\rangle = \|x\| = \|x^*\| = 1\}.$$

Lemma 2.4 ([9]). *Let B be a closed linear operator with dense domain in X. Let $S(B)$ be the numerical range of B and let Σ be the complement of $\overline{S(A)}$ in $\mathbb{C}$. If $\lambda \in \Sigma$ then $\lambda - B$ is one-to-one and has dense range. Moreover, if Σ_0 is a component of Σ satisfying $\rho(B) \cap \Sigma_0 \ne \emptyset$ then the spectrum of B is contained in the complement S_0 of Σ_0 and*

$$\|R(\lambda, B)\| \le \frac{1}{d(\lambda, \overline{S(B)})},$$

where $d(\lambda, \overline{S(B)})$ is the distance of λ from $\overline{S(B)}$.

Lemma 2.5 ([9]). *Let $A(x, D)$ satisfy (A1) and (A2). Then $-A$ generates an analytic C_0-semigroup on $L^2(\Omega)$.*

The main object of this section is to show that $-A$ also generates an α-times resolvent family for some $\alpha \in (1, 2)$. The proof follows the same lines of Theorem 2.7 in [9].

Theorem 2.6. *Let $A(x, D)$ satisfy (A1) and (A2). Then there is some $\alpha \in (1, 2)$, such that $-A \in \mathcal{A}^\alpha$.*

Proof. Let $A_{\lambda_0} = A + \lambda_0 I$, where λ_0 is given by Lemma 2.2, then we have

$$\mathrm{Re}(A_{\lambda_0}u, u)_0 \geq c_0\|u\|_{m,2}^2,$$

for $u \in D(A_{\lambda_0}) = H^{2m}(\Omega) \cap H_0^m(\Omega)$. By integration by parts, we have for some constant $b > 0$,

$$|\mathrm{Im}(A_{\lambda_0}u, u)_0| \leq |(A_{\lambda_0}u, u)_0| \leq b\|u\|_{m,2}^2.$$

So the numerical range of A_{λ_0}

$$S(A_{\lambda_0}) = \{(A_{\lambda_0}u, u)_0 : \ u \in D(A), \|u\|_2 = 1\} \subset \overline{\Sigma_{\theta_1}},$$

where $\theta_1 = \arctan(b/c_0) < \pi/2$, where c_0 is given by Lemma 2.2. Choose $\theta_1 < \theta < \pi/2$, then for $\lambda \in (\Sigma_\theta)^c - \{0\}$, here $(\Sigma_\theta)^c$ means the complement of Σ_θ, we have

$$d(\lambda, S(A_{\lambda_0})) \geq c_\theta|\lambda|.$$

By Lemma 2.3, $R(\mu - A_{\lambda_0}) = L^2(\Omega)$ for $\mu > 0$; and by Lemma 2.4, for $\mu > 0$, $\mu - A_{\lambda_0}$ is injective, so $(-\infty, 0) \subset \rho(A_{\lambda_0})$. Then $(\Sigma_\theta)^c - \{0\}$ is in $\overline{S(A_{\lambda_0})}^c$ and its intersection with $\rho(A_{\lambda_0})$ is not empty. Again by Lemma 2.4, we obtain that $\rho(A_{\lambda_0}) \supset (\Sigma_\theta)^c - \{0\}$ and for $\lambda \in (\Sigma_\theta)^c - \{0\}$,

$$\|R(\lambda, A_{\lambda_0})\| \leq d(\lambda, \overline{S(A_{\lambda_0})})^{-1} \leq \frac{1}{c_\theta|\lambda|}.$$

Since $2(1 - \theta_1/\pi) > 1$, if we pick $1 < \alpha < 2(1 - \theta_1/\pi)$, then there exists $\varepsilon > 0$ such that $\rho(-A_{\lambda_0}) \supset \Sigma_{\alpha(\pi/2+\varepsilon)}$ and

$$\|R(\lambda, -A_{\lambda_0})\| \leq \frac{1}{c_\alpha|\lambda|}, \qquad \lambda \in \Sigma_{\alpha(\pi/2+\varepsilon)},$$

for some constant c_α, so we have by Lemma 1.5 that $-A_{\lambda_0} \in \mathcal{A}^\alpha$, and thus $-A \in \mathcal{A}^\alpha$ by Lemma 1.6. □

From the proof of Theorem 2.6 and the subordination principle (Lemma 1.7), we have

Theorem 2.7. *Let $A(x, D)$ satisfy (A1) and (A2), and the numerical range of A, $S(A)$, be contained in $\omega + \overline{\Sigma_\theta}$ for some $\omega \in \mathbb{R}$ and $\theta \in (0, \pi/2)$. Then $-A \in \mathcal{A}^\alpha$ for all $0 < \alpha < 2(1 - \theta/\pi)$.*

If the operator A is of the form

$$Au = \sum_{|\mu|,|\nu| \leq m} (-1)^{|\mu|} D^\mu(a_{\mu\nu}(x)D^\nu u), \tag{2.3}$$

by using Gårding's inequality in a quadratic integral form (see [1]), the smoothing conditions can be improved.

Theorem 2.8. *If A is given by (2.3) and satisfies*

(a) $a_{\mu\nu}$ is uniformly continuous on Ω, $|\mu| = |\nu| = m$; and $a_{\mu\nu}$ is bounded and measurable for $|\mu| + |\nu| \leq 2m$;

(b) A is strongly elliptic, which in this case means that

$$\mathrm{Re}\Big(\sum_{|\mu|=|\nu|=m} a_{\mu\nu}(x)\xi^{\mu+\nu} \Big) \geq C|\xi|^{2m}, \quad \forall x \in \overline{\Omega},\ \xi \in \mathbb{R}^n,$$

for some constant $C > 0$;

(c) the numerical range of A is contained in $\omega + \overline{\Sigma_\theta}$ for some $\omega \in \mathbb{R}$ and $\theta \in (0, \pi/2)$. Then $-A \in \mathcal{A}^\alpha$ for all $0 < \alpha < 2(1 - \theta/\pi)$.

Next we consider the strongly elliptic operator of second order in divergence form:

$$Au = -\sum_{j,k=1}^{n} \mathrm{D}^j(a_{jk}(x)\mathrm{D}^k u) + \sum_{j=1}^{n} b_j(x)\mathrm{D}^j u + c(x)u, \tag{2.4}$$

where $\mathrm{D}^j u$ means the j-th derivative of u.

Fattorini ([4]) has the following result:

Lemma 2.9. *Let A be given by (2.4) and strongly elliptic. Suppose that $a_{jk}, b_j, c \in L^\infty(\Omega)$ and $(a_{jk}(x))$ are real symmetric matrices, then $-A$ generates a cosine function on $L^2(\Omega)$.*

As a direct consequence, we have

Corollary 2.10. *Under the assumptions of Lemma 2.9, $-A$ generates an α-times resolvent family for all $\alpha \in (0, 2]$.*

For second order elliptic operator with the highest coefficients not necessarily real generating α-times resolvent family, we make the following assumption. Let A be given by (2.4).

$(S1)$ A is strongly elliptic.

$(S2)$ the coefficients $a_{jk}(x) \in C(\overline{\Omega})$, and there exists $0 < \theta < \pi/2$ such that for all $x \in \Omega$, $a(x) := (a_{jk}(x)) = B(x) + D(x)$, where $D(x) = \mathrm{diag}(d_1(x), \cdots, d_n(x))$ is a diagonal matrix with elements $d_j(x) \in \overline{\Sigma_\theta}(j = 1, \cdots, n)$, and $B(x)$ is conjugate symmetric, i.e., $B(x) = \overline{B(x)}^{\top}$ and the eigenvalues of $B(x)$ are nonnegative.

$(S3)$ b_j, c are complex measurable functions and bounded on Ω.

Theorem 2.11. *Suppose that A satisfies $(S1) - (S3)$, then for all $0 < \alpha < 2(1 - \theta/\pi)$, $-A \in \mathcal{A}^\alpha$.*

Proof. Let $Pu = \sum_{j,k=1}^{n} \mathrm{D}^j(a_{jk}(x)\mathrm{D}^k u)$, then

$$(Pu, u)_0 = -\sum \int_\Omega a_{jk}(x)\mathrm{D}^k u \mathrm{D}^j \overline{u}\, dx.$$

For fixed $x \in \Omega$, let $B = (b_{jk}) = B(x)$, its eigenvalues are $\lambda_1, \cdots, \lambda_n \geq 0$, and the corresponding eigenvectors are $\mu_1, \cdots, \mu_n$, respectively. Let $\beta_j = \mu_j/|\mu_j|$, which has the form $\beta_j = (\beta_{j1}, \cdots, \beta_{jn})^\top$ for $j = 1, \cdots, n$, then $\beta_1, \cdots, \beta_n$ are orthogonal vectors and $B\beta_j = \lambda_j \beta_j$ for $j = 1, \cdots, n$. Let $T = (\beta_1, \cdots, \beta_n)$, then T is an invertible matrix and

$$\begin{aligned}\overline{T}^\top BT &= (\overline{\beta_1}, \cdots, \overline{\beta_n})^\top B(\beta_1, \cdots, \beta_n) \\ &= (\overline{\beta_1}, \cdots, \overline{\beta_n})^\top (\lambda_1\beta_1, \cdots, \lambda_n\beta_n) \\ &= \operatorname{diag}(\lambda_1, \cdots, \lambda_n) =: E,\end{aligned}$$

so that

$$\begin{aligned}\sum_{j,k=1}^{n} b_{jk}(x)\xi_j\overline{\xi_k} &= (\overline{\xi_1}, \cdots, \overline{\xi_n})B(\xi_1, \cdots, \xi_n)^\top \\ &= (\overline{\xi_1}, \cdots, \overline{\xi_n})(\overline{T}^\top)^{-1}ET^{-1}(\xi_1, \cdots, \xi_n)^\top \\ &= (\overline{T^{-1}(\xi_1, \cdots, \xi_n)^\top})^\top ET^{-1}(\xi_1, \cdots, \xi_n)^\top \\ &= \sum_{j=1}^{n} \lambda_j |y_j|^2 \geq 0\end{aligned}$$

where $(y_1, \cdots, y_n)^\top = T^{-1}(\xi_1, \cdots, \xi_n)^\top$. On the other hand, since

$$\begin{aligned}-(Pu, u)_0 &= \sum \int_\Omega a_{jk}(x)\mathrm{D}^k u \mathrm{D}^j \overline{u}\, dx \\ &= \sum \int_\Omega b_{jk}(x)\mathrm{D}^k u \mathrm{D}^j \overline{u}\, dx + \sum \int_\Omega d_j(x)|\mathrm{D}^j u|^2\, dx,\end{aligned}$$

we know the numerical range of $-P$ falls into $\overline{\Sigma_\theta}$, thus it follows from Theorem 2.8 that $P \in \mathcal{A}^\alpha$ when $\alpha < 2(1 - \theta/\pi)$.

Next, by Gårding's inequality,

$$|(Pu, u)_0| \geq c_1\|u\|_{1,2}^2 - c_2\|u\|_2^2,$$

so for $u \in H_0^1(\Omega) \cap H^2(\Omega)$ with $\|u\|_2 = 1$, we have

$$\begin{aligned}\|Pu\|_2 &= \sup\{|(Pu, v)_0| : v \in L^2(\Omega), \|v\|_2 = 1\} \\ &\geq |(Pu, u)_0| \geq c_1\|u\|_{1,2}^2 - c_2.\end{aligned}$$

Let $Qu = \sum_{j=1}^n b_j(x)\mathrm{D}^j u + c(x)u$, for every $\varepsilon > 0$,

$$\begin{aligned}\|Qu\|_2 &= \sup\{(Qu, v)_0 : v \in L^2(\Omega), \|v\|_2 = 1\} \\ &\leq \sup\{\sum \|b_j\|_\infty \cdot \|Du\|_2 \cdot \|v\|_2 \\ &\quad + \|c\|_\infty \cdot \|u\|_2 \cdot \|v\|_2 : v \in L^2(\Omega), \|v\|_2 = 1\} \\ &\leq \sum \|b_j\|_\infty \cdot \|u\|_{1,2} + \|c\|_\infty \\ &\leq c_1\varepsilon\|u\|_{1,2}^2 + \frac{1}{4c_1\varepsilon}(\sum \|b_j\|_\infty)^2 + \|c\|_\infty \\ &\leq \varepsilon\|Pu\|_2 + C(\varepsilon)\end{aligned}$$

which implies that

$$\|Qu\|_2 \leq \varepsilon\|Pu\|_2 + C(\varepsilon)\|u\|_2$$

for all $u \in H_0^1(\Omega) \cap H^2(\Omega)$, thus by Lemma 1.6, $-A = P + Q \in \mathcal{A}^\alpha$. □

Remark 2.12. The strongly elliptic condition $(S1)$ is equivalent to saying that the matrices $\mathrm{Re}\, a(x)$ are positive definite, while the eigenvalues condition in $(S2)$ also requires some information on the imaginary part of $a_{jk}(x)$. For example, let $A_1 = \begin{pmatrix} 2 & -i \\ i & 2 \end{pmatrix}$, $A_2 = \begin{pmatrix} 2 & -3i \\ 3i & 2 \end{pmatrix}$, then $\mathrm{Re}\, A_1 = \mathrm{Re}\, A_2 = \mathrm{diag}(2, 2)$ are positive definite, but the eigenvalues of A_1 are 3 and 1, while the eigenvalues of A_2 are 5 and -1.

References

[1] S. Agmon, Lectures on Elliptic Boundary Value Problems, *Van Nostrand Math. Studies* **2**, 1965.

[2] E. Bajlekova, Fractional Evolution Equations in Banach Spaces, *Ph.D. Dissertation*, Einhoven University of Technology, 2001.

[3] G.Da Prato and M. Iannelli, Linear integro-differential equation in Banach space, *Rend. Sem. Math. Padova* **62** (1980), 207-219.

[4] H.O. Fattorini, *Second Order Linear Differential Equations in Banach Spaces*, North-Holland, 1985.

[5] Y. Fujita, Integrodifferential equations which interpolate the heat equation and the wave equation, *Osaka J. Math.* **27** (1990), 309-321.

[6] C. Lizama, Regularized solutions for Volterra equations, *J. Math. Anal. Appl.* **243** (2000), 278-292.

[7] F. Mainardi, Fractional relaxation-oscillation and fractional diffusion-wave phenomena, *Chaos, Solitons and Fractals* **7** (1996), 1461-1477.

[8] H. Oka, Linear Volterra equations and integrated solution families, *Semigroup Forum* **53** (1996), 278-297.

[9] A. Pazy, *Semigroups of Linear Operators and Applications to Partial Differential Equations* , Springer-Verlag, Berlin, 1983.

[10] J. Prüss, Evolutionary Integral Equations and Applications, *Monographs in Math.* **87**, Birkhaüser Verlag, 1993.

[11] H.B. Stewart, Generation of analytic semigroups by strongly elliptic operators, *Trans. Amer. Math. Soc.* **199** (1974), 141-162.

[12] H.B. Stewart, Generation of analytic semigroups by strongly elliptic operators under general boundary conditions, *Trans. Amer. Math. Soc.* **259** (1980), 299-310.

[13] H. Tanabe, Equations of Evolution, *Monographs and Studies in Math.* **6**, 1979.

In: Research on Evolution Equation Compendium. Volume 2 ISBN: 978-1-60456-752-6
Editor: Gaston M. N'Guerekata

Existence Results for Pseudo Almost Periodic Differential, Functional, and Neutral Integral Equations

Toka Diagana*
Department of Mathematics, Howard University,
2441 6th Street N.W., Washington D.C. 20059, USA

This article is dedicated to the memory of my dear uncle El Hadj Bocar Baba Diagana for having taught us the basic knowledge and for having guided our first footsteps in the right direction.

Abstract

This paper is a survey devoted to the author's recent results related to the existence and uniqueness of pseudo almost periodic solutions to some abstract differential equations and partial differential equations. Some of those results will be slightly modified and applied to some concrete problems. As main tools, we will make extensive use of the method of analytic semigroups, and both the Banach and Zima's fixed-point principles. Applications include the study of pseudo almost periodic solutions to some boundary-value problems such as the heat equation with delay as well as the logistic differential equation.

Key Words: pseudo almost periodic function; almost periodic function; existence and uniqueness of a pseudo almost periodic solution; semilinear differential equation, analytic semigroup; Zima fixed-point theorem, Banach fixed-point theorem, logistic equation with delay, integral equation of advanced type, heat equation, functional differential equation

AMS Subject Classification: 43A60; 35B15; 47B55.

*E-mail address: tdiagana@howard.edu

1. Introduction

Let $(\mathbb{X}, \|\cdot\|)$ be a complex Banach space. An original question consists of studying pseudo almost periodic solutions to the semilinear differential equation

$$u'(t) = Au(t) + f(t, u(t)), \quad t \in \mathbb{R}, \tag{1.1}$$

where A is a densely defined closed linear operator on $\mathbb{X}$, and the forcing term $f : \mathbb{R} \times \mathbb{X} \mapsto \mathbb{X}$ is a jointly continuous function.

Now when A is the infinitesimal generator of an analytic semigroup (respectively, a Hille-Yosida operator) and $f : \mathbb{R} \times \mathbb{X} \mapsto \mathbb{X}$ is pseudo almost periodic, both the Banach fixed-point principle and the method of analytic semigroups were used in **[3, 9, 28]** as a main tool to determine sufficient conditions for the existence and uniqueness of pseudo almost periodic solutions to Eq. (1.1). In contrast with the above-mentioned contributions, we will make use of the so-called Zima's fixed-point principle **[42]** to deal with the existence of pseudo almost periodic solutions to Eq. (1.1).

A slightly different version of Eq. (1.1) which has some interesting applications, especially in the delay case, consists of the functional-differential equation

$$u'(t) = Au(t) + f(t, u(h(t))), \quad t \in \mathbb{R}, \tag{1.2}$$

where A is a densely defined closed (possibly unbounded) linear operator acting on $\mathbb{X}$, $h : \mathbb{R} \mapsto \mathbb{R}$ is a continuous function with $h(\mathbb{R}) = \mathbb{R}$, and $f : \mathbb{R} \times \mathbb{X} \mapsto \mathbb{X}$ is a jointly continuous function. In the case when A is the infinitesimal generator of an analytic semigroup and $f : \mathbb{R} \times \mathbb{X} \mapsto \mathbb{X}$ is pseudo almost periodic, the Zima's fixed-point principle will be used to study the existence of pseudo almost periodic solutions to Eq. (1.2). One should stress that the study of pseudo almost periodic solutions to Eq. (1.2) was recently considered in Diagana **[18]** in the case when A is the infinitesimal generator of an exponentially stable c_0-semigroup. Here, we consider the case when A is the infinitesimal generator of an analytic semigroup. Furthermore, various other formulations of Eq. (1.2) had been studied by several authors; among them are **[22]**, **[24]**, **[29]**, **[30]**, and **[37]** and the references therein. In the particular case when $h(t) = t - p$ with $p > 0$, one obtains the delay case, which is important in many applications.

As a generalization of both Eq. (1.1) and Eq. (1.2) we also consider the existence of pseudo almost periodic solutions to their perturbed semilinear equation

$$u'(t) = Au(t) + f(t, Bu(t)), \quad t \in \mathbb{R}, \tag{1.3}$$

where A is the infinitesimal generator of an analytic semigroup acting on $\mathbb{X}$, $B : D(B) \subset \mathbb{X} \mapsto \mathbb{X}$ is a densely defined closed linear operator, and $f : \mathbb{R} \times \mathbb{X} \mapsto \mathbb{X}$ is a jointly continuous function. As previously, under some additional assumptions on A, B, and f, the existence and uniqueness of a pseudo almost periodic solution to Eq. (1.3) is obtained by combining both semigroup methods and the Zima's fixed-point principle. One should indicate that Eq. (1.3) was recently considered by Diagana et al.[17] with the help of both fractional powers of operators and the Banach fixed-point theorem. Here, we consider its related abstract setting and use the Zima's fixed point principle rather than the Banach's one to deal with those existence results.

In addition to Eq. (1.1), Eq. (1.2), and Eq. (1.3) we also study pseudo almost periodic solutions to more general equation, which consists of the following abstract partial evolution equation

$$\frac{d}{dt}[u(t)+f(t,Bu(t))]=Au(t)+g(t,Cu(t)),\quad t\in\mathbb{R} \tag{1.4}$$

where A is the infinitesimal generator of an analytic semigroup acting on $\mathbb{X}$, B, C are arbitrary densely defined closed linear operators on $\mathbb{X}$, and f, g are some jointly continuous functions. Upon some appropriate assumptions, the existence and uniqueness of a pseudo almost periodic (mild) solution to Eq. (1.4) is obtained. One should mention that various formulations of Eq. (1.4) were extensively studied by Hernández et al. **[25, 26, 27]**. In particular, in **[27]**, both the method of fractional powers of operators and the Banach fixed-point principle were combined to study almost periodic and asymptotically almost periodic solutions to a slightly different version of Eq. (1.4) corresponding to the delay case. More recently, Diagana **[15]** had studied sufficient conditions for the existence and uniqueness of pseudo almost periodic solutions to Eq. (1.4) by combining both semigroup methods and the Banach fixed-point principle. However, there, the operator A was assumed to be the infinitesimal generator of an exponentially stable c_0-semigroup. Moreover, the assumptions made there require much more regularity for the linear operator A, i.e., it would have been more realistic for applications to assume that A is the infinitesimal generator of an analytic semigroup. Here, the latter assumption will be made.

One should also point out that Eq. (1.4) has various applications. More precisely, one can make use of it to characterize the existence and uniqueness of pseudo almost periodic solutions to the class of partial differential equations of the form

$$\frac{\partial}{\partial t}[u(t,x)+f(t,Bu(t,x))] = Au(t,x)+g(t,Cu(t,x)) \tag{1.5}$$

$$u(t,\cdot)_{|\partial\Omega} = 0 \tag{1.6}$$

for all $t \in \mathbb{R}$ with $\Omega \subset \mathbb{R}^n$ is a (bounded) subset.

Furthermore, Eq. (1.4) may also enable us to characterize pseudo almost periodic solutions to partial neutral functional differential equations [25], integro-differential equations **[10, 26]**, and many others.

At the end of this survey, we consider another problem, which is very interesting for applications, but not directly linked to the previous ones. Namely, we study some sufficient conditions for the existence of pseudo almost periodic solutions to the abstract integral equation of the form

$$u(t)=f(u(h_1(t)))+\int_t^{\infty} Q(s,u(s),u(h_2(s)))C(t-s)ds+g(t) \tag{1.7}$$

for each $t \in \mathbb{R}$, where $f, g, h_1, h_2, C : \mathbb{R} \mapsto \mathbb{R}$ are continuous functions with $h_i(\mathbb{R}) = \mathbb{R}$ for $i = 1, 2$, and $Q : \mathbb{R}\times\mathbb{R}\times\mathbb{R} \mapsto \mathbb{R}$ is jointly continuous.

Setting, $h_1(t) = h_2(t) = t - p$ where $p > 0$ is a constant, in Eq. (1.7), one obtains the so-called neutral delay integral equation of advanced type

$$u(t)=f(u(t-p))+\int_t^{\infty} Q(s,u(s),u(s-p)C(t-s)ds+g(t), \tag{1.8}$$

which was introduced in the literature by Burton [7] as an intermediate step while studying the existence and uniqueness of (periodic) bounded solutions to the logistic differential equation delay given by

$$u'(t) = au(t) + \alpha u'(t-p) - q(t, u(t), u(t-p)) \tag{1.9}$$

where $a > 0$, $0 \leq |\alpha| < 1$, and $p > 0$ are respectively constants.

Here, we study pseudo almost periodic solutions to Eq. (1.7) and use it to deal with pseudo almost periodic to Eq. (1.9). These results were recently obtained in Diagana et al.[19].

Applications include among others, the existence and uniqueness of pseudo almost periodic solutions to some boundary value problems related to the heat with delay.

The existence of almost automorphic, asymptotically almost automorphic, almost periodic, asymptotically almost periodic, and pseudo almost periodic solutions is among the most attractive topics in the qualitative theory of differential equations due to their significance and applications in physics, mathematical biology, and many others.

The concept of pseudo almost periodicity (p.a.p), which is the central question in this paper was first initiated by Zhang in **[39, 40, 41]** and is a natural generalization of the of the classical (Bochner) almost periodicity (a.p). Thus this new concept is welcome to implement another existing generalization of almost periodicity, the so-called notion of asymptotically almost periodicity due to Fréchet, see, e.g., **[23, 32, 33]**. Moreover, let us emphasis on the fact that the concept of pseudo almost periodicity is a special case of the well known Besicovitch almost periodicity, i.e., the space $B_1(\mathbb{R}, \mathbb{X})$, see, e.g., **[34]**. More details on those concepts and related issues can be found in **[1, 2, 3, 9, 11, 23, 28, 33, 38, 39, 40, 41]** and the references therein.

In addition to **[3, 9, 28]** one should mention that the study of pseudo almost periodic solutions to differential equations, partial differential equations, functional differential equations with delay, has been of a great interest for many mathematicians in the last few years, see, e.g., **[1, 2, 4, 11, 15, 16, 17, 19, 39, 40, 41]** and the references therein. Here, we examine some sufficient conditions, which do ensure the existence and uniqueness of a pseudo almost periodic solution to Eq. (1.1)-Eq. (1.2)- Eq. (1.3), Eq. (1.4), Eq. (1.7), and Eq. (1.8) using the above-mentioned techniques.

Let us recall some definitions and notations that we shall use in the sequel.

2. Preliminaries

2.1. Pseudo Almost Periodic Functions

Let $(\mathbb{X}, \|\cdot\|)$ be a Banach space and let $(BC(\mathbb{X}), \|\cdot\|_\infty)$ denote the Banach space of all bounded continuous functions $\phi : \mathbb{R} \mapsto \mathbb{X}$ endowed with the sup norm defined by

$$\|\phi\|_\infty := \sup_{t\in\mathbb{R}} \|\phi\|.$$

Similarly, $BC(\mathbb{R}\times\Omega)$ where $\Omega \subset \mathbb{X}$ is an open subset, denotes the space of all bounded continuous functions $\Phi : \mathbb{R} \times \Omega \mapsto \mathbb{X}$.

Definition 1. **[8]** *A function $f \in BC(\mathbb{X})$ is called almost periodic if for each $\varepsilon > 0$, there exists $l_\varepsilon > 0$ such that every interval of length l_ε contains a number τ with the following property:*

$$\|f(t+\tau) - f(t)\| < \varepsilon, \quad \forall t \in \mathbb{R}.$$

The number τ above is then called an *ε-translation number* of f, and the collection of such functions will be denoted $AP(\mathbb{X})$.

Example 1. *The function $f(t) = \sin t + \sin\sqrt{2}t$ is a classical example of an almost periodic function on $\mathbb{R}$.*

To construct more almost periodic functions on $\mathbb{R}$ one can use the following.

Proposition 1. *Let $f \in AP(\mathbb{R})$. If $g \in L^1(\mathbb{R})$, then $f * g$, the convolution of f with g on $\mathbb{R}$, is almost periodic.*

Proof. Since f is continuous and $g \in L^1(\mathbb{R})$, it is not hard to see that the function $t \mapsto (f * g)(t)$ is continuous. Moreover, $|(f * g)(t)| \leq \|f\|_\infty \|g\|_1$ for each $t \in \mathbb{R}$, where $\|g\|_1$ is the L^1-norm of g, and therefore $f * g \in BC(\mathbb{R})$.

It remains to prove that $f * g$ is almost periodic. First of all, note that when $g \equiv 0$ there is nothing to prove. From now on, we suppose $g \not\equiv 0$.

Since $f \in AP(\mathbb{R})$, for every $\varepsilon > 0$ there exists $l_\varepsilon > 0$ such that for all $\delta \in \mathbb{R}$ there exists $\tau \in [\delta, \delta + l_\varepsilon]$ with

$$|f(\sigma+\tau) - f(\sigma)| \leq \frac{\varepsilon}{\|g\|_1} \quad \text{for each } \sigma \in \mathbb{R}.$$

In particular, the following holds

$$|f(t-s+\tau) - f(t-s)| \leq \frac{\varepsilon}{\|g\|_1} \quad \text{for each } \sigma = t - s \in \mathbb{R}. \tag{2.10}$$

Now

$$(f * g)(t+\tau) - (f * g)(t) = \int_{-\infty}^{+\infty} \{f(t-\sigma+\tau) - f(t-\sigma)\} g(\sigma) d\sigma, \quad t \in \mathbb{R}.$$

Thus from Eq. (2.10) and the fact that $g \in L^1(\mathbb{R})$ it easily follows that

$$\|\sigma_\tau(f * g) - (f * g)\|_\infty \leq \varepsilon,$$

and hence $f * g \in AP(\mathbb{R})$. ∎

Example 2. *Combining Example 1 and Proposition 1 it easily follows that for each function $K \in L^1(\mathbb{R})$, the function defined by*

$$\mathcal{L}(t) := \int_{-\infty}^{+\infty} K(t-s)(\sin s + \sin\sqrt{2}s) ds, \quad (t \in \mathbb{R})$$

is almost periodic.

Similarly

Definition 2. *A function $F \in BC(\mathbb{R} \times \Omega)$ is called almost periodic in $t \in \mathbb{R}$ uniformly in any $K \subset \Omega$ bounded subset if for each $\varepsilon > 0$ there exists $l_\varepsilon > 0$ such that every interval of length $l_\varepsilon > 0$ contains a number τ with the following property*

$$\|F(t+\tau, x) - F(t,x)\| < \varepsilon, \quad (t \in \mathbb{R},\ x \in K).$$

Here again, the number τ above is called an ε-translation number of F, and the class of such functions will be denoted $AP(\mathbb{R} \times \Omega)$.

More details on properties of almost periodic functions $\phi : \mathbb{R} \mapsto \mathbb{X}$ and as well as those of the form $F : \mathbb{R} \times \mathbb{X} \mapsto \mathbb{X}$ can be found in the literature, especially in [8, 23, 32, 33, 38] and the references therein.

Throughout the rest of the paper, we suppose that $\Omega = \mathbb{X}$ and set

$$AP_0(\mathbb{X}) := \{f \in BC(\mathbb{X}) : \lim_{T \to \infty} \frac{1}{2T} \int_{-T}^{T} \|f(s)\| ds = 0\}.$$

Similarly, let $AP_0(\mathbb{R} \times \mathbb{X})$ denote the collection of functions $F \in BC(\mathbb{R} \times \mathbb{X})$ such that

$$\lim_{T \to \infty} \frac{1}{2T} \int_{-T}^{T} \|F(t,u)\| dt = 0$$

uniformly in $u \in \mathbb{X}$.

Example 3. *Both $f_\omega(t) = \dfrac{1}{t^2 + \omega^2}$ with $\omega \neq 0$, and $g(t) = e^{-|t|}$ belong to $AP_0(\mathbb{R})$.*

Proposition 2. *Let $f \in AP_0(\mathbb{R})$. If $g \in L^1(\mathbb{R})$, then $f * g$, the convolution of f with g on $\mathbb{R}$, belongs to $AP_0(\mathbb{R})$.*

Proof. From $f \in AP_0(\mathbb{R})$ and $g \in L^1(\mathbb{R})$ it is clear that $f * g \in BC(\mathbb{R})$. By assumption,

$$\lim_{T \to \infty} \frac{1}{2T} \int_{-T}^{T} |f(t)|\, dt = 0.$$

Now setting

$$J(T) := \frac{1}{2T} \int_{-T}^{T} \int_{-\infty}^{+\infty} |f(t-s)|\, |g(s)|\, ds\, dt$$

it follows that

$$\begin{aligned}
\frac{1}{2T} \int_{-T}^{T} |(f * g)(t)|\, dt &\leq J(T) \\
&= \frac{1}{2T} \int_{-T}^{T} \int_{-\infty}^{+\infty} |f(t-s)|\, |g(s)|\, ds\, dt \\
&= \int_{-\infty}^{+\infty} |g(s)| \left(\frac{1}{2T} \int_{-T}^{T} |f(t-s)|\, dt \right) ds \\
&= \int_{-\infty}^{+\infty} |g(s)| \left(\frac{1}{2T} \int_{-T-s}^{T-s} |f(r)|\, dr \right) ds \\
&= \int_{-\infty}^{+\infty} |g(s)|\, \phi_T(s)\, ds,
\end{aligned}$$

where $f_T(u) = \dfrac{1}{2T}\displaystyle\int_{-T-u}^{T-u} |f(r)|\, dr$.

Clearly, $f_T(u) \mapsto 0$ as $T \mapsto \infty$. Next, since f_T is bounded and $g \in L^1(\mathbb{R})$, using the Lebesgue dominated convergence theorem, it follows that $\lim_{T\to\infty} J(T) = 0$, and hence $f * g \in AP_0(\mathbb{R})$. ∎

Example 4. *Combining Example 3 and Proposition 2 it easily follows that for each function $K \in L^1(\mathbb{R})$, the function defined by*

$$\mathcal{M}_\omega(t) := \int_{-\infty}^{+\infty} \frac{K(t-s)}{s^2+\omega^2}\, ds, \quad (t \in \mathbb{R})$$

belongs to $AP_0(\mathbb{R})$.

Definition 3. *A function $f \in BC(\mathbb{X})$ is called pseudo almost periodic if it can be expressed as $f = g + \phi$, where $g \in AP(\mathbb{X})$ and $\phi \in AP_0(\mathbb{X})$. The collection of such functions will be denoted by $PAP(\mathbb{X})$.*

Note that the functions g and ϕ appearing in Definition 3 are respectively called the almost periodic and the ergodic perturbation components of f. Furthermore, the decomposition in Definition 3 is unique **[39, 40, 40]**

Example 5. *The function $f_\omega(t) = \mathcal{L}(t)+\mathcal{M}_\omega(t)$ ($\omega \neq 0$) where $\mathcal{L}$ and $\mathcal{M}_\omega$ are respectively the given in Example 1 and Example 3, is pseudo almost periodic.*

As a consequence of Proposition 1 and Proposition 2 the following holds.

Proposition 3. *Let $f \in PAP(\mathbb{R})$. If $g \in L^1(\mathbb{R})$, then $f * g$, the convolution of f with g on $\mathbb{R}$, belongs to $PAP(\mathbb{R})$.*

We now equip $PAP(\mathbb{X})$ the collection of all pseudo almost periodic functions from $\mathbb{R}$ into $\mathbb{X}$ with the sup norm. It is well known that $(PAP(\mathbb{X}), \|.\|_\infty)$ is a Banach space, see, e.g., **[28]**.

Definition 4. *A function $f \in BC(\mathbb{R} \times \mathbb{X})$ is called pseudo almost periodic in $t \in \mathbb{R}$ uniformly in $x \in \mathbb{X}$ if it can be expressed as*

$$f = g + \phi,$$

where $g \in AP(\mathbb{R} \times \mathbb{X})$ and $\phi \in AP_0(\mathbb{R} \times \mathbb{X})$. The collection of such functions will be denoted by $PAP(\mathbb{R} \times \mathbb{X})$.

2.2. The Zima's Fixed-Point Theorem

We now recall the so-called Zima's fixed-point theorem **[42]** that we shall use in the sequel. For that, consider a given Banach space $(\mathbb{Y}, \|\cdot\|_{\mathbb{Y}})$ equipped with both a binary relation $\prec$ and a mapping $m : \mathbb{Y} \mapsto \mathbb{Y}$ such that:

(i) The relation $\prec$ is transitive;

(ii) The norm $\|\cdot\|_{\mathbb{Y}}$ is monotonic, i.e., $\theta \prec u \prec v$, then $\|u\|_{\mathbb{Y}} \leq \|v\|_{\mathbb{Y}}$ for all $u, v \in \mathbb{Y}$;

(iii) $\theta \prec m(u)$ and $\|m(u)\|_{Y} = \|u\|_{\mathbb{Y}}$ for each $u \in \mathbb{Y}$.

Theorem 4. **[42]** *In the Banach space* $(\mathbb{Y}, \|\cdot\|_{\mathbb{Y}}, \prec, m)$ *given above, let* $\Gamma : \mathbb{Y} \mapsto \mathbb{Y}$ *and* $B : \mathbb{Y} \mapsto \mathbb{Y}$ *be mappings such that*

(iv) B *is a bounded linear operator with spectral radius* $r(B) < 1$;

(v) If $\theta \prec u \prec v$, *then* $Bu \prec Bv$ *for all* $u, v \in \mathbb{Y}$;

(vi) $m(\Gamma u - \Gamma v) \prec Bm(u - v)$ *for all* $u, v \in \mathbb{Y}$.

Then the equation $\Gamma u = u$ *has a unique solution.*

Observe that Theorem 4 is more general than the classical Banach fixed-point theorem and has recently found several applications, especially those dealing with the existence and uniqueness of solutions to differential equations, partial differential equations, functional differential equations, and integro-differential equations, see, e.g., Bugajewski and Zima **[5, 6]**.

3. Pseudo Almost Periodic Solutions

Throughout the rest of this paper $(\mathbb{X}, \|\cdot\|)$ stands for a (complex) Banach space.

3.1. Pseudo Almost Periodic Solutions To Eq. (1.1)

To deal with the existence and uniqueness of pseudo almost periodic solutions to Eq. (1.1), we require the following assumptions:

(H.1) The function $f : \mathbb{R} \times \mathbb{X} \mapsto \mathbb{X}$, $(t, u) \mapsto f(t, u)$ is pseudo almost periodic in $t \in \mathbb{R}$ uniformly in $u \in \mathbb{X}$, i.e., $f = g + \phi$ where $g \in AP(\mathbb{R} \times \mathbb{X})$ and $\phi \in AP_0(\mathbb{R} \times \mathbb{X})$; and f satisfies Lipschitz condition in $u \in \mathbb{X}$ for each $t \in \mathbb{R}$ in the following sense

$$\|f(t, u) - f(t, v)\| \leq L(t) \,.\, \|u - v\|, \ \ \forall t \in \mathbb{R}, \ u, v \in \mathbb{X},$$

where $L : R \mapsto [0, \infty)$ is a continuous function satisfying

$$\int_{-\infty}^{+\infty} L(\sigma) d\sigma < \infty;$$

(H.2) The operator A is the infinitesimal generator of an analytic semigroup $T(t)$ such that there exist $M > 0$ and $\omega > 0$ with

$$\|T(t)\| \leq M e^{-\omega t}$$

for all $t \geq 0$;

(H.3) If $r(\Pi)$ denotes the spectral radius of the bounded linear operator Π defined by

$$(\Pi u)(t) = M \,.\int_{-\infty}^{t} L(\sigma)u(\sigma)d\sigma, \;\; t \in \mathbb{R}, \;\; u \in PAP(\mathbb{X}),$$

then $r(\Pi) < 1$, where M is the constant appearing in **(H.2)**.

Theorem 5. *Suppose that assumptions* **(H.1)**-**(H.2)**-**(H.3)** *hold. Then Eq. (1.1) has a unique pseudo almost periodic solution.*

Proof. Let u be a bounded solution to Eq. (1.1). It is well known **[3]** that u can be explicitly expressed by

$$u(t) := \int_{-\infty}^{t} T(t-s)f(s, u(s)ds, \;\; t \in \mathbb{R}.$$

Now if $u \in PAP(\mathbb{X})$, then using **(H.1)** it follows that $u = \alpha(u) + \beta(u)$, where

$$\alpha(u)(t) := \int_{-\infty}^{t} T(t-s)g(s, u(s))ds, \quad \text{and}$$

$$\beta(u)(t) := \int_{-\infty}^{t} T(t-s)\phi(s, u(s))ds,$$

where $g \in AP(\mathbb{R} \times \mathbb{X})$ and $\phi \in AP_0(\mathbb{R} \times \mathbb{X})$.

First of all, note that since $f \in PAP(\mathbb{R} \times \mathbb{X})$ and satisfies the Lipschitz's condition **(H.1)**, for each $v \in PAP(\mathbb{X})$, $f(\cdot, v(\cdot))$ belongs to $PAP(\mathbb{X})$, see, e.g., [9]. Moreover, $g(\cdot, v(\cdot))$ and $\phi(\cdot, v(h(\cdot)))$ are respectively the almost periodic and ergodic perturbation components of f.

Next we show that $\alpha(u) \in AP(\mathbb{X})$. Since $g(\cdot, v(\cdot)) \in AP(\mathbb{X})$, therefore, for each $\varepsilon > 0$ there exists $\zeta > 0$ such that for all θ there is $\tau \in [\theta, \theta + \zeta]$ such that

$$\|g(s+\tau, u(s+\tau)) - g(s, u(s))\| < \left(\frac{\omega}{M}\right).\varepsilon, \;\; \forall s \in \mathbb{R}.$$

Considering $\alpha(u)(s+\tau) - \alpha(u)(s)$ and using **(H.2)** it easily follows that

$$\|\alpha(u)(t+\tau) - \alpha(u)(t)\| < \varepsilon, \;\; \forall t \in \mathbb{R},$$

and hence $t \mapsto \alpha(u)(t)$ is almost periodic.

The next step consists of showing that $t \mapsto \beta(u)(t)$ is in $AP_0(\mathbb{X})$. Clearly, $t \mapsto \beta(u)(t)$ is a bounded continuous function. Thus the remaining task consists of showing that

$$\lim_{T\to\infty} \frac{1}{2T} \int_{-T}^{T} \|\beta(u)(t)\| dt = 0.$$

Using **(H.2)** it follows that

$$\lim_{T\to\infty} \frac{1}{T} \int_{-T}^{T} \|\beta(u)(t)\| dt \leq I + J,$$

where

$$I = \lim_{T\to\infty} \frac{M}{2T} \int_{-T}^{T} dt \left(\int_{-T}^{t} e^{-\omega(t-s)} \|\phi(s, u(s))\| ds \right), \quad \text{and}$$

$$J = \lim_{T\to\infty} \frac{M}{2T} \int_{-T}^{T} dt \int_{-\infty}^{-T} e^{-\omega(t-s)} \|\phi(s, u(s))\| ds = 0.$$

It remains to show that $I = J = 0$. For that, we will make use of the following arguments:

(A1) $\phi(\cdot, u(\cdot)) \in AP_0(\mathbb{X})$, as we mentioned it above;

(A2) $K = \sup_{t\in\mathbb{R}} \|\phi(t, u(t))\| < \infty$.

Now

$$\begin{aligned} I &= \lim_{T\to\infty} \frac{M}{2T} \int_{-T}^{T} \|\phi(t, u(t))\| dt \left(\int_{-T}^{t} e^{-\omega(t-s)} ds \right) \\ &= \lim_{T\to\infty} \frac{M}{2T} \int_{-T}^{T} \|\phi(t, u(t))\| dt \left(\frac{1}{\omega}[1 - e^{-\omega(t+T)}] \right) \\ &\leq \frac{M}{\omega} \cdot \lim_{T\to\infty} \frac{1}{2T} \int_{-T}^{T} \|\phi(t, u(t))\| dt \\ &= 0, \end{aligned}$$

by **(A1)**.

Similarly,

$$\begin{aligned} J &= \lim_{T\to\infty} \frac{M}{2T} \int_{-\infty}^{-T} e^{\omega s} \|\phi(s, u(s))\| ds \int_{-T}^{T} e^{-\omega t} dt \\ &\leq \lim_{T\to\infty} \frac{M\,.\,K}{2T} \int_{-\infty}^{-T} e^{\omega s} ds \left(\frac{1}{\omega}[e^{\omega T} - e^{-\omega T}] \right) \\ &= \lim_{T\to\infty} \frac{M\,.\,K}{2T\omega^2} \left(1 - e^{-2\omega T}\right) \\ &= 0, \end{aligned}$$

by **(A2)**, and hence $\beta(u) \in AP_0(\mathbb{X})$.

Define the nonlinear operator

$$\Gamma(u)(t) := \int_{-\infty}^{t} T(t-s) f(s, u(s)) ds, \quad u \in PAP(\mathbb{X}), \ t \in \mathbb{R}.$$

In view of the above, Γ maps $PAP(\mathbb{X})$ into itself, and $\alpha(u)$ and $\beta(u)$ are respectively the almost periodic and ergodic perturbation components of $\Gamma(u)$. To complete the proof, we have to show that $\Gamma : PAP(\mathbb{X}) \mapsto PAP(\mathbb{X})$ has a unique fixed-point. For that, we consider the Banach space $\mathbb{Y} = (PAP(\mathbb{X}), \|\cdot\|_\infty)$ equipped with both the relation $\prec$ and the mapping $m : PAP(\mathbb{X}) \mapsto PAP(\mathbb{X})$ defined by: If $u, v \in PAP(\mathbb{X})$,

$$u \prec v \ \text{ if and ony if } \ \|u(t)\| \leq \|v(t)\|, \ \ \forall t \in \mathbb{R},$$

and $m(u) = \|u\|$, i.e., $(m(u))(t) = \|u(t)\|$ for all $t \in \mathbb{R}$.

Now

(1) The binary relation $\prec$ is transitive;

(2) $\theta \prec m(u)$ and $\|m(u)\|_\infty = \|u\|_\infty$ for each $u \in PAP(\mathbb{X})$; and

(3) The norm $\|\cdot\|_\infty$ is monotonic, that is, $\theta \prec u \prec v$, then

$$\|u\|_\infty \leq \|v\|_\infty, \quad \forall u, v \in PAP(\mathbb{X}).$$

For all $u, v \in PAP(\mathbb{X})$,

$$\begin{aligned}
\|\Gamma(u)(t) - \Gamma(v)(t)\| &\leq \int_{-\infty}^{t} \|T(t-s)\left(f(s,u(s)) - f(s,v(s))\right)\|ds \\
&\leq \int_{-\infty}^{t} Me^{-\omega(t-s)}L(s)\|u(s) - v(s)\|ds \\
&\leq M. \int_{-\infty}^{t} L(s)\|u(s) - v(s)\|ds,
\end{aligned}$$

and hence $m(\Gamma(u) - \Gamma(v)) \prec \Pi m(u - v)$.

We also observe that Π is increasing, that is, if $\theta \prec u \prec v$, then $\Pi u \prec \Pi v$ for all $u, v \in PAP(\mathbb{X})$. Therefore, using Theorem 4 it follows that the operator Γ has a unique fixed-point in $PAP(\mathbb{X})$, which obviously is the only pseudo almost periodic solution to Eq. (1.1). ∎

3.2. Pseudo Almost Periodic Solutions to Eq. (1.2)

To deal with the existence and the uniqueness of pseudo almost periodic solutions to Eq. (1.2) we need assumptions **(H.1)-(H.2)** given previously and the following:

(H.4) The function $h : \mathbb{R} \mapsto \mathbb{R}$ is continuous, $h(\mathbb{R}) = \mathbb{R}$, and $u(h) \in PAP(\mathbb{R})$ whenever $u \in PAP(\mathbb{X})$

Theorem 6. *Suppose that assumptions* **(H.1)-(H.2)-(H.3)-(H.4)** *hold. Then, Eq. (1.2) has a unique pseudo almost periodic solution.*

The proof of Theorem 6 is similar to that of Theorem 5. However, for the sake of clarity, we will give its main steps.

Proof. Let u be a bounded solution to Eq. (1.2). It is well known [3] that u can be explicitly expressed by

$$u(t) := \int_{-\infty}^{t} T(t-s)f(s, u(h(s)))ds, \quad t \in \mathbb{R}.$$

Now if $u \in PAP(\mathbb{X})$, then using **(H.1)** it follows that $u = \alpha(u) + \beta(u)$, where

$$\alpha_h(u)(t) := \int_{-\infty}^{t} T(t-s)g(s, u(h(s)))ds, \quad \text{and}$$

$$\beta_h(u)(t) := \int_{-\infty}^{t} T(t-s)\phi(s, u(h(s)))ds,$$

where $g \in AP(\mathbb{R} \times \mathbb{X})$ and $\phi \in AP_0(\mathbb{R} \times \mathbb{X})$.

As in the proof of Theorem 5, for each $v \in PAP(\mathbb{X})$, the function $f(\cdot, v(h(\cdot)))$ belongs to $PAP(\mathbb{X})$, by **(H.4)**. Moreover, the functions $g(\cdot, v(h(\cdot)))$ and $\phi(\cdot, v(h(\cdot)))$ are respectively the almost periodic and ergodic perturbation components of f.

Next we show that $\alpha(u) \in AP(\mathbb{X})$. Since $g(\cdot, v(h(\cdot))) \in AP(\mathbb{X})$, therefore, for each $\varepsilon > 0$ there exists $\zeta > 0$ such that for all θ there is $\tau \in [\theta, \theta + \zeta]$ such that

$$\|\alpha_h(u)(t+\tau) - \alpha_h(u)(t)\| < \varepsilon, \quad \forall t \in R,$$

and hence $t \mapsto \alpha(u)(t)$ is almost periodic.

The next step consists of showing that $t \mapsto \beta(u)(t)$ is in $AP_0(\mathbb{X})$. Clearly, $t \mapsto \beta(u)(t)$ is a bounded continuous function. Thus the remaining task consists of showing that

$$\lim_{T\to\infty} \frac{1}{2T} \int_{-T}^{T} \|\beta_h(u)(t)\| dt = 0.$$

Using **(H.2)** it follows that

$$\lim_{T\to\infty} \frac{1}{2T} \int_{-T}^{T} \|\beta(u)(t)\| dt \le I + J,$$

where

$$I = \lim_{T\to\infty} \frac{M}{2T} \int_{-T}^{T} dt \left(\int_{-T}^{t} e^{-\omega(t-s)} \|\phi(s, u(h(s)))\| ds \right), \quad \text{and}$$

$$J = \lim_{T\to\infty} \frac{M}{2T} \int_{-T}^{T} dt \int_{-\infty}^{-T} e^{-\omega(t-s)} \|\phi(s, u(h(s)))\| ds.$$

Using similar arguments as in the proof of Theorem 5 one can easily show that $I = J = 0$.

Define the nonlinear operator

$$\Gamma_h(u)(t) := \int_{-\infty}^{t} T(t-s) f(s, u(h(s))) ds, \quad u \in PAP(\mathbb{X}), \ t \in \mathbb{R}.$$

Clearly, Γ_h maps $PAP(\mathbb{X})$ into itself, and $\alpha_h(u)$ and $\beta_h(u)$ are respectively the almost periodic and ergodic perturbation components of $\Gamma_h(u)$. Consider the Banach space $Y = (PAP(\mathbb{X}), \|\cdot\|_\infty)$ equipped with both the relation $\prec$ and the mapping $m : PAP(\mathbb{X}) \mapsto PAP(\mathbb{X})$ defined by: If $u, v \in PAP(\mathbb{X})$,

$$u \prec v \ \text{ if and ony if } \ \|u(t)\| \le \|v(t)\|, \quad \forall t \in \mathbb{R},$$

and $m(u) = \|u\|$, i.e., $(m(u))(t) = \|u(t)\|$ for all $t \in \mathbb{R}$.

Proceeding as previously, for all $u, v \in PAP(\mathbb{X})$,

$$\begin{aligned}
\|\Gamma_h(u)(t) - \Gamma_h(v)(t)\| &\le \int_{-\infty}^{t} \|T(t-s)\left(f(s, u(h(s))) - f(s, v(h(s)))\right)\| ds \\
&\le \int_{-\infty}^{t} M e^{-\omega(t-s)} L(s) \|u(h(s)) - v(h(s))\| ds \\
&\le M. \int_{-\infty}^{t} L(s) \|u(h(s)) - v(h(s))\| ds,
\end{aligned}$$

and hence $m(\Gamma_h(u) - \Gamma_h(v)) \prec \Pi m(u - v)$.

Therefore, by Theorem 4 it follows that the operator Γ_h has a unique fixed-point in $PAP(\mathbb{X})$, which obviously is the only pseudo almost periodic solution to Eq. (1.2). ∎

Example 6. *For $p > 0$, consider the heat equation with delay given by*

$$\frac{\partial u}{\partial t}(t,x) = \frac{\partial^2 u}{\partial x^2}(t,x) + f\left(t, u(t-p,x)\right), \quad t \in \mathbb{R},\ x \in [0,\pi] \tag{3.11}$$

$$u(t,0) = u(t,\pi) = 0, \quad t \in \mathbb{R}, \tag{3.12}$$

where $f(t,u) = a(t)\sin u + b(t)\sin(\sqrt{2}u)$ with $a, b \in PAP(\mathbb{R}) \cap L^1(\mathbb{R})$.

Let $\mathbb{X} = L^2[0,\pi]$ equipped with its natural topology and let A be the operator defined by

$$D(A) = \{u \in L^2[0,\pi] : u'' \in L^2[0,\pi],\ u(0) = u(\pi) = 0\}$$
$$Au = u'', \ \forall u \in D(A).$$

It is well-known [27] that A is the infinitesimal generator of an analytic semigroup $T(t)$ on $L^2[0,\pi]$ with

$$\|T(t)\| \le e^{-t}, \quad t \ge 0.$$

Using **[9, Proposition 2.2]** *it can be easily checked that f satisfies assumption (***H.1***) with $L(t) = |a(t)| + \sqrt{2}|b(t)|$.*

Now set

$$(\Pi' u)(t) := \int_{-\infty}^{t} \Big(|a(s)| + \sqrt{2}|b(s)|\Big) u(s)ds$$

for all $t \in \mathbb{R}$ and $u \in PAP(\mathbb{R})$.

Corollary 7. *The boundary value problem Eq. (3.11)-Eq. (3.12) has a unique pseudo almost periodic solution whenever $r(\Pi') < 1$.*

4. Pseudo Almost Periodic Solutions To Some Partial Evolution Equations

Throughout the rest of the paper, $(\mathbb{Y}, \|\cdot\|_{\mathbb{Y}})$ denotes an arbitrary Banach space, which is continuously embedded into the ground space $(\mathbb{X}, \|\cdot\|)$. Furthermore, the constant $\vartheta > 0$ will denote the bound of such an embedding, i.e.,

$$\|u\| \le \vartheta \,.\, \|u\|_{\mathbb{Y}}, \quad \forall u \in \mathbb{Y}.$$

Examples of such $\mathbb{Y}$ include among others the following three cases:

(i) The domain $(D(A), \|\cdot\|_{D(A)})$ of a closed operator A where $\|\cdot\|_{D(A)}$ is the graph norm defined by

$$\|u\|_{D(A)} := \|u\| + \|Au\|$$

for each $u \in D(A)$.

(ii) Let $\alpha \in (0,1)$. Suppose that A is a densely defined closed invertible linear operator on $\mathbb{X}$. If A is the infinitesimal generator of an analytic semigroup, as we previously mentioned, the fractional powers A^α of A exist. In this event, one sets $\mathbb{Y} = (D(A^\alpha), \|\cdot\|_\alpha)$, where

$$\|u\|_\alpha := \|u\| + \|A^\alpha u\|$$

for each $u \in D(A^\alpha)$.

(iii) $\mathbb{Y} = \mathbb{X}$. In this event, the linear operators B appearing in Eq. (1.3) must be bounded linear operators on $\mathbb{X}$ (see assumption (H.6) below).

4.1. Pseudo Almost Periodic Solutions to Eq. (1.3)

This subsection is devoted to the existence and uniqueness of a pseudo almost periodic solution to (1.3).

In addition to assumption **(H.1)**, our present setting requires the following assumptions:

(H.1)' The function $f : \mathbb{R} \times \mathbb{Y} \mapsto \mathbb{X}$, $(t,u) \mapsto f(t,u)$ is pseudo almost periodic in $t \in \mathbb{R}$ uniformly in $u \in \mathbb{Y}$, and $f : \mathbb{R} \times \mathbb{Y} \mapsto \mathbb{Y}$ satisfies Lipschitz condition in $u \in \mathbb{Y}$ for each $t \in \mathbb{R}$ in the following sense

$$\|f(t,u) - f(t,v)\|_{\mathbb{Y}} \leq L(t)\,.\,\|u - v\|, \quad \forall t \in \mathbb{R},\ u, v \in \mathbb{Y},$$

where $L : \mathbb{R} \mapsto [0,\infty)$ is a continuous function satisfying

$$\int_{-\infty}^{+\infty} L(\sigma) d\sigma < \infty;$$

(H.5) The operator A is the infinitesimal generator of an analytic semigroup $T(t)$ such that there exist constants $M > 0$ and $\delta > 0$ with

$$\|T(t)\|_{B(\mathbb{Y})} \leq M e^{-\delta t}, \quad \forall t \geq 0.$$

(H.6) The linear operator $B \in B(\mathbb{Y}, \mathbb{X})$. Set $\|B\|_{B(\mathbb{Y},\mathbb{X})} = \varpi$.

(H.7) If $r(\Sigma)$ denotes the spectral radius of the bounded linear operator Σ defined by

$$(\Sigma u)(t) := M\,.\,\varpi\,.\,\vartheta\,.\int_{-\infty}^{t} L(s) u(s) ds,$$

$t \in \mathbb{R}, \quad u \in PAP(\mathbb{Y})$, then $r(\Sigma) < 1$.

Lemma 8. *Suppose that assumptions* **(H.1)'-(H.5)-(H.6)** *hold. Define the nonlinear operator Λ by: For each $\phi \in PAP(\mathbb{Y})$,*

$$(\Lambda\phi)(t) := \int_{-\infty}^{t} T(t-s) f(s, B\phi(s)) ds.$$

Then Λ maps $PAP(\mathbb{Y})$ into itself.

Proof. First of all, let us check that Λ is well-defined. Indeed, if $\phi \in PAP(\mathbb{Y})$, then $t \mapsto B\phi(t)$ is in $PAP(\mathbb{Y})$, by **(H.6)**. And hence, by the composition theorem of pseudo almost periodic functions [3], the function $\psi(\cdot) = f(\cdot, B\phi(\cdot))$ belongs to $PAP(\mathbb{Y})$ whenever $\phi \in PAP(\mathbb{Y})$. In particular, $N = \sup_{t\in\mathbb{R}} \|f(t, B\phi(t))\|_{\mathbb{Y}} < \infty$.

Now

$$\begin{aligned}
\|(\Lambda\phi)(t)\|_{\mathbb{Y}} &= \|\int_{-\infty}^{t} T(t-s)f(s, B\phi(s))ds\|_{\mathbb{Y}} \\
&\leq \int_{-\infty}^{t} \|T(t-s)f(s, B\phi(s))\|_{\mathbb{Y}}ds \\
&\leq M\,.\int_{-\infty}^{t} e^{-\omega(t-s)}\,.\,\|f(s, B\phi(s))\|_{\mathbb{Y}}ds \\
&= M\,.\,N\,.\int_{-\infty}^{t} e^{-\omega(t-s)}ds \\
&= M\,.\,N\,.\int_{0}^{+\infty} e^{-\omega\sigma}d\sigma \\
&= \frac{M\,.\,N}{\delta} \\
&< \infty,
\end{aligned}$$

and hence $t \mapsto (\Lambda\phi)(t)$ is well-defined.

Since $\psi \in PAP(\mathbb{Y})$, one can write $\psi = p + q$ where $p \in AP(\mathbb{Y})$ and $q \in AP_0(\mathbb{Y})$, and hence $\Lambda(\phi) = \Upsilon(p) + \Upsilon(q)$ where

$$\Upsilon p(t) := \int_{-\infty}^{t} T(t-s)p(s)ds, \text{ and}$$

$$\Upsilon q(t) := \int_{-\infty}^{t} T(t-s)q(s)ds.$$

Clearly, $\Upsilon(p) \in AP(\mathbb{Y})$. Indeed, since $p \in AP(\mathbb{Y})$, for every $\varepsilon > 0$ there exists $\varsigma > 0$ such that for all θ there is $\tau \in [\zeta, \zeta + \varsigma]$ with

$$\|\sigma_\tau p - p\|_{\mathbb{Y},\infty} < \mu\,.\,\varepsilon$$

where $\mu = \dfrac{\omega}{M}$.

Thus, using **(H.1)'** it easily follows that $\|\Upsilon(p)(t+\tau) - \Upsilon(p)(t)\|_{\mathbb{Y}} < \varepsilon$ for each $t \in \mathbb{R}$, and hence $\|\sigma_\tau\Upsilon(p) - \Upsilon(p)\|_{\mathbb{Y},\infty} < \varepsilon$, that is, $\Upsilon(p) \in AP(\mathbb{Y})$.

Next, we show that $t \mapsto \Upsilon(q)(t)$ is in $AP_0(\mathbb{Y})$. First, note that $s \mapsto \Upsilon(q)(s)$ is a bounded continuous function. Thus we have to show that

$$\lim_{T\to\infty} \frac{1}{2T} \int_{-T}^{T} \|\Upsilon(q)(t)\|_{\mathbb{Y}}\, dt = 0.$$

Clearly,

$$\lim_{T\to\infty} \frac{1}{2T} \int_{-T}^{T} \|\Upsilon(q)(t)\|_{\mathbb{Y}}dt \leq I + J$$

where

$$I := \lim_{T\to\infty} \frac{M}{2T} \int_{-T}^{T} dt \left(\int_{-r}^{t} e^{-\omega(t-s)} \, \|q(s)\|_{\mathbb{Y}} ds \right),$$

and

$$J := \lim_{r\to\infty} \frac{M}{2T} \int_{-T}^{T} dt \int_{-\infty}^{-T} e^{-\omega(t-s)} \, \|q(s)\|_{\mathbb{Y}} ds.$$

We must show that $I = J = 0$. To this end, notice that $C' = \|q\|_{\mathbb{Y},\infty} < \infty$. Clearly,

$$\begin{aligned} I &= \lim_{T\to\infty} \frac{M}{2T} \int_{-T}^{T} \|q(t)\|_{\mathbb{Y}} dt \left(\int_{-T}^{t} e^{-\omega(t-s)} ds \right) \\ &= \lim_{T\to\infty} \frac{M}{2T} \int_{-T}^{T} \|q(t)\|_{\mathbb{Y}} dt \left(\int_{0}^{t+T} e^{-\omega\sigma} d\sigma \right) \\ &\leq \lim_{T\to\infty} \frac{M}{2T} \int_{-T}^{T} \|q(t)\|_{\mathbb{Y}} dt \left(\int_{0}^{+\infty} e^{-\omega\sigma} d\sigma \right) \\ &= \lim_{r\to\infty} \frac{M}{2T\omega} \int_{-T}^{T} \|q(t)\|_{\mathbb{Y}} dt \\ &= \frac{M}{\omega} \cdot \lim_{T\to\infty} \frac{1}{2T} \int_{-T}^{T} \|q(t)\|_{\mathbb{Y}} dt \\ &= 0. \end{aligned}$$

Similarly,

$$\begin{aligned} J &\leq \lim_{T\to\infty} \frac{M\,.\,C'}{2T} \int_{-T}^{T} dt \int_{t+T}^{+\infty} e^{-\omega\sigma} d\sigma \\ &\leq \lim_{T\to\infty} \frac{M\,.\,C'}{2T} \int_{-T}^{T} dt \int_{2T}^{+\infty} e^{-\omega\sigma} d\sigma \\ &= \lim_{T\to\infty} \left[\frac{M\,.\,C'\,.\,e^{-2\omega T}}{\delta} \right] \\ &= 0, \end{aligned}$$

and hence $\Upsilon(q)$ belongs to $AP_0(\mathbb{Y})$.

∎

Theorem 9. *Suppose that assumptions* **(H.1)'-(H.5)-(H.6)-(H.7)** *hold. Then Eq. (1.3) has a unique pseudo almost periodic (mild) solution.*

Proof. From Lemma 8 it is clear that $\Lambda : PAP(\mathbb{Y}) \mapsto PAP(\mathbb{Y})$ where

$$(\Lambda u)(t) = \int_{-\infty}^{t} T(t-s) f(s, Bu(s)) ds$$

is well-defined.

Consider the Banach space $\mathcal{M} = (PAP(\mathbb{Y}), \|\cdot\|_{\mathbb{Y},\infty})$ equipped with both the relation $\prec$ and the mapping $m : PAP(\mathbb{Y}) \mapsto PAP(\mathbb{Y})$ defined by: If $u, v \in PAP(\mathbb{Y})$,

$$u \prec v \text{ if and ony if } \|u(t)\|_{\mathbb{Y}} \leq \|v(t)\|_{\mathbb{Y}}, \ \ \forall t \in \mathbb{R},$$

and $m(u) = \|u\|_{\mathbb{Y}}$, i.e., $(m(u))(t) = \|u(t)\|_{\mathbb{Y}}$ for all $t \in \mathbb{R}$.

Now, for all $\phi, \psi \in PAP(\mathbb{Y})$,

$$\begin{aligned}
\|(\Lambda\phi)(t) - (\Lambda\psi)(t)\|_{\mathbb{Y}} &= \left\| \int_{-\infty}^{t} T(t-s)\left[f(s, B\phi(s)) - f(s, B\psi(s))\right] ds \right\|_{\mathbb{Y}} \\
&\leq \int_{-\infty}^{t} \|T(t-s)\left[f(s, B\phi(s)) - f(s, B\psi(s))\right]\|_{\mathbb{Y}}\, ds \\
&\leq M\,.\,\varpi\,.\int_{-\infty}^{t} L(t) e^{-\omega(t-s)} \|\phi(s) - \psi(s)\| ds \\
&\leq M\,.\,\varpi\,.\,\vartheta\,.\int_{-\infty}^{t} L(t) e^{-\omega(t-s)} \|\phi(s) - \psi(s)\|_{\mathbb{Y}} ds \\
&\leq M\,.\,\varpi\,.\,\vartheta\,.\int_{-\infty}^{t} L(t) \|\phi(s) - \psi(s)\|_{\mathbb{Y}} ds
\end{aligned}$$

and hence $m(\Lambda(u) - \Lambda(v)) \prec \Sigma\, m(u - v)$.

We also observe that Δ is increasing, that is, if $\theta \prec u \prec v$, then $\Sigma u \prec \Sigma v$ for all $u, v \in PAP(\mathbb{Y})$. Therefore, by Theorem 4 it follows that the operator Λ has a unique fixed-point in $PAP(\mathbb{Y})$, which obviously is the only pseudo almost periodic solution to Eq. (1.3). ∎

Corollary 10. *Suppose that assumptions* **(H.1)'-(H.5)-(H.6)-(H.7)** *hold in the case where* $\mathbb{Y} = \mathbb{X}$, *and* $B \in B(\mathbb{X})$. *Then Eq. (1.3) has a unique pseudo almost periodic (mild) solution.*

4.2. Pseudo Almost Periodic Solutions to Eq. (1.4)

In what follows, $BC(\mathbb{R}, \mathbb{Y})$ will denote the collection of bounded continuous functions. It is equipped with the $\mathbb{Y}$-sup norm defined by $\|u\|_{\infty,\mathbb{Y}} := \sup\limits_{t\in\mathbb{R}} \|u(t)\|_{\mathbb{Y}}$.

Definition 5. *A function* $u \in BC(\mathbb{R}, \mathbb{X})$ *is said to be a bounded mild solution to Eq. (1.4) provided that the function* $s \to AT(t-s)f(s, Bu(s))$ *is integrable on* $(-\infty, t)$ *for each* $t \in \mathbb{R}$ *and*

$$u(t) = -f(t, Bu(t)) - \int_{-\infty}^{t} AT(t-s)f(s, Bu(s))ds + \int_{-\infty}^{t} T(t-s)g(s, Cu(s))ds$$

for each $\forall t \in \mathbb{R}$.

The existence result of this subsection requires assumptions **(H.8)-(H.9)-(H.10)** given below.

(H.8) The operator A is the infinitesimal generator of an exponentially stable semigroup $(T(t))_{t\geq 0}$ such that there exist constants $M > 0$ and $\delta > 0$ with

$$\|T(t)\|_{B(\mathbb{Y})} \leq Me^{-\delta t}, \ \ \forall t \geq 0.$$

Furthermore, the function $\sigma \to AT(\sigma)$ defined from $(0, \infty)$ into $B(\mathbb{Y})$ is strongly (Lebesgue) measurable and there exist a nonincreasing function $\gamma : \mathbb{R} \to [0, \infty)$ and a constant $\omega > 0$ with $\rho := \displaystyle\int_0^\infty e^{-\omega s}\gamma(s)ds < \infty$ such that

$$\|AT(s)\|_{B(\mathbb{Y},\mathbb{X})} \leq e^{-\omega s}\,.\,\gamma(s), \ \ s > 0.$$

(H.9) The functions $f, g : \mathbb{R}\times\mathbb{Y} \mapsto \mathbb{X}$, $(t, u) \mapsto f(t, u), g(t, u)$ are pseudo almost periodic in t uniformly in u, and $f, g : \mathbb{R} \times \mathbb{Y} \mapsto \mathbb{Y}$ are Lipschitz, i.e., there exist constants $K_f, K_g > 0$ such that

$$\|f(t, u) - f(t, v)\|_{\mathbb{Y}} \leq K_f\,.\,\|u - v\|, \ \text{ and}$$

$$\|g(t, u) - g(t, v)\|_{\mathbb{Y}} \leq K_g\,.\,\|u - v\|$$

for all $t \in \mathbb{R}$, and $\forall u, v \in \mathbb{Y}$.

(H.10) The linear operators $B, C \in B(\mathbb{Y}, \mathbb{X})$ with

$$\max(\|B\|_{B(\mathbb{Y},\mathbb{X})}, \ \|C\|_{B(\mathbb{Y},\mathbb{X})}) = \varpi.$$

In addition to Lemma 8 (where f and B should be respectively replaced by g and C), the proof of the main result of this subsection requires the following technical lemma:

Lemma 11. *Suppose that assumptions* **(H.8)-(H.9)-(H.10)** *hold. Define the nonlinear operator* Λ_1 *by: For each* $\phi \in PAP(\mathbb{Y})$,

$$(\Lambda'\phi)(t) = \int_{-\infty}^t AT(t - s)f(s, B\phi(s))ds$$

Then Λ' *maps* $PAP(\mathbb{Y})$ *into itself.*

Proof. Let $\phi \in PAP(\mathbb{Y})$. Since $B \in B(\mathbb{Y}, \mathbb{X})$ it follows that the function $t \mapsto B\phi(t)$ belongs to $PAP(\mathbb{Y})$. Again, using the composition theorem of pseudo almost periodic functions [3] it follows that $\psi(\cdot) = f(\cdot, B\phi(\cdot))$ is in $PAP(\mathbb{Y})$ whenever $\phi \in PAP(\mathbb{Y})$. In particular, $N = \sup\limits_{t\in R} \|f(t, B\phi(t))\|_{\mathbb{Y}} < \infty$.

Arguing as in the proof of Lemma 8 it can be easily shown that Λ_1 is well-defined. Now write $\psi = w + z$, where $w \in AP(\mathbb{Y})$ and $z \in AP_0(\mathbb{Y})$, and hence $\Lambda_1\phi = \Xi(w) + \Xi(z)$ where

$$\Xi w(t) := \int_{-\infty}^t AT(t - s)w(s)ds, \ \text{ and } \ \Xi z(t) := \int_{-\infty}^t AT(t - s)z(s)ds.$$

Clearly, $\Xi(w) \in AP(\mathbb{Y})$. Indeed, since $w \in AP(\mathbb{Y})$, for every $\varepsilon > 0$ there exists $\varsigma > 0$ such that for all ζ there is $\tau \in [\zeta, \zeta + \varsigma]$ with the property: $\|\sigma_\tau w - w\|_{\mathbb{Y}} < \nu \,.\, \varepsilon$ where $\nu = \dfrac{1}{\rho}$. Arguing as in the proof of Lemma 8 it can be easily follows that $\Xi(w) \in AP(\mathbb{Y})$.

Now

$$\begin{aligned}
\frac{1}{2r}\int_{-r}^{r} \|(\Xi z)(t)\|_{\mathbb{Y}}\, dt &\leq \frac{1}{2r}\int_{-r}^{r}\int_{-\infty}^{t} \|AT(t-s)\|_{B(\mathbb{Y},\mathbb{X})}\, \|z(s)\|_{\mathbb{Y}}\, ds\, dt \\
&\leq \frac{1}{2r}\int_{-r}^{r}\int_{-\infty}^{t} e^{-\omega(t-s)} \,.\, \gamma(t-s)\, \|z(s)\|_{\mathbb{Y}}\, ds\, dt \\
&\leq \frac{1}{2r}\int_{-\infty}^{-r}\int_{-r}^{r} e^{-\omega(t-s)}\, \gamma(t-s)\|z(s)\|_{\mathbb{Y}}\, dt\, ds \\
&\quad + \frac{1}{2r}\int_{-r}^{r}\int_{s}^{r} e^{-\omega(t-s)}\gamma(t-s)\|z(s)\|_{\mathbb{Y}} dt ds \\
&\leq \frac{\|z\|_{\infty,\mathbb{Y}}}{2r}\int_{-\infty}^{-r} \gamma(-r-s)\,.\int_{-r-s}^{r-s} e^{-\omega\tau}\, d\tau\, ds \\
&\quad + \frac{\rho}{2r}\int_{-r}^{r} \|z(s)\|_{\mathbb{Y}}\, dt \\
&\leq \frac{\rho\,.\,\|z\|_{\infty,\mathbb{Y}}}{2r\omega} + \frac{\rho}{2r}\int_{-r}^{r} \|z(s)\|_{\mathbb{Y}}\, dt.
\end{aligned}$$

Clearly, $\lim_{r\to\infty} \dfrac{1}{2r}\displaystyle\int_{-r}^{r} \|(\Xi z)(t)\|_{\mathbb{Y}} dt = 0$, and hence $\Xi z \in AP_0(\mathbb{Y})$. ∎

Theorem 12. *Suppose that assumptions* **(H.8)-(H.9)-(H.10)** *hold. Then Eq. (1.4) has a unique pseudo almost periodic (mild) solution whenever*

$$\Theta := \varpi\left[\frac{M\,.\,K_g}{\delta} + K_f\,.\,\left(1 + \int_0^\infty e^{-\omega s}\gamma(s) ds\right)\right] < 1.$$

Proof. Define the nonlinear operator $\Gamma : PAP(\mathbb{Y}) \mapsto PAP(\mathbb{Y})$ by

$$\begin{aligned}
(\Gamma u)(t) &= -f(t, Bu(t)) - \int_{-\infty}^{t} AT(t-s)f(s, Bu(s))ds \\
&\quad + \int_{-\infty}^{t} T(t-s)g(s, Cu(s))ds.
\end{aligned}$$

For all $\phi, \psi \in PAP(\mathbb{Y})$,

$$\begin{aligned}
\|f(t, B\phi(t)) - f(t, B\psi(t))\|_{\mathbb{Y}} &\leq K_f\,.\,\|B\phi(t) - B\psi(t)\| \\
&\leq K_f\,.\,\varpi\,.\,\|\phi(t) - \psi(t)\|_{\mathbb{Y}}, \\
&\leq K_f\,.\,\varpi\,.\,\|\phi - \psi\|_{\infty,\mathbb{Y}},
\end{aligned}$$

by **(H.9)-(H.10)**.

Using the composition theorem of pseudo almost periodic functions [3] it follows that the operator defined by $\Omega(u)(t) = f(t, Bu(t))$ maps $PAP(\mathbb{Y})$ into $PAP(\mathbb{Y})$.

In view of the above, and both Lemma 8 and Lemma 11 it easily follows that the nonlinear operator Γ is well-defined, that is, it maps $PAP(\mathbb{Y})$ into $PAP(\mathbb{Y})$.

To complete the proof, we will successively approximate the expressions $\Lambda\phi - \Lambda\psi$ and $\Lambda'\phi - \Lambda'\psi$ in norm $\|\cdot\|_{\mathbb{Y}}$ and next apply the classical Banach fixed-point principle.

$$\begin{aligned}
\|(\Lambda'\phi)(t) - (\Lambda'\psi)(t)\|_{\mathbb{Y}} &= \left\| \int_{-\infty}^{t} AT(t-s)\,[f(s, B\phi(s)) - f(s, B\psi(s))]\,ds \right\|_{\mathbb{Y}} \\
&\le \int_{-\infty}^{t} \|AT(t-s)\,[f(s, B\phi(s)) - f(s, B\psi(s))]\|_{\mathbb{Y}}\,ds \\
&\le K_f\,.\,\varpi\,.\,\|\phi-\psi\|_{\infty,\mathbb{Y}}\,.\int_{-\infty}^{t} e^{-\omega(t-s)}\,\gamma(t-s)ds \\
&= K_f\,.\,\varpi\,.\,\|\phi-\psi\|_{\infty,\mathbb{Y}}\,.\int_{0}^{+\infty} e^{-\omega\sigma}\gamma(\sigma)d\sigma \\
&= K_f\,.\,\varpi\,.\,\rho\,.\,\|\phi-\psi\|_{\infty,\mathbb{Y}}.
\end{aligned}$$

Similarly,

$$\begin{aligned}
\|(\Lambda\phi)(t) - (\Lambda\psi)(t)\|_{\mathbb{Y}} &= \left\| \int_{-\infty}^{t} T(t-s)\,[g(s, C\phi(s)) - g(s, C\psi(s))]\,ds \right\|_{\mathbb{Y}} \\
&\le \int_{-\infty}^{t} \|T(t-s)\,[g(s, C\phi(s)) - g(s, C\psi(s))]\|_{\mathbb{Y}}\,ds \\
&\le M\,.\,K_g\,.\,\varpi\,.\,\|\phi-\psi\|_{\infty,\mathbb{Y}}\,.\int_{-\infty}^{t} e^{-\delta(t-s)}ds \\
&= \frac{M\,.\,K_g\,.\,\varpi}{\delta}\,.\,\|\phi-\psi\|_{\infty,\mathbb{Y}}.
\end{aligned}$$

Clearly, for all $\phi, \psi \in PAP(\mathbb{Y})$,

$$\|\Gamma(\phi) - \Gamma(\psi)\|_{\infty,\mathbb{Y}} \le \Theta\,.\,\|\phi-\psi\|_{\infty,\mathbb{Y}}.$$

Consequently, if $\Theta < 1$, then the operator Γ has a unique fixed-point, which obviously is the only pseudo almost periodic solution to Eq. (1.4).

∎

Corollary 13. *Suppose that assumptions* **(H.8)-(H.9)** *hold in the case where* $\mathbb{Y} = \mathbb{X}$, *and* $B, C \in B(\mathbb{X})$. *Then Eq. (1.4) has a unique pseudo almost periodic (mild) solution whenever*

$$\Theta' := \varpi' \left[\frac{M\,.\,K_g}{\delta} + K_f\,.\,\left(1 + \int_{0}^{\infty} e^{-\omega s}\gamma(s)ds \right) \right] < 1$$

where $\varpi' = \max(\|B\|, \|C\|)$.

Example 7. $\mathbb{X} = L^2[0, \pi]$ *equipped with its natural topology. For* $p > 0$*, consider the boundary value problem with delay given by*

$$\begin{aligned} \frac{\partial}{\partial t}\left[u(t,x) + f\left(t, u(t-p,x)\right)\right] &= \frac{\partial^2 u}{\partial x^2}(t,x) + g\left(t, u(t-p,x)\right) \quad (4.13) \\ u(t,0) &= u(t,\pi) = 0, \quad t \in \mathbb{R}, \quad (4.14) \end{aligned}$$

where $f, g : \mathbb{R} \times L^2[0,\pi] \mapsto L^2[0,\pi]$ *are jointly continuous functions.*

To illustrate Theorem 12 we examine the existence and uniqueness of pseudo almost periodic solutions to the heat equation given by the system Eq. (4.13)- Eq. (4.14). For that, as in Example 6, suppose that A *is the linear operator defined by*

$$\begin{aligned} D(A) &= \{u \in L^2[0,\pi] : u'' \in L^2[0,\pi], \;\; u(0) = u(\pi) = 0\} \\ Au(\cdot) &= \Delta u = u''(\cdot), \;\; \forall u(\cdot) \in D(A). \end{aligned}$$

and take $\mathbb{Y} = D(A)$ *equipped with the graph norm of* A.

Note that the linear operator A *has many properties, in particular has a discrete spectrum with eigenvalues of the form* $\lambda_n = -n^2, n \in \mathbb{N}$ *with corresponding normalized eigenfunctions given by* $z_n(t) := \sqrt{\frac{2}{\pi}}\sin(nt)$*. Moreover, the following properties hold:*

(a) $\{z_n : n \in \mathbb{N}\}$ *is an orthonormal basis for* $L^2[0,\pi]$*;*

(b) $Au = -\sum_{n=1}^{\infty} n^2 \langle u, z_n\rangle z_n$*, for each* $u \in D(A)$*;*

(c) $T(t)u = \sum_{n=1}^{\infty} e^{-n^2 t}\langle u, z_n\rangle z_n$*, for each* $u \in L^2[0,\pi]$*;*

(d) Let $\mathbb{X}_\alpha$ *denote the space* $D((-A)^\alpha)$ *endowed with the graph norm* $\|\cdot\|_\alpha$*. It can be shown that* $\mathbb{X}_\alpha \mapsto \mathbb{X}_\beta$ *is continuous for* $0 < \beta \leq \alpha \leq 1$ *and there exist some constants* $C_\alpha, \delta_\alpha > 0$ *such that*

$$\|T(t)\|_{B(\mathbb{X}_\alpha,\mathbb{X})} \leq \frac{C_\alpha e^{-\delta_\alpha t}}{t^\alpha}, \quad t > 0.$$

Moreover, it is possible to define fractional powers of A *as a closed linear operator, see, e.g., [25, 26, 27, 31]. In particular,*

(e) For $u \in L^2[0,\pi]$ *and* $\alpha \in (0,1)$, $(-A)^{-\alpha}u = \sum_{n=1}^{\infty} \frac{1}{n^{2\alpha}}\langle u, z_n\rangle z_n$*;*

(f) The operator $(-A)^\alpha : D((-A)^\alpha) \subseteq L^2[0,\pi] \mapsto L^2[0,\pi]$ *is given by*

$$(-A)^\alpha u = \sum_{n=1}^{\infty} n^{2\alpha}\langle u, z_n\rangle z_n,$$

for each $u \in D((-A)^\alpha)$ *where*

$$D((-A)^\alpha) = \{u(\cdot) \in L^2[0,\pi] : \sum_{n=1}^{\infty} n^{2\alpha} \langle u, z_n \rangle z_n \in L^2[0,\pi]\}.$$

Theorem 14. *Under assumptions* **(H.9)**, *the heat-like equation with delay given by the system Eq. (4.13)-Eq. (4.14) has a unique solution whenever*

$$K_g + 2K_f < 1.$$

5. Case of the Logistic Equation With Delay

Throughout the rest of this subsection, we suppose that $\mathbb{X} = \mathbb{R}$ equipped the classical absolute value. Note however that when we deal with the pseudo almost periodicity of the kernel Q it would be more convenient to choose $\mathbb{X} = \mathbb{R} \times \mathbb{R}$, see **(H.13)**.

Our setting requires the following assumptions:

(H.11) The function $f, g : \mathbb{R} \mapsto \mathbb{R}$ are pseudo almost periodic and f satisfies,

$$|f(x) - f(y)| \leq \alpha \,.\, |x - y|, \qquad 0 \leq \alpha < 1,$$

for all $x, y \in \mathbb{R}$;

(H.12) The function $h_i : \mathbb{R} \mapsto \mathbb{R}$ is continuous, $h_i(\mathbb{R}) = \mathbb{R}$, and $u(h_i) \in PAP(\mathbb{R})$ $(i = 1, 2)$ whenever $u \in PAP(\mathbb{R})$;

(H.13) The function $Q : \mathbb{R} \times (\mathbb{R} \times \mathbb{R}) \mapsto \mathbb{R}$, $(t, x, y) \mapsto Q(t, x, y)$ is pseudo almost periodic in $t \in \mathbb{R}$ uniformly if $(x, y) \in \mathbb{R} \times \mathbb{R}$. Setting $Q = Q_1 + Q_2$ with $Q_1 \in AP(\mathbb{R} \times \mathbb{R} \times \mathbb{R}, \mathbb{R})$ and $Q_2 \in AP_0(\mathbb{R} \times \mathbb{R} \times \mathbb{R}, \mathbb{R})$, we suppose that $Q_2(\cdot, v(\cdot), v(h_2(\cdot))) \in L^1(\mathbb{R})$ for each $v \in PAP(\mathbb{R})$ where h_2 is the function appearing in **(H.12)**. Furthermore, there exists $0 \leq k \leq 1$ such that

$$|Q(t, x, y) - Q(t, w, z)| \leq (k \,.\, |x - w| + (1 - k) \,.\, |y - z|)$$

for all $x, y, z, w \in \mathbb{R}$;

(H.14) $0 < \int_0^\infty |C(-s)| ds = C_0 < \infty.$

Our main result requires the following technical lemma:

Lemma 15. *Under assumptions* **(H.12)-(H.13)-(H.14)**, *the function defined by*

$$\Gamma u(t) := \int_t^\infty Q(s, u(s), u(h_2(s))) C(t - s) ds$$

maps $PAP(\mathbb{R})$ *into itself.*

Proof. Let $u \in PAP(\mathbb{R})$. First of all, note that $t \mapsto u(h_2(t))$ is pseudo almost periodic, by **(H.12)**. Using **(H.13)** it follows that $s \mapsto Q(s, u(s), u(h_2(s)))$ is pseudo almost periodic, see, e.g., **[3, 9]**.

Now write $Q = Q_1 + Q_2$ where $Q_1 \in AP(\mathbb{R} \times \mathbb{R}^2, \mathbb{R})$ and $Q_2 \in AP_0(\mathbb{R} \times \mathbb{R}^2, \mathbb{R})$. Consequently, $\Gamma u = \Gamma_1(u) + \Gamma_2(u)$ where

$$\Gamma_1 u(t) := \int_t^\infty Q_1(s, u(s), u(h_2(s)))C(t-s)ds$$

and

$$\Gamma_2 u(t) := \int_t^\infty Q_2(s, u(s), u(h_2(s)))C(t-s)ds.$$

To complete the proof, it remains to prove that $\Gamma_1 u \in AP(\mathbb{R})$ and $\Gamma_2 u \in AP_0(\mathbb{R})$.

Since $Q_1(\cdot, u(\cdot), u(h_2(\cdot))) \in AP(\mathbb{R})$, for each $\varepsilon > 0$, there exists $\delta > 0$ such that for all γ, there is $\tau \in [\gamma, \gamma + \delta]$ with

$$|Q_1(s+\tau, u(s+\tau), u(h_2(s+\tau))) - Q_1(s, u(s), u(h_2(s)))| < \frac{\varepsilon}{C_0} \tag{5.15}$$

for each $s \in \mathbb{R}$.

Note that $\Gamma_1 u(t+\tau)) = \int_t^\infty Q_1(r+\tau, u(r+\tau), u(h_2(r+\tau)))C(t-r)dr$, by setting $r = s - \tau$. Considering $\Gamma_1 u(t+\tau) - \Gamma_1 u(t)$ it easily follows that

$$|\Gamma_1 u(t+\tau) - \Gamma_1 u(t)| < \varepsilon, \quad \forall t \in \mathbb{R},$$

by Eq. (5.15) and **(H.14)**, and hence $\Gamma_1(u) \in AP(\mathbb{R})$.

The next step consists of showing that $\Gamma_2 u \in AP_0(\mathbb{R})$. It is clear that $s \mapsto \Gamma_2(u)(s)$ is a bounded continuous function. Thus, it remains to show that

$$\lim_{r \to \infty} \frac{1}{2r} \int_{-r}^{r} |\Gamma_2 u(t)|\, dt = 0.$$

Clearly,

$$\lim_{r \to \infty} \frac{1}{2r} \int_{-r}^{r} |\Gamma_2 u(t)|\, dt \leq I + J,$$

where

$$I := \lim_{r \to \infty} \frac{1}{2r} \int_{-r}^{r} dt \left(\int_t^r |Q_2(s, u(s), u(h_2(s)))| \,.\, |C(t-s)| ds \right), \quad \text{and}$$

$$J := \lim_{r \to \infty} \frac{1}{2r} \int_{-r}^{r} dt \int_r^\infty |Q_2(s, u(s), u(h_2(s)))| \,.\, |C(t-s)| ds.$$

To show that $I = J = 0$, we make use of the following arguments:

(A0)' $\int_{-r}^{s} |C(t-s)| dt = \int_0^{r+s} |C(-v)| dv \leq C_0$ for all $r + s \geq 0$;

(A1)' $Q_2(\cdot, u(\cdot), u(h_2(\cdot))) \in AP_0(\mathbb{R})$;

(A2)' $Q_2(\cdot, u(\cdot), u(h_2(\cdot))) \in L^1(\mathbb{R})$.

Indeed, by changing the order of integration we obtain:

$$\begin{aligned} I &= \lim_{r\to\infty} \frac{1}{2r} \int_{-r}^{r} |Q_2(s, u(s), u(h_2(s)))|\, ds \left(\int_{-r}^{s} |C(t-s)| dt \right) \\ &\leq \lim_{r\to\infty} \frac{C_0}{2r} \int_{-r}^{r} |Q_2(s, u(s), u(h_2(s)))|\, ds \\ &= 0, \end{aligned}$$

by **(A0)'**-**(A1)'**.

Similarly,

$$\begin{aligned} J &= \lim_{r\to\infty} \frac{1}{2r} \int_{r}^{\infty} |Q_2(s, u(s), u(h_2(s)))|\, ds \int_{-r}^{r} |C(t-s)| dt \\ &= \lim_{r\to\infty} \frac{1}{2r} \int_{r}^{\infty} |Q_2(s, u(s), u(h_2(s)))| ds \int_{s-r}^{s+r} |C(-v)| dv \\ &= \lim_{r\to\infty} \int_{r}^{\infty} |Q_2(s, u(s), u(h_2(s)))| \,.\, \phi_r(s)\, ds, \end{aligned}$$

where $\phi_r(s) = \dfrac{1}{2r} \displaystyle\int_{s-r}^{s+r} |C(-v)| dv$.

Clearly, $\phi_r(s) \leq \dfrac{C_0}{2r}$, by $\displaystyle\int_{s-r}^{s+r} |C(-v)| dv \leq C_0$ for all $s \geq r$. And hence $\phi_r(s) \mapsto 0$ as $r \mapsto \infty$. Since $Q_2(\cdot, u(\cdot)\, u(h_2(\cdot))) \in L^1(\mathbb{R})$ it follows that

$$\lim_{r\to\infty} \int_{r}^{\infty} |Q_2(s, u(s), u(h_2(s)))| \,.\, \phi_r(s)\, ds = 0,$$

by **(A.2)'** and the Lebesgue dominated convergence theorem. And therefore $\Gamma_2 u \in AP_0(\mathbb{R})$. ∎

Theorem 16. *Under assumptions* **(H.11)-(H.12)-(H.13)-(H.14)***, Eq. (1.7) has a unique pseudo almost periodic solution whenever* $\alpha + C_0 < 1$.

Proof. Let $u \in PAP(\mathbb{R})$. Define the nonlinear operator

$$\Lambda(u)(t) := f(u(h_1(t))) + \int_{t}^{\infty} Q(s, u(s) u(h_2(s))) C(t-s) ds + g(t), \qquad t \in \mathbb{R}.$$

First of all, let us mention that $f(u(h_1(\cdot))) \in PAP(\mathbb{R})$, which follows immediately from the composition theorem of pseudo almost periodic functions in [3, 9]. Thus, in view of the previous facts and Lemma 15, it easily follows that Λ maps $PAP(\mathbb{R})$ into itself and that $\Gamma_1 u$ and $\Gamma_2 u$ are respectively the almost periodic and ergodic perturbation components of $\Lambda(u)$.

To complete the proof, we must show that $\Lambda : PAP(\mathbb{R}) \mapsto PAP(\mathbb{R})$ has a unique fixed-point.

For $u, v \in PAP(\mathbb{X})$,

$$|\Lambda(u)(t) - \Lambda(v)(t)|$$

$$\leq \alpha\|u - v\|_\infty + \int_t^\infty |(Q(s, u(s), u(h_2(s))) - Q(s, v(s), v(h_2(s))))| \,.\, |C(t-s)|ds$$

$$\leq \alpha\|u - v\|_\infty + \int_t^\infty [k|u(s) - v(s)| + (1-k)|u(h_2(s) - v(h_2(s))] \,.\, |C(t-s)|ds$$

$$\leq \alpha\|u - v\|_\infty + \|u - v\|_\infty \,.\, \int_t^\infty |C(t-s)|ds,$$

and hence

$$\|\Lambda(u) - \Lambda(v)\|_\infty \leq (\alpha + C_0) \,.\, \|u - v\|_\infty.$$

Therefore, by the Banach fixed-point principle, the operator Λ has a unique fixed point whenever $\alpha + C_0 < 1$, which obviously is the only pseudo almost periodic solution to Eq. (1.7).

∎

Setting $h_1(t) = h_2(t) = t - p$, one can easily see that (**H.12**) holds, and hence the next corollary is a straightforward consequence of Theorem 16. (In assumption (**H.13**), we suppose that the ergodic component Q_2 of Q is given such that $Q_2(\cdot, v(\cdot), v(\cdot - p)) \in L^1(\mathbb{R})$ for each $v \in PAP(\mathbb{R})$.)

Corollary 17. *Under assumptions* (**H.11**)-(**H.13**)-(**H.14**), *Eq. (1.8) has a unique pseudo almost periodic solution whenever* $\alpha + C_0 < 1$.

The rest of this subsection is devoted to the existence and uniqueness of pseudo almost periodic solutions to Eq. (1.9). In what follows we define the function

$$\tilde{q}(t, x, y) := q(t, x, y) - a\alpha y$$

for all $t \in \mathbb{R}$ and $x, y \in \mathbb{R}$, where a, α and q are respectively the constants and the function appearing in Eq. (1.9).

We require the following assumption:

(**H.5**) The function $\tilde{q} : \mathbb{R} \times (\mathbb{R} \times \mathbb{R}) \mapsto \mathbb{R}$, $(t, x, y) \mapsto \tilde{q}(t, x, y)$ is pseudo almost periodic in $t \in \mathbb{R}$ uniformly if $(x, y) \in \mathbb{R} \times \mathbb{R}$. Setting $\tilde{q} = q_1 + q_2$ where $q_1 \in AP(\mathbb{R} \times \mathbb{R} \times \mathbb{R}, \mathbb{R})$ and $q_2 \in AP_0(\mathbb{R} \times \mathbb{R} \times \mathbb{R}, \mathbb{R})$, we suppose that $q_2(\cdot, v(\cdot), v(\cdot - p))) \in L^1(\mathbb{R})$ for each $v \in PAP(\mathbb{R})$. Furthermore, suppose that $\alpha, a > 0$, and

$$|q(t, x, y) - q(t, w, z)| \leq (1 - a\alpha)|x - w|$$

for all $t, x, y, z, w \in \mathbb{R}$.

Theorem 18. *Under assumption* (**H.5**), *the logistic equation, Eq. (1.9), has a unique pseudo almost periodic solution whenever* $\alpha + \dfrac{1}{a} < 1$.

Proof. One follows along the same lines as in [7]. We are interested in bounded solutions only. Thus if u is a bounded solution to Eq. (1.9), then

$$\frac{d}{dt}[(u(t) - \alpha u(t-p))\, e^{-at}] = [a\alpha u(t-p) - q(t, u(t), u(t-p))]e^{-at}.$$

Clearly,

$$u(t) = \alpha u(t-p) + \int_t^{\infty} [q(s, u(s), u(s-p)) - a\alpha u(s-p)]e^{a(t-s)}ds, \quad (5.16)$$

for each $t \in \mathbb{R}$, by $\lim_{t\to\infty} [(u(t) - \alpha u(t-p))\, e^{-at}] = 0$ (u is bounded).

To complete the proof, in Eq. (1.7), take $f(t) = \alpha t$, $h_1(t) = h_2(t) = t-p$, $C(t) = e^{at}$, $g(t) = 0$, and

$$Q(t, u(t), u(t-p)) = \tilde{q}(t, u(t), u(t-p)), \quad \forall t \in \mathbb{R},$$

and follow along the same lines as in the proof of Theorem 16. ∎

References

[1] E. Ait Dads, K. Ezzinbi, and O. Arino, Pseudo Almost Periodic Solutions for Some Differential Equations in a Banach Space, *Nonlinear Anal* (TMA). **28** (1997), no. 7, pp. 1141–1155.

[2] E. Ait Dads and O. Arino, Exponential Dichotomy and Existence of Pseudo Almost Periodic Solutions of Some Differential Equations, *Nonlinear Anal* (TMA). **27** (1996), no. 4, pp. 369–386.

[3] B. Amir and L. Maniar, Composition of Pseudo-Almost Periodic Functions and Cauchy Problems with Operator of Nondense Domain, *Ann. Math. Blaise Pascal*. **6** (1999), no. 1, pp. 1–11.

[4] D. Bugajewski, T. Diagana, and C. M. Mahop, Asymptotic and Pseudo Almost Periodicity of Convolution Operator and Applications to Differential and Integral Equations, *Zeitschrift für Analysis und ihre Anwendungen*, 2005, to appear.

[5] D. Bugajewski and M. Zima, On the Darboux Problem of Neutral Type, *Bull. Austral. Math. Soc.* **54** (1996), no. 3, pp. 451–458.

[6] D. Bugajewski and M. Zima, On the Spectral Radius of Linearly Bounded Operators and Existence Results for Functional-Differential Equations. Dynamical systems and differential equations (Wilmington, NC, 2002). *Discrete Contin. Dyn. Syst.* 2003, suppl., pp. 147–155.

[7] T. A. Burton, Basic Neutral Integral Equations of Advanced Type, *Nonlinear Anal. (TMA)* **31** (1998), no. 3/4, pp. 295–310.

[8] C. Corduneanu, *Almost Periodic Functions*, Second Edition, Chelsea - New York, 1989.

[9] C. Cuevas and M. Pinto, Existence and Uniqueness of Pseudo Almost Periodic Solutions of Semilinear Cauchy Problems with Non-dense Domain, *Nonlinear Anal* (TMA). **45** (2001), pp. 73–83.

[10] W. Desch, R. Grimmer, Ronald, and W. Schappacher, Well-Posedness and Wave Propagation for a Class of Integro-differential Equations in Banach Space. *J. Differential Equations* **74** (1988), no. 2, pp. 391–411.

[11] T. Diagana, Pseudo Almost Periodic Solutions to Some Differential Equations, *Nonlinear Anal* (TMA). **60** (2005), no. 7, pp. 1277–1286.

[12] T. Diagana, Some Remarks on Some Second-Order Hyperbolic Differential Equations, *Semigroup Forum*. **68** (2004), no. 3, pp. 357–364.

[13] T. Diagana, C. M. Mahop and G. M. NGuérékata, Pseudo Almost Periodic Solution to Some Semilinear Differential Equations. *Mathematical and Computer Modelling* **43** (2006), No. 1-2., pp. 89–96.

[14] T. Diagana and G. M. N'Guérékata, Pseudo Almost Periodic Mild Solutions To Hyperbolic Evolution Equationa in Abstract Intermediate Banach Spaces. Applicable Analysis (2006), to appear.

[15] T. Diagana, Existence and Uniqueness of Pseudo Almost Periodic Solutions to Some Classes of Partial Evolution Equations. *Nonlinear Analysis* (TMA), 2005, to appear.

[16] T. Diagana, Existence and Uniqueness of Pseudo Almost Periodic Solutions to Semilinear Differential Equations and Applications. *Nonlinear Analysis* (TMA), 2005, to appear.

[17] T. Diagana, C. M. Mahop, G. M. N'Guérékata, and B. Toni, Existence and Uniqueness of Pseudo Almost Periodic Solutions to Some Classes of Semilinear Differential Equations and Applications. *Nonlinear Analysis (TMA)*, 2005, to appear.

[18] T. Diagana, Existence and Uniqueness of Pseudo Almost Periodic Solutions to Some Functional Differential Equations. *Functional Differential Equations*, 2005, to appear.

[19] T. Diagana, C. M. Mahop, Pseudo Almost Periodic Solutions to A Neutral Delay Integral Equation. *Cubo Journal*, 2005, to appear.

[20] T. Diagana, G. M. N'Guérékata, and N. V. Minh, Almost Automorphic Solutions of Evolution Equations, *Proc. Amer. Math. Soc*. **134** (2004), no. 11, pp. 3289–3298.

[21] T. Diagana and G. M. N'Guérékata, Almost Automorphic Solutions to Semilinear Evolution Equations, *Functional Differential Equations*, 2005, to appear.

[22] A. Feldstein, A. Iserles, D. Levin, Embedding of Delay Equations into an Infinite Dimensional ODE System, *J. Differential Equations*. **117** (1995), no.1, pp. 127–150.

[23] A. M. Fink, *Almost Periodic Differential Equations*, Lecture Notes in Mathematics **377**, Springer-Verlag, New York - Berlin, 1974.

[24] L. Fox, D. F. Mayers, J. R. Ockendon, A. B. Tayler, On a Functional Differential Equation, *J. Inst. Math. Appl.* **8** (1971), p. 271–307.

[25] E. Hernández and H. R. Henríquez, Existence of Periodic Solutions of Partial neutral Functional Differential Equations with Unbounded Delay. *J. Math. Anal. Appl* **221** (1998), no. 2, pp. 499–522.

[26] E. Hernández, Existence Results for Partial Neutral Integro-differential Equations with Unbounded Delay. *J. Math. Anal. Appl* **292** (2004), no. 1, pp. 194–210.

[27] E. Hernández M., M. L. Pelicer, and J. P. C. dos Santos , Asymptotically almost periodic and almost periodic solutions for a class of evolution equations, *Electron. J. Diff. Eqns* **2004**(2004), no. 61, pp. 1–15.

[28] H. X. Li, F. L. Huang, and J. Y. Li, Composition of pseudo almost-periodic functions and semilinear differential equations. *J. Math. Anal. Appl.* **255** (2001), no. 2, pp. 436–446.

[29] Z. Jackiewicz, A Note on Existence and Uniqueness of Solutions to Neutral Functional Differential Equations with State Dependent Delays, *Comment. Math. Univ. Carolin.* **36** (1995), no. 1, p. 15–17.

[30] T. Kato and J. B. McLeod, The Functional-Differential Equation $y'(x) = ay(\lambda x) + by(x)$, *Bull. Amer. Math. Soc.* **77** (1971), p. 891–937.

[31] A. Lunardi, *Analytic semigroups and optimal regularity in parabolic problems* , PNLDE Vol. *16*, Birkhäauser Verlag, Basel, 1995.

[32] G. M. N'Guérékata, *Almost Automorphic Functions and Almost Periodic Functions in Abstract Spaces*, Kluwer Academic / Plenum Publishers, New York-London-Moscow, 2001.

[33] G. M. N'Guérékata, *Topics in Almost Automorphy*, Springer, New York, 2005.

[34] A. A. Pankov, *Bounded and Almost Periodic Solutions of Nonlinear Operator Differential Equations*. (Translated from Russian by V. S. Zjackovski and A. A. Pankov). Mathematics and Applications (Russian Series), v. **55**. Kluwer Academic Publishers, 1985.

[35] A. Pazy, *Semigroups of Linear Operators and Applications to Partial Differential Equations*, Springer-Verlag, 1983.

[36] M. Renardy and R. C. Rogers, *An Introduction to Partial Differential Equations*, Texts in Appl. Math. **13**, Springer-Verlag, New York-Berlin- Heidelberg-London-Paris, 1992.

[37] B. Rzepecki, On Some Functional-Differential Equations. *Glas. Mat. Ser. III* **19(39)** (1984), no. 1, pp. 73–82.

[38] S. Zaidman, *Topics in Abstract Differential Equations* Pitman Research Notes in Mathematics Ser. II John Wiley and Sons, New York, 1994-1995.

[39] C. Y. Zhang, Pseudo Almost Periodic Solutions of Some Differential Equations, *J. Math. Anal. Appl.* **151** (1994), pp. 62–76.

[40] C. Y. Zhang, Pseudo Almost Periodic Solutions of Some Differential Equations II, *J. Math. Anal. Appl.* **192** (1995), pp. 543–561.

[41] C. Y. Zhang, Integration of Vector-Valued Pseudo Almost Periodic Functions, *Proc. Amer. Math. Soc.* **121** (1994), pp. 167–174.

[42] M. Zima, A Certain Fixed-Point Theorem and Its Application to Integral-Functional Equations, *Bull. Austral. Math. Soc.*.**48** (1992), pp. 179–186.

In: Research on Evolution Equation Compendium. Volume 2 ISBN: 978-1-60456-752-6
Editor: Gaston M. N'Guerekata

AN ASYMPTOTIC MODEL OF A NONLINEAR KELVIN-VOIGT VISCOELASTIC PLATE*

Robert P. Gilbert and Robert J. Ronkese
Department of Mathematical Sciences
University of Delaware, Newark, DE 19716

Abstract

In the 1970s, COWIN and HEGEDUS [**5, 6**] introduced an adaptive elasticity model for bone deposition and reabsorption (remodeling). In 1998, FIGUEIREDO and TRABUCHO [**8**] published a paper on an asymptotic model of an adaptive elastic rod. Afterwards, MONNIER and TRABUCHO [**9**] proved the existence and uniqueness of a solution for this model; TRABUCHO [**10**] himself extended this result to a model that contains non-linear strain terms along with linear strain terms in a rate remodeling equation. Here, it is proposed to introduce a *Kelvin-Voigt* model for the plate that includes non-linear terms rate remodelling terms with respect to strain and the time derivative of strain. As many biomechanists consider cancellous bone to be structure consisting of both rode and plates, adding the study of the plate permitsre a more realistic modeling of the trabeculae of spongy bone. Recent studies have suggested that the dissipation of acoustic energy in cancellous bone interrogation via ultrasound is due to the viscous properties of the trabeculae, rather than that of interstitial blood and marrow [**1**]. This implies use of a Kelvin-Voigt model for both the plate-like and rod-like trabeculae.

1. Introduction

The model of COWIN and HEGEDUS [**5, 6**] has a rate remodeling (deposition and reabsorption) equation of the form:

$$\dot{\phi} = a(\phi) + A_{km}(\phi)e_{km}$$

where $a(\phi)$ is a constitutive function, A_{km} is a rate remodeling coefficient and e_{km} is the strain tensor. This equation was coupled with equations describing the equilibrium of forces, both of the body and of the surfaces, to be found on the bone material. Later, an existence and uniqueness theorem was postulated in detail in a paper by MONNIER and TRABUCHO [**9**]. However, a subsequent paper by TRABUCHO [**10**] suggested adding a nonlinear term $c_{ijkl}e_{kl}e_{ij}$ for the strain tensor in the above equation and then proved existence and uniqueness. In this paper, a *Kelvin-Voigt* model of a nonlinear viscoelastic plate is considered with the more complicated rate remodeling equation:

$$\dot{\phi}^\varepsilon = a^\varepsilon(\phi^\varepsilon)+A^\varepsilon_{kl}(\phi^\varepsilon)e^\varepsilon_{kl}(u^\varepsilon)+\bar{A}^\varepsilon_{kl}(\phi^\varepsilon)\dot{e}^\varepsilon_{kl}(u^\varepsilon)+c^\varepsilon_{ijkl}(\phi^\varepsilon)e^\varepsilon_{kl}(u^\varepsilon)e^\varepsilon_{ij}(u^\varepsilon)+d^\varepsilon_{ijkl}(\phi^\varepsilon)\dot{e}^\varepsilon_{kl}(u^\varepsilon)e^\varepsilon_{ij}(u^\varepsilon)$$

*This research was supported in part by NSF grant INT 0438765 and by an Alexander v. Humboldt Senior Scientist Award at the Ruhr University Bochum.

with $\bar{A}_{km}$ as a rate remodeling coefficient and d_{ijkl} as a viscosity coefficient that are grouped with tensors that are the time derivatives of the strain. This model includes the time rate of change of strain, which we believe represents a truer picture of the remodeling process. This paper introduces the asymptotic model and shows how asymptotic expansions of both the displacement vector and the stress tensor can be simplified. Equations of the *displacement-stress model* are derived as well as a characterization of the first term of the displacement expansion. The first few terms of the asymptotic expansion of the time rate of change of the volume fraction of spongy bone material is derived.

2. Set-up of the Problem

The middle-surface of the three dimensional plate is the interior, ω, of a simply connected curve, γ, in the x_1x_2 plane. The plate, which is to have a thickness 2ε, will be designated by Ω^ε. The so called *reference plate* will have a thickness of L which will be designated by Ω. Thus, $\Omega^\varepsilon = \omega \times (-\varepsilon, \varepsilon)$ and $\Omega = \omega \times (0, L)$. The lateral boundary of Ω^ε is $\Gamma_0^\varepsilon = \gamma \times [-\varepsilon, \varepsilon]$. The lateral boundary of Ω is $\Gamma^0 = \gamma \times [0, L]$. The boundary on the top and bottom of Ω^ε is $\Gamma_+^\varepsilon = \omega \times \{\varepsilon\}$ and $\Gamma_-^\varepsilon = \omega \times \{-\varepsilon\}$, respectively. For the top and bottom of Ω, there is $\Gamma^+ = \omega \times \{L\}$ and $\Gamma^- = \omega \times \{0\}$. It is assumed that the constant $\varepsilon > 0$ and ε is much smaller than the dimensions of ω. The plate is assumed to be clamped so that a displacement vector $\mathbf{u}$ vanishes on the lateral boundary. The notation for subscripts

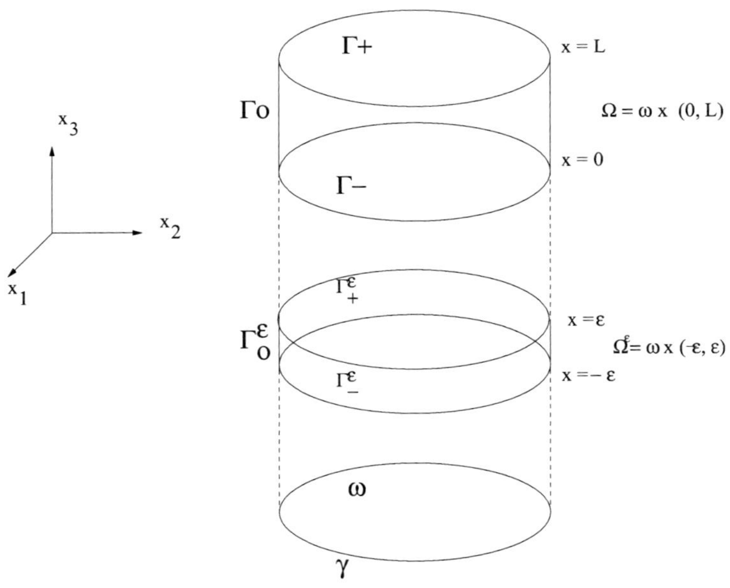

Plate Models in Different Domains

Figure 1. The plates Ω^ε and Ω.

is as follows: Latin indices take on values of {1,2,3} while Greek indices take on values of

{1,2}. For repeated indices, the summation convention will be employed.

A system of inertia associated with the coordinate system $Ox_1^\varepsilon x_2^\varepsilon x_3^\varepsilon$ is chosen so that the axis Ox_3^ε passes through the centroid of each cross section despite the thickness of 2ε. Also:

$$\int_{\omega^\varepsilon} x_\alpha^\varepsilon d\omega^\varepsilon = \int_{\omega^\varepsilon} x_1^\varepsilon x_2^\varepsilon d\omega^\varepsilon = 0, \quad \alpha = \{1, 2\}$$

Differential operators will be either represented by $\frac{\partial}{\partial x_i}$ or simply by ∂_i. The unit normal $\mathbf{n}^\varepsilon = (n_1^\varepsilon, n_2^\varepsilon, 0)$ on the lateral boundary, while $\mathbf{n}^\varepsilon = (0, 0, 1)$ on the top and $\mathbf{n}^\varepsilon = (0, 0, -1)$ on the bottom.

The plate Ω^ε is composed of a nonlinear, viscoelastic, homogenous material where the Lamé constants of isotropic materials, which are related to Young's modulus and Poisson's ratio [**2**, pp. 59, 60], are replaced in the anisotropic case by the fourth-order tensors c_{ijkm}^ε and d_{ijkm}^ε used in the definition of the stress tensor as seen below. Volume forces, denoted by $\mathbf{f}^\varepsilon = (f_i^\varepsilon) : \Omega^\varepsilon \Rightarrow \mathbf{R}^3$, act in the interior of Ω^ε and surface forces, denoted by $\mathbf{g}^\varepsilon = (g_i^\varepsilon) : \Omega^\varepsilon \Rightarrow \mathbf{R}^3$, act on the boundaries $\Gamma^\varepsilon = \Gamma_0^\varepsilon \cup \Gamma_+^\varepsilon \cup \Gamma_-^\varepsilon$.

The following equations of equilibrium for the nonlinear viscoelastic clamped plate as given in Cowin and Hegedus [**5, 6**], are used to find a $\mathbf{u}^\varepsilon = (u_i^\varepsilon) : \bar{\Omega} \Rightarrow \mathbf{R}^3$ such that:

$$-\partial_j^\varepsilon \sigma_{ij}^\varepsilon(\mathbf{u}^\varepsilon) = \gamma(\xi_0 + M_\rho \phi(\mathbf{u}^\varepsilon)) f_i^\varepsilon, \ \textit{on}\ \Omega^\varepsilon$$

$$\sigma_{ij}(\mathbf{u}^\varepsilon) n_j^\varepsilon = g_i^\varepsilon, \ \textit{on}\ \Gamma^\varepsilon$$

where $\Gamma^\varepsilon = \Gamma_0^\varepsilon \cup \Gamma_+^\varepsilon \cup \Gamma_-^\varepsilon$; Γ_0^ε is the lateral boundary of the plate; $\Gamma_+^\varepsilon \cup \Gamma_-^\varepsilon$ is the boundary of the plate on the top and bottom; γ = density of the material; ξ_0 = the reference change in the volume fraction of material; M_ρ = mollification operator from $C^0(\bar{\Omega}^\varepsilon)$ into $C^\infty(R^3)$; P_η = truncation operator of class C^1; f_i = body forces; g_i = surface forces;

$$e_{ij}^\varepsilon(\mathbf{u}^\varepsilon) = \frac{1}{2}(\partial_i^\varepsilon u_j^\varepsilon + \partial_j^\varepsilon u_i^\varepsilon)$$

is the strain tensor; the stress tensor for a Kelvin-Voigt material has the form:

$$\sigma^\varepsilon(\mathbf{u}^\varepsilon) = \sigma_{ij}^\varepsilon(\mathbf{u}^\varepsilon) = c_{ijkm}^\varepsilon(\phi^\varepsilon) e_{ij}^\varepsilon(\mathbf{u}) + d_{ijkm}^\varepsilon(\phi^\varepsilon) \dot{e}_{ij}^\varepsilon(\mathbf{u})$$

and the clamping conditions of $\mathbf{u}^\varepsilon = \mathbf{0}$ apply on the lateral boundary Γ_0^ε. This is a *Kelvin-Voigt model* for homogeneous materials.

It is assumed, as in a paper by Figueiredo and Trabucho [**8**, pp. 3, 4], that the definition of the stress tensor contains the modified elasticity coefficient $c_{ijkl}^\varepsilon(\phi^\varepsilon)$ and the modified viscosity coefficient $d_{ijkl}(\phi^\varepsilon)$ which are dependent upon ϕ, the change in the volume fraction of the elastic material. a_{ijkl} is a elasticity coefficient and b_{ijkl} is a viscosity coefficient.

$$c_{ijkl}^\varepsilon = (\xi_0 + M_\rho \circ P_\eta(\phi(\mathbf{u})))\, a_{ijkl}, \quad d_{ijkl}^\varepsilon = (\xi_0 + M_\rho \circ P_\eta(\phi(\mathbf{u})))\, b_{ijkl}$$

ξ_0 is the original change in the volume fraction of elastic material. M_ρ is a mollifier. P_η is a truncation operator. Due to symmetry about the Ox_3 axis,

$$c_{ijkl}^\varepsilon(\phi^\varepsilon) = c_{jikl}^\varepsilon(\phi^\varepsilon) = c_{klij}^\varepsilon(\phi^\varepsilon), \ c_{\alpha\beta\gamma 3}(\phi^\varepsilon) = c_{\alpha 333}(\phi^\varepsilon) = 0$$

$$d^{\varepsilon}_{ijkl}(\phi^{\varepsilon}) = d^{\varepsilon}_{jikl}(\phi^{\varepsilon}) = d^{\varepsilon}_{klij}(\phi^{\varepsilon}),\ d_{\alpha\beta\gamma 3}(\phi^{\varepsilon}) = d_{\alpha 333}(\phi^{\varepsilon}) = 0$$

It is also assumed that the coefficients c_{ijkl} are positive definite for all symmetric tensors τ_{ij}, and so there is an elliptic condition that is satisfied:
$\exists c > 0 : c^{\varepsilon}_{ijkl}(\phi^{\varepsilon})\tau_{ij}\tau_{kl} \geq c\,\tau_{ij}\tau_{kl}\ \forall \tau_{ij} \in \mathbf{R}^{3x3},\ \tau_{ij} = \tau_{ji}$
Hence, we obtain the following proposed remodeling rate equation:

$$\dot{\phi}^{\varepsilon} = a^{\varepsilon}(\phi^{\varepsilon}) + A^{\varepsilon}_{kl}(\phi^{\varepsilon})e^{\varepsilon}_{kl}(u^{\varepsilon}) + \bar{A}^{\varepsilon}_{kl}(\phi^{\varepsilon})\dot{e}^{\varepsilon}_{kl}(u^{\varepsilon}) + c^{\varepsilon}_{ijkl}(\phi^{\varepsilon})e^{\varepsilon}_{kl}(u^{\varepsilon})e^{\varepsilon}_{ij}(u^{\varepsilon}) + d^{\varepsilon}_{ijkl}(\phi^{\varepsilon})\dot{e}^{\varepsilon}_{kl}(u^{\varepsilon})e^{\varepsilon}_{ij}(u^{\varepsilon})$$

$a^{\varepsilon}(\phi^{\varepsilon})$ is a constitutive function. $A^{\varepsilon}_{kl}(\phi^{\varepsilon})$ is a rate remodeling coefficient. $c^{\varepsilon}_{ijkl}(\phi^{\varepsilon})e^{\varepsilon}_{kl}(u^{\varepsilon})e^{\varepsilon}_{ij}(u^{\varepsilon})$ is a nonlinear strain term. $\bar{A}^{\varepsilon}_{kl}(\phi^{\varepsilon})\dot{e}^{\varepsilon}_{kl}(u^{\varepsilon})$ is a linear time derivative of strain term, and $d^{\varepsilon}_{ijkl}(\phi^{\varepsilon})\dot{e}^{\varepsilon}_{kl}(u^{\varepsilon})e^{\varepsilon}_{ij}(u^{\varepsilon})$ is a nonlinear time derivative of strain term. The modified elasticity coefficients c_{ijkl} may also be considered as rate remodeling coefficients. The same could be construed for the modified viscosity coefficient d_{ijkl}. FIGUEIREDO and TRABUCHO justify this by mentioning that the driving mechanism associated with the remodeling rate equation is the strain energy per unit volume $\sigma^{\varepsilon}_{ij}e^{\varepsilon}_{ij}(u^{\varepsilon})$ as some evidence suggests [**8**, pp. 4, 5].

For any function $g \in C^0[0,T] \cap C^0(\bar{\Omega}^{\varepsilon})$ whose extension to R^3 is $\bar{g}(t)$, the mollification operator is defined as follows:

$$M_{\rho}(g(t))(x^{\varepsilon}) = \omega_{\rho} \star \bar{g}(t)(x^{\varepsilon}) = \int_{R^3} \omega_{\rho}(x^{\varepsilon} - y)\bar{g}(t)(y)dy$$

$\omega_{\rho}(x^{\varepsilon})$ is a mollifier derived from the function ω:

$$\omega_{\rho}(x^{\varepsilon}) = \left(\int_{R^3} \omega(y)dy\right)^{-1}\frac{1}{\rho^n}\omega\left(\frac{x^{\varepsilon}}{\rho}\right); \quad \begin{aligned} \omega(x) &= exp(1/(\|x\|^2_{R^3} - 1))\ \ if\|x\|_{R^3} < 1 \\ &= 0,\ otherwise \end{aligned}$$

For d^{ε} = the change in volume fraction and real numbers $0 < \eta << 1, 0 < \rho << 1$, the truncation operator P_{η} of class C^1 is defined as:

$$P_{\eta}(d^{\varepsilon})(x^{\varepsilon}) = \begin{cases} -\xi_0(x^{\varepsilon}) + \frac{\eta}{2} & \text{if } d^{\varepsilon}(x^{\varepsilon}) \leq -\xi_0(x^{\varepsilon}) + \frac{\eta}{2} \\ d^{\varepsilon}(x^{\varepsilon}) & \text{if } \eta - \xi_0(x^{\varepsilon}) \leq d^{\varepsilon}(x^{\varepsilon}) \leq 1 - \xi_0(x^{\varepsilon}) - \eta \\ 1 - \xi_0(x^{\varepsilon}) & \text{if } d^{\varepsilon}(x^{\varepsilon}) \geq 1 - \xi_0(x^{\varepsilon}) \end{cases}$$

with $0 < \frac{\eta}{2} \leq (\xi_0 + P_{\eta}(d^{\varepsilon}))(x^{\varepsilon}) \leq 1$

There is a variational formulation of this problem. One seeks to find a solution $\mathbf{u}^{\varepsilon} = (u^{\varepsilon}_i)$ such that

$$\mathbf{u}^{\varepsilon} \in V(\Omega^{\varepsilon}) = \{\mathbf{v}^{\varepsilon} = (v^{\varepsilon}_i) \in [H^1(\Omega^{\varepsilon})]^3 : \mathbf{v}^{\varepsilon} = \mathbf{0} \in \Gamma^{\varepsilon}_{+} \cup \Gamma^{\varepsilon}_{-}\},$$

$$\int_{\Omega^{\varepsilon}} \sigma^{\varepsilon}_{ij}(\mathbf{u}^{\varepsilon})e^{\varepsilon}_{ij}(\mathbf{v}^{\varepsilon})d\mathbf{x}^{\varepsilon} = \int_{\Omega^{\varepsilon}} f^{\varepsilon}_i v^{\varepsilon}_i d\mathbf{x}^{\varepsilon} + \int_{\Gamma^{\varepsilon}_0} g^{\varepsilon}_i v^{\varepsilon}_i ds^{\varepsilon} + \int_{\Gamma^{\varepsilon}_{+} \cup \Gamma^{\varepsilon}_{-}} h^{\varepsilon}_i v_i d\omega$$

for all $\mathbf{v}^{\varepsilon} \in V(\Omega^{\varepsilon})$.

It is assumed that $f^{\varepsilon}_i \in L^2(\Omega^{\varepsilon})$ and $g^{\varepsilon}_i \in L^2(\Gamma^{\varepsilon})$ which makes the problem well-posed.The left side of the above equation is symmetric and continuous. The right side is continuous as well.

3. Scalings of Unknowns

For the scaling of various elements in the equations, the basic relationship of functions in the reference domain, Ω, to the those in the other, Ω^ε, comes from CIARLET [**3**, pp. 32, 33]:

$x^\varepsilon = (x_i^\varepsilon) = (x_1, x_2, \varepsilon x_3)$

$\partial_\alpha^\varepsilon = \partial_\alpha$ and $\partial_3^\varepsilon = \frac{1}{\varepsilon}\partial_3$

$v_\alpha^\varepsilon(x^\varepsilon) = v_\alpha(x)$ and $v_3^\varepsilon(x^\varepsilon) = \frac{1}{\varepsilon}v_3(x)\ \forall x^\varepsilon \in \overline{\Omega}$

$f_\alpha^\varepsilon(x^\varepsilon) = f_\alpha(x)$ and $f_3^\varepsilon(x^\varepsilon) = \varepsilon f_3(x)\ \forall x^\varepsilon \in \Omega^\varepsilon$

$g_\alpha^\varepsilon(x^\varepsilon) = \varepsilon g_\alpha(x)$ and $g_3^\varepsilon(x^\varepsilon) = \varepsilon^2 g_3(x)\ \forall x^\varepsilon \in \Gamma_0^\varepsilon$

$h_\alpha^\varepsilon(x^\varepsilon) = h_\alpha(x)$ and $h_3^\varepsilon(x^\varepsilon) = \varepsilon h_3(x)\ \forall x^\varepsilon \in \Gamma_+^\varepsilon \cup \Gamma_-^\varepsilon$

$\sigma_{\alpha\beta}^\varepsilon(x^\varepsilon) = \sigma_{\alpha\beta}(\mathbf{u}(\varepsilon)(x)), \sigma_{\alpha 3}^\varepsilon(x^\varepsilon) = \varepsilon\sigma_{\alpha\beta}(\mathbf{u}(\varepsilon)(x)), \sigma_{33}^\varepsilon(x^\varepsilon)(x^\varepsilon) = \varepsilon^2\sigma_{33}(\mathbf{u}(\varepsilon)(x))$

$e_{\alpha\beta}^\varepsilon(\mathbf{u}(\varepsilon)) = e_{\alpha\beta}(\mathbf{u}(\varepsilon))(x), e_{\alpha\beta}^\varepsilon(\mathbf{u}^\varepsilon) = \frac{1}{\varepsilon}e_{\alpha 3}(\mathbf{u}(\varepsilon))(x), e_{33}^\varepsilon(\mathbf{u}(\varepsilon)) = \frac{1}{\varepsilon^2}e_{33}(\mathbf{u}(\varepsilon))(x)$

where Γ_+^ε and Γ_-^ε represent the top and the bottom of the plate respectively.

First, the aforementioned scalings convert the terms in the proposed rate remodeling equation to their counterparts in the domain Ω which are then substituted back into the equation:

$$
\begin{aligned}
\dot{\phi}(\varepsilon) &= a^\varepsilon(\phi^\varepsilon) + A_{\alpha\beta}e_{\alpha\beta}(\varepsilon) + c_{\alpha\beta\gamma\mu}e_{\gamma\mu}(\varepsilon)e_{\alpha\beta}(\varepsilon) + \tfrac{2}{\varepsilon}[A_{3\beta}e_{3\beta}(\varepsilon) + \bar{A}_{3\beta}\dot{e}_{3\beta}(\varepsilon)] \\
&+ \tfrac{1}{\varepsilon^2}[A_{33}e_{33}(\varepsilon) + \bar{A}_{33}\dot{e}_{33}(\varepsilon)] + 4c_{3\alpha 3\beta}e_{3\beta}(\varepsilon)e_{3\alpha}(\varepsilon) + 4d_{3\alpha 3\beta}\dot{e}_{3\beta}(\varepsilon)e_{3\alpha}(\varepsilon) \\
&+ c_{\alpha\beta 33}e_{33}(\varepsilon)e_{\alpha\beta}(\varepsilon) + d_{\alpha\beta 33}\dot{e}_{33}(\varepsilon)e_{\alpha\beta}(\varepsilon)] \\
&+ \tfrac{1}{\varepsilon^4}[c_{3333}e_{33}(\varepsilon)e_{33}(\varepsilon) + d_{3333}\dot{e}_{33}(\varepsilon)e_{33}(\varepsilon)]
\end{aligned}
$$

Next, a Kelvin-Voigt model for the plate will be derived. A general approach will be attempted with fourth-order tensors and time derivatives in the definition of the stress tensor. Using Ω^ε as the volume of integration,

$$
\begin{aligned}
&\int_{\Omega^\varepsilon} \sigma_{ij}^\varepsilon e_{ij}^\varepsilon d\vec{x} = \\
&= \int_{\Omega^\varepsilon} [c_{ijkl}e_{kl}^\varepsilon e_{ij}^\varepsilon + d_{ijkl}\dot{e}_{kl}^\varepsilon e_{ij}^\varepsilon] d\vec{x} \\
&= \int_{\Omega^\varepsilon} [c_{\alpha\beta\gamma\mu}e_{\gamma\mu}^\varepsilon e_{\alpha\beta}^\varepsilon + d_{\alpha\beta\gamma\mu}\dot{e}_{\gamma\mu}^\varepsilon e_{\alpha\beta}^\varepsilon] d\vec{x} \\
&+ \int_{\Omega^\varepsilon} [c_{\alpha\beta 33}e_{33}^\varepsilon e_{\alpha\beta}^\varepsilon + d_{\alpha\beta 33}\dot{e}_{33}^\varepsilon e_{\alpha\beta}^\varepsilon] d\vec{x} \\
&+4 \int_{\Omega^\varepsilon} [c_{3\alpha 3\beta}e_{3\beta}^\varepsilon e_{3\alpha}^\varepsilon + d_{3\alpha 3\beta}\dot{e}_{3\beta}^\varepsilon e_{3\alpha}^\varepsilon] d\vec{x}
\end{aligned}
$$

$$+\int_{\Omega^\varepsilon}[c_{33\alpha\beta}e^\varepsilon_{\alpha\beta}e^\varepsilon_{33}+d_{33\alpha\beta}\dot{e}^\varepsilon_{\alpha\beta}e^\varepsilon_{33}]d\vec{x}$$

$$+\int_{\Omega^\varepsilon}[c_{3333}e^\varepsilon_{33}e^\varepsilon_{33}+d_{3333}\dot{e}^\varepsilon_{33}e^\varepsilon_{33}]d\vec{x}$$

Now, reverting back to the reference domain Ω as the volume of integration:

$$\int_\Omega \sigma_{ij}e_{ij}d\vec{x}=$$

$$=\int_\Omega[c_{\alpha\beta\gamma\mu}e_{\gamma\mu}e_{\alpha\beta}+d_{\alpha\beta\gamma\mu}\dot{e}_{\gamma\mu}e_{\alpha\beta}]d\vec{x}$$

$$+\frac{1}{\varepsilon^2}\int_\Omega[c_{\alpha\beta33}e_{33}e_{\alpha\beta}+d_{\alpha\beta33}\dot{e}_{33}e_{\alpha\beta}]d\vec{x}$$

$$+\frac{4}{\varepsilon^2}\int_\Omega[c_{3\alpha3\beta}e_{3\beta}e_{3\alpha}+d_{3\alpha3\beta}\dot{e}_{3\beta}e_{3\alpha}]d\vec{x}$$

$$+\frac{1}{\varepsilon^2}\int_\Omega[c_{33\alpha\beta}e_{\alpha\beta}e_{33}+d_{33\alpha\beta}\dot{e}_{\alpha\beta}e_{33}]d\vec{x}$$

$$+\frac{1}{\varepsilon^4}\int_\Omega[c_{3333}e_{33}e_{33}+d_{3333}\dot{e}_{33}e_{33}]d\vec{x}$$

Asymptotic expansions of the following:

$$\mathbf{u}(\varepsilon)=\mathbf{u}^0+\varepsilon^1\mathbf{u}^1+\varepsilon^2\mathbf{u}^2+...$$

$$e_{ij}(\mathbf{u}(\varepsilon))=e^0_{ij}+e^1_{ij}+e^2_{ij}+...$$

$$\dot{e}_{ij}(\mathbf{u}(\varepsilon))=\dot{e}^0_{ij}+\dot{e}^1_{ij}+\dot{e}^2_{ij}+...$$

$$\sigma_{\alpha\beta}(\varepsilon)=\varepsilon^{-2}\sigma^{-2}_{\alpha\beta}+\varepsilon^{-1}\sigma^{-1}_{\alpha\beta}+\sigma^{-2}_{\alpha\beta}+\varepsilon^1\sigma^1_{\alpha\beta}+...$$

$$\sigma_{3\beta}(\varepsilon)=\varepsilon^{-2}\sigma^{-2}_{3\beta}+\varepsilon^{-1}\sigma^{-1}_{3\beta}+\sigma^{-2}_{3\beta}+\varepsilon^1\sigma^1_{3\beta}+...$$

$$\sigma_{33}(\varepsilon)=\varepsilon^{-4}\sigma^{-4}_{33}+\varepsilon^{-3}\sigma^{-3}_{33}+\varepsilon^{-2}\sigma^{-2}_{33}+\varepsilon^{-1}\sigma^{-1}_{33}+...$$

can be inserted into the force balance equation:

$$\int_\Omega \sigma_{ij}(\varepsilon)e_{ij}(\mathbf{v})d\vec{x}=\int_\Omega f_iv_id\vec{x}+\int_{\Gamma^0}g_iv_id\vec{x}+\int_{\Gamma^+\cup\Gamma^-}h_iv_id\vec{x} \tag{1}$$

where upon the comparison of this equation with the above equation of fourth-order tensors and negative powers of ε yields:

$\sigma^{-4}_{33}=c_{3333}e^0_{33}+d_{3333}\dot{e}^0_{33}$ and $\sigma^{-3}_{33}=c_{3333}e^1_{33}+d_{3333}\dot{e}^1_{33}$

$$\sigma^p_{33}=c_{33\alpha\beta}e^{p+2}_{\alpha\beta}+d_{33\alpha\beta}\dot{e}^{p+2}_{\alpha\beta}+c_{3333}e^{p+4}_{33}+d_{3333}\dot{e}^{p+4}_{33},\ p\geq -2$$

$$\sigma^p_{3\alpha}=2c_{3\alpha3\beta}e^{p+2}_{3\beta}+2d_{3\alpha3\beta}\dot{e}^{p+2}_{3\beta},\ p\geq -2$$

$$\sigma_{\alpha\beta}^{-2} = c_{\alpha\beta 33}e_{33}^{0} + d_{\alpha\beta 33}\dot{e}_{33}^{0} \text{ and } \sigma_{\alpha\beta}^{-1} = c_{\alpha\beta 33}e_{33}^{1} + d_{\alpha\beta 33}\dot{e}_{33}^{1}$$

$$\sigma_{\alpha\beta}^{p} = c_{\alpha\beta 33}e_{33}^{p+2} + d_{\alpha\beta 33}\dot{e}_{33}^{p+2} + c_{\alpha\beta\gamma\mu}e_{\gamma\mu}^{p} + d_{\alpha\beta\gamma\mu}\dot{e}_{\gamma\mu}^{p},\ p \geq 0$$

Setting the non-zero powers of ε equal to zero, one gets the following equations:

$$\varepsilon^{-4}: \int_{\Omega} \sigma_{33}^{-4} e_{33}(\mathbf{v}) d\vec{x} = 0, \quad \varepsilon^{-3}: \int_{\Omega} \sigma_{33}^{-3} e_{33}(\mathbf{v}) d\vec{x} = 0$$

For $p = \{-2, -1, 1, 2, 3, ...\}$, $\varepsilon^{p}: \int_{\Omega} \sigma_{ij}^{p} e_{ij}(\mathbf{v}) d\vec{x} = 0$

$$\varepsilon^{0}: \int_{\Omega} \sigma_{ij}^{0} e_{ij}(\mathbf{v}) d\vec{x} = \int_{\Omega} f_i v_i d\vec{x} + \int_{\Gamma}^{0} g_i v_i d\vec{x} + \int_{\Gamma^{+}\cup\Gamma^{-}} h_i v_i d\vec{x}$$

One is now in a position to characterize many of the quantities in the above equations leading to significant simplifications in the asymptotic expansions.

Lemma 1. *Given* $\psi(t) = \frac{1}{2}\int_{\Omega} d_{ijkl} e_{kl} e_{kl} d\vec{x}$ *with* $d_{ijkl} > 0$, *if* $\frac{\partial}{\partial t}\psi(t) \leq 0$ *and* $e_{kl}(0) = 0$, *then* $e_{kl}(t) = 0$.

Proof.

$$\frac{1}{2}\frac{\partial}{\partial t}\int_{\Omega} d_{ijkl} e_{kl} e_{kl} d\vec{x} = \frac{\partial}{\partial t}\psi(t) \leq 0 \Rightarrow \psi(t) \leq K$$

However, $e_{ij}(0) = 0 \Rightarrow \psi(0) = 0 \Rightarrow K = 0$ and $\psi(t) \leq 0$; but assuming $d_{ijkl} > 0$, $\psi(t) \geq 0$. Combining the two inequalities concerning ψ leads to: $\psi(t) = 0 \Rightarrow e_{ij}(t) = 0$ □

Theorem 1. *If* $\int_{\Omega} \sigma_{33}^{-4} e_{33}(\mathbf{v}) d\vec{x} = 0$, *then* $e_{33}^{0} = 0, \sigma_{33}^{-4} = 0, \sigma_{\alpha\beta}^{-2} = 0$

Proof. Choose $\mathbf{v} = \mathbf{u}^{0}$.

$$\int_{\Omega} \sigma_{33}^{-4} e_{33}(\mathbf{v}) d\vec{x} = 0 = \int_{\Omega} \sigma_{33}^{-4} e_{33}^{0} dx$$

$$= \int_{\Omega} [c_{3333} e_{33}^{0} e_{33}^{0} + d_{3333} \dot{e}_{33}^{0} e_{33}^{0}] d\vec{x}$$

$$= \int_{\Omega} [c_{3333} e_{33}^{0} e_{33}^{0} + d_{3333} \frac{1}{2}\frac{\partial}{\partial t}(e_{33}^{0} e_{33}^{0})] d\vec{x}$$

Using *Lemma 1*, $e_{33}^{0} = 0 \Rightarrow \dot{e}_{33}^{0} = 0 \Rightarrow \sigma_{33}^{-4} = 0, \sigma_{\alpha\beta}^{-2} = 0$ □

Theorem 2. *If* $\int_{\Omega} \sigma_{33}^{-3} e_{33}(\mathbf{v}) d\vec{x} = 0$, *then* $e_{33}^{1} = 0, \sigma_{33}^{-3} = 0,\ \sigma_{\alpha\beta}^{-1} = 0$

Proof. Choose $\mathbf{v} = \mathbf{u}^{1}$.

$$\int_{\Omega} \sigma_{33}^{-3} e_{33}(\mathbf{v}) d\vec{x} = 0 = \int_{\Omega} \sigma_{33}^{-3} e_{33}^{1} d\vec{x}$$

$$= \int_{\Omega} [c_{3333} e_{33}^{1} e_{33}^{1} + d_{3333} \dot{e}_{33}^{1} e_{33}^{1}] d\vec{x}$$

$$= \int_{\Omega} [c_{3333} e_{33}^{1} e_{33}^{1} + d_{3333} \frac{1}{2}\frac{\partial}{\partial t}(e_{33}^{1} e_{33}^{1})] d\vec{x}$$

Using *Lemma 1*, $e_{33}^{1} = 0 \Rightarrow \dot{e}_{33}^{1} = 0 \Rightarrow \sigma_{33}^{-3} = 0,\ \sigma_{\alpha\beta}^{-1} = 0$ □

Theorem 3. *If* $\sigma^{-2}_{\alpha\beta} = 0, e^0_{33} = 0$ *and* $\int_\Omega \sigma^{-2}_{ij} e_{ij}(\mathbf{v})dx = 0$, *then*

$$e_{3\beta} = 0,\ \sigma^{-2}_{3\beta} = 0,\ \int_\Omega \sigma^{-2}_{33} e_{33}(\mathbf{v})d\vec{x} = 0$$

Proof. Choose $\mathbf{v} = \mathbf{u}^0$.

$$\int_\Omega \sigma^{-2}_{ij} e_{ij}(\mathbf{v})d\vec{x} = 0 = \int_\Omega \sigma^{-2}_{ij} e^0_{ij} d\vec{x}$$

$$\int_\Omega [\sigma^{-2}_{\alpha\beta} e^0_{\alpha\beta} + \sigma^{-2}_{3\beta} e^0_{3\beta} + \sigma^{-2}_{33} e^0_{33}]d\vec{x} = 0$$

From *Theorem 1*, $\sigma^{-2}_{\alpha\beta} = 0 = e^0_{33} \Rightarrow \int_\Omega \sigma^{-2}_{3\beta} e^0_{3\beta} d\vec{x} = 0$

$$2\int_\Omega [c_{3\alpha 3\beta} e^0_{3\beta} e^0_{3\beta} + d_{3\alpha 3\beta} \dot{e}^0_{3\beta} e^0_{3\beta}]d\vec{x} = 0$$

$$2\int_\Omega [c_{3\alpha 3\beta} e^0_{3\beta} e^0_{3\beta} + d_{3\alpha 3\beta} \frac{1}{2}\frac{\partial}{\partial t}(e^0_{3\beta} e^0_{3\beta})]d\vec{x} = 0$$

Using *Lemma 1*, $e^0_{3\beta} = 0 \Rightarrow \dot{e}^0_{3\beta} = 0 \Rightarrow \sigma^{-2}_{3\beta} = 0 \Rightarrow$

$$\int_\Omega \sigma^{-2}_{ij} e_{ij}(\mathbf{v})d\vec{x} = 0 \Rightarrow \int_\Omega \sigma^{-2}_{33} e_{33}(\mathbf{v})d\vec{x} = 0$$

□

Theorem 4. *If* $\sigma^{-1}_{\alpha\beta} = 0,\ e^1_{33} = 0$ *and* $\int_\Omega \sigma^{-1}_{ij} e^1_{ij} dx = 0$, *then*

$$e^1_{3\beta} = 0, \sigma^{-1}_{3\beta} = 0,\ \int_\Omega \sigma^{-1}_{33} e_{33}(\mathbf{v})d\vec{x} = 0$$

Proof. Choose $\mathbf{v} = \mathbf{u}^1$ so that $\int \sigma^{-1}_{ij} e^1_{ij} d\vec{x} = 0$.
From *Theorem 2*, $\sigma^{-1}_{\alpha\beta} = 0 = e^1_{33} \Rightarrow \int_\Omega \sigma^1_{3\beta} e^1_{3\beta} d\vec{x} = 0$

$$2\int_\Omega [c_{3\alpha 3\beta} e^1_{3\beta} e^1_{3\beta} + d_{3\alpha 3\beta} \dot{e}^1_{3\beta} e^1_{3\beta}]d\vec{x} = 0$$

$$2\int_\Omega c_{3\alpha 3\beta} e^1_{3\beta} e^1_{3\beta} + d_{3\alpha 3\beta} \frac{1}{2}\frac{\partial}{\partial t}(e^1_{3\beta} e^1_{3\beta})]d\vec{x} = 0$$

Using *Lemma 1*, $e^1_{3\beta} = 0 \Rightarrow \dot{e}^1_{3\beta} = 0 \Rightarrow \sigma^{-1}_{3\beta} = 0$

$$\int_\Omega \sigma^{-1}_{ij} e_{ij}(\mathbf{v})d\vec{x} = 0 \Rightarrow \int_\omega \sigma^{-1}_{33} e_{33}(\mathbf{v})d\vec{x} = 0$$

□

To find out more about σ^{-2}_{33} and σ^{-1}_{33}, a technique used by CIARLET and DESTUYNDER, [**4**, p. 330] in the case for the linear elastic plate is employed here and can work in general.

Theorem 5. *For $n = \{-2, -1\}$, if $\sigma^n_{\alpha\beta} = 0 = \sigma^n_{3\beta}$, then $\sigma^n_{33} = 0$*

Proof. For $n = \{-2, -1\}$,

$$\int_\Omega \sigma^n_{ij} e_{ij} d\vec{x} = 0 = \int_\Omega (\sigma^n_{\alpha\beta} e_{\alpha\beta} + \sigma^n_{3\beta} e_{3\beta} + \sigma^n_{33} e_{33}) d\vec{x}$$

¿From *Theorem 1* and *Theorem 2*, $\sigma^n_{\alpha\beta} = 0 \Rightarrow$

$$\int_\Omega (\sigma^n_{33} e_{33} + \sigma^n_{3\beta} e_{3\beta}) d\vec{x} = 0$$

Using Green's formula,

$$0 = \int_\Omega (\sigma^n_{33} \frac{\partial v_3}{\partial x_3} + \sigma^n_{3\beta} \frac{\partial v_3}{\partial x_\beta}) d\vec{x} = -\int_\Omega (\frac{\partial \sigma^n_{33}}{\partial x_3} + \frac{\partial \sigma^n_{3\beta}}{\partial x_\beta}) v_3 d\vec{x} + \int_{\Gamma^+ \cup \Gamma^-} \sigma^n_{33} v_3 d\omega + \int_{\Gamma^+ \cup \Gamma^-} \sigma^n_{3\beta} v_3 d\omega$$

Since $\int_\Omega (\sigma^n_{33} e_{33} + \sigma^n_{3\beta} e_{3\beta}) d\vec{x} = 0$, one looks for a solution where $\frac{\partial \sigma^n_{33}}{\partial x_3} + \frac{\partial \sigma^n_{3\beta}}{\partial x_\beta} = 0$ in Ω and $\sigma_{33} = 0$ on $\Gamma^+ \cup \Gamma^-$. From *Theorem 3* and *Theorem 4* for $n = \{-2, -1\}$, $\sigma^n_{3\beta} = 0$, so $\frac{\partial \sigma^n_{3\beta}}{\partial x_\beta} = 0 \Rightarrow \frac{\partial \sigma^n_{33}}{\partial x_3} = 0$. Thus, σ^n_{33} is a constant with respect to x_3.

$$0 = \int_{\Gamma^+ \cup \Gamma^-} \sigma^n_{33} v_3 d\vec{x} = \int_{\Gamma^+} \sigma^n_{33} v_3 d\vec{x} - \int_{\Gamma^-} \sigma^n_{33} v_3 d\vec{x} \;\; \forall v_3 (\Gamma^+ \cup \Gamma^-)$$

So $\sigma^n_{33} = 0$ on $\Gamma^+ \cup \Gamma^-$. However, $\frac{\partial \sigma^n_{33}}{\partial x_3} = 0$, so $\sigma^n_{33} = 0$ in Ω. □

Corollary 1. *For the viscoelastic case, all negative powers of σ_{ij} are zero.*

Corollary 2. *For any positive odd integer n, if $\sigma^n_{3\beta} = 0 = \sigma^n_{\alpha\beta}$, then $\sigma^n_{33} = 0$*

Proof. For any positive odd integer n, $\int_\Omega \sigma^n_{ij} e_{ij}(\mathbf{v}) d\vec{x} = 0$. Given that $\sigma^n_{\alpha\beta} = 0$, this simplifies to $\int_\Omega (\sigma^n_{33} e_{33} + \sigma^n_{3\beta} e_{3\beta}) d\vec{x} = 0$. Now, Green's formula can be applied to this equation just as it was above in the cases of $n = \{-2, -1\}$. □

Theorem 6. *Since $\sigma^{-2}_{33} = 0$, there exists a $\mathbf{u}^2$ such that*

$$e^2_{33} = \frac{\int_0^t [(-\frac{c_{33\alpha\beta}}{d_{3333}} e^0_{\alpha\beta} - \frac{d_{33\alpha\beta}}{d_{3333}} \dot{e}^0_{\alpha\beta}) exp(\int_0^s \frac{c_{3333}}{d_{3333}} dq)] ds}{exp(\int_0^t \frac{c_{3333}}{d_{3333}} ds)}$$

Proof. ¿From *Theorem 5*, $\sigma^{-2}_{33} = 0 \Rightarrow \sigma^{-2}_{ij} = 0$ and there would exist a $\mathbf{u}^2$ such that

$$\sigma^{-2}_{33} = c_{33\alpha\beta} e^0_{\alpha\beta} + d_{33\alpha\beta} \dot{e}^0_{\alpha\beta} + c_{3333} e^2_{33} + d_{3333} \dot{e}^2_{33} = 0$$

$$\dot{e}^2_{33} + \frac{c_{3333}}{d_{3333}} e^2_{33} = -\frac{c_{33\alpha\beta}}{d_{3333}} e^0_{\alpha\beta} - \frac{d_{33\alpha\beta}}{d_{3333}} \dot{e}^0_{\alpha\beta}$$

$$e^2_{33} = \frac{\int_0^t [(-\frac{c_{33\alpha\beta}}{d_{3333}} e^0_{\alpha\beta} - \frac{d_{33\alpha\beta}}{d_{3333}} \dot{e}^0_{\alpha\beta}) exp(\int_0^s \frac{c_{3333}}{d_{3333}} dq)] ds}{exp(\int_0^t \frac{c_{3333}}{d_{3333}} ds)}$$

□

Theorem 7. *Since* $\sigma_{33}^{-1} = 0$*, there exists a* $\mathbf{u}^3$ *such that*

$$e_{33}^3 = \frac{\int_0^t [(-\frac{c_{33\alpha\beta}}{d_{3333}} e_{\alpha\beta}^1 - \frac{d_{33\alpha\beta}}{d_{3333}} \dot{e}_{\alpha\beta}^1) exp(\int_0^s \frac{c_{3333}}{d_{3333}} dq)] ds}{exp(\int_0^t \frac{c_{3333}}{d_{3333}} ds)}$$

Proof. ¿From *Theorem 5*, $\sigma_{33}^{-1} = 0 \Rightarrow \sigma_{ij}^{-1} = 0$ and there exists a $\mathbf{u}^3$ such that

$$\sigma_{33}^{-1} = c_{33\alpha\beta} e_{\alpha\beta}^1 + d_{33\alpha\beta} \dot{e}_{\alpha\beta}^1 + c_{3333} e_{33}^3 + d_{3333} \dot{e}_{33}^3 = 0$$

$$\dot{e}_{33}^3 + \frac{c_{3333}}{d_{3333}} e_{33}^3 = -\frac{c_{33\alpha\beta}}{d_{3333}} e_{\alpha\beta}^1 - \frac{d_{33\alpha\beta}}{d_{3333}} \dot{e}_{\alpha\beta}^1$$

$$e_{33}^3 = \frac{\int_0^t [(-\frac{c_{33\alpha\beta}}{d_{3333}} e_{\alpha\beta}^1 - \frac{d_{33\alpha\beta}}{d_{3333}} \dot{e}_{\alpha\beta}^1) exp(\int_0^s \frac{c_{3333}}{d_{3333}} dq)] ds}{exp(\int_0^t \frac{c_{3333}}{d_{3333}} ds)}$$

□

Now, a simplification of the asymptotic expansions of both the displacement vector $\mathbf{u}$ and the stress tensor σ will be derived.

Theorem 8. $e_{\gamma\mu}^1 = 0, \;\; \mathbf{u}^1 = 0$

Proof. ¿From *Theorem 2*, $e_{33}^1 = 0$ and from *Theorem 4*, $e_{3\beta}^1 = 0$. Insert this into $\int_\Omega \sigma_{ij}^{-1} e_{ij}^1 d\vec{x} = 0. \Rightarrow \int_\Omega \sigma_{\alpha\beta}^{-1} e_{\gamma\mu}^1 d\vec{x} = 0$

$$\int_\Omega (c_{\alpha\beta\gamma\mu} e_{\gamma\mu}^1 + d_{\alpha\beta\gamma\mu} \dot{e}_{\gamma\mu}^1) e_{\gamma\mu}^1 d\vec{x} = 0$$

$$\int_\Omega [c_{\alpha\beta\gamma\mu} e_{\gamma\mu}^1 e_{\gamma\mu}^1 + d_{\alpha\beta\gamma\mu} \frac{1}{2}\frac{\partial}{\partial t}(e_{\gamma\mu}^1 e_{\gamma\mu}^1)] d\vec{x} = 0$$

Using *Lemma 1*, $e_{\gamma\mu}^1 = 0 \Rightarrow \dot{e}_{\gamma\mu}^1 = 0 \Rightarrow e_{ij}^1 = 0 \Rightarrow \mathbf{u}^1 = 0$ □

Theorem 9. $e_{33}^3 = 0, \;\; \sigma_{\alpha\beta}^1 = 0, \;\; e_{3\beta}^3 = 0, \; \sigma_{3\beta}^1 = 0$

Proof. ¿From *Theorem 5*, $\sigma_{33}^{-1} = 0$ and from *Theorem 8*, $e_{\alpha\beta}^1 = 0$. Together, both lead to $e_{33}^3 = 0$. Combining this with $e_{\alpha\beta}^1 = 0$ leads to $\sigma_{\alpha\beta}^1 = 0$. Now, choose $\mathbf{v} = \mathbf{u}^3$, and insert it into the equation $\int_\Omega \sigma_{ij}^1 e_{ij}(\mathbf{v}) d\vec{x} = 0$. Since $\sigma_{\alpha\beta}^1 = 0 = e_{33}^3$, this equation becomes:

$$4\int_\Omega [c_{3\beta 3\mu} e_{3\mu}^3 + d_{3\beta 3\mu} \dot{e}_{3\mu}^3] e_{3\beta}^3 d\vec{x} = 0$$

$$4\int_\Omega [c_{3\beta 3\mu} e_{3\mu}^3 e_{3\mu}^3 + d_{3\beta 3\mu} \frac{1}{2}\frac{\partial}{\partial t}(e_{3\mu}^3 e_{3\mu}^3)] d\vec{x} = 0$$

Using *Lemma 1*, $e_{3\mu}^3 = e_{3\beta}^3 = 0 \Rightarrow \dot{e}_{3\mu}^3 = 0 \Rightarrow \sigma_{3\beta}^1 = 0$ □

A characterization of e_{33}^{n+4} in terms of $e_{\alpha\beta}^{n+2}$ for any odd integer $n \geq 1$ suggests further steps in evaluating positive odd powers of the asymptotic expansions of the displacement vector and the stress tensor.

Theorem 10. *The positive odd powers in the asymptotic expansions of both the displacement vector* $\mathbf{u}$ *and the stress tensor* σ *are all zero.*

Proof. ¿From *Theorem 9*, $\sigma^1_{\alpha\beta} = 0 = \sigma^1_{3\beta}$ and

$$\int_\Omega \sigma^1_{ij} e_{ij}(\mathbf{v}) d\vec{x} = 0 \Rightarrow \int_\Omega \sigma^1_{33} e_{33}(\mathbf{v}) d\vec{x} = 0$$

Due to *Corollary 2* to *Theorem 5*, one can set $\sigma^n_{33} = 0$ for $n = 1$. However, keep the notation with the exponent n.

$$\sigma^n_{33} = c_{33\alpha\beta} e^{n+2}_{\alpha\beta} + d_{33\alpha\beta} \dot{e}^{n+2}_{\alpha\beta} + c_{3333} e^{n+4}_{33} + d_{3333} \dot{e}^{n+4}_{33} = 0$$

Now, take the Laplace transforms of all terms.

$$c_{33\alpha\beta} L\{e^{n+2}_{\alpha\beta}\} + d_{33\alpha\beta}[sL\{e^{n+2}_{\alpha\beta}\} - e^{n+2}_{\alpha\beta}(0)] + c_{3333} L\{e^{n+4}_{33}\} + d_{3333}[sL\{e^{n+4}_{33}\} - e^{n+4}_{33}(0)] = 0$$

Since $e^{n+2}_{\alpha\beta}(0) = 0 = e^{n+4}_{33}(0)$

$$\begin{aligned}
L\{e^{n+4}_{33}\} &= -\frac{[c_{33\alpha\beta} + d_{33\alpha\beta} s]}{[c_{3333} + d_{3333} s]} L\{e^{n+2}_{\alpha\beta}\} \\
&= -\frac{d_{33\alpha\beta}[s + c_{33\alpha\beta}/d_{33\alpha\beta}]}{d_{3333}[s + c_{3333}/d_{3333}]} L\{e^{n+2}_{\alpha\beta}\} \\
&= -\frac{d_{33\alpha\beta}}{d_{3333}} \frac{s}{s + c_{3333}/d_{3333}} L\{e^{n+2}_{\alpha\beta}\} - \frac{c_{33\alpha\beta}}{d_{3333}} \frac{1}{s + c_{3333}/d_{3333}} L\{e^{n+2}_{\alpha\beta}\} \\
&= = -\frac{d_{33\alpha\beta}}{d_{3333}} L\{e^{n+2}_{\alpha\beta}\} + (\frac{d_{33\alpha\beta} c_{3333}}{(d_{3333})^2} - \frac{c_{33\alpha\beta}}{d_{3333}}) \frac{1}{s + c_{3333}/d_{3333}} L\{e^{n+2}_{\alpha\beta}\}
\end{aligned}$$

$$e^{n+4}_{33}(t) = -\frac{d_{33\alpha\beta}}{d_{3333}} e^{n+2}_{\alpha\beta}(t) + (\frac{d_{33\alpha\beta} c_{3333}}{(d_{3333})^2} - \frac{c_{33\alpha\beta}}{d_{3333}}) \int_0^t e^{n+2}_{\alpha\beta}(\tau) exp\{-\frac{c_{3333}}{d_{3333}}(t-\tau)\} d\tau$$

Thus, when $n = -1$, $e^1_{\alpha\beta} = 0$ leads to $e^3_{33} = 0$. Now, consider the case of $n = 1$. Due to *Corollary 2* to *Theorem 5*, $\sigma^1_{3\beta} = 0 = \sigma^1_{\alpha\beta}$ leads to $\sigma^1_{33} = 0$. This would mean that $\sigma^1_{\mathbf{ij}} = \mathbf{0}$ and that:

$$e^5_{33}(t) = -\frac{d_{33\alpha\beta}}{d_{3333}} e^3_{\alpha\beta}(t) + (\frac{d_{33\alpha\beta} c_{3333}}{(d_{3333})^2} - \frac{c_{33\alpha\beta}}{d_{3333}}) \int_0^t e^3_{\alpha\beta}(\tau) exp\{-\frac{c_{3333}}{d_{3333}}(t-\tau)\} d\tau$$

This can be put to good use in the following way. Since $e^3_{3\beta} = 0 = e^3_{33}$,

$$\int_\Omega \sigma^3_{ij} e^3_{ij} d\vec{x} = 0 \Rightarrow \int_\Omega \sigma^3_{\alpha\beta} e^3_{\alpha\beta} d\vec{x} = 0$$

$$= \int_\Omega [c_{\alpha\beta\gamma\mu} e^3_{\gamma\mu} + d_{\alpha\beta\gamma\mu} \dot{e}^3_{\gamma\mu} + c_{\alpha\beta 33} e^5_{33} + d_{\alpha\beta 33} \dot{e}^5_{33}] e^3_{\gamma\mu} d\vec{x} = 0$$

Substitute for e^5_{33}:

$0 = \int_\Omega \{c_{\alpha\beta\gamma\mu} e^3_{\gamma\mu} + d_{\alpha\beta\gamma\mu} \dot{e}^3_{\gamma\mu}$

$+ c_{\alpha\beta 33}[-\frac{d_{33\alpha\beta}}{d_{3333}} e^3_{\alpha\beta}(t) + (\frac{d_{33\alpha\beta} c_{3333}}{(d_{3333})^2} - \frac{c_{33\alpha\beta}}{d_{3333}}) \int_0^t e^3_{\alpha\beta}(\tau) exp\{-\frac{c_{3333}}{d_{3333}}(t-\tau)\} d\tau]$

$$+ d_{\alpha\beta 33}[-\tfrac{d_{33\alpha\beta}}{d_{3333}}\dot{e}^3_{\alpha\beta}(t) - (\tfrac{d_{33\alpha\beta}c_{3333}}{(d_{3333})^2} - \tfrac{c_{3333}}{d_{33\alpha\beta}})e^3_{\gamma\mu}exp\{-\tfrac{c_{3333}}{d_{3333}}(0)\}$$
$$- \tfrac{c_{3333}}{d_{3333}}(\tfrac{d_{33\alpha\beta}c_{3333}}{(d_{3333})^2} - \tfrac{c_{33\alpha\beta}}{d_{3333}}) \int_0^t e^3_{\gamma\mu}(\tau)exp\{\tfrac{c_{3333}}{d_{3333}}(t-\tau)\}d\tau]\}e^3_{\gamma\mu}d\vec{x}$$

$$0 = \int_\Omega\{c_{\alpha\beta\gamma\mu}e^3_{\gamma\mu}e^3_{\gamma\mu} + d_{\alpha\beta\gamma\mu}\tfrac{1}{2}\tfrac{\partial}{\partial t}(e^3_{\gamma\mu}e^3_{\gamma\mu})$$
$$+ c_{\alpha\beta 33}[-\tfrac{d_{33\alpha\beta}}{d_{3333}}e^3_{\gamma\mu}e^3_{\gamma\mu} - (\tfrac{d_{33\alpha\beta}c_{3333}}{(d_{3333})^2} - \tfrac{c_{33\alpha\beta}}{d_{3333}}) \int_0^t e^3_{\gamma\mu}(\tau)e^3_{\gamma\mu}(\tau)exp\{-\tfrac{c_{3333}}{d_{3333}}(t-\tau)\}d\tau]$$
$$+ d_{\alpha\beta 33}[-\tfrac{d_{33\alpha\beta}}{d_{3333}}\tfrac{1}{2}\tfrac{\partial}{\partial t}(e^3_{\gamma\mu}e^3_{\gamma\mu}) + (\tfrac{d_{33\alpha\beta}c_{3333}}{(d_{3333})^2} - \tfrac{c_{33\alpha\beta}}{d_{3333}})(e^3_{\gamma\mu}e^3_{\gamma\mu})$$
$$- \tfrac{c_{3333}}{d_{3333}}(\tfrac{d_{33\alpha\beta}c_{3333}}{(d_{3333})^2} - \tfrac{c_{33\alpha\beta}}{d_{3333}}) \int_0^t e^3_{\gamma\mu}(\tau)e^3_{\gamma\mu}(\tau)exp\{-\tfrac{c_{3333}}{d_{3333}}(t-\tau)\}d\tau]\}d\vec{x}$$

$$e^3_{\gamma\mu} = e^3_{\alpha\beta} = 0 \Rightarrow \dot{e}^3_{\gamma\mu} = 0 \Rightarrow e^5_{33} = 0 \Rightarrow \dot{e}^5_{33} = 0 \Rightarrow \mathbf{u}^3 = 0,\ \sigma^3_{\alpha\beta} = 0$$

Substitute $\mathbf{v} = \mathbf{u}^5$ into $\int_\Omega \sigma^3_{ij}e_{ij}(\mathbf{v})d\vec{x} = 0$ to get:

$$\int_\Omega \sigma^3_{3\beta}e^5_{3\beta}d\vec{x} = 0 = 4\int_\Omega [c_{3\beta 3\mu}e^5_{3\mu} + d_{3\beta 3\mu}\dot{e}^5_{3\mu}]e^5_{3\mu}d\vec{x}$$

$$4\int_\Omega [c_{3\beta 3\mu}e^5_{3\mu}e^5_{3\mu} + d_{3\beta 3\mu}\frac{1}{2}\frac{\partial}{\partial t}(e^5_{3\beta}e^5_{3\beta})]d\vec{x} = 0$$

Using *Lemma 1*, $e^5_{3\mu} = 0 \Rightarrow \dot{e}^5_{3\mu} = 0 \Rightarrow \sigma^3_{3\beta} = 0$

Since $\sigma^3_{\alpha\beta} = 0 = \sigma^3_{3\beta}$, $\int_\Omega \sigma^3_{ij}e_{ij}(\mathbf{v})d\vec{x} = 0$ becomes $\int_\Omega \sigma^3_{33}e_{33}(\mathbf{v})d\vec{x} = 0$. Due to *Corollary 2* to *Theorem 5*, one can set $\sigma^3_{33} = 0$ which leads to $\boldsymbol{\sigma^3_{ij} = 0}$ and to the same situation as before when $n = 1$ was inserted into:

$$e^{n+4}_{33}(t) = -\frac{d_{33\alpha\beta}}{d_{3333}}e^{n+2}_{\alpha\beta}(t) + (\frac{d_{33\alpha\beta}c_{3333}}{(d_{3333})^2} - \frac{c_{33\alpha\beta}}{d_{3333}}) \int_0^t e^{n+2}_{\alpha\beta}(\tau)exp\{-\frac{c_{3333}}{d_{3333}}(t-\tau)\}d\tau$$

to produce:

$$e^5_{33}(t) = -\frac{d_{33\alpha\beta}}{d_{3333}}e^3_{\alpha\beta}(t) + (\frac{d_{33\alpha\beta}c_{3333}}{(d_{3333})^2} - \frac{c_{33\alpha\beta}}{d_{3333}}) \int_0^t e^3_{\alpha\beta}(\tau)exp\{-\frac{c_{3333}}{d_{3333}}(t-\tau)\}d\tau$$

Now, set $n = 3$ and one obtains:

$$e^7_{33}(t) = -\frac{d_{33\alpha\beta}}{d_{3333}}e^5_{\alpha\beta}(t) + (\frac{d_{33\alpha\beta}c_{3333}}{(d_{3333})^2} - \frac{c_{33\alpha\beta}}{d_{3333}}) \int_0^t e^5_{\alpha\beta}(\tau)exp\{-\frac{c_{3333}}{d_{3333}}(t-\tau)\}d\tau$$

The same pattern of calculations starts anew. Therefore, by induction on the odd integer $n \geq 1$, the positive odd powers of the asymptotic expansions of the displacement vector $\mathbf{u}$ and the stress tensor σ are equal to zero. □

The constitutive equation for the general Kelvin-Voigt model on the domain Ω^ε is:

$$\sigma^\varepsilon_{ij} = c^\varepsilon_{ijkl}(\phi^0)e^\varepsilon_{kl} + d^\varepsilon_{ijkl}(\phi^0)\dot{e}^\varepsilon_{kl}$$

Next, the coefficients $D^\varepsilon_{ijkl}(\phi^0)$ of the inverse matrix defined by the viscoelastic coefficients $d^\varepsilon_{ijkl}(\phi^0)$ are used to rearrange the constitutive equation into a first order differential equation:

$$\frac{de^\varepsilon_{kl}}{dt} + D^\varepsilon_{ijkl}(\phi^0)c^\varepsilon_{ijkl}(\phi^0)e^\varepsilon_{kl} = D^\varepsilon_{ijkl}(\phi^0)\sigma^\varepsilon_{ij}$$

One solves for e_{kl} in the above system of differential equations, noting that $e^{\varepsilon}_{kl}(0) = 0$, and obtains:

$$e^{\varepsilon}_{kl} = \frac{\int_0^t D^{\varepsilon}_{ijkl}\sigma^{\varepsilon}_{ij} exp(\int_t^s D^{\varepsilon}_{ijkl}c^{\varepsilon}_{ijkl}d\tau)ds}{exp(\int_0^t D^{\varepsilon}_{ijkl}c^{\varepsilon}_{ijkl}d\tau)}$$

Inside the above equation, the following can be defined:

$$T_{klij}(t,s) = \frac{D^{\varepsilon}_{ijkl}(s) exp(\int_t^s D^{\varepsilon}_{ijkl}c^{\varepsilon}_{ijkl}d\tau)}{exp(\int_0^t D^{\varepsilon}_{ijkl}c^{\varepsilon}_{ijkl}d\tau)}$$

from which $e^{\varepsilon}_{kl}(t) = \int_0^t T_{ijkl}(t,s)\sigma^{\varepsilon}_{ij}(s)$

For any tensor τ^{ε}_{kl}, this last equation can be put into the following form:

$$\int_{\Omega^\varepsilon}[\int_0^t T_{klnq}\sigma^{\varepsilon}_{nq}ds]\tau^{\varepsilon}_{kl}d\vec{x}^{\varepsilon} - \int_{\Omega^\varepsilon} e^{\varepsilon}_{ij}(\mathbf{u}^{\varepsilon})\tau^{\varepsilon}_{kl}d\vec{x}^{\varepsilon} = 0$$

$$\begin{aligned}
&= \int_{\Omega^\varepsilon}[\int_0^t T_{\alpha\beta\gamma\mu}\sigma^{\varepsilon}_{\gamma\mu}ds + \int_0^t T_{\alpha\beta 33}\sigma^{\varepsilon}_{33}ds]\tau^{\varepsilon}_{\alpha\beta}d\vec{x}^{\varepsilon} \\
&+ \int_{\Omega^\varepsilon}[4\int_0^t T_{3\beta 3\mu}\sigma^{\varepsilon}_{3\mu}ds]\tau^{\varepsilon}_{3\beta}d\vec{x}^{\varepsilon} \\
&+ \int_{\Omega^\varepsilon}[\int_0^t T_{33\gamma\mu}\sigma^{\varepsilon}_{\gamma\mu}ds + \int_0^t T_{3333}\sigma^{\varepsilon}_{33}ds]\tau^{\varepsilon}_{33}d\vec{x}^{\varepsilon} \\
&- \int_{\Omega^\varepsilon} e^{\varepsilon}_{ij}\tau^{\varepsilon}_{kl}d\vec{x}^{\varepsilon} = 0
\end{aligned}$$

Now, convert from Ω^ε to Ω making the appropriate changes in variables and scaling:

$$\begin{aligned}
&= \int_{\Omega}[\int_0^t T_{\alpha\beta\gamma\mu}\sigma_{\gamma\mu}ds]\tau_{\alpha\beta}d\vec{x} + \varepsilon^2\int_{\Omega}[\int_0^t T_{\alpha\beta 33}\sigma_{33}ds]\tau_{\alpha\beta}d\vec{x} \\
&+ 4\varepsilon^2\int_{\Omega}[\int_0^t T_{3\beta 3\mu}\sigma_{3\mu}ds]\tau_{3\beta}d\vec{x} + \varepsilon^2\int_{\Omega}[\int_0^t T_{33\gamma\mu}\sigma_{\gamma\mu}ds]\tau_{33}d\vec{x} \\
&+ \varepsilon^4\int_{\Omega}[\int_0^t T_{3333}\sigma_{33}ds]\tau_{33}d\vec{x} - \int_{\Omega} e_{ij}\tau_{kl}d\vec{x} = 0
\end{aligned}$$

This equation is coupled with:

$$\int_{\Omega}\sigma_{ij}e_{ij}(\mathbf{v})d\vec{x} = \int_{\Omega}\gamma(\xi_0 + P_\eta(\phi_0))f_i v_i d\vec{x} + \int_{\Gamma^0} g_i v_i d\vec{x} + \int_{\Gamma^+\cup\Gamma^-} h_i v_i d\vec{x}$$

for all test functions v_i, and the two equations characterize the mixed displacement-stress approach.

Theorem 11. *The first term* $u^0 = (u^0_1, u^0_2, u^0_3)$ *in the asymptotic expansion* $u(\epsilon) = u^0 + \epsilon^1 u^1 + \epsilon^2 u^2 + ...$ *satisfies:* $u^0_3 = u^0_3(x_1, x_2, t)$, $u^0_\beta = \zeta^0_\beta(x_1, x_2, t) - x_3\frac{\partial u^0_3}{\partial x_\beta}$ *Additionally,* $\sigma^0_{\gamma\mu} = \frac{1}{2}T^{-1}_{\alpha\beta\gamma\mu}(\frac{\partial^2 u^0_\alpha}{\partial t\partial x_\beta} + \frac{\partial^2 u^0_\beta}{\partial t\partial x_\alpha})$

Proof. One takes the ε^0 terms in the first of the two equations of the mixed displacement-stress approach and creates the following equations:

$$\int_\Omega [\int_0^t T_{\alpha\beta\gamma\mu}\sigma^0_{\gamma\mu}ds]\tau_{\alpha\beta}d\vec{x} - \int_\Omega e^0_{\alpha\beta}\tau_{\alpha\beta}d\vec{x} = 0$$

$$\int_\Omega e^0_{3\beta}\tau_{3\beta}d\vec{x} = 0, \int_\Omega e^0_{33}\tau_{33}d\vec{x} = 0$$

¿From which $e^0_{\alpha\beta} = \int_0^t T_{\alpha\beta\gamma\mu}\sigma^0_{\gamma\mu}ds$ and $e^0_{33} = 0\ \forall\tau_{33}$ and $e^0_{3\beta} = 0\ \forall\tau_{3\beta}$

Now $e^0_{33} = 0 \Rightarrow \frac{\partial u^0_3}{\partial x_3} = 0 \Rightarrow u^0_3 = u^0_3(x_1, x_2, t)$

Furthermore, $e^0_{3\beta} = 0 \Rightarrow \frac{\partial u^0_\beta}{\partial x_3} = -\frac{\partial u^0_3}{\partial x_\beta} \Rightarrow u^0_\beta(x_1, x_2, t) = \zeta^0_\beta(x_1, x_2, t) - x_3\frac{\partial u^0_3}{\partial x_\beta}$

Finally, using the fundamental theorem of calculus, $e^0_{\alpha\beta} = \int_0^t T_{\alpha\beta\gamma\mu}\sigma^0_{\gamma\mu}ds$ becomes

$$T_{\alpha\beta\gamma\mu}\sigma^0_{\gamma\mu} = \frac{\partial}{\partial t}e^0_{\alpha\beta} = \frac{1}{2}(\frac{\partial^2 u^0_\alpha}{\partial t\partial x_\beta} + \frac{\partial^2 u^0_\beta}{\partial t\partial x_\alpha})$$

$$\sigma^0_{\gamma\mu} = \frac{1}{2}T^{-1}_{\alpha\beta\gamma\mu}(\frac{\partial^2 u^0_\alpha}{\partial t\partial x_\beta} + \frac{\partial^2 u^0_\beta}{\partial t\partial x_\alpha})$$

□

Theorem 12. *The second term* $u^1 = (u^1_1, u^1_2, u^1_3)$ *in the asymptotic expansion* $u(\epsilon) = u^0 + \epsilon^1 u^1 + \epsilon^2 u^2 + \ldots$ *satisfies:* $u^1_3 = u^1_3(x_1, x_2, t), u^1_\beta = \zeta^1_\beta(x_1, x_2, t) - x_3\frac{\partial u^0_3}{\partial x_\beta}$
Additionally, $\sigma^1_{\gamma\mu} = \frac{1}{2}T^{-1}_{\alpha\beta\gamma\mu}(\frac{\partial^2 u^1_\alpha}{\partial t\partial x_\beta} + \frac{\partial^2 u^1_\alpha}{\partial t\partial x_\beta})$

Proof. One takes the previous proof and substitutes the superscript 1 for the superscript 0 wherever the superscript 0 is to be found. The theorem for terms with the superscript 1 results. Considering that the odd terms in the asymptotic expansion for the displacement are zero, this is a moot point. □

Theorem 13.

$$\sigma^0_{3\mu} = \frac{1}{4}T^{-1}_{3\beta3\mu}\frac{\partial}{\partial t}e^2_{3\beta},\ \sigma^0_{\gamma\mu} = T^{-1}_{33\gamma\mu}\frac{\partial}{\partial t}e^2_{33},\ T^{-1}_{\alpha\beta\gamma\mu}\frac{\partial}{\partial t}e^0_{\alpha\beta} = T^{-1}_{33\gamma\mu}\frac{\partial}{\partial t}e^2_{33}$$

Proof. The ε^2 equation leads to three different equations according to the subscripts involved:

$$\int_\Omega [\int_0^t T_{\alpha\beta\gamma\mu}\sigma^2_{\gamma\mu}ds + \int_0^t T_{\alpha\beta33}\sigma^0_{33}ds]\tau_{\alpha\beta}dx - \int_\Omega e^2_{\alpha\beta}\tau_{\alpha\beta}dx = 0$$

$$4\int_\Omega [\int_0^t T_{3\beta3\mu}\sigma^0_{3\mu}ds]\tau_{\alpha\beta}dx - \int_\Omega e^2_{3\beta}\tau_{3\beta}dx = 0$$

$$\int_\Omega [\int_0^t T_{33\gamma\mu}\sigma^0_{\gamma\mu}ds]\tau_{33}dx - \int_\Omega e^2_{33}\tau_{33}dx = 0$$

The ε^4 equation adds a last equation to those above:

$$\int_\Omega [\int_0^t T_{33\gamma\mu}\sigma^2_{\gamma\mu} ds]\tau_{33} dx + \int_\Omega [\int_0^t T_{3333}\sigma^2_{33} ds]\tau_{33} dx - \int_\Omega e^4_{33}\tau_{33} dx = 0$$

One should recall that the stress tensor has its own asymptotic expansion of:

$$\sigma(\epsilon) = \sigma^0 + \epsilon^2\sigma^2 + \epsilon^4\sigma^4 + ...$$

which with the above equations leads to a further characterization of some terms in the expansion by way of the fundamental theorem of calculus:

$$\sigma^0_{3\mu} = \frac{1}{4} T^{-1}_{3\beta 3\mu}\frac{\partial}{\partial t} e^2_{3\beta}, \quad \sigma^0_{\gamma\mu} = T^{-1}_{33\gamma\mu}\frac{\partial}{\partial t} e^2_{33}$$

Combining with an earlier expression for $\sigma^0_{\gamma\mu}$, one has:

$$\sigma^0_{\gamma\mu} = T^{-1}_{\alpha\beta\gamma\mu}\frac{\partial}{\partial t} e^0_{\alpha\beta} = T^{-1}_{33\gamma\mu}\frac{\partial}{\partial t} e^2_{33}$$

Assuming that $e^0_{\alpha\beta} \neq 0$, one sees that $e^2_{33} \neq 0$ while $e^0_{33} = 0 = e^1_{33}$ □

Other expressions with $\sigma^0_{\alpha\beta}$ can be obtained from the force balance equation as follows:

Lemma 2. *For $v_\beta \in C^1(\Omega)$ and $v_3 \in C^1(\Omega)$,*

$$\int_\Omega \sigma_{\alpha\beta}\partial_\alpha v^0_\beta d\vec{x} = \int_\Omega \gamma' f_\beta v^0_\beta d\vec{x} + \int_{\Gamma^0} g_\beta v^0_\beta d\omega + \int_{\Gamma^+\cup\Gamma^-} h_\beta v^0_\beta d\omega$$

$$0 = \int_\Omega \gamma' f_3 v^0_3 d\vec{x} + \int_{\Gamma^0} g_3 v^0_3 d\omega + \int_{\Gamma^+\cup\Gamma^-} h_3 v^0_3 d\omega$$

Proof. Let $\gamma' = \gamma(\xi_0 + P_\eta(\phi_0))$. ¿From the ε^0 equation, one can derive the following:
Substitute $\sigma_{3\beta}$ for $\tau_{3\beta}$ in $\int_\Omega \tau_{3\beta} e^0_{3\beta} d\vec{x} = 0$ to get $\int_\Omega \sigma_{3\beta} e^0_{3\beta} d\vec{x} = 0$.
Substitute $\sigma_{\alpha\beta}$ for $\tau_{\alpha\beta}$ in $\int_\Omega \tau_{\alpha\beta} e^0_{\alpha\beta} d\vec{x} = 0$ to get $\int_\Omega \sigma_{\alpha\beta} e^0_{\alpha\beta} d\vec{x} = 0$.

Now, two variants of the force balance equation (1) on page 6 are:

$$\int_\Omega (\sigma_{\alpha\beta}\partial_\alpha v^0_\beta + \sigma_{3\beta}\partial_3 v^0_\beta) d\vec{x} = \int_\Omega \gamma' f_\beta v^0_\beta d\vec{x} + \int_{\Gamma^0} g_\beta v^0_\beta d\omega + \int_{\Gamma^+\cup\Gamma^-} h_\beta v^0_\beta d\omega$$

$$\int_\Omega (\sigma_{33}\partial_3 v^0_3 + \sigma_{3\beta}\partial_\beta v^0_3) d\vec{x} = \int_\Omega \gamma' f_3 v^0_3 d\vec{x} + \int_{\Gamma^0} g_3 v^0_3 d\omega + \int_{\Gamma^+\cup\Gamma^-} h_3 v^0_3 d\omega$$

The two equations can be added together:

$\int_\Omega (\sigma_{\alpha\beta}\partial_\alpha v^0_\beta + \sigma_{3\beta} e^0_{3\beta} + \sigma_{33} e^0_{33}) d\vec{x} =$
$\int_\Omega \gamma'(f_\beta v^0_\beta + f_3 v^0_3) d\vec{x} + \int_{\Gamma^0}(g_\beta v^0_\beta + g_3 v^0_3) d\omega + \int_{\Gamma^+\cup\Gamma^-}(h_\beta v^0_\beta + h_3 v^0_3) d\omega$

The second and third terms of the left side are zero due to the beginning of this proof. The v^0_β terms on the right side can be grouped with $\int_\Omega \sigma_{\alpha\beta}\partial_\alpha v^0_\beta d\vec{x}$ to form one equation. The v^0_3 terms on the right side can be grouped with zero to form another. □

Theorem 14. *For $v_\beta \in C^1(\Omega)$ and $\sigma_{\alpha\beta} \in C^1(\Omega)$,*

$$-\int_\omega \partial_\alpha \sigma_{\alpha\beta} d\omega = \int_\omega \gamma' f_\beta d\omega + \int_{\partial\omega} g_\beta ds$$

with boundary conditions given as:

$$[\sigma_{\alpha\beta} v_\beta^0]_{\Gamma^+\cup\Gamma^-} = \int_{\Gamma^+\cup\Gamma^-} h_\beta v_\beta^0 d\omega$$

Proof. ¿From the above lemma, the first of the resulting equations is:

$$\int_\Omega \sigma_{\alpha\beta} \partial_\alpha v_\beta^0 d\vec{x} = \int_\Omega \gamma' f_\beta v_\beta^0 d\vec{x} + \int_{\Gamma^0} g_\beta v_\beta^0 d\omega + \int_{\Gamma^+\cup\Gamma^-} h_\beta v_\beta^0 d\omega$$

Integrate by parts on the left side noting that $v_\beta = 0$ on Γ^0 to get $[\sigma_{\alpha\beta} v_\beta^0]_{\Gamma^+\cup\Gamma^-} - \int_\Omega \partial_\alpha \sigma_{\alpha\beta} v_\beta^0 d\vec{x}$.
Decompose each integral in the middle three terms as follows:

$$-\int_0^L [\int_\omega \partial_\alpha \sigma_{\alpha\beta} d\omega] v_\beta^0 dx_3 = \int_0^L [\int_\omega \gamma' f_\beta d\omega] v_\beta^0 dx_3 + \int_0^L [\int_{\partial\omega} g_\beta ds] v_\beta^0 dx_3$$

with boundary conditions given as:

$$[\sigma_{\alpha\beta} v_\beta^0]_{\Gamma^+\cup\Gamma^-} = \int_{\Gamma^+\cup\Gamma^-} h_\beta v_\beta^0 d\omega$$

The quantities in the square brackets of the equation with the decomposed integrals can form an equation which is stated as follows:

$$-\int_\omega \partial_\alpha \sigma_{\alpha\beta} d\omega = \int_\omega \gamma' f_\beta d\omega + \int_{\partial\omega} g_\beta ds$$

This is the strong formulation of the equation at the start of this proof, which is the first of the two equations derived from the force balance equation (1) in *Lemma 2*. It is also a statement of this theorem. □

Theorem 15. *For $v_\beta \in C^1(\Omega)$, $f_3 \in C^1(\Omega)$, $g_3 \in C^1(\Omega)$, $h_3 \in C^1(\Omega)$,*

$$0 = \int_\omega \gamma' x_\beta \partial_3 f_3 d\omega + \int_{\partial\omega} x_\beta \partial_3 g_3 ds$$

with boundary conditions given as:

$$0 = \int_{\Gamma^+\cup\Gamma^-} x_\beta (\partial_3 h_3) v_\beta^0 d\omega$$

Proof. ¿From the above lemma, the second of the resulting equations is:

$$0 = \int_\Omega \gamma' f_3 v_3^0 d\vec{x} + \int_{\Gamma^0} g_3 v_3^0 d\omega + \int_{\Gamma^+\cup\Gamma^-} h_3 v_3^0 d\omega$$

¿From *Theorem 3*, $e^0_{3\beta} = 0 \Rightarrow v^0_3 = \zeta^0_3(x_\alpha, x_3, t) - x_\beta \frac{\partial v^0_\beta}{\partial x_3}$. Choose $\zeta^0_3(x_\alpha, x_3, t) = 0$ so that $v^0_3 = -x_\beta \frac{\partial v^0_\beta}{\partial x_3}$.

$$0 = \int_\Omega \gamma' x_\beta f_3 \partial_3 v^0_\beta d\vec{x} + \int_{\Gamma^0} x_\beta g_3 \partial_3 v^0_\beta ds + \int_{\Gamma^+ \cup \Gamma^-} x_\beta h_3 \partial_3 v^0_\beta ds$$

Integrate by parts noting that $v_\beta = 0$ on Γ^0.

$$0 = \int_\Omega \gamma' x_\beta (\partial_3 f_3) v^0_\beta d\vec{x} + \int_{\Gamma^0} x_\beta (\partial_3 g_3) v^0_\beta d\omega + \int_{\Gamma^+ \cup \Gamma^-} x_\beta (\partial_3 h_3) v^0_\beta d\omega$$

Decompose each integral as follows:

$$0 = \int_0^L [\int_\omega \gamma' x_\beta \partial_3 f_3 d\omega] v^0_\beta dx_3 + \int_0^L [\int_{\partial\omega} x_\beta \partial_3 g_3 ds] v^0_\beta dx_3$$

with boundary conditions given as:

$$0 = \int_{\Gamma^+ \cup \Gamma^-} x_\beta (\partial_3 h_3) v^0_\beta d\omega$$

The quantities in the square brackets of the equation with the decomposed integrals can form an equation which is stated as follows:

$$0 = \int_\omega \gamma' x_\beta \partial_3 f_3 d\omega + \int_{\partial\omega} x_\beta \partial_3 g_3 ds$$

This is the strong formulation of the equation at the start of this proof, which is the second of the two equations derived from the force balance (1) in *Lemma 2*. It is also a statement of this theorem. □

Next, one considers the time rate of change in the volume fraction and its asymptotic expansion.

Theorem 16. *The asymptotic expansion for the time rate of change in the volume fraction is given by:* $\dot{\phi} = \dot{\phi}^0 + \varepsilon^1 \dot{\phi}^1 + \varepsilon^2 \dot{\phi}^2 + \ldots$
where $\dot{\phi}^0 = c_{3333} e^2_{33} e^2_{33} + d_{3333} \dot{e}^2_{33} \dot{e}^2_{33} + A_{33} e^2_{33} + \tilde{A}_{33} \dot{e}^2_{33} + 2c_{\alpha\beta 33} e^0_{\alpha\beta} e^2_{33} + 2d_{\alpha\beta 33} \dot{e}^0_{\alpha\beta} \dot{e}^2_{33} + A_{\alpha\beta} e^0_{\alpha\beta} + \tilde{A}_{\alpha\beta} \dot{e}^0_{\alpha\beta} + c_{\alpha\beta\gamma\mu} e^0_{\alpha\beta} e^0_{\alpha\beta} + d_{\alpha\beta\gamma\mu} \dot{e}^0_{\alpha\beta} \dot{e}^0_{\alpha\beta} + a$

with $\phi^0(0) = \bar{\phi}^0$ *and* $\dot{\phi}^1 = 2A_{3\beta} e^2_{3\beta} + 2\tilde{A}_{3\beta} \dot{e}^2_{3\beta}$

and $\dot{\phi}^2 = 2c_{3333} e^2_{33} e^4_{33} + 2d_{3333} \dot{e}^2_{33} \dot{e}^4_{33} + A_{33} e^4_{33} + \tilde{A}_{33} \dot{e}^4_{33} + 2c_{\alpha\beta 33}(e^0_{\alpha\beta} e^4_{33} + 2e^2_{\alpha\beta} e^2_{33}) + 2d_{\alpha\beta 33}(\dot{e}^0_{\alpha\beta} \dot{e}^4_{33} + 2\dot{e}^2_{\alpha\beta} \dot{e}^2_{33}) + 8c_{3\beta 3\mu} e^2_{3\beta} e^2_{3\mu} + 8d_{3\beta 3\mu} \dot{e}^2_{3\beta} \dot{e}^2_{3\mu} + A_{\alpha\beta} e^2_{\alpha\beta} + \tilde{A}_{\alpha\beta} \dot{e}^2_{\alpha\beta} + c_{\alpha\beta\gamma\mu}(e^0_{\alpha\beta} e^2_{\gamma\mu} + + e^2_{\alpha\beta} e^0_{\gamma\mu}) + d_{\alpha\beta\gamma\mu}(\dot{e}^0_{\alpha\beta} \dot{e}^2_{\gamma\mu} + \dot{e}^2_{\alpha\beta} \dot{e}^0_{\gamma\mu})$

Proof. In the domain Ω_ε, the time rate of change in volume fraction is given by: $\dot{\phi}^\varepsilon = a^\varepsilon + A^\varepsilon_{kl} e^\varepsilon_{kl} + \tilde{A}^\varepsilon_{kl} \dot{e}^\varepsilon_{kl} + c^\varepsilon_{ijkl} e^\varepsilon_{ij} e^\varepsilon_{kl} + d^\varepsilon_{ijkl} \dot{e}^\varepsilon_{ij} \dot{e}^\varepsilon_{kl}$
$= a^\varepsilon + A^\varepsilon_{\alpha\beta} e^\varepsilon_{\alpha\beta} + 2A^\varepsilon_{3\beta} e^\varepsilon_{3\beta} + A^\varepsilon_{33} e^\varepsilon_{33} + \tilde{A}^\varepsilon_{\alpha\beta} \dot{e}^\varepsilon_{\alpha\beta} + 2\tilde{A}^\varepsilon_{3\beta} \dot{e}^\varepsilon_{3\beta} + \tilde{A}^\varepsilon_{33} \dot{e}^\varepsilon_{33} + c^\varepsilon_{\alpha\beta\gamma\mu} e^\varepsilon_{\alpha\beta} e^\varepsilon_{\gamma\mu} +$

$2c^{\varepsilon}_{\alpha\beta 33}e^{\varepsilon}_{\alpha\beta}e^{\varepsilon}_{33} + 4c^{\varepsilon}_{3\beta 3\mu}e^{\varepsilon}_{3\beta}e^{\varepsilon}_{3\mu} + c^{\varepsilon}_{3333}e^{\varepsilon}_{33}e^{\varepsilon}_{33} + d_{\alpha\beta\gamma\mu}\dot{e}^{\varepsilon}_{\alpha\beta}\dot{e}^{\varepsilon}_{\gamma\mu} + 2d^{\varepsilon}_{\alpha\beta 33}\dot{e}^{\varepsilon}_{\alpha\beta}\dot{e}^{\varepsilon}_{33} + 4d^{\varepsilon}_{3\beta 3\mu}\dot{e}^{\varepsilon}_{3\beta}\dot{e}^{\varepsilon}_{3\mu} + d^{\varepsilon}_{3333}\dot{e}^{\varepsilon}_{33}\dot{e}^{\varepsilon}_{33}$

The scalings:

$e_{\alpha\beta}(u(x)) = e^{\varepsilon}_{\alpha\beta}(u(x^{\varepsilon}))$, $e_{3\beta}(u(x)) = \varepsilon e^{\varepsilon}_{3\beta}(u(x^{\varepsilon}))$, $e_{33}(u(x)) = \varepsilon^2 e^{\varepsilon}_{33}(u(x^{\varepsilon}))$

convert the expression for $\dot{\phi}^{\varepsilon}$ into the expression for $\dot{\phi}$ on the domain Ω, which is the first expression mentioned in the above theorem.

Now, the expression for $\dot{\phi}$ in the domain Ω can be multiplied by ε^4 and asymptotic expansions in the form: $e_{ij} = e^0_{ij} + \varepsilon^2 e^2_{ij} + \varepsilon^4 e^4_{ij} + ...$ can be substituted for the various strain tensors. From *Theorems 1,3,10*, one may note that

$0 = e^0_{33} = e^0_{3\beta} = e^{2n+1}_{ij}$ for integer $n \geq 0$.

$$\varepsilon^4\dot{\phi}^0 + \varepsilon^5\dot{\phi}^1 + \varepsilon^6\dot{\phi}^2 + ... =$$

$$c_{3333}(e^0_{33} + \varepsilon^2 e^2_{33} + \varepsilon^4 e^4_{33} + ...)(e^0_{33} + \varepsilon^2 e^2_{33} + \varepsilon^4 e^4_{33} + ...) + d_{3333}(\dot{e}^0_{33} + \varepsilon^2 \dot{e}^2_{33} + \varepsilon^4 \dot{e}^4_{33} + ...)(\dot{e}^0_{33} + \varepsilon^2 \dot{e}^2_{33} + \varepsilon^4 \dot{e}^4_{33} + ...)$$

$$+ \varepsilon^{\mathbf{2}}[A_{33}(e^0_{33} + \varepsilon^2 e^2_{33} + \varepsilon^4 e^4_{33} + ...) + \tilde{A}_{33}(\dot{e}^0_{33} + \varepsilon^2 \dot{e}^2_{33} + \varepsilon^4 \dot{e}^4_{33} + ...) + 2c_{\alpha\beta 33}(e^0_{\alpha\beta} + \varepsilon^2 e^2_{\alpha\beta} + \varepsilon^4 e^4_{\alpha\beta} + ...)(e^0_{33} + \varepsilon^2 e^2_{33} + \varepsilon^4 e^4_{33} + ...) + 4c_{3\beta 3\mu}(e^0_{3\beta} + \varepsilon^2 e^2_{3\beta} + \varepsilon^4 e^4_{3\beta} + ...)(e^0_{3\mu} + \varepsilon^2 e^2_{3\mu} + \varepsilon^4 e^4_{3\mu} + ...) + 2d_{\alpha\beta 33}(\dot{e}^0_{\alpha\beta} + \varepsilon^2 \dot{e}^2_{\alpha\beta} + \varepsilon^4 \dot{e}^4_{\alpha\beta} + ...)(\dot{e}^0_{33} + \varepsilon^2 \dot{e}^2_{33} + \varepsilon^4 \dot{e}^4_{33} + ...) + 4d_{3\beta 3\mu}(\dot{e}^0_{3\beta} + \varepsilon^2 \dot{e}^2_{3\beta} + \varepsilon^4 \dot{e}^4_{3\beta} + ...)(\dot{e}^0_{3\mu} + \varepsilon^2 \dot{e}^2_{3\mu} + \varepsilon^4 \dot{e}^4_{3\mu} + ...)]$$

$$+ \varepsilon^{\mathbf{3}}[2A_{3\beta}(e^0_{3\beta} + \varepsilon^2 e^2_{3\beta} + \varepsilon^4 e^4_{3\beta} + ...) + 2\tilde{A}_{3\beta}(\dot{e}^0_{3\beta} + \varepsilon^2 \dot{e}^2_{3\beta} + \varepsilon^4 \dot{e}^4_{3\beta} + ...)]$$

$$+ \varepsilon^{\mathbf{4}}[a + A_{\alpha\beta}(e^0_{\alpha\beta} + \varepsilon^2 e^2_{\alpha\beta} + \varepsilon^4 e^4_{\alpha\beta} + ...) + \tilde{A}_{\alpha\beta}(\dot{e}^0_{\alpha\beta} + \varepsilon^2 \dot{e}^2_{\alpha\beta} + \varepsilon^4 \dot{e}^4_{\alpha\beta} + ...) + c_{\alpha\beta\gamma\mu}(e^0_{\alpha\beta} + \varepsilon^2 e^2_{\alpha\beta} + \varepsilon^4 e^4_{\alpha\beta} + ...)(e^0_{\gamma\mu} + \varepsilon^2 e^2_{\gamma\mu} + \varepsilon^4 e^4_{\gamma\mu} + ...) + d_{\alpha\beta\gamma\mu}(\dot{e}^0_{\alpha\beta} + \varepsilon^2 \dot{e}^2_{\alpha\beta} + \varepsilon^4 \dot{e}^4_{\alpha\beta} + ...)(\dot{e}^0_{\gamma\mu} + \varepsilon^2 \dot{e}^2_{\gamma\mu} + \varepsilon^4 \dot{e}^4_{\gamma\mu} + ...)]$$

The terms with the lowest power of ε are those with ε^4, and these will be designated as $\dot{\phi}^0$. In likewise fashion, those with ε^5 as a coefficient will be $\dot{\phi}^1$, and those with ε^6 as coefficent will be $\dot{\phi}^2$. The asymptotic expansion for the time rate of change in the volume fraction is given by: $\dot{\phi} = \dot{\phi}^0 + \varepsilon^1\dot{\phi}^1 + \varepsilon^2\dot{\phi}^2 + ...$

with $\dot{\phi}^0$, $\dot{\phi}^1$ and $\dot{\phi}^2$ as given in the statement of this theorem. □

References

[1] Buchanan, J.L., Gilbert, R. P. and Khashanah, K.; "Determination of Parameters of Cancellous Bone Using Low-Frequency Acoustic Measurements", *Journal of Computational Acoustics*, Vol. 12, No. 2, pp. 99-126, 2004.

[2] Chou, P.C. and Pagano, N.J.; Elasticity-Tensor, *Dyadic and Engineering Approaches*, Dover, New York, 1992.

[3] Ciarlet, P.G.; *Mathematical Elasticity, Volume II: Theory of Plates*, North-Holland, Amsterdam, 1997.

[4] Ciarlet, P.G. and Destuynder, P.; "A justification of the two-dimensional linear plate mdifferential equationl ", *Journal de Mècanique*, vol. 18, no.2, 1979.

[5] Cowin, S.C. and Hegedus, D.H., "Bone remodeling I: theory of adaptive elasticity", *Journal of Elasticity*, vol.6, no.3, pp. 313-326, 1976.

[6] Cowin, S.C. and Hegedus, D.H., "Bone remodeling II: small strain adaptive elasticity", *Journal of Elasticity*, vol.6, no.4, pp. 337-352, 1976.

[7] Cowin, S.C. and Nachlinger, R.R., "Bone remodeling III: uniqueness and stability in adaptive elasticity theory", *Journal of Elasticity*, vol.8, no.3, pp. 285-295, 1978.

[8] Figueiredo, I. and Trabucho, L.;"Asymptotic Model of a Nonlinear Adaptive Elastic Rod", *Mathematics and Mechanics of Solids*, vol. 9, no. 4, 2004.

[9] Monnier, J. and Trabucho, L.; "An Existence and Uniqueness Result in Bone Remodeling Theory", *Computer Methods in Applied Mechanics and Engineering*, vol. 151, pp. 539-544, 1998.

[10] Trabucho, L.; "Non-linear Bone Remodelling: An Existence and Uniqueness Result", *Mathematical Methods in the Applied Sciences*, vol. 23, pp. 1331-1346, 2000.

[11] Trabucho, L. and Viaño, J.M.;"Mathematical Modeling of Rods", found in the *Handbook of Numerical Analysis*, Ciarlet, P.G. and Lions, J.L.(editors); Volume IV, North-Holland, Amsterdam, 1996.

In: Research on Evolution Equation Compendium. Volume 2 ISBN: 978-1-60456-752-6
Editor: Gaston M. N'Guerekata

MATHEMATICAL ANALYSIS OF A BILATERAL OBSTACLE PROBLEM FOR A CLASS OF SECOND-ORDER OPERATORS

Laurent Lévi and Guy Vallet
Université de Pau et des Pays de l'Adour
Laboratoire de Mathématiques Appliquées - UMR 5142 CNRS
BP 1155 - 64013 PAU Cedex - FRANCE

Abstract

We investigate some inner bilateral obstacle problems for a class of strongly degenerate parabolic-hyperbolic quasilinear operators associated with homogeneous Dirichlet data in a multidimensional bounded domain. We first introduce the concept of an entropy process solution, more convenient and generalizing the notion of an entropy solution. Moreover, the boundary conditions are expressed by using the background of *Divergence Measure Fields*. We ensure that proposed definition warrants uniqueness. The existence of an entropy process solution is obtained through the vanishing viscosity and penalization methods.

1. Introduction

1.1. Mathematical Framework

Obstacle problems for conservation laws in physics and mechanics have been studied by many authors [6, 12, 16]. In this work, we are especially interested in the second-order quasilinear operator stemming from the theory of fluid flows through porous media:

$$\mathbb{P}(t,x,.) : u \to \partial_t u + \sum_{i=1}^{p} \partial_{x_i}\varphi_i(t,x,u) + \psi(t,x,u) - \Delta\phi(u),$$

where ϕ is a nondecreasing function; especially ϕ' may vanish on nonempty intervals of $\mathbb{R}$. Let us given two thresholds a and b that are fixed reals such that $a \leq 0 \leq b$. For any positive T, the bilateral obstacle problem for $\mathbb{P}$ on the bounded domain Ω of $\mathbb{R}^p$, $p \geq 1$, may be formally described: find a measurable function u on $]0,T[\times\Omega$ satisfying the formulation of free boundary $\partial[a < u < b]$,

$$\begin{cases} a \leq u \leq b \text{ in }]0,T[\times\Omega, \\ \mathbb{P}(t,x,u) = 0 \text{ on the (unknown) zone } [a < u < b], \\ \mathbb{P}(t,x,u) \leq 0 \text{ on the (unknown) zone } [a < u = b], \\ \mathbb{P}(t,x,u) \geq 0 \text{ on the (unknown) zone } [a = u < b], \\ u = 0 \text{ on }]0,T[\times\partial\Omega, u(0,.) = u_0 \text{ on } \Omega, \end{cases} \tag{1}$$

where u_0 is a measurable function on Ω such that $a \leq u_0 \leq b$ a.e. in Ω.

On the one hand, we emphasize that without any assumption on the sign of the source term for $\mathbb{P}$ a bilateral obstacle condition on initial data does not *a priori* pass on to the associated solution. On the other hand, entropy formulations for the Dirichlet problem to strongly degenerate parabolic-hyperbolic operators have been introduced by J.Carrillo in 1999 ([2]). Since then, numerous works have been achieved on this matter [1, 13, 14, 17, 18]. Here, we first provide in Subsection 1.3. the definition of an entropy solution to (1). However, since we release the smoothness assumptions on the convective term φ and the reactive one ψ it is more convenient to introduce the concept of an *entropy process solution* to (1). This notion, highlighted in [18] for the Dirichlet to $\mathbb{P}$, may be considered as an extension to the second order of the framework due to R.Eymard, T.Gallouët & R.Herbin in [7]. The uniqueness of an entropy process solution to (1) is stated in Section 2. and its existence is obtained in Section 3. *via* the vanishing viscosity method and by relaxing the bilateral obstacle condition. These two results provide the existence on an entropy solution to (1) and warrant the strong convergence in $L^q(]0,T[\times\Omega)$, $1 \leq q < +\infty$, of the sequence of approximate solutions.

1.2. Main Notations and Assumptions on Data

The reaction term ψ is a continuous function on $[0,T] \times \bar{\Omega} \times [a,b]$. In addition ψ is Lipschitzian with respect to its third variable with a constant M_ψ, uniformly with respect to (t,x) in $]0,T[\times\Omega$. The flux term φ is a $W^{1,+\infty}$-class vector-valued function on $]0,T[\times\Omega\times]a,b[$ such that for all i in $\{1,..,p\}$, $\partial_{x_i}\varphi_i$ is Lipschitzian with respect to its third variable with a constant $M_{\partial_{x_i}\varphi_i}$, uniformly with respect to (t,x) in $]0,T[\times\Omega$. Thus we can set $M_\varphi = \max_{i\in\{1,..,p\}} M_{\partial_{x_i}\varphi_i}$ and define for any t of $[0,T]$,

$$M(t) = \max(-a,b)e^{\mathcal{K}_1 t} + \frac{\mathcal{K}_2}{\mathcal{K}_1}\left(e^{\mathcal{K}_2 t} - 1\right), \tag{2}$$

where $\mathcal{K}_1 = M_\psi + M_\varphi$ and $\mathcal{K}_2 = \max_{[0,T]\times\bar{\Omega}} |\psi(t,x,0) + Div\varphi(t,x,0)|$.

The diffusive term ϕ is a nondecreasing $W^{1,+\infty}(]a,b[)$-class function such that $\phi(0) = 0$. We note $E = \{l \in \mathbb{R}, \{l\} = \phi^{-1}(\{\phi(l)\})\}$. Hence, if ϕ_0^{-1} denotes a generalized inverse of ϕ, the hypo-inverse for example defined through

$$\forall r \in Im\phi,\ \phi_0^{-1}(r) = Inf\phi^{-1}(\{r\}),$$

then $\phi(E)$ is the set of points where ϕ_0^{-1} is continuous ([2]).

We assume that Ω is a bounded subset of $\mathbb{R}^p$ such that $\Gamma = \partial\Omega$ is Lipschitz-deformable (see [1, 3, 4, 13]). Then the space of L^2-Divergence Measure fields on Q - denoted $\mathcal{DM}_2(Q)$ - is given by

$$\mathcal{DM}_2(Q) = \left\{V = (v_0, v_1, .., v_p) \in (L^2(Q))^{p+1}, Div_{(t,x)}V \in \mathcal{M}_b(Q)\right\},$$

where $\mathcal{M}_b(Q)$ is the space of bounded Radon measures on Q. For any V in $\mathcal{DM}_2(Q)$, it is useful to define a linear mapping Λ_V on $H^1(Q) \cap L^\infty(Q) \cap C(Q)$ through the following generalized Gauss-Green formula coming from [4]:

$$\Lambda_V(\xi) := \langle V, \xi \rangle_\partial = \int_Q V.(\partial_t \xi, \nabla \xi) dx dt + \int_Q \xi d[Div_{(t,x)} V], \tag{3}$$

where $d[Div_{(t,x)}V]$ denotes the Borelian measure on Q associated with the bounded Radon measure $Div_{(t,x)}V$. In addition the next property holds (see [13]): let V be an element of $\mathcal{DM}_2(Q)$ such that v_0 is continuous at $t = 0$ and $t = T$ with respect to the $L^1(\Omega)$-norm then, for any ξ in $H^1(Q) \cap L^\infty(Q) \cap C(Q)$,

$$\begin{aligned}
&\lim_{\rho \to 0^+} \int_Q V\xi.(0, \nabla \rho_\rho) dx dt \\
&= \int_\Omega v_0(T,x)\xi(0,x)dx - \int_\Omega v_0(0,x)\xi(0,x)dx - \langle V, \xi \rangle_\partial,
\end{aligned} \tag{4}$$

for any boundary-layer sequence $(\rho_\rho)_{\rho>0}$, *i.e.* a sequence of $C^2(\Omega) \cap C(\overline{\Omega})$-class functions such that $\lim_{\rho \to 0^+} \rho_\rho = 1$ pointwise in Ω, $0 \le \rho_\rho \le 1$ in Ω, $\rho_\rho = 0$ on Γ. In (4), if ξ belongs to $L^2(0,T; H_0^1(\Omega))$ the right-hand side is equal to 0.

We will consider "sgn_λ" the approximation of the function "sgn" given for any positive parameter λ and nonnegative real x by:

$$sgn_\lambda(x) = \min\left(\frac{x}{\lambda}, 1\right) \text{ and } sgn_\lambda(-x) = -sgn_\lambda(x).$$

To simplify the writing, we refer to the notations

$$\begin{aligned}
\mathbf{F}(u,k) &= sgn(u-k)\{\varphi(t,x,u) - \varphi(t,x,k)\}, \\
G(u,k) &= sgn(u-k) Div_x \varphi(t,x,k) + \psi(t,x,u), \\
U_k = (|u-k|, -\nabla|\phi(u) - \phi(k)| &+ \mathbf{F}(u,k)), \overline{\nabla}\zeta = (\partial_t \zeta, \nabla \zeta).
\end{aligned}$$

Eventually, for any n in $\mathbb{N}^\star$, $\mathcal{H}^n$ stands for the n-dimensional Hausdorff measure and for all s of $]0,T]$, Q_s is the cylinder $]0,s[\times\Omega$, $\Sigma_s =]0,s[\times\Gamma$ with the convention $Q = Q_T$ and $\Sigma = \Sigma_T$. The outer normal of Ω is denoted ν.

1.3. Two Concepts of Weak Solutions

The existence of possible internal and boundary layers leads us to propose a mathematical formulation for (1) through an entropy inequality inside the studied field - using the classical Kruzhkov entropy pairs - and on its boundary; the latter is viewed as an extension to the second order of the F.Otto's formulation provided in [15] for hyperbolic first-order operators and uses - as in [1, 13, 18] - the mathematical framework of *Divergence-Measure Fields*. In addition, to take into account the bilateral obstacle condition we only consider the Kruzhkov pairs for a parameter k that has to belong to the bounded interval $[a,b]$. That is why, by referring to a preliminary study of the positiveness obstacle problem for $\mathbb{P}$ in [11], it will be said that:

Definition 1. *A measurable function u on Q is an entropy solution to (1) if:*

$$a \leq u \leq b \text{ a.e. in } Q, \tag{5}$$

$$\begin{gathered}\phi(u) \in L^2(0,T;H_0^1(\Omega)),\ \partial_t\phi(u) \in L^2_{loc}(0,T;L^2(\Omega)),\\ \partial_t u \in L^2(0,T;H^{-1}(\Omega)),\end{gathered} \tag{6}$$

$$\forall k \in [a,b],\ U_k \in \mathcal{DM}_2(Q), \tag{7}$$

$\forall k \in [a,b]$, $\forall \xi \in \mathcal{D}(]-\infty,T[\times\Omega)$, $\xi \geq 0$,

$$\int_Q \left(U_k.\overline{\nabla}\xi dxdt - G(u,k)\xi\right)dxdt + \int_\Omega |u_0-k|\xi(0,.)dx \geq 0, \tag{8}$$

$\forall \xi \in L^\infty(Q)\cap H^1(Q)\cap C(Q)$, $\xi \geq 0$, $\xi(T,.) = \xi(0,.) = 0$, $\forall k \in [a,b]$,

$$\int_\Sigma \mathbf{F}(k,0).\nu\xi d\mathcal{H}^p \leq \langle U_k,\xi\rangle_\partial + \langle U_0,\xi\rangle_\partial. \tag{9}$$

Remark 1. *Considering $k = a$ and ξ in $\mathcal{D}(Q)$, $\xi \geq 0$, in (8) leads to:*

$$\int_Q ((u-a)\partial_t\xi - \nabla\phi(u).\nabla\xi + \mathbf{F}(u,a).\nabla\xi - G(u,a)\xi)dxdt \geq 0.$$

We observe that a.e. in Q, $\mathbf{F}(u,a) = \varphi(t,x,u) - \varphi(t,x,a)$. So thanks to some integrations by parts one gets:

$$-\langle \mathbb{P}(t,x,u),\xi\rangle_{\mathcal{D}'(Q),\mathcal{D}(Q)} + \int_Q (1-sgn(u-a))(Div\varphi(t,x,a)+\psi(t,x,u))\xi dxdt \geq 0.$$

Assume now that u is smooth enough so that $\mathbb{P}(t,x,u)$ belongs to $L^1(Q)$ and let us choose $\xi = (u-a)^+\zeta$ with ζ in $\mathcal{D}(Q)$, $\zeta \geq 0$. Since $(1-sgn(u-a))(Div\varphi(t,x,a)+\psi(t,x,u))(u-a)^+ = 0$ a.e. on Q, we obtain

$$\int_Q \mathbb{P}(t,x,u)(u-a)^+\zeta dxdt \leq 0.$$

That means $\mathbb{P}(t,x,u)(u-a)^+ \leq 0$ a.e. on Q. The same reasoning with $k = b$ leads to $\mathbb{P}(t,x,u)(u-b)^- \geq 0$ a.e. in Q that formally provides (1).

However, one of the feature of this work in comparison with [11] is to release the assumptions of regularity for φ and ψ. So for the existence property stated in Section 3. through the vanishing viscosity and penalization methods, we can only refer to an $L^\infty(Q)$-estimate of approximate solutions. That is why we lean on the original presentation of R.Eymard, T.Gallouët & R.Herbin in [7] for first-order quasilinear hyperbolic operators that consists - by using the basic tools of Young measure [5] - in introducing a free variable α living on $]0,1[$ and a new measurable and bounded unknown $]0,1[\times Q$, $\pi \equiv \pi(\alpha,t,x)$, that fulfills an entropy formulation on $]0,1[\times Q$ and called an *entropy process solution*. We adapt this concept when dealing with second order quasilinear operators associated with a forced bilateral constraint. From now, to simplify the writing, we set $\underset{\sim}{Q} =]0,1[\times Q$ and $dq = d\alpha dxdt$:

Definition 2. *Let* π *be in* $L^\infty(]0,1[\times Q)$ *and* $u(t,x) = \int_0^1 \pi(\alpha,t,x)d\alpha$. *We say that* π *is an entropy process solution to (1) if:*

$$a \leq \pi \leq b \text{ a.e. in } Q, \tag{10}$$

$$\text{for a.e. } (t,x) \text{ in } Q, \phi(\pi(\alpha,t,x)) = \phi(u(t,x)) \text{ for a.e. } \alpha \text{ in }]0,1[, \tag{11}$$

the smoothness properties (6) hold and:

$$\forall k \in [a,b], \Pi_k \in \mathcal{DM}_2(Q), \tag{12}$$

$\forall k \in [a,b]$, $\forall \xi \in \mathcal{D}(]-\infty,T[\times\Omega)$, $\xi \geq 0$,

$$\int_Q \Pi_k.\overline{\nabla}\xi dxdt - \int_Q G(\pi,k)\xi dq + \int_\Omega |u_0 - k|\xi(0,.)dx \geq 0. \tag{13}$$

$\forall \xi \in L^\infty(Q) \cap H^1(Q) \cap C(Q)$, $\xi(T,.) = \xi(0,.) = 0$, $\xi \geq 0$, $\forall k \in [a,b]$,

$$\int_\Sigma \mathbf{F}(k,0).\nu\xi d\mathcal{H}^p \leq \langle \Pi_k, \xi\rangle_\partial + \langle \Pi_0, \xi\rangle_\partial, \tag{14}$$

where

$$\Pi_k = \left(\int_0^1 |\pi(\alpha,.,.) - k|d\alpha, \int_0^1 (-\nabla|\phi(\pi)(\alpha,.,.) - \phi(k)| + \mathbf{F}(\pi(\alpha,.,.),k))d\alpha\right).$$

The concept of an entropy process solution generalize that of an entropy solution since an entropy process solution independent from the variable α is an entropy solution. But the first notion is more convenient that the second one, mainly because it requires a few estimates of approximate solutions and so an existence result is easier to obtain. Moreover, the existence and uniqueness of an entropy solution results from the existence and uniqueness of an entropy process solution. First by reasoning as F.Otto in [15], we may announce:

Proposition 1. *An entropy process solution* π *to (1) fulfills:*

$$ess \lim_{t\to 0^+} \int_{]0,1[\times\Omega} |\pi(\alpha,t,x) - u_0(x)|d\alpha dx = 0. \tag{15}$$

2. The Uniqueness Theorem

The proof basically relies on a inner comparison property which is an extension to second-order operators of the usual hyperbolic method of doubling variables due to S.N.Kuzhkov [9]; the contribution of diffusive terms being controlled thanks to an *energy inequality* in the same spirit as in the original paper of J.Carrillo [2]. To do so, we need some preliminary lemma. The first one takes into account that ϕ' may vanish on a nonnegligible subset of $\mathbb{R}$:

Lemma 1.
(i) $\forall k \in E$ *and a.e. on* Q,

$$sgn(u-k) = sgn(\phi(u)-\phi(k)) = sgn(\pi(\alpha,.,.)-k) \textit{ for a.e. } \alpha \textit{ in }]0,1[.$$

(ii) $\nabla\phi(\pi) = 0$ *a.e. on* $Q_\theta^\pi \equiv \{(\alpha,t,x) \in Q, \pi(\alpha,t,x) \notin E\}$

Proof.
For the first equality in (i) we remark that when k belongs to E, if $u(t,x) > k$ then $\phi(u(t,x)) > \phi(k)$. This way the second equality is a consequence of (11) and uses the same reasoning as for the first equality with π in the place of u. The point (ii) is already emphasized ([2]) with the setting of entropy solutions. It uses the monotony of ϕ_0^{-1} so that $Im\phi \setminus \phi(E)$ - the set of points where ϕ_0^{-1} is discontinuous - is a countable (and thus a Lebesgue-negligible) subset of $\mathbb{R}$. ■

The second lemma may be viewed as an *inequality version* of the energy equality stated by J.Carrillo in [2]. It permits to determine the sign of diffusive terms appearing in the method of doubling variables. We prove that this inequality is fulfilled by any entropy process solution to (1) but only for k in E. We emphasize that in [11] the forthcoming relation (16) results from an underlying formulation of the unilateral obstacle problem for $\mathbb{P}$ through a strong variational inequality (in the sense of J.L.Lions in [12]). But here, in the context of the bilateral obstacle problem, we have not been able to establish such a formulation and so we directly start from (13). Indeed:

Lemma 2. *Let* π *be an entropy process solution to (1). Then, for any real* k *of* $[a,b]\cap E$, *for any nonnegative function* ζ *of* $\mathcal{D}(Q)$,

$$\int_Q \Pi_k.\overline{\nabla}\zeta dxdt - \int_Q G(\pi,k)\zeta dq \geq \limsup_{\lambda\to 0^+} \int_Q sgn'_\lambda(\phi(\pi)-\phi(k))[\nabla\phi(\pi)]^2\zeta dq.$$

Proof. We consider (13) for any nonnegative ξ in $\mathcal{D}(Q)$ and, thanks to a density argument, for any nonnegative ξ in $H_0^1(Q)$ so as to choose the test-function $\xi = \zeta|sgn_\lambda(\phi(u)-\phi(k))|$ with ζ in $\mathcal{D}(Q)$, $\zeta \geq 0$. Let us perform the following transformations (by setting $\mathcal{S}_\lambda(v) = sgn_\lambda(\phi(v)-\phi(k))$):
• For the evolution term,

$$\int_Q |\pi-k|\partial_t\xi dq = I_1 + I_2 \text{ where,}$$

$$I_1 = \int_Q |\pi-k||\mathcal{S}_\lambda(u)|\partial_t\zeta dq,$$

$$I_2 = \int_Q |\pi-k|\zeta sgn'_\lambda(\phi(u)-\phi(k))sgn(\phi(u)-\phi(k))\partial_t\phi(u)dq.$$

In I_2 we have taken into account that $sgn(\mathcal{S}_\lambda(u)) = sgn(\phi(u)-\phi(k))$ a.e. on Q when k belongs to E. But due to Lemma 1 (i), $|\pi-k| = sgn(u-k)(\pi-k)$ a.e. on Q and $sgn(\phi(u)-$

$\phi(k)) = sgn(u-k)$ a.e. on Q. So the integration with respect to α may be calculated. We get:

$$\begin{aligned} I_2 &= \int_Q \zeta sgn'_\lambda(\phi(u)-\phi(k))(u-k)\partial_t\phi(u)dxdt \\ &= \int_Q \partial_t(\zeta \mathcal{S}_\lambda(u))(u-k)dxdt - \int_Q \partial_t\zeta \mathcal{S}_\lambda(u)(u-k)dxdt. \end{aligned}$$

Since $\partial_t u$ is an element of $L^2(0,T;H^{-1}(\Omega))$,

$$\int_Q \partial_t(\zeta \mathcal{S}_\lambda(u))(u-k)dxdt = -\int_0^T \langle \partial_t u, \mathcal{S}_\lambda(u)\zeta\rangle_{H^{-1},H_0^1} dt.$$

An integration by parts in time using the Mignot-Bamberger formula ([8]) gives

$$-\int_0^T \langle \partial_t u, \mathcal{S}_\lambda(u)\zeta\rangle_{H^{-1},H_0^1} dt \int_Q \left(\int_k^u \mathcal{S}_\lambda(\tau)d\tau \right) \partial_t\zeta dxdt.$$

We now take the λ-limit through the Lebesgue dominated convergence Theorem in I_1 and I_2. Since k belongs to E, I_2 goes to 0^+ (Lemma 1 (i)).

• For the diffusive term, considering that $sgn(\mathcal{S}_\lambda(\pi)) = sgn(\phi(\pi)-\phi(k))$ a.e. on Q provides

$$\begin{aligned} & -\int_Q \nabla|\phi(\pi)-\phi(k)|.\nabla\xi dq \\ = & -\int_Q [\nabla|\phi(\pi)-\phi(k)|.\nabla\zeta]\,|\mathcal{S}_\lambda(\pi)|dq - \int_Q sgn'_\lambda(\phi(\pi)-\phi(k))[\nabla\phi(\pi)]^2\zeta dq. \end{aligned}$$

• Concerning the convective term, as a consequence of (11), we write:

$$\int_Q \mathbf{F}(\pi,k).\nabla\xi dq = \int_Q [\mathbf{F}(\pi,k).\nabla\zeta]\,|\mathcal{S}_\lambda(\pi)|dq + \int_Q \zeta\mathbf{F}(\pi,k).\nabla|\mathcal{S}_\lambda(\pi)|dq.$$

To take the λ-limit, we transform the second term in the right-hand side by by taking into account that due to Lemma 1 (i), $sgn(\mathcal{S}_\lambda(\pi)) = sgn(\phi(\pi)-\phi(k)) = sgn(\pi-k)$ a.e. on Q. There follows

$$\begin{aligned} & \int_Q \zeta\mathbf{F}(\pi,k).\nabla|\mathcal{S}_\lambda(\pi)|dq \\ = & \int_Q \zeta sgn'_\lambda(\phi(\pi)-\phi(k))\{\varphi(t,x,\pi)-\varphi(t,x,k)\}.\nabla\phi(\pi)dq. \end{aligned}$$

Now, we introduce the vector-valued function

$$\mathbf{H}_\lambda(t,x,r) = \int_{\phi(k)}^r [\varphi(t,x,\phi_0^{-1}(\tau)) - \varphi(t,x,k)]sgn'_\lambda(\tau-\phi(k))d\tau,$$

so that thanks to Lemma 1 (ii),

$$\int_Q \zeta sgn'_\lambda(\phi(\pi)-\phi(k))\{\varphi(t,x,\pi)-\varphi(t,x,k)\}\nabla\phi(\pi)dq$$
$$= \int_Q \zeta Div\mathbf{H}_\lambda(t,x,\phi(\pi))dq$$
$$- \int_Q \zeta \int_{\phi(k)}^{\phi(\pi)} \left(Div_x\varphi(t,x,\phi_0^{-1}(\tau))-Div_x\varphi(t,x,k)\right) sgn'_\lambda(\tau-\phi(k))d\tau dq,$$

The first term is integrated by parts and for the second one we come back to the definition of sgn'_λ to obtain:

$$\int_Q \zeta\mathbf{F}(\pi,k).\nabla|sgn_\lambda(\phi(\pi)-\phi(k))|dq$$
$$= -\int_Q \frac{1}{\lambda}\int_I \left([\varphi(t,x,\phi_0^{-1}(\tau))-\varphi(t,x,k)].\nabla\zeta d\tau\right) dq$$
$$-\int_Q \frac{1}{\lambda}\int_I \left(Div_x[\varphi(t,x,\phi_0^{-1}(\tau))-\varphi(t,x,k)]\zeta d\tau\right) dq,$$

where $I = I(\phi(\pi),\phi(k))\cap[\phi(k)-\lambda,\phi(k)+\lambda]$ and $I(r,p) = [min(r,p),max(r,p)]$ for all (r,p) of $\mathbb{R}^2$. Since k belongs to E, the generalized function ϕ_0^{-1} is continuous in $\phi(k)$ ([2]). This way, the previous integral goes to 0 when λ tends to 0^+. The study of the reaction term does not present any difficulty. ■

This way the next Kuzhkov-type relation holds:

Proposition 2. *Let π_1 and π_2 satisfying (6,10,11,13). Then for any nonnegative Ψ of $\mathcal{D}(Q\times Q)$:*

$$-\int_{Q\times Q} |\pi_1-\tilde{\pi}_2|(\partial_t+\partial_{\tilde{t}})\Psi-(\nabla_x+\nabla_{\tilde{x}})|\phi(u_1)-\phi(\tilde{u}_2)|.(\nabla_x+\nabla_{\tilde{x}})\Psi dqd\tilde{q}$$
$$-\int_{Q\times Q} \left(\mathbf{F}(\pi_1,\tilde{\pi}_2).\nabla_x\Psi+\tilde{\mathbf{F}}(\tilde{\pi}_2,\pi_1).\nabla_{\tilde{x}}\Psi\right) dqd\tilde{q}$$
$$+\int_{Q\times Q} (G(\pi_1,\tilde{\pi}_2)+\tilde{G}(\tilde{\pi}_2,\pi_1))\Psi dqd\tilde{q}\le 0.$$

Proof. To simplify we set $d\tilde{q} = d\tilde{\alpha}d\tilde{x}d\tilde{t}$ and we add a "tilde" superscript to any function in "tilde" variables. On the one hand in (16) written in variables (α,t,x) for π_1 we take $k(\tilde{\alpha},\tilde{t},\tilde{x}) = \tilde{\pi}_2$ for a.e. $(\tilde{\alpha},\tilde{t},\tilde{x})$ in $Q\setminus Q_\theta^{\tilde{\pi}_2}$, so that $k(\tilde{\alpha},\tilde{t},\tilde{x})$ belongs to E. On the other hand in (13) written in variables (α,t,x) for π_1, we take $k(\tilde{\alpha},\tilde{t},\tilde{x}) = \tilde{\pi}_2$ for a.e. $(\tilde{\alpha},\tilde{t},\tilde{x})$ in $Q_\theta^{\tilde{u}_2}$. Each inequality obtained is integrated with respect to the variables $\tilde{\alpha}$, $\tilde{t}$ and $\tilde{x}$ on the corresponding domain. We add and use a version of the Fatou's Lemma to deal with the "lim sup" in the right-hand side. Indeed we observe by coming back to the proof of Lemma 2 for λ fixed, that the function

$$k \mapsto \int_{Q} sgn'_{\lambda}(\phi(\pi) - \phi(k))[\nabla\phi(\pi)]^2\zeta dq$$

is uniformly bounded with respect to λ and k in $[a,b]$. Finally we obtain for π_1:

$$\begin{aligned}
&- \int_{Q\times Q} |\pi_1 - \tilde{\pi}_2|\Psi_t \, dqd\tilde{q} + \int_{Q\times Q} [\nabla_x|\phi(\pi_1) - \phi(\tilde{\pi}_2)| - \mathbf{F}(\pi_1,\tilde{\pi}_2)].\nabla_x\Psi \, dqd\tilde{q} \\
&+ \int_{Q\times Q} G(\pi_1,\tilde{\pi}_2)\Psi \, dqd\tilde{q} \\
&\leq -\limsup_{\lambda\to 0^+} \int_{\mathbb{Q}} sgn'_{\lambda}(\phi(\pi_1) - \phi(\tilde{\pi}_2))[\nabla\phi(\pi_1)]^2\Psi \, dqd\tilde{q},
\end{aligned}$$

where in the right-had side we have used the lemma 1 (ii) to rewrite the integration field under the form $\mathbb{Q} = Q \setminus Q_{\theta}^{\pi_1} \times Q \setminus Q_{\theta}^{\tilde{\pi}_2}$. Besides, we integrate over Q the Gauss-Green formula:

$$\int_{Q} \nabla_x\phi(u_1).\nabla_{\tilde{x}}[sgn_{\lambda}(\phi(u_1) - \phi(\tilde{u}_2))\Psi] \, d\tilde{t}d\tilde{x} = 0 \text{ a.e. } (t,x) \text{ in } Q.$$

By developing the partial derivatives and taking into account that $\phi(u_2)$ is an element of $L^2(0,T;H_0^1(\Omega))$, the λ-limit provides the next equality:

$$\begin{aligned}
&\int_{Q\times Q} sgn(\phi(\pi_1) - \phi(\tilde{\pi}_2))\nabla_x\phi(\pi_1).\nabla_{\tilde{x}}\Psi \, dqd\tilde{q} \\
&= \lim_{\lambda\to 0^+} \int_{Q\times Q} sgn'_{\lambda}(\phi(\pi_1) - \phi(\tilde{\pi}_2))\nabla_x\phi(\pi_1).\nabla_{\tilde{x}}\phi(\tilde{\pi}_2)\Psi \, dqd\tilde{q} \\
&= \lim_{\lambda\to 0^+} \int_{\mathbb{Q}} sgn'_{\lambda}(\phi(\pi_1) - \phi(\tilde{\pi}_2))\nabla_x\phi(\pi_1).\nabla_{\tilde{x}}\phi(\tilde{\pi}_2)\Psi \, dqd\tilde{q}.
\end{aligned}$$

We apply the same reasoning with $\tilde{\pi}_2$ and bring the results together to obtain:

$$\begin{aligned}
&- \int_{Q\times Q} |\pi_1 - \tilde{\pi}_2|(\partial_t + \partial_{\tilde{t}})\Psi - (\nabla_x + \nabla_{\tilde{x}})|\phi(\pi_1) - \phi(\tilde{\pi}_2)|.(\nabla_x + \nabla_{\tilde{x}})\Psi dqd\tilde{q} \\
&- \int_{Q\times Q} \left(\mathbf{F}(\pi_1,\tilde{\pi}_2).\nabla_x\Psi + \tilde{\mathbf{F}}(\tilde{\pi}_2,\pi_1).\nabla_{\tilde{x}}\Psi\right) dqd\tilde{q} \\
&+ \int_{Q\times Q} (G(\pi_1,\tilde{\pi}_2) + \tilde{G}(\tilde{\pi}_2,\pi_1))\Psi dqd\tilde{q} \\
&\leq -\limsup_{\lambda\to 0^+} \int_{\mathbb{Q}} sgn'_{\lambda}(\phi(\pi_1) - \phi(\tilde{\pi}_2)) \, [\nabla_x\phi(\pi_1) - \nabla_{\tilde{x}}\phi(\tilde{\pi}_2)]^2 \Psi \, dqd\tilde{q}.
\end{aligned}$$

The desired inequality follows. ■

¿From Proposition 2 we may state the main result of this section, that is the T-Lipschitzian dependence in L^1 of an entropy process solution to (1) with respect to the corresponding initial data. For the treatment of boundary terms, the proof follows C.Mascia, A.Porretta & A.Terracina's one ([13]) but it needs to be transcribed in the framework of entropy process solutions. It leads to:

Theorem 1. *If π_1 and π_2 are two entropy process solutions to (1) corresponding to initial data $u_{0,1}$ and $u_{0,2}$ respectively then for a.e. t of $[0,T]$,*

$$\int_{]0,1[^2\times\Omega} |\pi_1(\alpha,t,x)-\pi_2(\tilde{\alpha},t,x)|\,dxd\alpha d\tilde{\alpha} \leq e^{M_\psi t}\int_\Omega |u_{0,1}-u_{0,2}|\,dx.$$

Proof. In Proposition 2 we choose $\Psi=\xi(t,x,\tilde{t},\tilde{x})\rho_l(x)\rho_m(\tilde{x})$ for any boundary-layer sequences $(\rho_l)_{l>0}$ and $(\rho_m)_{m>0}$ and any nonnegative ξ in $\mathcal{D}((]0,T[\times\overline{\Omega})^2)$. We develop the partial derivatives and we argue that due to (3) and (4),

$$\begin{aligned}
&\lim_{m\to0^+}\left(\lim_{l\to0+}\left(\int_{Q\times Q}\Big(\nabla_x|\phi(\pi_1)-\phi(\tilde{\pi}_2)|-\mathbf{F}(\pi_1,\tilde{\pi}_2)\Big)\xi\rho_m\nabla_x\rho_l dqd\tilde{q}\right)\right)\\
=&\int_Q\langle\Pi^1_{\tilde{\pi}_2},\xi\rangle_\partial d\tilde{q},
\end{aligned}$$

where $\Pi^1_{\tilde{\pi}_2}$ refers to $\Pi_{\tilde{\pi}_2}$ with $\pi=\pi_1$. Similarly

$$\begin{aligned}
&\lim_{m\to0^+}\left(\lim_{l\to0+}\left(\int_{Q\times Q}(\nabla_{\tilde{x}}|\phi(\pi_1)-\phi(\tilde{\pi}_2)|-\tilde{\mathbf{F}}(\tilde{\pi}_2,\pi_1))\,\xi\rho_l\nabla_{\tilde{x}}\rho_m dqd\tilde{q}\right)\right)\\
=&\int_Q\langle\tilde{\Pi}^2_{\pi_1},\xi\rangle_\partial dq,
\end{aligned}$$

and we obtain:

$$\begin{aligned}
&-\int_{Q\times Q}\Pi^1_{\tilde{\pi}_2}.\overline{\nabla}_{(t,x)}\xi dxdtd\tilde{x}d\tilde{t}+\int_{Q\times Q}G(\pi_1,\tilde{\pi}_2)\xi dqd\tilde{q}\\
&+\int_{Q\times Q}(\nabla_x|\phi(\pi_1)-\phi(\tilde{\pi}_2)|.\nabla_{\tilde{x}}\xi+\nabla_{\tilde{x}}|\phi(\pi_1)-\phi(\tilde{\pi}_2)|.\nabla_x\xi)dqd\tilde{q}\\
&-\int_{Q\times Q}\tilde{\Pi}^2_{\pi_1}.\overline{\nabla}_{(\tilde{t},\tilde{x})}\xi dxdtd\tilde{x}d\tilde{t}+\int_{Q\times Q}\tilde{G}(\tilde{\pi}_2,\pi_1)\xi dqd\tilde{q}\\
\leq&\ -\int_Q\langle\Pi^1_{\tilde{\pi}_2},\xi\rangle_\partial d\tilde{q}-\int_Q\langle\tilde{\Pi}^2_{\pi_1},\xi\rangle_\partial dq-I_1-I_2,
\end{aligned}\tag{16}$$

where

$$I_1 = \lim_{m\to 0^+} \int_{Q\times Q} \xi \nabla_x |\phi(\pi_1) - \phi(\tilde{\pi}_2)|.\nabla_{\tilde{x}}\rho_m dq d\tilde{q},$$

$$I_2 = \lim_{l\to 0^+} \int_{Q\times Q} \xi \nabla_{\tilde{x}} |\phi(\pi_1) - \phi(\tilde{\pi}_2)|.\nabla_{x}\rho_l dq d\tilde{q}.$$

An integration by parts with respect to x and then with respect to $\tilde{x}$ allows us to express the limit with respect to m in I_1. Indeed:

$$\lim_{m\to 0^+} \int_{Q\times Q} \xi \nabla_x |\phi(\pi_1) - \phi(\tilde{\pi}_2)|.\nabla_{\tilde{x}}\rho_m dq d\tilde{q} = \int_{Q\times Q} Div_{\tilde{x}}[|\phi(\pi_1) - \phi(\tilde{\pi}_2)|.\nabla_x \xi] dq d\tilde{q}.$$

An integration by parts with respect to $\tilde{x}$ and then with respect to x provides:

$$I_1 = -\int_{Q\times\Sigma} \nabla_x |\phi(\pi_1)|.\tilde{\nu}\xi dq d\mathcal{H}^p_{(\tilde{t},\tilde{x})}.$$

With the same arguments,

$$I_2 = -\int_{Q\times\Sigma} \nabla_{\tilde{x}} |\phi(\tilde{\pi}_2)|.\nu\xi d\tilde{q} d\mathcal{H}^p_{(t,x)}.$$

We take now into account (14) for π_1 and $\tilde{\pi}_2$ to have a majoration of the right-hand side of (16) in terms of:

$$\begin{aligned} &\int_{Q\times\Sigma} \left(\nabla_x|\phi(\pi_1)| - \tilde{\mathbf{F}}(\pi_1,0)\right).\tilde{\nu}\xi dq d\mathcal{H}^p_{(\tilde{t},\tilde{x})} + \int_Q \langle \Pi^1_0, \xi\rangle_\partial d\tilde{q} \\ + &\int_{Q\times\Sigma} \left(\nabla_{\tilde{x}}|\phi(\tilde{\pi}_2)| - \mathbf{F}(\tilde{\pi}_2,0)\right).\nu\xi d\tilde{q} d\mathcal{H}^p_{(t,x)} + \int_Q \langle \tilde{\Pi}^2_0, \xi\rangle_\partial dq. \end{aligned} \tag{17}$$

We choose $\xi_n = W_n(x-\tilde{x})w_n(t-\tilde{t})\gamma$ where γ is a nonnegative element of $\mathcal{D}(]0,T[)$, $(W_n)_n$ and $(w_n)_n$ are the standard mollifier sequences on $\mathbb{R}^p$ and $\mathbb{R}$. In addition, n is large enough so that for any $\tilde{t}$ in $]0,T[$, $t \mapsto \gamma(t)w_n(t-\tilde{t})$ and for any t in $]0,T[$, $\tilde{t} \mapsto \gamma(t)w_n(t-\tilde{t})$ belongs to $\mathcal{D}(]0,T[)$. We take the limit with respect to n in (17) by referring to [13] providing that each line goes to 0. For the left-hand side of (16), we use classical techniques and the fact that

$$(\partial_t + \partial_{\tilde{t}})\xi_n = \gamma'(t)W_n(x-\tilde{x})w_n(t-\tilde{t}) \text{ and } (\nabla_x + \nabla_{\tilde{x}})\xi_n = 0.$$

Eventually one gets,

$$-\int_{Q\times]0,1[} |\pi_1 - \pi_2|\gamma'(t) dq d\tilde{\alpha} \le M_\psi \int_{Q\times]0,1[} |\pi_1 - \pi_2|\gamma dq d\tilde{\alpha},$$

for any nonnegative γ in $\mathcal{D}(]0,T[)$ and so, by density, for any nonnegative γ in $H^1_0(]0,T[)$. Now the conclusion is classical: it uses for γ a Lipschitzian and piecewise linear approximation of $\mathbb{I}_{[0,\tau]}$, for a.e. τ in $]0,T[$, the initial condition (15) for π_1 and π_2 and the Gronwall

Lemma. ■

When $u_{0,1} = u_{0,2}$ a.e. on Ω, we deduce (see [7] or [5] in the framework of Young measure solutions) the existence of a measurable function χ on Q, such that $\pi_1(\alpha,t,x) = \pi_2(\tilde{\alpha},t,x) = \chi(t,x)$ for a.e. α and $\tilde{\alpha}$ in $]0,1[$ and for a.e. (t,x) in Q. Thus in Definition 2 the integrations with respect to the Lebesgue measure on $]0,1[$ may be performed. So that $\chi = u$ a.e. on Q and u is namely an entropy solution to (1) in the sense of Definition 1. As a consequence,

Corollary 1. *If (1) has an entropy process solution then it has an entropy solution. In addition, if u_1 and u_2 are two entropy solutions corresponding to initial data $u_{0,1}$ and $u_{0,2}$ respectively, then for a.e. t in $]0,T[$:*

$$\int_\Omega |u_1(t,x) - u_2(t,x)|\,dx \le e^{M_\psi t} \int_\Omega |u_{0,1} - u_{0,2}|\,dx.$$

3. The Viscosity and Penalization Methods

Our aim is to obtain an existence result for (1) by introducing some diffusion on the whole domain and by relaxing the bilateral obstacle condition. This leads us to introduce first, for any function f (or $f(t,x,.)$), the Lipschitz bounded extension $f^\star$ of f outside $[a,b]$ defined through:

$$f^\star(u) = f(u) \text{ if } u \in [a,b], f^\star(u) = f(a) \text{ if } u \le a, f^\star(u) = f(b) \text{ if } u \ge b,$$

and similarly for $f^\star(t,x,.)$ (in this case, observe that $\partial_{x_i} f^\star = (\partial_{x_i} f)^\star$). Now for any positive parameter δ intended to tend to 0^+ - so that it will be supposed less or equal than a fixed value δ_0 - we define $\phi^\star_\delta = \phi^\star + \delta Id_{\mathbb{R}}$ a bilipschitzian function and we consider the nondegenerate penalized parabolic operator

$$\mathbb{P}_\delta(t,x,.) : u \to \partial_t u + \sum_{i=1}^{p} \partial_{x_i} \varphi_i^\star(t,x,u) + \psi^\star(t,x,u) - \Delta\phi^\star_\delta(u) + \beta_\delta(u),$$

with $\beta_\delta(z) = \frac{1}{\delta}(-(z-a)^- + (z-b)^+)$. We consider the resulting problem: find a measurable and bounded function u_δ on Q satisfying

$$\begin{cases} \mathbb{P}_\delta(t,x,u_\delta) = 0 \text{ on } Q, \\ u_\delta = 0 \text{ on }]0,T[\times\partial\Omega,\ u_\delta(0,.) = u_0^\delta \text{ on } \Omega, \end{cases} \tag{18}$$

where u_0^δ is a regularization of u_0 obtained by the mean of mollifiers, so that:

$$\forall\delta > 0,\ u_0^\delta \in \mathcal{D}(\Omega),\ a \le u_0^\delta \le b \text{ a.e. in } \Omega,\ \lim_{\delta\to 0^+} u_0^\delta = u_0 \text{ in } L^1(\Omega).$$

3.1. Some *a priori* Estimates

We know (see for example [10]) that for a fixed positive δ, the problem (18) has a unique weak solution u_δ in $L^\infty(Q)\cap H^1(Q)\cap L^\infty(0,T;H_0^1(\Omega))$ with $\phi_\delta^\star(u_\delta)$ in $L^2(0,T;H^2(\Omega))$. In this section, we are mainly interested in the δ-uniform *a priori* estimations for the sequence $(u_\delta)_{\delta>0}$. Indeed,

Proposition 3. *There exist some constants C_1, C_2 and C_3 independent from δ such that*

$$\forall t\in[0,T],\ |u_\delta(t,.)|\le M(t) \text{ a.e. in } \Omega, \tag{19}$$

$$\frac{1}{\delta}\|(u_\delta-a)^-\|_{L^2(Q)}+\frac{1}{\delta}\|(u_\delta-b)^+\|_{L^2(Q)}\le C_1 \tag{20}$$

$$\|\widehat{\phi_\delta^\star}(u_\delta)\|_{L^2(0,T;H_0^1(\Omega))}+\|\partial_t u_\delta\|_{L^2(0,T;H^{-1}(\Omega))}\le C_2, \tag{21}$$

$$\|\sqrt{t}\,\partial_t\widehat{\phi_\delta^\star}(u_\delta)\|_{L^2(Q)}\le C_3, \tag{22}$$

where $M(t)$ is defined by (2) and $\widehat{\phi^\star}_\delta(x)=\int_0^x(\phi_\delta^{\star\prime}(\tau))^{1/2}d\tau.$

Remark 2. *We have not been able to establish any δ-uniform estimates for $(u_\delta)_{\delta>0}$ in $W^{1,1}(Q)$ or even in $BV(Q)$ as in [11] for the positiveness obstacle problem for $\mathbb{P}$ when φ is a $W^{2,+\infty}$-class vector-valued function and ψ a $W^{1,+\infty}$-class function. Furthermore even if we have enough compactness for $(\phi_\delta^\star(u_\delta))_{\delta>0}$, we cannot take advantage of it since $(\phi^\star)^{-1}$ may not exist. So to describe the behavior of $(u_\delta)_{\delta>0}$ when δ goes to 0^+ we can only refer to (19), which leads to consider the notion of an entropy process solution to (1).*

Proof of Proposition 3 - With the notations of Section 1.2., $M_{\psi^\star}=M_\psi$ and $M_{\partial_{x_i}\varphi_i^\star}=M_{\partial_{x_i}\varphi_i}$. So the standard maximum principle for u_δ ensures (19), the independence with respect to δ resulting from the monotonicity of the penalized operator β_δ. For (20), we treat each term separately: to estimate $\frac{1}{\delta}(u_\delta-b)^+$ we take the $L^2(Q)$-scalar product between (18) and $\frac{1}{\delta}(u_\delta-b)^+$. We observe that

$$\frac{1}{\delta}\int_Q\partial_t u_\delta(u_\delta-b)^+dxdt=\frac{1}{2\delta}\int_Q\partial_t[(u_\delta-b)^+]^2dxdt=\frac{1}{2\delta}\int_\Omega[(u_\delta-b)^+]^2(T,x)dx,$$

since $a\le u_0^\delta\le b$ a.e. on Ω. So the contribution of the evolution term is nonnegative and it is the same for the diffusive one, thanks to the Green formula. For the convective and reactive integrals, due to the definition of $\varphi^\star$ and $\psi^\star$,

$$\begin{aligned}&\frac{1}{\delta}\int_Q(Div\varphi^\star(t,x,u_\delta)+\psi(t,x,u_\delta))(u_\delta-b)^+dxdt\\ =\ &\frac{1}{\delta}\int_Q(Div\varphi(t,x,b)+\psi(t,x,b))(u_\delta-b)^+dxdt\\ \le\ &\frac{1}{2}\|\varphi(t,x,b)+\psi(t,x,b)\|^2_{L^2(Q)}+\frac{1}{2}\|\frac{1}{\delta}(u_\delta-b)^+\|^2_{L^2(Q)}dxdt.\end{aligned}$$

Eventually, we remark that

$$\frac{1}{\delta}\int_Q \beta_\delta(u_\delta)(u_\delta - b)^+ dxdt = \|\frac{1}{\delta}(u_\delta - b)^+\|^2_{L^2(Q)}.$$

So we conclude the existence of a constant C, independent from δ, such that The previous techniques with $-\frac{1}{\delta}(u_\delta - a)^-$ provide (20).

The energy estimate (21) results from the $L^2(Q)$-scalar product between (18) and u_δ. Since $\beta_\delta(u_\delta)u_\delta$ is nonnegative a.e. in Q, it warrants thanks to (19) a δ-uniform bound for $\nabla\widehat{\phi^\star_\delta}(u_\delta)$ in $L^2(Q)^p$. Then by coming back to the definition of the norm in $L^2(0,T;H^{-1}(\Omega))$ and referring to (20) we derive the estimation of $\partial_t u_\delta$. Let us focus on (22). We take the $L^2(Q)$-scalar product between (18) and $t\partial_t\phi^\star_\delta(u_\delta)$. This way,

$$\int_Q t\partial_t u_\delta \partial_t \phi^\star_\delta(u_\delta)dxdt = \frac{1}{2}\|\sqrt{t}\partial_t\widehat{\phi^\star_\delta}(u_\delta)\|^2_{L^2(Q)}.$$

For the diffusive integral, since $w = \phi^\star_\delta(u_\delta)$ belongs to the functional space $\mathcal{W}(0,T) \equiv \{v \in L^2(0,T;H^1_0(\Omega)\cap H^2(\Omega)); \partial_t v \in L^2(Q)\}$, we use the density of $\mathcal{D}([0,T];H^1_0(\Omega)\cap H^2(\Omega))$ into $\mathcal{W}(0,T)$ to carry out calculations with a sequence $(w_k)_k$ of mollified functions with respect to the time variable. One has:

$$\begin{aligned} -\int_Q t\Delta w_k \partial_t w_k dxdt &= \int_Q t\nabla w_k.\partial_t \nabla w_k dxdt \int_Q \frac{t}{2}\partial_t \left[\nabla w_k\right]^2 dxdt \\ &= -\frac{1}{2}\int_Q \left[\nabla w_k\right]^2 dxdt + \int_\Omega \frac{T}{2}\left[\nabla w_k\right]^2 (T,x)dx. \end{aligned}$$

Then, we pass to the limit with k to obtain

$$\begin{aligned} \int_Q t\Delta\phi^\star_\delta(u_\delta)\partial_t\phi^\star_\delta(u_\delta)dxdt &= \frac{1}{2}\int_Q \left[\nabla\phi^\star_\delta(u_\delta)\right]^2 dxdt - \int_\Omega \frac{T}{2}\left[\nabla\phi^\star_\delta(u_\delta)\right]^2 (T,x)dx \\ &\leq \frac{1}{2}\int_Q \left[\nabla\phi^\star_\delta(u_\delta)\right]^2 dxdt \leq C, \end{aligned}$$

where C is a constant independent from δ thanks to (19) and (21).

We develop the partial derivatives in the convective term to write:

$$\begin{aligned} &\int_Q tDiv_x\varphi^\star(t,x,u_\delta)\partial_t\phi^\star_\delta(u_\delta)dxdt \\ = &\int_Q \sum_{i=1}^p \left(\partial_u\varphi^\star_i(t,x,u_\delta)\sqrt{t}\partial_{x_i}\widehat{\phi^\star_\delta}(u_\delta)\sqrt{t}\partial_t\widehat{\phi^\star_\delta}(u_\delta)\right)dxdt \\ &+\int_Q \sum_{i=1}^p \left(\partial_{x_i}\varphi^\star_i(t,x,u_\delta)\sqrt{t}\sqrt{(\phi^\star_\delta)'(u_\delta)}\sqrt{t}\partial_t\widehat{\phi^\star_\delta}(u_\delta)\right)dxdt. \end{aligned}$$

The Young inequality with $p=2$ and (19) prove that:

$$\int_Q tDiv_x\varphi_i^\star(t,x,u_\delta)\partial_t\phi_\delta^\star(u_\delta)dxdt \leq C\|\nabla\widehat{\phi_\delta^\star}(u_\delta)\|^2_{L^2(Q)^p}+\frac{1}{4}\|\sqrt{t}\partial_t\widehat{\phi_\delta^\star}(u_\delta)\|^2_{L^2(Q)}.$$

Again, with the Young inequality and (19):

$$\int_Q t\psi^\star(t,x,u_\delta)\partial_t\phi_\delta^\star(u_\delta)dxdt \leq C+\frac{1}{4}\|\sqrt{t}\partial_t\widehat{\phi_\delta^\star}(u_\delta)\|^2_{L^2(Q)},$$

where C doest not depend on δ thanks to (19). It is the same for the penalized term since, due to (20), $\frac{1}{\delta}(u_\delta-a)^-$ and $\frac{1}{\delta}(u_\delta-b)^+$ are bounded in $L^2(Q)$ uniformly with respect to δ, that completes the proof of Proposition 3. ■

We give now a formulation of boundary conditions for the solutions to the nondegenerate relaxed problem that will be the starting point to derive the formulation of boundary conditions for the solution to (1) by taking the limit with respect to δ. In what follows $\mathbf{F}^\star$, $G^\star$ and $U^\star_{k,\delta}$ refer to $\mathbf{F}$, G and U_k with $\varphi^\star$, $\psi^\star$ and $\phi_\delta^\star$ in the place of φ, ψ and ϕ_δ respectively. This way, by arguing as in [11, 18], we prove that:

Proposition 4. *For any δ, the next compatibility condition holds on Σ:*

$$\int_\Sigma \mathbf{F}(0,k).\nu\xi d\mathcal{H}^p \leq \left\langle U^\star_{k,\delta},\xi\right\rangle_\partial+\left\langle U^\star_{0,\delta},\xi\right\rangle_\partial, \tag{23}$$

for all nonnegative ξ in $L^\infty(Q)\cap H^1(Q)\cap C(Q)$, $\xi(T,.)=\xi(0,.)=0$ and any real k. Moreover there exists a constant C_4, independent from δ such that

$$\|Div_{(t,x)}U^\star_{k,\delta}\|_{\mathcal{M}_b(Q)}\leq C_4. \tag{24}$$

Proof.

First point - We observe that $\Delta|\phi_\delta^\star(u_\delta)-\phi_\delta^\star(k)|$ is a bounded Radon measure on Q, for any real k. As in [17], this assertion comes from the Kato's Inequality and uses the fact that $\phi_\delta^\star(u_\delta)$ is an element of $L^2(0,T;H^2(\Omega))$. As a consequence, $U^\star_{k,\delta}$ and $(0,\nabla|\phi_\delta^\star(u_\delta)-\phi_\delta^\star(k)|)$ belong to $\mathcal{DM}_2(Q)$ for any real k. Thus, by using the boundary condition for u_δ and the relation (4) with $\xi(T,.)=\xi(0,.)=0$, we may successively write:

$$\begin{aligned}
&\left\langle U^\star_{k,\delta},\xi\right\rangle_\partial\\
&=-\lim_{\rho\to0+}\int_Q U^\star_{k,\delta}\xi.(0,\nabla\rho_\rho)\,dxdt\\
&=\lim_{\rho\to0+}\int_Q \xi\nabla|\phi_\delta^\star(u_\delta)-\phi_\delta^\star(k)|.\nabla\rho_\rho dxdt-\lim_{\rho\to0+}\int_Q \xi\mathbf{F}^\star(u_\delta,k).\nabla\rho_\rho\,dxdt\\
&=\lim_{\rho\to0+}\int_Q \xi(0,\nabla|\phi_\delta^\star(u_\delta)-\phi_\delta^\star(k)|).(0,\nabla\rho_\rho)\,dxdt+\int_\Sigma \mathbf{F}(0,k).\nu\xi d\mathcal{H}^p\\
&=-\langle(0,\nabla|\phi_\delta^\star(u_\delta)-\phi_\delta^\star(k)|),\xi\rangle_\partial+\int_\Sigma \mathbf{F}(0,k).\nu\xi d\mathcal{H}^p.
\end{aligned}$$

Let us conclude through the technical property:

Claim 1. *([18]) Let w be in $L^2(0,T;H_0^1(\Omega))$ such that $w \geq 0$ a.e. on Q and Δw belongs to $\mathcal{M}_b(Q)$. Then, for any nonnegative ξ of $L^\infty(Q)\cap H^1(Q)\cap C(Q)$*

$$\langle (0,\nabla w),\xi\rangle_\partial \leq 0.$$

With $w = |\phi_\delta^\star(u_\delta)-\phi_\delta^\star(k)| + |\phi_\delta^\star(u_\delta)| - |\phi_\delta^\star(k)|$ provides (23) since, due to (4),

$$-\langle (0,\nabla|\phi_\delta^\star(u_\delta))|,\xi\rangle_\partial = \left\langle U_{0,\delta}^\star,\xi\right\rangle_\partial$$

We remark now that u_δ fulfills (8) with $\phi_\delta^\star$ in the place of ϕ. Indeed for any nonnegative function ξ of $\mathcal{D}(]-\infty,T[\times\Omega)$, we take the $L^2(Q)$-scalar product between (18) and the test-function $sgn_\lambda(u_\delta - k)\xi$. As soon as k belongs to $[a,b]$, $\beta_\delta(u_\delta)sgn_\lambda(u_\delta-k)\xi$ is nonnegative. After some integration by parts we obtain (with $S_\lambda(v) = sgn_\lambda(v-k)$ and $I_\lambda(v) = \int_k^v S_\lambda(\tau)d\tau$):

$$\begin{aligned}
&\int_Q I_\lambda(u_\delta)\partial_t\xi dxdt + \int_Q\left(\int_k^{u_\delta}\partial_u\varphi^\star(t,x,\tau)S_\lambda(\tau)d\tau\right).\nabla\xi dxdt\\
+&\int_\Omega I_\lambda(u_0^\delta)\xi(0,.)dx - \int_Q (Div\varphi^\star(t,x,k)+\psi^\star(t,x,u_\delta))\,S_\lambda(u_\delta)\xi dxdt\\
-&\int_Q\left(\int_k^{u_\delta}[Div\varphi^\star(t,x,\tau)-Div\varphi^\star(t,x,k)]S_\lambda'(\tau)d\tau\right)\xi dxdt\\
\geq&\int_Q S_\lambda(u_\delta)\nabla\phi_\delta^\star(u_\delta).\nabla\xi dxdt.
\end{aligned}\tag{25}$$

For a.e. x, $sgn_\lambda'(x) = \frac{1}{\lambda}\mathbb{I}_{]-\lambda,\lambda[}(x)$ and since the function $z\mapsto Div\varphi^\star(t,x,z)$ is continuous uniformly with respect to (t,x) in Q, the third term in the left-hand side vanishes when λ goes to 0^+. Thus (8) for u_δ is obtained at the λ–limit thanks to the Lebesgue dominated convergence Theorem. We deduce that for any real k of $[a,b]$,

$$\theta_{k,\delta} \equiv -Div_{(t,x)}U_{k,\delta}^\star - G^\star(u_\delta,k)\tag{26}$$

is a nonnegative Radon measure on Q such that

$$||\theta_{k,\delta}||_{\mathcal{M}_b(Q)} = \int_Q d\theta_{k,\delta} = -\int_Q d[Div_{(t,x)}U_{k,\delta}^\star] - \int_Q G^\star(u_\delta,k)\,dxdt.$$

But by using (3) and (4):

$$\begin{aligned}
&-\int_Q d[Div_{(t,x)}U_{k,\delta}^\star]\\
=&\ \langle(0,\nabla|\phi_\delta^\star(u_\delta)-\phi_\delta^\star(k)|),1\rangle_\partial - \int_\Sigma \mathbf{F}(0,k).\nu d\mathcal{H}^p + \int_\Omega |u_0^\delta - k|dx\\
-&\int_\Omega |u_\delta(T,x)-k|dx.
\end{aligned}$$

Then as a consequence of (19) and since u_0^δ is uniformly bounded with respect to δ, we claim the existence of a constant C independent from δ such that:

$$-\int_Q d[Div_{(t,x)}U_{k,\delta}^\star] \leq \langle (0, \nabla|\phi_\delta^\star(u_\delta) - \phi_\delta^\star(k)|), 1\rangle_\partial + C.$$

On the one hand, the previous inequality with $k = 0$ and Claim 1 with $w = |\phi_\delta^\star(u_\delta)|$ provide

$$-\int_Q d[Div_{(t,x)}U_{0,\delta}^\star] \leq C,$$

that means that $(\theta_{0,\delta})_{\delta>0}$ given by (26) is a bounded sequence in $\mathcal{M}_b(Q)$ and $(Div_{(t,x)}U_{0,\delta}^\star)_{\delta>0}$ too. On the other hand for any k in $[a,b]$, Claim 1 with $w = |\phi_\delta^\star(u_\delta) - \phi_\delta^\star(k)| + |\phi_\delta^\star(u_\delta)| - |\phi_\delta^\star(k)|$ gives:

$$-\int_Q d[Div_{(t,x)}U_{k,\delta}^\star] \leq \langle (0, -\nabla|\phi_\delta^\star(u_\delta)|), 1\rangle_\partial + C.$$

But by referring to (3) and (4) and using the homogeneous boundary condition for u_δ, we observe that:

$$\langle (0, -\nabla|\phi_\delta^\star(u_\delta)|), 1\rangle_\partial = \left\langle U_{0,\delta}^\star, 1\right\rangle_\partial + \int_\Omega |u_0^\delta|dx - \int_\Omega |u_\delta(T,x)|dx.$$

Thus, due to the uniform boundedness for u_0^δ, there exists a constant C independent from δ such that:

$$|\langle (0, -\nabla|\phi_\delta^\star(u_\delta)|), 1\rangle_\partial| \leq |\left\langle U_{0,\delta}^\star, 1\right\rangle_\partial| + C \leq \|Div_{(t,x)}U_{0,\delta}^\star\|_{\mathcal{M}_b(Q)} + C.$$

Eventually the estimate of $Div_{(t,x)}U_{0,\delta}^\star$ highlighted previously ensures that the sequence $(\theta_{k,\delta})_{\delta>0}$ is a uniformly bounded in $\mathcal{M}_b(Q)$ and so $(Div_{(t,x)}U_{k,\delta}^\star)_{\delta>0}$ too. Relation (24) follows, which completes the proof of Proposition 4. ■

3.2. Convergence Toward an Entropy Process Solution

Due to Proposition 3, there exists a measurable function u on Q such that, up to a subsequence when δ goes to 0^+, $(u_\delta)_{\delta>0}$ converges toward u in $L^\infty(Q)$ weak $\star$ and $(\partial_t u_\delta)_{\delta>0}$ weakly converges toward $\partial_t u$ in $L^2(0,T;H^{-1}(\Omega))$. Besides there exists a function Φ such that, up to a subsequence, $(\phi_\delta^\star(u_\delta))_{\delta>0}$ goes to Φ in $L^2(0,T;H_0^1(\Omega))$−weak. But since $(t\phi_\delta^\star(u_\delta))_{\delta>0}$ is uniformly bounded in $H^1(Q)$ with respect to δ, we may be sure that $(\phi_\delta^\star(u_\delta))_{\delta>0}$ strongly converges (up to a subsequence) toward Φ in $L^q(Q)$ for any q in $[1,+\infty[$. In order to connect $\phi(u)$ and Φ, we first remind some properties of bounded sequences in L^∞:

Claim 2. *([7]) Let $(u_n)_{n>0}$ be a sequence of measurable functions on an open bounded subset O such that $(u_n)_n$ is uniformly bounded in the $L^\infty(Q)$-norm the by a constant M. Then, there exists a subsequence $(u_{\varphi(n)})_{n>0}$ and a measurable and bounded function π on $]0,1[\times O$ such that for all continuous and bounded functions h on $O\times]-M,M[$,*

$$\forall \xi \in L^1(O),\ \lim_{n\to+\infty}\int_O h(\omega,u_{\varphi(n)})\xi dx = \int_{]0,1[\times O} h(\omega,\pi(\alpha,\omega))d\alpha\xi d\omega.$$

Such a result has found its first application in the approximation through the artificial viscosity method of the Cauchy problem in $\mathbb{R}^p$ for a scalar conservation law, as one can establish a uniform L^∞-control of approximate solutions. It has also been applied to the numerical analysis of transport equations since "Finite-Volume" schemes mainly give an L^∞-estimate uniformly with respect to the mesh length of the numerical solution. Here, we refer to this concept when the approximating sequence is $(u_\delta)_{\delta>0}$ and so there exists a function π in $L^\infty(]0,1[\times Q)$ such that thanks to (20), $a\leq\pi\leq b$ a.e. in $]0,1[\times Q$. Of course, we also have for a.e. (t,x) in Q,

$$u(t,x)=\int_0^1 \pi(\alpha,t,x)d\alpha \text{ and } \Phi(t,x)=\int_0^1 \phi(\pi(t,x,\alpha))d\alpha.$$

Furthermore we remark that

$$\int_{]0,1[\times Q} |\phi(\pi(\alpha,t,x))-\Phi(t,x)|d\alpha dxdt = \lim_{\delta\to 0^+}\int_Q |\phi_\delta^\star(u_\delta)-\Phi(t,x)|dxdt = 0.$$

Thus we even have, for a.e. (α,t,x) of $]0,1[\times Q$, $\phi(\pi(\alpha,t,x))=\Phi(t,x)$. But since ϕ is a nondecreasing function, for a.e. (t,x) in Q, $\phi^{-1}(\{\Phi(t,x)\})=[i_1,i_2]$, where $i_1\leq\pi(.,t,x)\leq i_2$ a.e. in $]0,1[$. This way, by integrating from 0 to 1 we prove that $\phi(i_1)=\phi(u(t,x))=\phi(i_2)=\Phi(t,x)$ that is namely (11). The previous developments guide us toward the next statement:

Theorem 2. - *The obstacle problem (1) admits an entropy process solution.*

Proof. We have highlighted a function u and a process π such that (10,11,6) hold. Moreover due to (24) there exists an element of $\mathcal{M}_b(Q)$ - identified in the sense of distributions on Q to $Div_{(t,x)}\Pi_k$ - such that up to a subsequence when δ goes to 0^+, $(Div_{(t,x)}U^\star_{k,\delta})_{\delta>0}$ converges toward $Div_{(t,x)}\Pi_k$ in $\mathcal{M}_b(Q)$ weak $\star$ and (12) follows. To establish (13) we start from (25). For the left hand-side of the δ-limit only refers to the claim 2 while for the right-hand side of (25) we use the smoothness of u_δ and the Green formula:

$$\begin{aligned}\int_Q S_\lambda(u_\delta)\nabla\phi_\delta^\star(u_\delta).\nabla\xi dxdt &= \int_Q \nabla\left(\int_k^{u_\delta} S_\lambda(\tau)(\phi_\delta^\star)'(\tau)d\tau\right).\nabla\xi dxdt\\ &= -\int_Q K_\delta(u_\delta)\Delta\xi dxdt,\end{aligned}$$

where K_δ is the the continuous function $x \to \int_k^x S_\lambda(\tau)(\phi_\delta^\star)'(\tau)d\tau$. We pass to the limit with δ by arguing that $|H_\delta(x)-H(x)| \leq \delta|x-k|$ and referring to Claim 2. It follows

$$\begin{aligned}
& \int_Q \left(\int_k^\pi S_\lambda(\tau)d\tau\right)\partial_t\xi dq + \int_Q\int_k^\pi \partial_u\varphi(t,x,\tau)S_\lambda(\tau)d\tau.\nabla\xi dq \\
+ & \int_\Omega \left(\int_k^{u_0} S_\lambda(\tau)d\tau\right)\xi(0,.)dx - \int_Q (Div\varphi(t,x,\pi)+\psi(t,x,\pi))S_\lambda(\pi)d\tau\xi dq \\
- & \int_Q\int_k^\pi [Div\varphi(t,x,\tau)-Div\varphi(t,x,k)]S'_\lambda(\tau)d\tau\xi dq \\
\geq & -\int_Q \left(\int_k^\pi S_\lambda(\tau)\phi'(\tau)d\tau\right)\Delta\xi dq.
\end{aligned}$$

The λ-limit relies on the Lebesgue dominated convergence Theorem and uses the continuity, uniformly with respect to (t,x) in Q, of $z \mapsto Div\varphi(t,x,z)$ to deal with the third line in the left-hand side. Note that for the right-hand side:

$$\lim_{\lambda\to 0^+} \int_{Q\times]0,1[} \left(\int_k^\pi S_\lambda(\tau)\phi'(\tau)d\tau\right)\Delta\xi d\alpha dxdt = \int_{Q\times]0,1[} |\phi(\pi)-\phi(k)|\Delta\xi d\alpha dxdt.$$

An integration by parts gives (13).

We establish (14) by passing to the δ-limit in (23) and using that:

Claim 3. - *([18]) - Let $(\mu_\rho)_{\rho>0}$ be a sequence of $\mathcal{M}_b(Q)$, $\mu_\rho \geq 0$, converging toward μ in $\mathcal{M}_b(Q)$ weak $\star$. Then, for any nonnegative ξ of $L^\infty(Q)\cap C(Q)$,*

$$\int_Q \xi d\mu \leq \liminf_{\rho\to 0^+}\int_Q \xi d\mu_\rho.$$

This claim with the nonnegative Radon measure $\theta_{k,\delta}$ defined by (26) provide that for any nonnegative ξ of $H^1(Q)\cap L^\infty(Q)\cap C(Q)$, with $\xi(T,.)=\xi(0,.)=0$, up to a subsequence,

$$\int_Q \xi d\theta_k \leq \liminf_{\delta\to 0^+}\int_Q \xi d\theta_{k,\delta},$$

where in the sense of the bounded Radon measures on Q, $\theta_k = Div_{(t,x)}\Pi_k + G$ with $G = \lim_{\delta\to 0+} G^\star(u_\delta,k)$ in $L^\infty(Q)$ weak $\star$ (up to a subsequence). This way,

$$\limsup_{\delta\to 0^+}\int_Q \xi d[Div_{(t,x)}U^\star_{k,\delta}] \leq \int_Q \xi d[Div_{(t,x)}\Pi_k].$$

Eventually, because $(U^{\star}_{k,\delta})_{\delta>0}$ weakly converges in $L^2(Q)^{p+1}$ toward Π_k,

$$\limsup_{\delta\to 0^+}\langle U^{\star}_{k,\delta},\xi\rangle_{\partial} = \int_Q \Pi_k.\overline{\nabla}\xi dxdt + \limsup_{\delta\to 0^+}\int_Q \xi d[Div_{(t,x)}U^{\star}_{k,\delta}],$$

and this way, for any real k of $[a,b]$,

$$\limsup_{\delta\to 0^+}\langle U^{\star}_{k,\delta},\xi\rangle_{\partial} \le \langle \Pi_k,\xi\rangle_{\partial}.$$

Relation (14) follows, that completes the proof of Theorem 2. ■

As pointed out in R.Diperna's paper [5] within the framework work of Young measure solutions, the strong convergence of approximate solutions occurs if and only if the process π may be identified to a function independent from the variable α, so that the associated Young measure reduces to a Dirac mass at almost all points of the physical domain. So, as a consequence of Theorem 2 and Corollary 1 we claim that:

Theorem 3. - *The bilateral obstacle problem (1) admits a unique entropy solution that is the limit of the whole sequence* $(u_\delta)_{\delta>0}$ *of solutions to (18)*$_{\delta>0}$ *when* δ *goes to* 0^+ *in* $L^q(Q)$, $1 \le q < +\infty$.

References

[1] R. Bürger, H. Frid & K.H. Karlsen, On a free boundary problem for a strongly degenerate quasi-linear parabolic equation with an application to a model of pressure filtration. *SIAM J. Math. Anal.* **34**, No.3, 611-635 (2003).

[2] J.Carrillo, Entropy solutions for nonlinear degenerate problems. *Arch. Ration. Mech. Anal.* **147**, No.4, 269-361 (1999).

[3] G.Q.Chen & H.Frid, Divergence-measure fields and hyperbolic conservation laws. *Arch. Ration. Mech. Anal.* **147**, No.2, 89-118 (1999).

[4] G.Q.Chen & H.Frid, On the theory of divergence-measure fields and its applications. *Bol. Soc. Bras. Mat., Nova Sr.* **32**, No.3, 401-433 (2001).

[5] R.J.Diperna, Measure-valued solutions to conservation laws. *Arch. Ration. Mech. Anal.* **88**, 223-270 (1985).

[6] G.Duvaut & J.L.Lions, Les inéquations en mécanique et en physique. *Travaux et recherches mathématiques.* **21**. Paris: Dunod. XX. (1972).

[7] R.Eymard, T.Gallouët & R.Herbin, Existence and uniqueness of the entropy solution to a nonlinear hyperbolic equation. *Chin. Ann. Math., Ser. B* **16**, No.1, 1-14 (1995).

[8] G.Gagneux & M.Madaune-Tort, Analyse mathématique de modèles non linéaires de l'ingénierie pétrolière. *Mathématiques & Applications (Paris).* **22**. Paris: Springer-Verlag. xiv, 188 p. (1995).

[9] S.N.Kuzhkov, First order quasilinear equations in several independent variables. *Math. USSR, Sb.* **10**, 217-243 (1970).

[10] O.A.Ladyzhenskaya, V.A.Solonnikov, N.N.Ural'tseva, Linear and quasi-linear equations of parabolic type. *Translations of Mathematical Monographs.* **23**. Providence (AMS). XI, 648 p. (1968).

[11] L. Lévi, The Positiveness Problem for a Class of Degenerate Parabolic-Hyperbolic Operators *Adv. Math. Sci. Appl.* **15**, No 1, 307-333 (2005)

[12] J.L.Lions, Quelques méthodes de résolution des problèmes aux limites non linéaires. Etudes mathématiques. Paris: Dunod; Paris: Gauthier-Villars. XX, 554 p. (1969).

[13] C.Mascia, A.Porretta & A.Terracina, Nonhomogeneous Dirichlet problems for degenerate parabolic-hyperbolic equations. *Arch. Ration. Mech. Anal.* **163**, No.2, 87-124 (2002).

[14] A.Michel & J.Vovelle, Entropy formulation for parabolic degenerate equations with general Dirichlet boundary conditions and application to the convergence of FV methods. *SIAM J. Numer. Anal.* **41**, No.6, 2262-2293 (2003).

[15] F.Otto, Initial-boundary value problem for a scalar conservation law. *C. R. Acad. Sci., Paris, Sr. I* **322**, No.8, 729-734 (1996).

[16] J.F.Rodrigues, Obstacle problems in mathematical physics. *North-Holland Mathematics Studies,* **134**, *Notas de Matemtica* **(114)**. Amsterdam: North-Holland. XV, 352 p. (1987).

[17] E.Rouvre & G.Gagneux, Formulation forte entropique de lois scalaires hyperboliques-paraboliques dégénérées. *Ann. Fac. Sci. Toulouse, VI. Sér., Math.* **10**, No.1, 163-183 (2001).

[18] G.Vallet, Dirichlet problem for a degenerated hyperbolic-parabolic equations. *Adv. Math. Sci. Appl.* **15**, 423-450 (2005)

In: Research on Evolution Equation Compendium. Volume 2 ISBN: 978-1-60456-752-6
Editor: Gaston M. N'Guerekata

APPLICATION OF THE PICARD OPERATORS TO SECOND ORDER ODE'S

Alexandru Bica[1]*, *Loredana Galea*[2]† *and Sorin Muresan*[1]‡
[1]Department of Mathematics and Informatics, University of Oradea
Str. Universităţii No.1, 410087, Oradea, Romania
[2]Faculty of Law and Economics, The Agora University of Oradea
Piaţa Tineretului no.8, 410526, Oradea, Romania

Abstract

Using the Perov's fixed point theorem and the fiber contractions theorem are obtained sufficient conditions for the smooth dependence by the end points of the solution of the two point boundary value problem for second order ODE's. Finally, an application in the metabolism control is presented.

2000 AMS Mathematics Subject Classification: 47H10.

Keywords and phrases:: two point boundary value problem, Picard operators, fiber contractions theorem, smooth dependence by the end points.

1. Introduction

Consider the following two point boundary value problem corresponding to the nonlinear second order ODE:

$$\begin{cases} x''(t) = f(t,x(t),x'(t)),\ t \in [a,b] \\ x(a) = \alpha,\ x(b) = \beta,\ \alpha,\beta \in \mathbb{R} \end{cases} \tag{1}$$

The existence and uniqueness of the solution of (1) is studied in [1] and [2]. In [2] is presented a new approach for the problem (1) using the Perov's fixed point theorem and considering the equivalence of the problem (1), in $C^2[a,b]$, with the following system of Fredholm integral equations:

$$\begin{cases} x(t) = \frac{t-a}{b-a} \cdot \beta + \frac{b-t}{b-a} \cdot \alpha - \int_a^b G(t,s) \cdot f(s,x(s),y(s))\,ds \\ y(t) = \frac{\beta-\alpha}{b-a} - \int_a^b \frac{\partial G}{\partial t}(t,s) \cdot f(s,x(s),y(s))\,ds \end{cases}, \tag{2}$$

---

*E-mail address: abica@uoradea.ro
†E-mail address: loredana.galea@univagora.ro
‡E-mail address: smuresan@uoradea.ro

where was considered $y(t) = x'(t)$. Here, $G(t,s)$ is the well known Green function. The Perov's fixed point theorem (see [2], [4], [6], [7], [9], [10], [11] and [13]) generates an efficient technique to approach existence, uniqueness and approximation of the solution of the system of equations (2) (and consequently for the problem (1)).

Theorem 1 ***(Perov,*** *[2], [5], [7]) Let (X,d) be a generalized metric space (with the metric taking values in $\mathbb{R}_+^n$) and $A : X \to X$ a mapping for which there exist a matrix $Q \in M_n(\mathbb{R}_+)$ such that:*

$$d(A(x), A(y)) \leqslant Q d(x,y), \ \forall x,y \in X \tag{3}$$

If all eigenvalues of Q lies in the open unit ball from the complex plane, then:

(i) the operator A has a unique fixed point $x^* \in X$

(ii) for any $x_0 \in X$, the sequence of successive approximations $(x_m)_{m\in\mathbb{N}} \subset X$ defined by $x_m = A(x_{m-1})$, $\forall m \in \mathbb{N}^*$, is convergent to x^*

(iii) the following inequality holds:

$$d(x_m, x^*) \leqslant Q^m (I_n - Q)^{-1} \cdot d(x_0, x_1), \ \forall m \in \mathbb{N}^*.$$

Here, I_n is the unity matrix in $M_n(\mathbb{R})$ and in the inequality (3) the order relation is defined in $\mathbb{R}_+^n$ component-wise, $\mathbb{R}_+^n = \left\{x = (x_1, ..., x_n) \in \mathbb{R}^n | x_i \geqslant 0, \forall i = \overline{1,n}\right\}$.

In the study of smooth dependence of the solution of the problem (1), by data and by parameters, are very useful the notions of Picard and weakly Picard operators (see [8], [11]), and the fiber contractions theorem (see [9], [10], [11], [6]). Below, we investigate the dependence of the solution of (1) by the end points a and b.

2. Preliminaries

Definition 1 *([8], [10], [11]) Let (X,d) be a metric space. An operator $A : X \to X$ is Picard operator if there exists $x^* \in X$ such that:*

(i) x^* is the unique fixed point of A .

(ii) the sequence $(A^n(x_0))_{n\in\mathbb{N}}$ converges to x^* , for all $x_0 \in X$, where $A^0 = Id(X)$ and $A^{n+1} = A \circ A^n$, $\forall n \in \mathbb{N}$.

Definition 2 *([8], [10], [11]) Let (X,d) be a metric space. An operator $A : X \to X$ is weakly Picard operator if the sequence $(A^n(x_0))_{n\in\mathbb{N}}$ converges for all $x_0 \in X$ and the limit (which may depend on x_0) is a fixed point of A.*

Applications of the technique of Picard and weakly Picard operators can be viewed in [4], [11], [13]. The above definitions can be extended in the same way for operators on generalized metric spaces (with the metric having vectorial values).

Using the Perov's fixed point theorem I.A. Rus have obtained the result:

Theorem 2 ***(of fiber generalized contractions, Rus*** *[9], [10]) Let (X,d) be a metric space (generalized or not) and (Y,ρ) be a complete generalized metric space $\left(\rho(x,y) \in \mathbb{R}_+^n\right)$. Let the operators $B : X \to X$, $C : X \times Y \to Y$ and the continuous operator $A : X \times Y \to X \times Y$. Suppose that:*

(i) $B : X \to X$ is weakly Picard operator
(ii) $A(x,y) = (B(x), C(x,y))$, for all $x \in X, y \in Y$
(iii) there exist a matrix $Q \in M_n(\mathbb{R}_+)$ with $Q^m \to 0$ as $m \to \infty$, such that:

$$\rho(C(x,y_1), C(x,y_2)) \leqslant Q \cdot \rho(y_1, y_2),$$

for all $x \in X$ and $y_1, y_2 \in Y$.

Then, the operator A is weakly Picard operator. Moreover, if B is Picard operator, then A is Picard operator.

Remark 1 *(see [2]) For a matrix $Q \in M_n(\mathbb{R}_+)$, the following properties are equivalent:*
(i) $Q^m \to 0$ as $m \to \infty$
(ii) all eigenvalues of Q lies in the open unit ball from the complex plane.

3. The Main Result

In the following, we will study the smooth dependence by the end points a and b of the solution of the problem (1). In this aim we will use the Picard operator's technique. Other way to approach data dependence of the solution of differential equations can be found in [12].

Consider $c, d \in \mathbb{R}$ such that $c \leqslant a < b \leqslant d$ and the two point boundary value problem,

$$\begin{cases} x''_{t^2}(t;a,b) = f(t, x(t;a,b), x'_t(t;a,b)) \\ x(a;a,b) = \alpha,\ x(b;a,b) = \beta \end{cases} \tag{4}$$

This problem is equivalent with the following integro-differential equation:

$$x(t;a,b) = \frac{t-a}{b-a} \cdot \beta + \frac{b-t}{b-a} \cdot \alpha - \int_a^b G(t,s) \cdot f\left(s; x(s;a,b), x'_t(s;a,b)\right) ds \tag{5}$$

for all $t, a, b \in [c,d]$ with $a \leqslant t \leqslant b$, where G is the well-known Green function.

We see that, denoting $y = x'_t$, the study of the equation (5) it reduces to the study of the following system of integral equations:

$$\begin{cases} x(t;a,b) = \frac{t-a}{b-a} \cdot \beta + \frac{b-t}{b-a} \cdot \alpha - \int_a^b G(t,s) \cdot f(s, x(s;a,b), y(s;a,b))\, ds \\ y(t;a,b) = \frac{\beta-\alpha}{b-a} - \int_a^b \frac{\partial G}{\partial t}(t,s) \cdot f(s, x(s;a,b), y(s;a,b))\, ds \end{cases} \tag{6}$$

where

$$G(t,s) = \begin{cases} \dfrac{(s-a)(b-t)}{b-a}, & \text{if } s \leqslant t \\ \dfrac{(t-a)(b-s)}{b-a}, & \text{if } s \geqslant t \end{cases},$$

$$\frac{\partial G}{\partial t}(t,s)=\begin{cases}-\dfrac{s-a}{b-a}, & \text{if } s\leqslant t\\ \dfrac{b-s}{b-a}, & \text{if } s\geqslant t\end{cases}.$$

Let $C\left([c,d]^3\right)=\left\{u:[c,d]^3\to\mathbb{R}\middle| u \text{ continuous}\right\}$ and $\|\cdot\|$ be the Chebychev's norm on $C\left([c,d]^3\right)$,

$$\|u\|=\max\left\{|u(x,y,z)|:(x,y,z)\in[c,d]^3\right\}.$$

Consider the space $X=C\left([c,d]^3\right)\times C\left([c,d]^3\right)$ with the generalized metric $d:X\times X\to\mathbb{R}^2$,

$$d((x_1,y_1),(x_2,y_2))=(\|x_1-x_2\|,\|y_1-y_2\|),\quad \forall(x_1,y_1),(x_2,y_2)\in X.$$

Define the operator $B:X\to X$, $B(x,y)=(B_1(x,y),B_2(x,y))$ with

$$B_1(x,y)(t;a,b)=\frac{t-a}{b-a}\cdot\beta+\frac{b-t}{b-a}\cdot\alpha-\int_a^b G(t,s)\cdot f(s,x(s;a,b),y(s;a,b))\,ds \tag{7}$$

$$B_2(x,y)(t;a,b)=\frac{\beta-\alpha}{b-a}-\int_a^b\frac{\partial G}{\partial t}(t,s)\cdot f(s,x(s;a,b),y(s;a,b))\,ds \tag{8}$$

in the condition $f\in C([c,d]\times\mathbb{R}^2)$. In the supplementary condition $f(s,\cdot,\cdot)\in C^1(\mathbb{R}^2)$, $\forall s\in[c,d]$ we can define the operator $C:X\times X\to X$,
$C((x,y),(u,v))=(C_1((x,y),(u,v)),C_2((x,y),(u,v)))$ with

$$C_1((x,y),(u,v))(t;a,b)=\frac{(b-t)(\alpha-\beta)}{(b-a)^2}+G(t,a;a,b)\cdot \tag{9}$$

$$\cdot f(a,x(a;a,b),y(a;a,b))-\int_a^b\Big(\frac{\partial G}{\partial a}(t,s;a,b)\cdot f(s,x(s;a,b),y(s;a,b))+G(t,s;a,b)\cdot$$

$$\cdot\left[\frac{\partial f}{\partial x}(s,x(s;a,b),y(s;a,b))u(s;a,b)+\frac{\partial f}{\partial y}(s,x(s;a,b),y(s;a,b))v(s;a,b)\right]\Big)ds$$

$$C_2((x,y),(u,v))(t;a,b)=\frac{(\beta-\alpha)}{(b-a)^2}+\frac{\partial G}{\partial t}(t,a;a,b)\cdot \tag{10}$$

$$\cdot f(a,x(a;a,b),y(a;a,b))-\int_a^b\Big(\frac{\partial^2 G}{\partial t\partial a}(t,s;a,b)\cdot f(s,x(s;a,b),y(s;a,b))+\frac{\partial G}{\partial t}(t,s;a,b)\cdot$$

$$\cdot\left[\frac{\partial f}{\partial x}(s,x(s;a,b),y(s;a,b))u(s;a,b)+\frac{\partial f}{\partial y}(s,x(s;a,b),y(s;a,b))v(s;a,b)\right]\Big)ds.$$

Finally, let $A:X\times X\to X\times X$ be the operator defined by $A((x,y),(u,v))=(B(x,y),C((x,y),(u,v)))$.

For $(x,y) \in X$ fixed, consider the system,

$$\begin{cases} u(t;a,b) := \text{second part of (9)} \\ v(t;a,b) := \text{second part of (10)} \end{cases}. \tag{11}$$

We will suppose that the following conditions holds:
(S1)(continuity): $f \in C\left([c,d] \times \mathbb{R}^2\right)$;
(S2) (Lipschitz): there exist $L_1 > 0$, $L_2 > 0$ such that:

$$|f(s,z_1,w_1) - f(s,z_2,w_2)| \leqslant L_1 |z_1 - z_2| + L_2 |w_1 - w_2|, \quad \forall s \in [c,d], \forall z_1, z_2, w_1, w_2 \in \mathbb{R}$$

(S3)(smoothness): $f(s,\cdot,\cdot) \in C^1\left(\mathbb{R}^2\right)$, $\forall s \in [c,d]$.

Theorem 3 *a) In the conditions (S1), (S2) if $L_1(d-c)^2 < 1$ and $L_2(d-c) < \frac{3}{4}$ then the system (6) of integral equations has in X an unique solution (x^*,y^*) such that $x^*,y^* \in C^1\left([c,d]^3\right)$, $x^*(\cdot;a,b) \in C^2([c,d]), \forall a,b \in [c,d]$ and $\frac{\partial x^*}{\partial t} = y^*$. Moreover, for all $(x_0,y_0) \in X$ the sequence $(x_m,y_m)_{m \in \mathbb{N}}$ defined by*

$$x_{m+1}(t;a,b) = \frac{t-a}{b-a} \cdot \beta + \frac{b-t}{b-a} \cdot \alpha - \int_a^b G(t,s) \cdot f(s, x_m(s;a,b), y_m(s;a,b))\, ds \tag{12}$$

$$y_{m+1}(t;a,b) = \frac{\beta - \alpha}{b-a} - \int_a^b \frac{\partial G}{\partial t}(t,s) \cdot f(s, x_m(s;a,b), y_m(s;a,b))\, ds \tag{13}$$

uniformly converges to (x^*,y^*) for all $t,a,b \in [c,d]$ and

$$d((x^*,y^*),(x_m,y_m)) \leqslant \frac{\left[\frac{1}{4}L_1(d-c)^2 + L_2(d-c)\right]^{m-1}}{1 - \frac{1}{4}L_1(d-c)^2 - L_2(d-c)}. \tag{14}$$

$$\cdot \begin{pmatrix} \frac{1}{4}L_1(d-c)^2 & \frac{1}{4}L_2(d-c)^2 \\ L_1(d-c) & L_2(d-c) \end{pmatrix} \cdot d((x_0,y_0),(x_1,y_1)), \forall m \in \mathbb{N}^*.$$

b) In the conditions (S1), (S2), (S3), if $L_1(d-c)^2 < 1$ and $L_2(d-c) < \frac{3}{4}$, then $x^*(t,\cdot,\cdot), y^*(t,\cdot,\cdot) \in C^1\left([c,d]^2\right)$ for all $t \in [c,d]$ and the pair $\left(\frac{\partial x^*}{\partial a}; \frac{\partial y^*}{\partial a}\right)$ is the unique solution of the system (11) for fixed $(x,y) = (x^*,y^*)$.

Proof 4 *a) From (S1) we infer that $B(X) \subset X$. Let $(x_1,y_1),(x_2,y_2) \in X$. We have:*

$$|B_1(x_1,y_1)(t;a,b) - B_1(x_2,y_2)(t;a,b)| \leqslant$$

$$\leqslant \int_a^b |G(t,s)| \cdot (L_1 |x_1(s;a,b) - x_2(s;a,b)| + L_2 |y_1(s;a,b) - y_2(s;a,b)|) ds \leqslant$$

$$\leqslant \frac{(d-c)^2}{4} (L_1 \|x_1 - x_2\| + L_2 \|y_1 - y_2\|)$$

$$|B_2(x_1,y_1)(t;a,b)-B_2(x_2,y_2)(t;a,b)|\leqslant$$

$$\leqslant\int_a^b\left|\frac{\partial G}{\partial t}(t,s)\right|\cdot|f(s,x_1(s;a,b),y_1(s;a,b))-f(s,x_2(s;a,b),y_2(s;a,b))|ds\leqslant$$

$$\leqslant(d-c)(L_1\|x_1-x_2\|+L_2\|y_1-y_2\|).$$

Consequently,

$$d(B(x_1,y_1),B(x_2,y_2))\leqslant Q\cdot d((x_1,y_1),(x_2,y_2)),$$

where

$$Q=\begin{pmatrix}\frac{(d-c)^2}{4}L_1 & \frac{(d-c)^2}{4}L_2\\ (d-c)L_1 & (d-c)L_2\end{pmatrix}.$$

The eigenvalues of Q are $\lambda_1=0$ and $\lambda_2=\frac{1}{4}(d-c)^2L_1+(d-c)L_2$. From hypothesis, follows $\lambda_2<1$ and therefore $Q^m\to 0$, as $m\to\infty$. Then B is Picard operator. From the Perov's fixed point theorem follows that the operator B have an unique fixed point $(x^*,y^*)\in X$ and the sequence of successive approximations given in (12) and (13) uniformly converges to (x^*,y^*) for all $t,a,b\in[c,d]$. Moreover, we have the inequality (14). On the other hand, $B(x^*,y^*)=(x^*,y^*)$ means,

$$x^*(t;a,b)=\frac{t-a}{b-a}\cdot\beta+\frac{b-t}{b-a}\cdot\alpha-\int_a^b G(t,s)\cdot(f(s,x^*(s;a,b),y^*(s;a,b)))ds \tag{15}$$

$$y^*(t;a,b)=\frac{\beta-\alpha}{b-a}-\int_a^b\frac{\partial G}{\partial t}(t,s)\cdot f(s,x^*(s;a,b),y^*(s;a,b))ds \tag{16}$$

and from condition (S1) we have that $x^*,y^*\in C^1\left([c,d]^3\right)$.

If we derive the equality (15) by t, from (16) follows $\frac{\partial x^*}{\partial t}=y^*$ and $x^*(\cdot,a,b)\in C^2([c,d])$.

Elementary calculus lead to $\frac{\partial^2x^*}{\partial t^2}(t;a,b)=f\left(t,x^*(t;a,b),\frac{\partial x^*}{\partial t}(t;a,b)\right)$ and $x^*(a;a,b)=\alpha$, $x^*(b;a,b)=\beta$.

Then x^* is the unique solution of (4).

b) From (S1) and (S3) we have $C(X\times X)\subset X$. For $(x,y)\in X$ arbitrary, we consider $(u_1,v_1),(u_2,v_2)\in X$ and have:

$$|C_1((x,y),(u_1,v_1))(t;a,b)-C_1((x,y),(u_2,v_2))(t;a,b)|\leqslant$$

$$\leqslant\int_a^b|G(t,s;a,b)|\cdot(\left|\frac{\partial f}{\partial x}(s,x(s;a,b),y(s;a,b))\right|\cdot|u_1(s;a,b)-u_2(s;a,b)|+$$

$$+\left|\frac{\partial f}{\partial y}(s,x(s;a,b),y(s;a,b))\right|\cdot|v_1(s;a,b)-v_2(s;a,b)|)ds\leqslant$$

$$\leqslant\frac{(d-c)^2}{4}(L_1\|u_1-u_2\|+L_2\|v_1-v_2\|)$$

because from (S2) we infer that:

$$\left|\frac{\partial f}{\partial x}(s,x(s;a,b),y(s;a,b))\right| \leqslant L_1$$

and

$$\left|\frac{\partial f}{\partial y}(s,x(s;a,b),y(s;a,b))\right| \leqslant L_2, \quad \forall s,a,b \in [c,d].$$

Analogous we obtain:

$$|C_2((x,y),(u_1,v_1))(t;a,b) - C_2((x,y),(u_2,v_2))(t;a,b)| \leqslant$$

$$\leqslant (d-c)(L_1\|u_1-u_2\| + L_2\|v_1-v_2\|).$$

Then,

$$d(C((x,y),(u_1,v_1)),C((x,y),(u_2,v_2))) \leqslant Q \cdot d((u_1,v_1),(u_2,v_2)).$$

We have $Q^m \to 0$ as $m \to \infty$ and therefore for fixed $(x,y) = (x^*,y^*)$, the operator $C((x^*,y^*),\cdot)$ have an unique fixed point (u^*,v^*). Then,

$$A((x^*,y^*),(u^*,v^*)) = ((x^*,y^*),(u^*,v^*))$$

and so, A is Picard operator.

If we choose $x_0 \in C^2\left([c,d]^3\right)$, $y_0 = \frac{\partial x_0}{\partial t}$, $u_0 = \frac{\partial x_0}{\partial a}$, $v_0 = \frac{\partial y_0}{\partial a}$ then the sequence given by $(A^n((x_0,y_0),(u_0,v_0)))_n$ uniformly converges on $X \times X$ to $((x^*,y^*),(u^*,v^*))$, that is x_n uniformly converges to x^*, $y_n = \frac{\partial x_n}{\partial t}$ uniformly converges to y^*, $u_n = \frac{\partial x_n}{\partial a}$ uniformly converges to u^*, $v_n = \frac{\partial y_n}{\partial a}$ uniformly converges to v^*.

Then $u^* = \frac{\partial x^*}{\partial a}$, $v^* = \frac{\partial y^*}{\partial a}$ and the pair $\left(\frac{\partial x^*}{\partial a}, \frac{\partial y^*}{\partial a}\right)$ is the unique fixed point of the system (11).

An analogous study can be realized for the dependence of the solution of the problem (4) by b.

Corollary 5 *In the conditions of the above theorem, the solution of the problem (4) and his derivative are smooth dependent by a and b.*

4. Application

We will apply the above result to study the dependence of the glycaemia profile by the moment of the "a jeun" blood-glucose measurement. In this sense, the system modeling blood-glucose homeostasis proposed and studied in [3],

$$\begin{cases} x' = -a\dfrac{xy}{x+G_0} - bx - my \\ y' = cx - dy \end{cases}, a,b,c,d,m,G_0 > 0 \tag{17}$$

is equivalent with the second order ODE:

$$x'' = -(b+d)x' - \frac{x'^2}{x+G_0} - \frac{x'(bx+aG_0)}{x+G_0} - \tag{18}$$

$$-\frac{[(a+m)c+bd]x^2 + G_0(ab+bd+mc)x}{x+G_0}, \quad t \in [t_0, T].$$

We investigate here the dependence of the solution of equations (18) and of his derivative by the initial moment t_0. So, we consider an interval $[\gamma, \delta]$ such that $\gamma \leqslant t_0 < T \leqslant \delta$.

For the equation (18) the conditions (S1), (S2), (S3) are fulfilled with:

$$L_1 = \frac{1}{25} + \frac{(b-a)G_0}{250} + 7(mc+bd) + 4ab + 3ac$$

and

$$L_2 = \frac{2}{5} + \frac{9}{5}b + d + \frac{aG_0}{50}$$

in the realistic hypothesis $x \in [-40, 40]$, $G(t) \in [50, 130]$ (in case of the presence of the blood-glucose homeostasis phenomenon, considering $G_0 \in (80, 100)$). Here $G(t)$ represent the blood-glucose level at the moment t.

We remember that in the system (17) x represent the deviation of the blood-glucose level from the homeostasis equilibrium G_0 and x' represent the speed of modification of the blood-glucose levels.

Applying the results from the above section we obtain:

Corollary 6 *If*

$$\left(\frac{2}{5} + \frac{9}{5}b + d + \frac{aG_0}{50}\right)(\delta - \gamma) < \frac{3}{4} \tag{19}$$

and

$$\left[\frac{1}{25} + \frac{(b-a)G_0}{250} + 7(mc+bd) + 4ab + 3ac\right](\delta - \gamma)^2 < 1 \tag{20}$$

then the deviation of the blood-glucose levels from the homeostasis equilibrium and the speed of modification of the blood-glucose levels are smooth dependent by the moment of the "a jeun" glycaemia measurement.

Proof 7 *Follows from Theorem 3.*

Remark 2 *In the conditions of the above Corollary, small deviations in the first moment of measurement t_0 lead to not significant difference in the measured blood-glucose levels. Consequently, the choice of the "a jeun" measurement moment is not so strictly for the patients for which the conditions (19) and (20) holds. Generally, for healthy persons the conditions (19) and (20) holds, since $\delta - \gamma < 1$ (for the glycaemia profile the unity of time is the day and for instance γ, δ can be $\gamma = 6AM$ and $\delta = 4AM$ in the next day) and the positive coefficients a, b, c, d, m have usually small values. Therefore, the physician have the possibility to choose the moment t_0 how he want between the hours 6:30 and 8 AM (according to the particularities of the patients and considering the diet paradigms).*

References

[1] P.B.Bailey, L. F. Shampine, P.E. Waltman, Nonlinear Two Point Boundary Value Problems , *Acad. Press,* New York, 1968.

[2] S.R. Bernfeld and V. Laksmikantham, Introduction to nonlinear boundary value problems , *Acad. Press*, New York, 1974.

[3] A.Bica, Lyapunov function for a bidimensional system-model in the glucose homeostasy and clinical interpretations, *Int.J.of Evolution Equations,* Vol.1, no.1, 69-79, 2005.

[4] G.Dezso, Continuous dependence and derivability with respect of the parameters of the fixed point, *Bull.Appl.and Computer Math.,Tech. Univ. Budapest*, **84**,pp.245-254,1998.

[5] G.Dezso, Fixed point theorems in generalized metric space, *PUMA Pure Math. Appl.* **11**, No.2, 183-186, 2000.

[6] V.Muresan, Some application of the fiber contraction theorem, *Studia Univ. Babes-Bolyai,* **45**(4), pp.87-96, 2000.

[7] A.I.Perov and A.V.Kibenko, On a general method to study the boundary value problems, *Iz.Akad. Nauk.*,**30**, pp249-264, 1966.

[8] I.A.Rus, Weakly Picard mappings, *Comment. Math. Univ*. Caroline, **34** (3), pp.769-773, 1993.

[9] I.A.Rus, A fiber generalized contractions theorem and applications, *Mathematica*, **41**(1), pp.85-90, 1999.

[10] I.A.Rus, Fiber generalized operators on generalized metric spaces and an application, *Scripta Sci. Math.*, **1**(2), pp.355-363, 1999.

[11] I.A.Rus, Picard operators and applications, , *Sci. Math.Japon.*, **58**(1), pp.191-219,2003.

[12] J.Sotomayor, Smooth dependence solutions of differential equation on initial data: a simple proof, *Bol. Soc. Brasil.,* **4** (1), pp.55-59, 1973.

[13] A. Tamasan, Differentiability with respect to lag for nonlinear pantograph equations, *PUMA*, **9**, pp.215-220, 1998.

In: Research on Evolution Equation Compendium. Volume 2 ISBN: 978-1-60456-752-6
Editor: Gaston M. N'Guerekata

Doubly Nonlinear Degenerate Parabolic Equations on Carnot Groups*

Junqiang Han [†] ***and Pengcheng Niu*** [‡]
Department of Applied Mathematics, Northwestern Polytechnical University,
Xi'an, Shaanxi 710072, People's Republic of China

Abstract

In this paper we use variational methods to study the nonexistence of positive solutions for the following doubly nonlinear degenerate parabolic equations on Carnot groups:

$$\begin{cases} \frac{\partial u}{\partial t} = \nabla_G \cdot \left(u^{m-1} |\nabla_G u|^{p-2} \nabla_G u \right) + V u^{m+p-2} & \text{in} \quad \Omega \times (0,T), \\ u(x,0) = u_0(x) \geq 0 & \text{in} \quad \Omega, \\ u(x,t) = 0 & \text{on} \quad \partial\Omega \times (0,T). \end{cases}$$

Here Ω is a bounded domain with smooth boundary in Carnot group G, ∇_G is the horizontal gradient on G, $T > 0$, $V \in L^1_{loc}(\Omega)$, u_0 is not identically zero, $m \in \mathrm{R}$, $1 < p < Q$ and $m + p - 2 > 0$.

Key Words: doubly nonlinear degenerate parabolic equations, nonexistence, Carnot group.

AMS Subject Classification: 35K55, 35K65

1. Introduction

We study the following doubly nonlinear degenerate parabolic equations:

$$\begin{cases} \frac{\partial u}{\partial t} = \nabla_G \cdot \left(u^{m-1} |\nabla_G u|^{p-2} \nabla_G u \right) + V u^{m+p-2} & \text{in} \quad \Omega \times (0,T), \\ u(x,0) = u_0(x) \geq 0 & \text{in} \quad \Omega, \\ u(x,t) = 0 & \text{on} \quad \partial\Omega \times (0,T). \end{cases} \tag{1.1}$$

Here Ω is a bounded domain with smooth boundary in Carnot group G, ∇_G is the horizontal gradient on G, $T > 0$, $V \in L^1_{loc}(\Omega)$, u_0 is not identically zero, $m \in \mathrm{R}$, $1 < p < Q$ and $m + p - 2 > 0$.

It is well known that Euclidean space R^n with its usual Abelian group structure is a trivial Carnot group and that the Heisenberg group H^n is the simplest non-commutative

*The project supported by National Natural Science Foundation of China, Grant No. 10371099.
[†]E-mail address: southhan@163.com
[‡]E-mail address: pengchengniu@yahoo.com.cn

Carnot group. The problem (1.1) on R^n and H^n has been studied by Kombe [11]. Goldstein and Kombe [6], [7], Kombe [12] investigated the problem (1.1) on R^n, H^n and G when $m = 1$, respectively.

The equation on R^n

$$\frac{\partial u}{\partial t} = div\left(u^{m-1}|\nabla u|^{p-2}\nabla u\right) \tag{1.2}$$

($m \in R$) is classified as "doubly nonlinear" (see [14]) or with "implicit nonlinearity"(see [10]). When $p = 2$, (1.2) becomes

$$\frac{\partial u}{\partial t} = \frac{1}{m}\Delta u^m. \tag{1.3}$$

This is the linear heat equation if $m = 1$, the porous medium equation if $m > 1$, the fast diffusion equation if $0 < m < 1$ and the very fast diffusion equation (super-diffusion) if $m \leq 0$, respectively. The case $m = 1$ in (1.2) corresponds to the p-Laplace heat equation

$$\frac{\partial u}{\partial t} = div\left(|\nabla u|^{p-2}\nabla u\right). \tag{1.4}$$

To perturb (1.2) by a suitable potential function V, we have

$$\frac{\partial u}{\partial t} = div\left(u^{m-1}|\nabla u|^{p-2}\nabla u\right) + Vu^{m+p-2}. \tag{1.5}$$

Indeed, it is a good combination of the porous medium equation, the fast diffusion equation, the very fast diffusion equation and the p-Laplace heat equation with a singular lower order term.

In this paper we are concerned with the nonexistence of positive solutions of the problem (1.1) on general Carnot groups G and our results recover the previous results in [6], [7], [11], [12]. In particular, we consider the singular potential $V(x) = \gamma\frac{|\nabla_G N|^p}{N^p}$ ($\gamma > 0$, N is a homogeneous norm in Carnot group) which appears in Hardy's inequality on polarizable Carnot groups (see [2]). It is well known that singularities of this type belong to a border line case where both the strong maximum principle and Gaussian bounds in [1], [13] fail. Such Valso lies outside the Kato class potentials which have been studied extensively. We also consider the highly singular, oscillating potential $V(x) = \gamma\frac{|\nabla_G N|^p}{N^p} + \beta\frac{|\nabla_G N|^p}{N^p}\sin\frac{1}{N^\alpha}$ ($\gamma > 0$, $\beta \in R\backslash\{0\}$ and $\alpha > 0$). We will show that nonexistence of positive solutions is intimately related to Hardy's inequality.

The plan of this paper is as follows. In Section 2. we introduce some notation and basic facts about Carnot groups. In Section 3. we study problem (1.1).

2. Notation and Preliminary Result

In this section we recall some notation and basic results about Carnot groups.

A Carnot group is a simply connected Lie group, G, with a Lie algebra g, which admits a stratification, $g = V_1 \oplus \cdots \oplus V_r$ satisfying $[V_1, V_j] = V_{j+1}$ for $j = 1, \ldots, r-1$ and $[V_1, V_r] = \{0\}$. The homogeneous dimension of G is $Q = \sum_{j=1}^{r} j(\dim V_j)$.

We assume that $\dim V_1 = m \geq 2$ and that $X = \{X_1, X_2, \ldots, X_m\}$is an orthonormal basis of V_1. The horizontal gradient on G is $\nabla_G = (X_1, \cdots, X_m)$.

The related doubly nonlinear degenerate operator on G is

$$\Delta_{G,m,p} = \nabla_G \cdot \left(u^{m-1} |\nabla_G u|^{p-2} \nabla_G u\right). \tag{2.1}$$

When $m = 1$, it becomes the p-sub-Laplacian on G:

$$\Delta_{G,p} u = \nabla_G \cdot \left(|\nabla_G u|^{p-2} \nabla_G u\right); \tag{2.2}$$

When $m = 1$ and $p = 2$, it becomes the sub-Laplacian on G:

$$\Delta_G = \sum_{j=1}^{m} X_j^2. \tag{2.3}$$

Folland [3] proved that in any Carnot group G, there exists a homogeneous norm N smooth on $G \backslash \{0\}$ such that $\Gamma = N^{2-Q}$ is a fundamental solution of $-\Delta_G$ at 0.

The following embedding result due to Folland and Stein [3], [4] holds on G:

Theorem 2.1. *Let $\Omega \subset G$ be an open set. For any $1 < p < Q$ there exists $S_p = S_p(G) > 0$ such that for $\phi \in C_0^\infty(\Omega)$,*

$$\left(\int_\Omega |\phi|^{p*} dx\right)^{\frac{1}{p*}} \leq S_p \left(\int_\Omega |\nabla_G \phi|^p dx\right)^{\frac{1}{p}}, \tag{2.4}$$

where $p* = \frac{pQ}{Q-p}$ is the Sobolev exponent relative to p.

It plays a crucial role in this paper.

We say that a Carnot group G is polarizable if the homogeneous norm $N = u^{\frac{1}{2-Q}}$ associated to Folland's solution u for the sub-Laplacian Δ_G is ∞-harmonic in $G \backslash \{0\}$.

Recently, D'Ambrosio [2] has proved the following L^p-Hardy's inequality on polarizable Carnot groups.

Theorem 2.2. *Let $1 < p < Q$ and $\Omega \subset G$ an open set. Assume that G is polarizable. Then we have for $\phi \in C_0^\infty(\Omega)$,*

$$\left(\frac{Q-p}{p}\right)^p \int_\Omega |\phi|^p \frac{|\nabla_G N|^p}{N^p} dx \leq \int_\Omega |\nabla_G \phi|^p \, dx. \tag{2.5}$$

It is an open problem whether (2.5) still holds for general Carnot groups. The Hardy inequality on polarizable Carnot groups says that the nonlinear operator $L_p u = \Delta_p u - \gamma \frac{|\nabla_G N|^p}{N^p} u^{p-1}$ is a positive operator if $\gamma \leq \left(\frac{Q-p}{p}\right)^p$. If $\gamma > \left(\frac{Q-p}{p}\right)^p$ then the operator L_p is unbounded from below. From this point of view it is natural to consider a sample potential such as $V(x) = \gamma \frac{|\nabla_G N|^p}{N^p}$.

For further discussion on Carnot groups, see [3], [5], [15, Chapter XIII, Section 5], [9], [8, Section 11.3] and the reference therein.

3. Nonexistence Results

Throughout this section, Ω is a bounded domain with smooth boundary in Carnot group G, $1<p<Q$ and $m+p-2>0$. We define the positive solutions in the following sense.

Definition 3.1. By a positive local solution continuous off of K, we mean

(i) K is a closed Lebesgue null subset of Ω;

(ii) $u:[0,T)\to L^1(\Omega)$ is continuous for some $T>0$;

(iii) $(x,t)\to u(x,t)\in C((\Omega\backslash K)\times(0,T))$;

(iv) $u(x,t)>0$ on $(\Omega\backslash K)\times(0,T)$;

(v) $\lim\limits_{t\to 0} u(\cdot,t)=u_0$ in the sense of distributions;

(vi) $\nabla_G u\in L^p_{loc}(\Omega)$ and u is a solution in the sense of distributions of the PDE.

Remark 3.2. If $0<a<b<T$ and K_0 is a compact subset of $\Omega\backslash K$, then $u(x,t)\geq\varepsilon_1>0$ for $(x,t)\in K_0\times[a,b]$ for some $\varepsilon_1>0$. We can weaken (iii), (iv) to be

(iii)' $u(x,t)$ is positive and locally bounded on $(\Omega\backslash K)\times(0,T)$;

(iv)' $\frac{1}{u(x,t)}$ is locally bounded on $(\Omega\backslash K)\times(0,T)$.

If a solution satisfies (i), (ii), (iii)', (iv)', (v) and (vi) then we call it a "general positive local solution off of K". This is more general than a positive local solution continuous off of K. If $K=\emptyset$, we simply call u "general positive local solution".

Lemma 3.3. *Let Ω be a bounded domain in G and $1<p<Q$, $0\leq M\in L^{\frac{Q}{p}}(\Omega)$, $\phi\in W_0^{1,p}(\Omega)$. Then for every $\varepsilon>0$ there exists a positive constant $C(\varepsilon)$ such that*

$$\int_\Omega M|\phi|^p\,dx\leq\varepsilon\int_\Omega|\nabla_G\phi|^p\,dx+C(\varepsilon)\int_\Omega|\phi|^p\,dx. \tag{3.1}$$

The proof is similar to them in [6] and [7].

Theorem 3.4. *Let $m\in\mathbb{R}$, $1<p<Q$, $3-\frac{p}{Q}\leq m+p<3$, $\lambda=\left(\frac{m+p-2}{p-1}\right)^{p-1}$, $V\in L^1_{loc}(\Omega\backslash K)$ where K is a closed Lebesgue null subset of Ω. If*

$$\sigma^p_{\inf}((1-\varepsilon_1)\lambda V):=\inf_{0\neq\phi\in C_0^\infty(\Omega\backslash K)}\frac{\int_\Omega|\nabla_G\phi|^p\,dx-\int_\Omega(1-\varepsilon_1)\lambda V|\phi|^p\,dx}{\int_\Omega|\phi|^p\,dx}=-\infty \tag{3.2}$$

for some $\varepsilon_1>0$, then the problem (1.1) has no general positive local solution off of K.

Proof. We argue by contradiction. Given any $T>0$, let $u:[0,T)\to L^1(\Omega)$ be a general positive local solution off of K to (1.1) in $(\Omega\backslash K)\times(0,T)$ with $u_0\geq 0$ but not identically zero. Multiply both sides of (1.1) by the test function $\frac{|\phi|^p}{u^{m+p-2}}$ where $\phi\in C_0^\infty(\Omega\backslash K)$, and integrate over Ω, to get

$$\begin{aligned}&\frac{1}{3-m-p}\frac{d}{dt}\int_\Omega u^{3-m-p}|\phi|^p\,dx-\int_\Omega\nabla_G\cdot\left(u^{m-1}|\nabla_G u|^{p-2}\nabla_G u\right)\frac{|\phi|^p}{u^{m+p-2}}dx\\ =&\int_\Omega V|\phi|^p\,dx.\end{aligned} \tag{3.3}$$

Integration by parts gives

$$\begin{aligned} L &= \int_\Omega \nabla_G \cdot \left(u^{m-1} |\nabla_G u|^{p-2} \nabla_G u\right) \frac{|\phi|^p}{u^{m+p-2}} dx \\ &= -\int_\Omega u^{m-1} |\nabla_G u|^{p-2} \nabla_G u \cdot \nabla_G \left(\frac{|\phi|^p}{u^{m+p-2}}\right) dx. \end{aligned} \tag{3.4}$$

A direct computation shows that

$$\begin{aligned} &\nabla_G \left(\frac{|\phi|^p}{u^{m+p-2}}\right) \\ = &\frac{1}{u^{2(m+p-2)}} \left(p |\phi|^{p-2} \phi \nabla_G \phi u^{m+p-2} - |\phi|^p (m+p-2) u^{m+p-3} \nabla_G u\right) \\ = &p \frac{|\phi|^{p-2} \phi}{u^{m+p-2}} \nabla_G \phi - (m+p-2) \frac{|\phi|^p}{u^{m+p-1}} \nabla_G u. \end{aligned}$$

Therefore

$$\begin{aligned} L &= -p \int_\Omega u^{1-p} |\nabla_G u|^{p-2} |\phi|^{p-2} \phi \nabla_G u \cdot \nabla_G \phi dx \\ &\quad + (m+p-2) \int_\Omega u^{-p} |\nabla_G u|^p |\phi|^p dx \\ &\geq (m+p-2) \int_\Omega u^{-p} |\nabla_G u|^p |\phi|^p dx \\ &\quad - p \int_\Omega u^{1-p} |\nabla_G u|^{p-1} |\phi|^{p-1} |\nabla_G \phi| dx. \end{aligned} \tag{3.5}$$

Here we can use the following elementary inequality: Let $p > 1$ and $w_1 \neq w_2$ be two positive different real numbers. Then

$$(p-1) w_2^p - p w_2^{p-1} w_1 > -w_1^p. \tag{3.6}$$

Choosing $w_1 = \lambda^{-\frac{1}{p}} |\nabla_G \phi|$ and $w_2 = \lambda^{\frac{1}{p(p-1)}} u^{-1} |\nabla_G u| |\phi|$, we have

$$\begin{aligned} &(m+p-2) \int_\Omega u^{-p} |\nabla_G u|^p |\phi|^p dx - p \int_\Omega u^{1-p} |\nabla_G u|^{p-1} |\phi|^{p-1} |\nabla_G \phi| dx \\ &> -\lambda^{-1} \int_\Omega |\nabla_G \phi|^p dx. \end{aligned} \tag{3.7}$$

Substituting (3.7) into (3.5) yields

$$L = \int_\Omega \nabla_G \cdot \left(u^{m-1} |\nabla_G u|^{p-2} \nabla_G u\right) \frac{|\phi|^p}{u^{m+p-2}} dx > -\lambda^{-1} \int_\Omega |\nabla_G \phi|^p dx. \tag{3.8}$$

Merging (3.8) into (3.3) and integrating from t_1 to t_2, with $0 < t_1 < t_2 < T$, we obtain

$$\begin{aligned} &\int_\Omega V |\phi|^p dx - \lambda^{-1} \int_\Omega |\nabla_G \phi|^p dx \\ < &\frac{1}{(3-m-p)(t_2 - t_1)} \int_\Omega \left[u(x,t_2)^{3-m-p} - u(x,t_1)^{3-m-p}\right] |\phi|^p dx. \end{aligned} \tag{3.9}$$

We know that $f(y) = y^{\frac{(3-m-p)Q}{p}}$ is concave when $3 - \frac{p}{Q} \leq m + p$. Applying Jensen's inequality for concave functions, we obtain

$$\begin{aligned}\int_\Omega u(x,t_i)^{\frac{(3-m-p)Q}{p}}\,dx &\leq |\Omega| \left(\frac{\int_\Omega u(x,t_i)\,dx}{|\Omega|}\right)^{\frac{(3-m-p)Q}{p}} \\ &= C(|\Omega|)\left(\int_\Omega u(x,t_i)\,dx\right)^{\frac{(3-m-p)Q}{p}} < \infty.\end{aligned}$$

Here we use the fact that Ω is bounded, whence $|\Omega|$ is finite. Therefore

$$u(x,t_i)^{3-m-p} \in L^{\frac{Q}{p}}(\Omega), \quad i = 1,2.$$

By Lemma 3.3, we have for every $\varepsilon > 0$, there exists a positive constant $C(\varepsilon)$ such that

$$\begin{aligned}&\frac{1}{(3-m-p)(t_2-t_1)}\int_\Omega \left[u(x,t_2)^{3-m-p} - u(x,t_1)^{3-m-p}\right]|\phi|^p\,dx \\ \leq\ & \varepsilon\int_\Omega |\nabla_G\phi|^p\,dx + C(\varepsilon)\int_\Omega |\phi|^p\,dx. \qquad (3.10)\end{aligned}$$

Replacing (3.9) by (3.10), we get

$$\int_\Omega V|\phi|^p\,dx - \lambda^{-1}\int_\Omega |\nabla_G\phi|^p\,dx \leq \varepsilon\int_\Omega |\nabla_G\phi|^p\,dx + C(\varepsilon)\int_\Omega |\phi|^p\,dx.$$

Therefore

$$\inf_{0\neq\phi\in C_0^\infty(\Omega\backslash K)} \frac{\int_\Omega |\nabla_G\phi|^p\,dx - \int_\Omega (1-\varepsilon_1)\lambda V|\phi|^p\,dx}{\int_\Omega |\phi|^p\,dx} \geq -\lambda(1-\varepsilon_1)C(\varepsilon_1) > -\infty. \qquad (3.11)$$

This contradicts our assumption (3.2). This completes the proof of Theorem 3.4. □

We now focus our attention on some singular potentials. First we treat the following positive singular potential:

$$V(x) = \gamma\frac{|\nabla_G N|^p}{N^p} \qquad (3.12)$$

where $\gamma > 0$. It appears in Hardy's inequality on polarizable Carnot groups (see [2]). Theorem 3.4 gives the following corollary.

Corollary 3.5. *Let $0 \in \Omega$, $m \in \mathrm{R}$, $1 < p < Q$, $\lambda = \left(\frac{m+p-2}{p-1}\right)^{p-1}$ and $V(x)$ be defined by (3.12). Then the problem (1.1) has no general positive local solution off of K if $3 - \frac{p}{Q} \leq m + p < 3$ and $\gamma > \lambda^{-1}\left(\frac{Q-p}{p}\right)^p$.*

We consider the following highly singular, oscillating potential for the next corollary:

$$V(x) = \gamma\frac{|\nabla_G N|^p}{N^p} + \beta\frac{|\nabla_G N|^p}{N^p}\sin\frac{1}{N^\alpha}, \qquad (3.13)$$

where $\gamma > 0$, $\beta \in \mathrm{R}\backslash\{0\}$ and $\alpha > 0$. The oscillating part has very large positive and negative oscillating parts. In particular, it oscillates widely, but important cancellations occur between the positive and the negative parts in the quadratic form. Therefore, nonexistence of positive solutions only depends on the size of γ.

Corollary 3.6. *Let* $0 \in \Omega$, $m \in \mathrm{R}$, $1 < p < Q$, $\lambda = \left(\frac{m+p-2}{p-1}\right)^{p-1}$ *and* $V(x)$ *be defined by (3.13). Then the problem (1.1) has no general positive local solution off of* K *if* $3 - \frac{p}{Q} \leq m + p < 3$ *and* $\gamma > \lambda^{-1}\left(\frac{Q-p}{p}\right)^{p}$.

To prove Corollary 3.5 and Corollary 3.6, we need to construct a sequence $\{\phi_n\}$ of test functions which satisfies

$$\frac{\int_\Omega |\nabla_G \phi_n|^p\,dx - \int_\Omega (1-\varepsilon_1)\lambda V\,|\phi_n|^p\,dx}{\int_\Omega |\phi_n|^p\,dx} \longrightarrow -\infty \text{ as } n \to \infty.$$

This can be done as in [6].

In the sequel, we study the problem (1.1) through another approach. We transform (1.1) to the following problem via change of variable $v = u^{\frac{m+p-2}{p-1}}$:

$$\begin{cases} \frac{\partial}{\partial t} v^{\frac{p-1}{m+p-2}} = \left(\frac{p-1}{m+p-2}\right)^{p-1} \Delta_{G,p} v + V v^{p-1} & \text{in } \Omega \times (0,T), \\ v(x,0) = v_0(x) \geq 0 & \text{in } \Omega, \\ v(x,t) = 0 & \text{on } \partial\Omega \times (0,T), \end{cases} \tag{3.14}$$

where $v_0(x) = u_0(x)^{\frac{m+p-2}{p-1}}$. The main point applying this method is that we obtain a different lower bound for $m+p$.

Theorem 3.7. *Let* $m \in \mathrm{R}$, $1 < p < Q$, $3 - \frac{p}{p+Q(p-1)} \leq m + p < 3$, $\lambda = \left(\frac{m+p-2}{p-1}\right)^{p-1}$, $V \in L^1_{loc}(\Omega \backslash K)$ *where* K *is a closed Lebesgue null subset of* Ω. *If*

$$\sigma^p_{\inf}((1-\varepsilon_1)\lambda V) := \inf_{0 \neq \phi \in C_0^\infty(\Omega\backslash K)} \frac{\int_\Omega |\nabla_G \phi|^p\,dx - \int_\Omega (1-\varepsilon_1)\lambda V\,|\phi|^p\,dx}{\int_\Omega |\phi|^p\,dx} = -\infty \tag{3.15}$$

for some $\varepsilon_1 > 0$, *then the problem (1.1) has no general positive local solution off of* K.

Proof. The proof is by contradiction. Given any $T > 0$, let $v : [0,T) \to L^1(\Omega)$ be a general positive local solution off of K to (3.14) in $(\Omega\backslash K) \times (0,T)$ with $v_0 \geq 0$ but not identically zero.

Multiply both sides of (3.14) by the test function $\frac{|\phi|^p}{v^{p-1}}$ where $\phi \in C_0^\infty(\Omega\backslash K)$, and integrate over Ω, to get

$$\frac{1}{3-m-p}\frac{d}{dt}\int_\Omega v^{\frac{(p-1)(3-m-p)}{m+p-2}}\,|\phi|^p\,dx = \lambda^{-1}\int_\Omega \Delta_{G,p} v\left(\frac{|\phi|^p}{v^{p-1}}\right)dx + \int_\Omega V\,|\phi|^p\,dx. \tag{3.16}$$

Applying integration by parts for $\int_\Omega \Delta_{G,p} v\left(\frac{|\phi|^p}{v^{p-1}}\right)dx$ and using (3.6), which is similar to the proof of Theorem 3.4, we have

$$\int_\Omega \Delta_{G,p} v\left(\frac{|\phi|^p}{v^{p-1}}\right)dx > -\int_\Omega |\nabla_G \phi|^p\,dx. \tag{3.17}$$

Substituting (3.17) into (3.16) and integrating from t_1 to t_2, where $0 < t_1 < t_2 < T$, we obtain

$$\frac{1}{(3-m-p)(t_2-t_1)}\int_\Omega \left[v(x,t_2)^{\frac{(p-1)(3-m-p)}{m+p-2}} - v(x,t_1)^{\frac{(p-1)(3-m-p)}{m+p-2}}\right]|\phi|^p\,dx$$
$$> \ -\lambda^{-1}\int_\Omega |\nabla_G\phi|^p\,dx + \int_\Omega V|\phi|^p\,dx. \tag{3.18}$$

We know that $f(y) = y^{\frac{(p-1)(3-m-p)Q}{(m+p-2)p}}$ is concave when $3 - \frac{p}{p+Q(p-1)} \le m+p$. Using Jensen's inequality for concave functions, we obtain

$$\int_\Omega v(x,t_i)^{\frac{(p-1)(3-m-p)Q}{(m+p-2)p}}\,dx \le |\Omega|\left(\frac{\int_\Omega v(x,t_i)\,dx}{|\Omega|}\right)^{\frac{(p-1)(3-m-p)Q}{(m+p-2)}}$$
$$= \ C(|\Omega|)\left(\int_\Omega v(x,t_i)\,dx\right)^{\frac{(p-1)(3-m-p)Q}{(m+p-2)p}} < \infty.$$

Here we use the fact that Ω is bounded, whence $|\Omega|$ is finite. Therefore

$$v(x,t_i)^{\frac{(p-1)(3-m-p)}{m+p-2}} \in L^{\frac{Q}{p}}(\Omega), \quad i = 1,2.$$

By Lemma 3.3, we have for every $\varepsilon > 0$, there exists a positive constant $C(\varepsilon)$ such that

$$\frac{1}{(3-m-p)(t_2-t_1)}\int_\Omega \left[v(x,t_2)^{\frac{(p-1)(3-m-p)}{m+p-2}} - v(x,t_1)^{\frac{(p-1)(3-m-p)}{m+p-2}}\right]|\phi|^p\,dx$$
$$\le \ \varepsilon\int_\Omega |\nabla_G\phi|^p\,dx + C(\varepsilon)\int_\Omega |\phi|^p\,dx. \tag{3.19}$$

Merging (3.19) into (3.18) gives

$$\int_\Omega V|\phi|^p\,dx - \lambda^{-1}\int_\Omega |\nabla_G\phi|^p\,dx < \varepsilon\int_\Omega |\nabla_G\phi|^p\,dx + C(\varepsilon)\int_\Omega |\phi|^p\,dx,$$

and so for ε_1 as in the hypothesis,

$$\inf_{0\neq\phi\in C_0^\infty(\Omega\setminus K)} \frac{\int_\Omega |\nabla_G\phi|^p\,dx - \int_\Omega (1-\varepsilon_1)\lambda V|\phi|^p\,dx}{\int_\Omega |\phi|^p\,dx} \ge -\lambda(1-\varepsilon_1)C(\varepsilon_1) > -\infty, \tag{3.20}$$

which contradicts the assumption (3.15). The proof of Theorem 3.7 is now complete. □

Corollary 3.8. *Let $0 \in \Omega$, $m \in \mathrm{R}$, $1 < p < Q$, $\lambda = \left(\frac{m+p-2}{p-1}\right)^{p-1}$ and $V(x)$ be defined by (3.12). Then the problem (1.1) has no general positive local solution off of K if $3 - \frac{p}{p+Q(p-1)} \le m+p < 3$ and $\gamma > \lambda^{-1}\left(\frac{Q-p}{p}\right)^p$.*

Corollary 3.9. *Let $0 \in \Omega$, $m \in \mathrm{R}$, $1 < p < Q$, $\lambda = \left(\frac{m+p-2}{p-1}\right)^{p-1}$ and $V(x)$ be defined by (3.13). Then the problem (1.1) has no general positive local solution off of K if $3 - \frac{p}{p+Q(p-1)} \le m+p < 3$ and $\gamma > \lambda^{-1}\left(\frac{Q-p}{p}\right)^p$.*

Remark 3.10. The methods we applied above gave us two different lower bound for $m+p$. The better lower bound for $m+p$ in each method depends on p.

(i) If $1 < p < \frac{2Q}{Q+1}$ then $3 - \frac{p}{Q} \leq m+p$ is the better lower bound;

(ii) If $p > \frac{2Q}{Q+1}$ then $3 - \frac{p}{p+Q(p-1)} \leq m+p$ is the better lower bound;

(iii) If $p = \frac{2Q}{Q+1}$ then both methods give the same lower bound for $m+p$.

References

[1] D. G. Aronson, Non-negative solutions of linear parabolic equations. *Ann. Scuola Norm. Sup. Pisa*, Vol. 22 (1968), 607–694.

[2] L. D'Ambrosio, Hardy type inequalities related to second order degenerate differential operators, to appear.

[3] G. B. Folland, Subelliptic estimates and function spaces on nilpotent Lie groups, *Ark. Math.*. Vol. **13** (1975), 161–207.

[4] G. B. Folland, E. M. Stein, Estimates for the $\overline{\partial}_b$ complex and analysis on the Heisenberg group, *Comm. Pure Appl. Math.*. Vol. 27 (1974), 429–522.

[5] G. B. Folland, E. M. Stein, *Hardy spaces on homogeneous groups, volume 28 of Mathematical Notes*, Princeton University Press, Princeton, New Jersey, 1982.

[6] J. A. Goldstein, I. Kombe, Nonlinear degenerate parabolic equations with singular lower order term, *Advances in Differential Equations* . Vol. 8 (2003), no. 10, 1153–1192.

[7] J. A. Goldstein, I. Kombe, Nonlinear degenerate parabolic equations on the Heisenberg group, *International Journal of Evolution Equations* . Vol. 1 (2005), 1–22.

[8] P. Hajlasz, P. Koskela, Sobolev met Poincaré, *Memoirs Amer. Math. Soc.*. Vol. 145 (2000), 688.

[9] J. Heinonen, Calculus on Carnot groups, In Fall School in Analysis (Jyväskylä, 1994), *Ber. Univ. Jyväskylä Math. Inst., Jyväskylä,* Vol. 68 (1995), 1–31.

[10] A. S. Kalashnikov, Some problems of the qualitative theory of second-order nonlinear degenerate parabolic equations, *Russian Math. Surveys*. Vol. 42 (1987), 169–222.

[11] I. Kombe, Doubly nonlinear parabolic equations with singular lower order term, *Nonlinear Analysis*. Vol. 56 (2004), 185–199.

[12] I. Kombe, The Hardy inequality and nonlinear parabolic equations on Carnot groups, to appear.

[13] P. Li, S. T. Yau, On the parabolic kernel of the Schrödinger operator, *Acta Math.* Vol. 156 (1986), 153–201.

[14] J. L. Lions, *Quelque methodes de resolution des problemes aux limites nonlineaire*, Springer, Berlin.

[15] E. M. Stein, *Harmonic analysis: real-variable methods, orthogonality, and oscillatory integrals*, Princeton University Press, Princeton, NJ, 1993. With the assistance of Timothy S. Murphy, Monographs in Harmonic Analysis, III.

In: Research on Evolution Equation Compendium. Volume 2 ISBN: 978-1-60456-752-6
Editor: Gaston M. N'Guerekata

SOME SCALAR CONSERVATION LAWS WITH DISCONTINUOUS FLUX

Julien Jimenez
Université de Pau et des Pays de l'Adour
Laboratoire de Mathématiques Appliquées - UMR 5142 CNRS
BP 1155 - 64013 - PAU Cedex - FRANCE

Abstract

We deal with the scalar conservation law in a one dimensional bounded domain Ω: $\partial_t u + \partial_x(k(x)g(u)) = 0$, associated with a bounded initial value u_0. The function k is supposed to be bounded, discontinuous at $\{x_0 = 0\}$, and with bounded variation. A weak entropy formulation for the Cauchy problem has been introduced by J.D Towers in [12]. In [11] the existence and the uniqueness is proved by N. Seguin and J. Vovelle through a regularization of the function k. We generalize the definition of J.D Towers and we adapt the method developed in [11] to establish an existence and uniqueness property in the case of the homogeneous Dirichlet boundary conditions.

1. Introduction

We are interested in the existence and uniqueness properties for a scalar conservation law made of an hyperbolic first-order quasilinear equation set in a one-dimensional bounded domain Ω, and for any positive finite real T, that can be formally described:

Find a bounded measurable function u on $Q =]0,T[\times\Omega$ such that

$$\left\{\begin{array}{rll} \dfrac{\partial u}{\partial t} + \dfrac{\partial}{\partial x}(k(x)g(u)) = & 0 & \text{in } Q =]0,T[\times\Omega, \\ u(0,x) = & u_0(x) & \text{on } \Omega, \\ u = & 0 & \text{on (a part of) }]0,T[\times\partial\Omega, \end{array}\right. \tag{1}$$

where k is a discontinuous function at a point x_0 of Ω.

Such an equation arises in the modelling of continuous sedimentation of solid particles in a liquid ([3]) or when one considers a two-phase flow in an heterogeneous porous medium without capillarity effects ([5], [4]).

By normalization, we suppose that $\Omega =]-1,1[$.

The initial condition u_0 belongs to $L^\infty(\Omega)$ and takes values in $[m,M]$ where m and M are two fixed reals, $m < M$.

The function k is discontinuous at $x_0 = 0$ and $k_{|[-1,0[}$ is an element of

$W^{1,+\infty}(]-1,0[)$ while $k_{|]0,1]}$ belongs to $W^{1,+\infty}(]0,1[)$.

Thus, thanks to a Cauchy criterion, we can define:

$$k_L = \lim_{x\to 0^-} k(x) \qquad \text{and} \qquad k_R = \lim_{x\to 0^+} k(x).$$

The flux function g is Lipschitzian on $\mathbb{R}$. We suppose also that:

$$g \text{ changes no more than once its monotony} \tag{2}$$

Eventually, we suppose that equation (1) is non degenerate in the sense of Panov (see [9], [10]), that is to say:

$$g \text{ is not linear on nondegenerate intervals} \tag{3}$$

and

$$\mathcal{L}\{x \in \Omega,\ k(x) = 0\} = 0. \tag{4}$$

where $\mathcal{L}$ denotes the Lebesgue measure.

The mathematical formulation for (1) is given in Section 2. through an entropy inequality on the whole Q, using the classical Kruzkov entropy pairs (see [7]) and involving a term that takes into account the jump of k along $\{x_0 = 0\}$. As soon as we are able to transcript in Section 3. the transmission conditions along the interface included in Definition 1, we are able to state, in Section 4., the uniqueness. To do so strong traces for u along the interface $\{x_0 = 0\}$ will be needed. Finally Section 5. is devoted to the existence property for (1) through a suitable regularization of the function k.

2. Definition of an Entropy Solution

We propose a definition extending that of J.D. Towers ([12]) - also used by N. Seguin and J. Vovelle ([11]) or F. Bachmann ([1]) - to the case where k depends on the space variable and for the homogeneous Dirichlet problem in a bounded interval of $\mathbb{R}$. So we say that:

Definition 1. *A function u of $L^\infty(Q)$ is an entropy solution to problem 1 if:*
(i) $\forall\kappa \in \mathbb{R}$, $\forall\varphi \in C_c^\infty([0,T[\times\Omega)$, $\varphi \geq 0$,

$$\left\{\begin{aligned} &\int_Q (|u(t,x)-\kappa|\varphi_t(t,x) + k(x)\Phi(u,\kappa)\varphi_x(t,x))dxdt \\ &\qquad\qquad - \int_Q k'(x)sgn(u-\kappa)g(\kappa)\varphi dxdt \\ &+ \int_\Omega |u_0-\kappa|\varphi(0,x)dx + |(k_L-k_R)g(\kappa)|\int_0^T \varphi(t,0)dt \quad \geq 0\,, \end{aligned}\right. \tag{5}$$

where

$$\Phi(u,\kappa) = sgn(u-\kappa)(g(u)-g(\kappa)),$$

(ii) for a.e. t in $]0,T[$, for any real κ,

$$k(1)(sgn(u_1^\tau(t)-\kappa)+sgn(\kappa))(g(u_1^\tau(t))-g(\kappa)) \;\geq\; 0, \tag{6}$$

$$k(-1)(sgn(u_{-1}^\tau(t)-\kappa)+sgn(\kappa))(g(u_{-1}^\tau(t))-g(\kappa)) \;\leq\; 0. \tag{7}$$

In this definition u_1^τ and u_{-1}^τ denote the traces of u respectively in $(+1)^-$ and $(-1)^+$ in the sense of A. Vasseur [13] (see also Y. Panov [10]). Indeed it follows from [13],

Lemma 1. *Let u be an entropy solution to (1). If for each* $(\alpha,\beta)\neq(0,0)$, *for a.e.* $x\in[-1,1]$, $\mathcal{L}(\{\lambda \mid \alpha+\beta.k(x)g'(\lambda)=0\})=0$, *there exists two functions* $u_{\pm 1}^\tau$ *in* $L^\infty(]0,T[)$ *such as, for every compact set K of* $]0,T[$,

$$\operatorname*{ess\,lim}_{x\to\pm 1}\int_K |u(t,x)-u_{\pm 1}^\tau(t)|dt = 0. \tag{8}$$

In [10], E. Panov proved the existence of these strong traces with a continuous flux function, when the boundary is not a characteristic hypersurface. The latter condition is satisfied here under (3) and (4), when we consider the problem (1) separately on $]-1,0[$ and on $]0,1[$.

Remark 1. *Of course, the statement of Lemma 1 also ensures the existence of strong traces for u,* γu^+ *and* γu^- *in* $L^\infty(]0,T[)$ *along* $\{x_0=0\}$ *respectively at right and at left.*

3. Conditions at the Interface $\{x_0=0\}$

Let us establish that the previous definition ensures the uniqueness. The proof is based on that proposed in [11] and relies essentially on the transmission conditions along $\{x_0=0\}$ underlying to entropy inequality (5). Indeed the existence of strong traces for u permits us to state first:

Lemma 2. *Let u in* $L^\infty(Q)$ *be an entropy solution to (1). So,*

for a.e. t in $]0,T[$, *for all real* κ,

$$k_L\Phi(\gamma u^-(t),\kappa)-k_R\Phi(\gamma u^+(t),\kappa)+|(k_L-k_R)g(\kappa)| \geq 0\ . \tag{9}$$

Proof. Let φ be a nonnegative element of $C_c^\infty(Q)$. We refer to the cut-off function on $\mathbb{R}$, ω_ε, for $\varepsilon>0$, introduced in [11]:

$$\omega_\varepsilon(x)=\begin{cases} 0 & \text{if} \quad 2\varepsilon<|x| \quad , \\ \dfrac{-|x|+2\varepsilon}{\varepsilon} & \text{if} \quad \varepsilon\leq|x|\leq 2\varepsilon \quad , \\ 1 & \text{if} \quad |x|<\varepsilon \quad . \end{cases}$$

such that $\omega_\varepsilon(x)\to 0$ if $x\neq 0$, and $\omega_\varepsilon(0)=1$ for all ε.

Thanks to a density argument we may choose $\varphi\omega_\varepsilon$ as test-function in (5). We pass to the limit when ε goes to 0^+ by using the Lebesgue dominated convergence Theorem providing that all the terms tend to 0 except $|k_L-k_R|g(\kappa)\int_0^T\varphi(t,0)dt$ (which does not depend on ε) and:

$$I_\varepsilon=\int_Q k(x)\Phi(u,\kappa)\varphi\,\omega'_\varepsilon\,dxdt\ .$$

By definition of ω_ε,

$$I_\varepsilon = \int_0^T \frac{1}{\varepsilon}\int_{-2\varepsilon}^{-\varepsilon} k(x)\Phi(u,\kappa)\varphi\,dxdt + \int_0^T -\frac{1}{\varepsilon}\int_{\varepsilon}^{2\varepsilon} k(x)\Phi(u,\kappa)\varphi\,dxdt,$$

and, by setting $L_\varepsilon = \int_0^T \frac{1}{\varepsilon}\int_{-2\varepsilon}^{-\varepsilon} |k(x)(\Phi(u,\kappa)\varphi(t,x) - k_L\Phi(\gamma u^-,\kappa)\varphi(t,0))|dtdx$,

we prove that $\lim_{\varepsilon\to 0^+} L_\varepsilon = 0$ because $\Phi(.,\kappa)$ is Lipschitzian on $[0,1]$, and due to the definition of k_L and γu^-. As a consequence, we obtain (9). □

As in [11], a Rankine-Hugoniot condition may be deduced from (9). To do so we need an additional hypothesis on the function g. So we suppose that:

$$\left\{\begin{array}{ll} \exists\ \kappa_1 \in \mathbb{R}, \kappa_1 \geq \operatorname{ess\,sup} u, & g(\kappa_1)(k_L - k_R) \leq 0 \\ \exists\ \kappa_2 \in \mathbb{R}, \kappa_2 \leq \operatorname{ess\,inf} u, & g(\kappa_2)(k_L - k_R) \geq 0 \end{array}\right. \tag{10}$$

Lemma 3. *Under (10), for a.e. t in $]0,T[$, the following Rankine-Hugoniot condition holds:*

$$k_L g(\gamma u^-(t)) = k_R g(\gamma u^+(t)) \,. \tag{11}$$

Proof. We choose $\kappa = \kappa_1$ in (9) to obtain:

$$k_R g(\gamma u^+) - k_L g(\gamma u^-) + g(\kappa_1)(k_L - k_R) + |g(\kappa_1)(k_L - k_R)| \geq 0$$

From (10), we deduce that $k_R g(\gamma u^+) \geq k_L g(\gamma u^-)$.

By choosing $\kappa = \kappa_2$ in (9), and using (10), we obtain the reverse inequality. □

4. The Uniqueness Theorem

First we recall that

Lemma 4. *If a bounded mapping u satisfies (5), then:*

$$\operatorname*{esslim}_{t\to 0^+} \int_\Omega |u(t,x) - u_0(x)|dx = 0. \tag{12}$$

We are now able to state an uniqueness property for (1) through a T-Lipschitzian dependence in $L^1(Q)$ of a weak entropy solution with respect to corresponding initial data.

Theorem 1. *Let u and v be two entropy solutions to (1) for initial conditions (u_0, v_0) in $(L^\infty(]-1,1[)^2$. Then, under (10):*

$$\int_0^T \int_{-1}^1 |u(t,x) - v(t,x)|dxdt \leq T\int_{-1}^1 |u_0(x) - v_0(x)|dx. \tag{13}$$

Proof. We use the method of doubling variables due to S. N. Kruzkov (see [7]) by reasoning in two steps: we consider first some test-functions vanishing on a vicinity of $\{x_0 = 0\}$. That provides a Kruzkov-type inequality between two entropy solutions from which one the former vanishing hypothesis is released by using (11).

Lemma 5. *Let u and v be two entropy solutions in $L^\infty(Q)$ to (1) associated with initial conditions u_0 and v_0 in $L^\infty(]-1,1[)$. For any nonnegative function φ in $C_c^\infty([0,T[\times\Omega)$, vanishing in a neighborhood of $\{x_0 = 0\}$,*

$$\begin{aligned}&\int_Q (|u(t,x)-v(t,x)|\varphi_t(t,x)+k(x)\Phi(u(t,x),v(t,x))\varphi_x(t,x))dxdt\\&+\int_\Omega |u_0(x)-v_0(x)|\varphi(0,x)dx \geq 0.\end{aligned} \tag{14}$$

Proof. Let $(\rho_j)_{j\in\mathbb{N}^*}$ be a classical sequence of mollifiers in $\mathbb{R}$, such that $\rho_j(x)\rho_j(-x)$, φ an element of $C_c^\infty([0,T[\times\Omega)$ satisfying the hypotheses of Lemma 5. For $j\in\mathbb{N}^*$ and $(t,x,s,y)\in Q\times Q$, we set:

$$\psi_j(t,x,s,y) = \varphi(\tfrac{t+s}{2},\tfrac{x+y}{2})\rho_j(t-s)\rho_j(x-y).$$

To simplify, we denote $w=\frac{t+s}{2}$, $z=\frac{x+y}{2}$, $u=u(t,x)$, $v=v(t,x)$, $\tilde{v}=v(s,y)$, $q=(t,x)$, $\tilde{q}=(s,y)$. By choosing $\kappa=\tilde{v}$ in (5) for u (respectively $\kappa=u$ in (5) for $\tilde{v}$) against the test-function ψ_j and integrating over Q with respect to $\tilde{q}$ (respectively q), it comes:

$$\begin{aligned}&\int_{Q\times Q}|u-\tilde{v}|\varphi_t(w,z)\rho_j(t-s)\rho_j(x-y)dqd\tilde{q}\\&-\int_{Q\times Q} sgn(u-\tilde{v})(k'(x)g(\tilde{v})-k'(y)g(u))\psi_j dqd\tilde{q}\\&+2\int_{Q\times\Omega}|u_0(x)-v_0(y)|\varphi(\tfrac{t}{2},z)\rho_j(x-y)\rho_j(t)dqdy\\&+\int_{Q\times\Omega}(|u-u_0|+|\tilde{v}-\tilde{v}_0|)\varphi(\tfrac{t}{2},z)\rho_j(x-y)\rho_j(t)dqdy\\&+\int_{Q\times Q}\Phi(u,\tilde{v})k(x)(\partial_x\varphi)(w,z)\rho_j(t-s)\rho_j(x-y)dqd\tilde{q}\\&+\int_{Q\times Q}\Phi(u,\tilde{v})(k(y)-k(x))\partial_y\psi_j(q,\tilde{q})\,dqd\tilde{q} \geq 0.\end{aligned} \tag{15}$$

We will just focus on the second and the sixth line. Indeed there is no difficulty to pass to the limit when j goes to $+\infty$ in the other lines by referring to the notion of Lebesgue points for an integrable function on Q (and by using (12) for the forth line). Let's study first the sixth line, denoted I_j. Coming back to the definition of ψ_j yields

$$I_j = I_{1,j}+I_{2,j},$$

where:

$$I_{1,j} = \int_{Q\times Q} \Phi(u,\tilde{v})(k(y)-k(x))\partial_y(\varphi(w,z))\rho_j(t-s)\rho_j(x-y)dqd\tilde{q},$$

$$I_{2,j} = \int_{Q\times Q} \Phi(u,\tilde{v})(k(y)-k(x))\varphi(w,z)\rho_j(t-s)\partial_y(\rho_j(x-y))dqd\tilde{q}.$$

By using the notion of Lebesgue points, we state that

$$\lim_{j\to+\infty} I_{1,j} = 0.$$

Next we write $I_{2,j} = I_a + I_b$ with:

$$I_a = \int_{Q\times Q} \{\Phi(u,\tilde{v}) - \Phi(u,v)\}(k(y)-k(x))\varphi(w,z)\rho_j(t-s)\partial_y(\rho_j(x-y))dqd\tilde{q}$$

and,

$$I_b = \int_{Q\times Q} \Phi(u,v)(k(y)-k(x))\varphi(w,z)\rho_j(t-s)\partial_y(\rho_j(x-y))dqd\tilde{q}.$$

Let us first consider I_b. We denote

$$T(q,\tilde{q}) = \Phi(u,v)(k(y)-k(x))\varphi(w,z)\rho_j(t-s)\partial_y(\rho_j(x-y)),$$

$$Q_- =]0,T[\times]-1,0[\text{ and } Q_+ =]0,T[\times]0,1[,$$

$$I_{b,1} = \int_{Q_-\times Q_-} T(q,\tilde{q})dqd\tilde{q}\ ,\ I_{b,2} = \int_{Q_-\times Q_+} T(q,\tilde{q})dqd\tilde{q}$$

$$I_{b,3} = \int_{Q_+\times Q_-} T(q,\tilde{q})dqdq \text{ and } I_{b,4} = \int_{Q_+\times Q_+} T(q,\tilde{q})dqd\tilde{q}.$$

Then, $I_b = I_{b,1} + I_{b,2} + I_{b,3} + I_{b,4}$, so we just need to study $I_{b,1}$ and $I_{b,2}$, the arguments for $I_{b,3}$ and $I_{b,4}$ being similar. We integrate by parts $I_{b,1}$ with respect to y to obtain:

$$\begin{aligned} I_{b,1} = \; & -\int_{Q_-\times Q_-} \Phi(u,v)k'(y)\varphi(w,z)\rho_j(t-s)\rho_j(x-y)dqd\tilde{q} \\ & -\tfrac{1}{2}\int_{Q_-\times Q_-} \Phi(u,v)(k(y)-k(x))\varphi_y(w,z)\rho_j(t-s)\rho_j(x-y)dqd\tilde{q} \\ & +\int_{Q_-}\int_0^T \Phi(u,v)(k_L-k(x))\varphi(w,\frac{x}{2})\rho_j(t-s)\rho_j(x)dqds \end{aligned}$$

When j goes to $+\infty$, the two last terms tend to 0, owing to the continuity of k on $]-1,0[$ and to the definition of k_L. Moreover, since $k_{|[-1,0[}$ belongs to $W^{1,+\infty}([-1,0[)$ and φ is continuous, the first term tends to:

$$-\int_{Q_-} \Phi(u(t,x),v(t,x))k'(x)\varphi(t,x)dq.$$

Similarly, $\lim_{j\to+\infty} I_{b,4} = -\int_{Q_+} \Phi(u(t,x),v(t,x))k'(x)\varphi(t,x)dq$.
By definition of ρ_j, $I_{b,2}$ is equal to:

$$\int_0^T \int_{-\frac{1}{j}}^0 \int_0^T \int_0^{\frac{1}{j}} \Phi(u,v)(k(y)-k(x))\varphi(w,z)\rho_j(t-s)\partial_y(\rho_j(x-y))dqd\tilde{q}$$

As φ vanishes on a neighborhood of $\{x_0 = 0\}$, from a certain j_0, $I_{b,2}$ vanishes and it is the same for $I_{b,3}$. Eventually:

$$\lim_{j\to+\infty} I_b = -\int_Q \Phi(u,v)k'(x)\varphi(t,x)dq$$

We study now I_a. By using the same decomposition as for I_b, it appears four integrals whose two vanish (because φ vanishes on a vicinity of $\{x_0 = 0\}$) and it only leads to consider the term, denoting by $I_{a,1}$:

$$\int_{Q_-\times Q_-} \{\Phi(u,\tilde{v}) - \Phi(u,v)\}(k(y)-k(x))\varphi(w,z)\rho_j(t-s)\partial_y(\rho_j(x-y))dqd\tilde{q}$$

By using the Lipschitz condition for ϕ and k, we highlight a nonnegative constant C_1 independent from j, such that:

$$|I_{a,1}| \leq C_1 \int_{Q_-\times Q_-} |v(s,y)-v(t,x)||x-y|\rho_j(t-s)|\partial_y(\rho_j(x-y))|dqd\tilde{q}$$

This way, due to the definition of ρ_j, there exists a nonnegative constant C_2 such that:

$$|I_{a,1}| \leq C_2 j^2 \int_{\{|t-s|\leq\frac{1}{j},|x-y|\leq\frac{1}{j}\}} |v(t,x)-v(s,y)|dqd\tilde{q},$$

so that, $\lim_{j\to+\infty} I_{a,1} = 0$, and as a consequence

$$\lim_{j\to+\infty} I_a = 0.$$

To sum up:

$$\lim_{j\to+\infty} I_j = \lim_{j\to+\infty} (I_{1,j} + I_a + I_b) = -\int_Q \Phi(u(t,x),v(t,x))k'(x)\varphi(t,x)dq.$$

We study now the j-limit of the second line in (15) that is:

$$L_j = -\int_{Q\times Q} sgn(u-\tilde{v})(k'(x)g(\tilde{v}) - k'(y)g(u))\psi_j dqd\tilde{q}.$$

We write

$$L_j = L_{1,j} - L_{2,j}$$

with

$$L_{1,j} = \int_{Q\times Q} k'(x)\Phi(u,\tilde{v})\psi_j dqd\tilde{q} \text{ and}$$

and

$$L_{2,j} = \int_{Q\times Q} g(u)sgn(u-\tilde{v})(k'(x)-k'(y))\psi_j dq d\tilde{q}.$$

On the one hand, it is clear that:

$$\lim_{j\to+\infty} L_{1,j} = \int_Q k'(x)\Phi(u(t,x),v(t,x))\varphi(t,x)dq.$$

On the other hand, as for the study of I_a and I_b we share $L_{2,j}$ into four terms whose two vanishes (φ vanishing on a neighborhood of $\{x_0=0\}$) so that we only consider:

$$L_{2,a} = \int_{Q_-\times Q_-} g(u(t,x))sgn(u(t,x)-v(s,y))(k'(x)-k'(y))\psi_j dq d\tilde{q}$$

and,

$$L_{2,b} = \int_{Q_+\times Q_+} g(u(t,x))sgn(u(t,x)-v(s,y))(k'(x)-k'(y))\psi_j dq d\tilde{q}.$$

We observe that:

$$|L_{2,a}| \le C\|g\|_\infty\|\varphi\|_\infty \int_{\Omega^-\times\Omega^-} |k'(x)-k'(y)|\rho_j(x-y)dxdy,$$

where $\Omega^- =]-1,0[$.

So that, since k' belongs to $L^\infty([-1,0[)$, $\lim_{j\to+\infty} L_{2,a} = 0$ and it is the same for $L_{2,b}$.

To summarize, $\lim_{j\to+\infty} L_j = \int_Q k'(x)\Phi(u(t,x),v(t,x))\varphi(t,x)dq$, and (14) follows that completes the proof of Lemma 5. □

Now we state that:

Lemma 6. *Under (10), the Kruzkov inequality (14) still holds for* φ *in* $C_c^\infty([0,T[\times\Omega)$, $\varphi\ge 0$.

Proof. Thanks to a density argument we can choose in (14) the test function $\varphi(1-\omega_\varepsilon)$ where ω_ε is defined in the proof of Lemma 2. By taking the ε-limit, it comes:

$$\int_Q (|u-v|\varphi_t + k(x)\Phi(u,v)\varphi_x)dxdt + \int_\Omega |u_0-v_0|\varphi(0,x)dx \ge J,$$

with:

$$J = \int_0^T (k_L\Phi(\gamma u^-,\gamma v^-) - k_R\Phi(\gamma u^+,\gamma v^+))\varphi(t,0)dt.$$

Inequality (9) shows that J is nonnegative. Indeed let us study, for *a.e.* t of $]0,T[$, the sign of:

$$I = k_L\Phi(\gamma u^-,\gamma v^-) - k_R\Phi(\gamma u^+,\gamma v^+)\ .$$

We just focus on the case when $\gamma u^+ - \gamma v^+$ and $\gamma u^- - \gamma v^-$ have an opposite sign. Otherwise due to (11), that is satisfied because of (10), $I = 0$. When $sgn(\gamma u^+ - \gamma v^+) = -sgn(\gamma u^- - \gamma v^-) \neq 0$, by using (11), we have:

$$I = 2k_L \Phi(\gamma u^-, \gamma v^-) = -2k_R \Phi(\gamma u^+, \gamma v^+)$$

We suppose that $k_L - k_R > 0$, $\gamma v^+ < \gamma u^+$ and $\gamma u^- < \gamma v^-$, the study of the other cases being similar. So,

$$I = -2k_L(g(\gamma u^-) - g(\gamma v^-)) = -2k_R(g(\gamma u^+) - g(\gamma v^+))$$

Here we must consider some different situations.

1. $\gamma u^- < \gamma v^- < \gamma v^+ < \gamma u^+$.
   Then (9) can be written, for any κ of $[\gamma u^-, \gamma u^+]$,

$$-k_L(g(\gamma u^-) - g(\kappa)) - k_R(g(\gamma u^+) - g(\kappa)) + (k_L - k_R)|g(\kappa)| \geq 0. \qquad (16)$$

   . if $g(\gamma v^-) \geq 0$, by choosing $\kappa = \gamma v^-$ in (16), we have:

$$-2k_L(g(\gamma u^-) - g(\gamma v^-)) \geq 0, \text{ so } I \geq 0.$$

   . if $g(\gamma v^+) \leq 0$, by choosing $\kappa = \gamma v^+$ in (16), we obtain:

$$-2k_R(g(\gamma u^+) - g(\gamma v^+)) \geq 0, \text{ so } I \geq 0.$$

   . if $g(\gamma v^-) < 0$ and $g(\gamma v^+) > 0$, we deduce from (11), as $k_L - k_R > 0$, that $k_R < 0$ and $k_L > 0$. If we suppose that $I = -2k_L(g(\gamma u^-) - g(\gamma v^-)) < 0$, then $g(\gamma u^-) > g(\gamma v^-)$. But we also have $I = -2k_R(g(\gamma u^+) - g(\gamma v^+))$, that implies that $g(\gamma u^+) < g(\gamma v^+)$. Consequently g changes at least twice its monotony that contradicts the assumption (2).

2. $\gamma u^- < \gamma v^+ < \gamma v^- < \gamma u^+$.

   As in the previous case, if $g(\gamma v^-) \geq 0$ or $g(\gamma v^+) \leq 0$, we have $I \geq 0$. If $g(\gamma v^-) < 0$ and $g(\gamma v^+) > 0$, there exists α in $]\gamma v^+, \gamma v^-[$ such that $g(\alpha) = 0$. Choosing $\kappa = \alpha$ in (9) written successively for u and v yields to $-2k_R g(\gamma u^+) \geq 0$ and $2k_R g(\gamma v^+) \geq 0$. Then $I \geq 0$.

3. $\gamma u^- < \gamma v^+ < \gamma u^+ < \gamma v^-$.
   . if $g(\gamma v^+) \geq 0$ or $g(\gamma u^+) \leq 0$, from (9), $I \geq 0$.
   . if $g(\gamma v^+) < 0$ and $g(\gamma u^+) > 0$, there exists β in $]\gamma v^+, \gamma u^+[$, such that $g(\beta) = 0$. By choosing $\kappa = \beta$ in (9), written for u and v, we show that $I \geq 0$.

All the others cases may be reduced to one of the previous situations. □

Now, in order to prove (13), thanks to Lemma 6 we choose in (14) the test-function φ such that, for any (t,x) in $[0,T[\times\Omega$,

$$\varphi(t,x)=\theta(t)\alpha_\varepsilon(x),\ \varepsilon>0$$

where $\theta\in C_c^\infty([0,T[)$ and α_ε is an element of $C_c^\infty(\Omega)$ such that $\alpha_\varepsilon=1$ on $]-1+\varepsilon,1-\varepsilon[$ and $|\alpha'_\varepsilon|\le\frac{2}{\varepsilon}$. We obtain:

$$\int_Q\{|u-v|\theta'(t)\alpha_\varepsilon(x)+k(x)\Phi(u,v)\theta(t)\alpha'_\varepsilon(x)\}dxdt+\int_\Omega|u_0-v_0|\theta(0)\alpha_\varepsilon(x)dx\ge 0.$$

There is no difficulty to pass to the limit when ε goes to 0^+. We just point out that due to the properties of $(\alpha_\varepsilon)_\varepsilon$ and to the definition of $v^\tau_{\pm1}$ and $u^\tau_{\pm1}$,

$$\lim_{\varepsilon\to0^+}\int_Q k(x)\Phi(u,v)\theta(t)\alpha'_\varepsilon(x)dq=\int_0^T\{k(-1)\Phi(u^\tau_{-1},v^\tau_{-1})-k(1)\Phi(u^\tau_1,v^\tau_1)\}\theta(t)dt.$$

It comes

$$\int_Q|u-v|\theta'(t)dxdt+\int_\Omega|u_0-v_0|\theta(0)dx\ge$$
$$\int_0^T k(1)\Phi(u^\tau_1,v^\tau_1)\theta(t)dt-\int_0^T k(-1)\Phi(u^\tau_{-1},v^\tau_{-1})\theta(t)dt.$$

Let's prove now that:

$$\int_0^T k(1)\Phi(u^\tau_1,v^\tau_1)\theta(t)dt-\int_0^T k(-1)\Phi(u^\tau_{-1},v^\tau_{-1})\theta(t)dt\ \ge\ 0\,.$$

By coming back to Definition 1, we know that, for *a.e* t in $]0,T[$, for all κ in $\mathbb{R}$:

$$\begin{aligned}k(1)(sgn(u^\tau_1(t)-\kappa)+sgn(\kappa))(g(u^\tau_1(t))-g(\kappa))\ &\ge\ 0,\\ k(1)(sgn(v^\tau_1(t)-\kappa)+sgn(\kappa))(g(v^\tau_1(t))-g(\kappa))\ &\ge\ 0.\end{aligned}$$

Thus, *a.e.* on $]0,T[$,

- if $u^\tau_1(t)$ and $v^\tau_1(t)$ have the same sign, we choose $\kappa=v^\tau_1$ in the first inequality (or $\kappa=u^\tau_1(t)$ in the second one) to obtain:

$$k(1)\Phi(u^\tau_1(t),v^\tau_1(t))\ge0,$$

- if $u^\tau_1(t)$ and $v^\tau_1(t)$ have an opposite sign, choosing $\kappa=0$ in the two inequalities gives:

$$k(1)\Phi(u^\tau_1(t),v^\tau_1(t))\ge0.$$

Hence, for *a.e.* t in $]0,T[$,

$$k(1)\Phi(u^\tau_1(t),v^\tau_1(t))\theta(t)\ge0.$$

Similarly, for *a.e.* t in $]0,T[$,

$$k(-1)\Phi(u_{-1}^{\tau}, v_{-1}^{\tau})\theta(t) \leq 0.$$

This way,

$$\int_{Q} |u - v|\theta_t(t)dxdt + \int_{\Omega} |u_0 - v_0|\theta(0)dx \geq 0.$$

The conclusion follows from classical arguments which completes the proof of Theorem 1. □

5. Existence of an Entropy Solution

The proof relies on a suitable regularization k_ε, $\varepsilon > 0$, of the function k and uses a compactness argument for the sequence $(k_\varepsilon \Phi(u_\varepsilon, \kappa))_{\varepsilon>0}$, where u_ε is the weak entropy solution to the corresponding mollified problem. We will consider two different situations:

$$\begin{cases} \exists \alpha \in \mathbb{R}^-, \ \forall x \leq \alpha, \ (k_L - k_R)g(x) \geq 0, \\ \exists \beta \in \mathbb{R}^+, \ \forall x \geq \beta, \ (k_L - k_R)g(x) \leq 0 \end{cases} \tag{17}$$

$$g(m) = g(M) = 0. \tag{18}$$

Remark 2. *As well (17) as (18) implies (10).*

Remark 3. *When we take into account (2) and (3), we observe that (18) includes the class of srictly convex (or strictly concave) functions that vanish at m and M. In addition (17) is fulfilled as soon as g is strictly monotone and vanishes at a point.*

In this framework we establish that:

Theorem 2. *The following assertions hold:*

(i) *Under (17) Problem (1) admits at least one entropy solution u.*

(ii) *Under (18) the problem (1) admits at least one entropy solution u such that, for a.e. (t,x) in Q, $m \leq u(t,x) \leq M$.*

We suppose first that the initial condition u_0 is smooth. Then through a Cauchy criterion in $L^1(Q)$ we come back to the situation of u_0 in $L^\infty(\Omega)$.

5.1. First Step: $u_0 \in C_c^\infty(\Omega)$

We apply the ideas introduced in [11] (also used in [1]) that is to consider a regular approximation of the function k. Let $(k_\varepsilon)_\varepsilon$ be a sequence of smooth functions such as, for every positive ε, $k_\varepsilon = k$ out of $]-\varepsilon,\varepsilon[$ and k_ε is monotone on $[-\varepsilon,\varepsilon]$ (depending on the sign of $k_L - k_R$). That implies:

$$\forall\, x \in \Omega^*, k_\varepsilon(x) \;\longrightarrow\; k(x) \;\text{ and } |k_\varepsilon|_{BV(\Omega)} \le |k|_{BV(\Omega)}.$$

Then we denote u_ε the unique entropy solution (see [2]) to the regularized problem:

Find a measurable and bounded function u in $BV(Q) \cap C([0,T];L^1(\Omega))$ such that formally

$$\begin{cases} \dfrac{\partial u_\varepsilon}{\partial t} + \dfrac{\partial}{\partial x}(k_\varepsilon(x)g(u_\varepsilon)) = 0 & \text{on } Q, \\ u_\varepsilon(0,x) = u_0(x) & \text{on } \Omega, \\ u = 0 & \text{on a part of }]0,T[\times\partial\Omega. \end{cases} \tag{19}$$

Lemma 7. (i) *Under (17), for any t in $]0,T[$, we set:*

$$R(t) = (\|u_0\|_\infty + \max(|\alpha|,\beta))e^{M_k M_g t} + \frac{e^{M_k M_g t} - 1}{M_g}|g(0)|$$

where $M_k = \max(\|k'\|_{L^\infty(]-1,0[)}, \|k'\|_{L^\infty(]0,1[)})$ and $M_g = Lip(g)$.

Then

$$|u_\varepsilon(t,x)| \le R(t), \text{ a.e. on } \Omega.$$

(ii) *Under (18), $m \le u_\varepsilon(t,x) \le M$, for a.e.$(t,x) \in Q$.*

(iii) *The sequence $(u_\varepsilon)_{\varepsilon>0}$ is strongly precompact in $L^1(Q)$.*

Proof. **(i)** Let $\mu > 0$. We introduce the viscous problem associated with (19):

Find $u_{\varepsilon,\mu}$ in $L^2(0,T;H^2(\Omega)) \cap C([0,T];H^1(\Omega))$, $\partial_t u_{\varepsilon,\mu} \in L^2(Q)$ such that,

$$\begin{cases} \partial_t u_{\varepsilon,\mu} + \partial_x(k_\varepsilon(x)g(u_{\varepsilon,\mu})) = \mu\partial_{xx}u_{\varepsilon,\mu} & \text{a.e. in } Q, \\ u_{\varepsilon,\mu} = u_0 & \text{on } \Omega, \\ u(t,1) = u(t,-1) = 0 & \text{for any } t \in]0,T[. \end{cases} \tag{20}$$

It is known that (20) admits an unique solution that converges toward u_ε in $C([0,T];L^1(\Omega))$ when μ tends to 0. We multiply (20) with $(u_{\varepsilon,\mu} - R(t))^+$, we integrate over $Q_s =]0,s[\times\Omega$, $s \in]0,T[$, and perform the following transformations:

$$\mu\int_{Q_s} \partial_{xx}u_{\varepsilon,\mu}(u_{\varepsilon,\mu} - R(t))^+ dxdt = -\mu\int_{Q_s}[\partial_x(u_{\varepsilon,\mu} - R(t))^+]^2,$$

$$\int_{Q_s}\partial_t u_{\varepsilon,\mu}(u_{\varepsilon,\mu} - R(t))^+ dtdx = \frac{1}{2}\|(u_{\varepsilon,\mu}(s,.) - R(s))^+\|^2_{L^2(\Omega)} + \int_{Q_s} R'(t)(u_{\varepsilon,\mu} - R(t))^+ dtdx,$$

since $u_{\varepsilon,\mu}(0,.) = u_0 \leq R(0)$. Moreover

$$\int_{Q_s} \partial_x(k_\varepsilon(x)g(u_{\varepsilon,\mu}))(u_{\varepsilon,\mu} - R(t))^+ dtdx = \int_{Q_s} k'_\varepsilon(x)g(R(t))(u_{\varepsilon,\mu} - R(t))^+ dtdx + \int_{Q_s} \partial_x(k_\varepsilon(x)(g(u_{\varepsilon,\mu}) - g(R(t)))(u_{\varepsilon,\mu} - R(t))^+ dtdx$$

where we use an integration by parts in the last term on the right-hand side. Then the Young inequality gives:

$$\left|\int_{Q_s} \partial_x(k_\varepsilon(x)(g(u_{\varepsilon,\mu}) - g(R(t)))(u_{\varepsilon,\mu} - R(t))^+ dtdx\right| \leq \mu \int_{Q_s} [\partial_x(u_{\varepsilon,\mu} - R(t))^+]^2 + \frac{1}{4\mu}\int_{Q_s} (k_\varepsilon M_g)^2((u_{\varepsilon,\mu} - R(t))^+)^2.$$

Gathering all terms yields to:

$$\frac{1}{2}\|(u_{\varepsilon,\mu}(s,.) - R(s))^+\|^2_{L^2(\Omega)} + \int_{Q_s} [R'(t) + k'_\varepsilon g(R(t))](u_{\varepsilon,\mu} - R(t))^+ dtdx \leq \frac{(\|k\|_\infty M_g)^2}{4\mu}\int_{Q_s} ((u_{\varepsilon,\mu} - R(t))^+)^2 dxdt.$$

Now let's show that the term

$$\Psi(t,x) = R'(t) + k'_\varepsilon g(R(t))$$

is nonnegative. Therefore, thanks to the Gronwall's Lemma the conclusion will follow. On $[-\varepsilon,\varepsilon]$, by definition of $R(t)$, $R(t) \geq \beta$ and $R'(t) \geq 0$. So:

if $k_L \leq k_R$, $k'_\varepsilon \geq 0$, and by (17), $g(R(t)) \geq 0$. Then $\Psi(t,x) \geq 0$.

if $k_R \leq k_L$, $k'_\varepsilon \leq 0$, and by (17), $g(R(t)) \leq 0$. Then $\Psi(t,x) \geq 0$.

Besides, on $]-1,\varepsilon[\cup[\varepsilon,1[$, $k_\varepsilon = k$. Moreover:

$R'(t) = M_g M_k R(t) + M_k|g(0)|$, and $k'g(R(t)) \geq -M_k M_g R(t) + k'g(0)$.

So

$\Psi(t,x) \geq k'g(0) + M_k|g(0)| \geq 0$.

To show that $u_\varepsilon(t,x) \geq -R(t)$, we multiply (20) with $(u_{\varepsilon,\mu} + R(t))^-$ and we use the same techniques as before, especially the first line in (17).

(ii) The proof refers to **(i)** and basically lies on the fact that $g(m) = g(M) = 0$. Indeed, the arguments of the proof of **(i)** still hold for $R(t) = M$ and, in this case, $\Psi(t,x) = R'(t) + k'_\varepsilon g(R(t)) = 0$. Then we apply the same method with $R(t) = m$. The conclusion follows.

(iii) Let $\varepsilon_0 > 0$. From **(i)** and **(ii)**, the sequence $(u_\varepsilon)_{\varepsilon>0}$ is bounded in $L^\infty(Q)$. So there exists a subsequence, still denoted by $(u_\varepsilon)_{\varepsilon>0}$, such that u_ε converges for the weak-$*$ topology in $L^\infty(Q)$ to a function u. Moreover, for each $\varepsilon < \varepsilon_0$, by definition of k_ε, $k_\varepsilon g(u) = kg(u)$ on $]-1,\varepsilon_0[\cup]\varepsilon_0,1[$. Then, under (3) and (4) it follows from a result of E. Yu. Panov in [9] that a subsequence $(u_\varepsilon)_{\varepsilon<\varepsilon_0}$ converges to u strongly in $L^1(]0,T[\times]-1,\varepsilon_0[\cup]\varepsilon_0,1[)$. By using a diagonal process we obtain that $(u_\varepsilon)_{\varepsilon>0}$ is strongly precompact in $L^1(Q)$. □

Remark 4. *To prove* **(iii)***, one could use the method developed in [11] that lies on the following BV-estimate:*

$$|k_\varepsilon(\Phi(u_\varepsilon,\kappa))|_{BV(Q)} \leq C(|u_0|_{BV(\Omega)} + |k|_{BV(\Omega)}),$$

where C is a positive real.
So, (4) and the compactness of the embedding from BV (Q) into $L^1(Q)$ ensure that a subsequence $(\Phi(u_\varepsilon,\kappa))_\varepsilon$ converges to a bounded and BV-function χ, a.e on Q. Besides we notice that, under (2) and (3), the assumption (17) or (18) on g implies that, for any bounded interval of $\mathbb{R}$, there exists κ_0 in $\mathbb{R}$ such that $\Phi(.,\kappa_0)$ is strictly monotone on this interval. That yields to the convergence of a subsequence $(u_\varepsilon)_\varepsilon$ in $L^1(Q)$.

Let u be the L^1-limit of a subsequence $(u_\varepsilon)_\varepsilon$. Now we have to establish that u is an entropy solution to (1). First we prove that u fulfills (5). To this purpose, we introduce the regularized entropy pairs, for any $\kappa\in\mathbb{R}$, and any real τ:

$$\Phi_\eta(\tau)=\int_\kappa^\tau sgn_\eta(r-\kappa)g'(r)dr \text{ and } I_\eta(\tau)=\int_\kappa^\tau sgn_\eta(r-\kappa)dr,$$

where sgn_η denotes the Lipschitzian approximation of the function sgn given for any positive η and any nonnegative real x by $sgn_\eta(x)=\min(\frac{x}{\eta},1)$ and $sgn_\eta(-x)=-sgn_\eta(x)$.
By coming back to (20) and considering the test-function $v=sgn_\eta(u_{\varepsilon,\mu}-\kappa)\varphi$, $\varphi\in C_c^\infty([0,T[\times\Omega)$, $\varphi\geq 0$, we can take the limit on μ with classical arguments. So we establish that u_ε fulfills the regularized entropy inequality for all φ in $C_c^\infty([0,T[\times\Omega)$,

$$\begin{aligned}&\int_Q I_\eta(u_\varepsilon)\varphi_t dxdt+\int_Q k_\varepsilon(x)\Phi_\eta(u_\varepsilon)\varphi_x dxdt\\ &+\int_Q k'_\varepsilon(x)(\Phi_\eta(u_\varepsilon)-I'_\eta(u_\varepsilon)g(u_\varepsilon))\varphi dxdt+\int_\Omega I_\eta(u_0)\varphi(0,x)dx \;\geq\; 0.\end{aligned}\tag{21}$$

We want to pass to the limit in (21), first with respect to ε and then with respect to η. The difficulty is only concentrated in the first term of the second line. That is why we write (with $dq=dxdt$):

$$\begin{aligned}&\int_Q k'_\varepsilon(x)(\Phi_\eta(u_\varepsilon)-I'_\eta(u_\varepsilon)g(u_\varepsilon))\varphi dq \int_0^T\int_{-1}^{-\varepsilon} k'_\varepsilon(x)(\Phi_\eta(u_\varepsilon)-I'_\eta(u_\varepsilon)g(u_\varepsilon))\varphi dq\\ &+\int_0^T\int_{-\varepsilon}^{\varepsilon} k'_\varepsilon(x)(\Phi_\eta(u_\varepsilon)-I'_\eta(u_\varepsilon)g(u_\varepsilon))\varphi dq+\int_0^T\int_{\varepsilon}^{1} k'_\varepsilon(x)(\Phi_\eta(u_\varepsilon)-I'_\eta(u_\varepsilon)g(u_\varepsilon))\varphi dq.\end{aligned}\tag{22}$$

However, owing to the definition of k_ε,

$$\int_0^T\int_{-1}^{-\varepsilon} k'_\varepsilon(x)(\Phi_\eta(u_\varepsilon)-I'_\eta(u_\varepsilon)g(u_\varepsilon))\varphi dq=\int_0^T\int_{-1}^{-\varepsilon} k'(x)(\Phi_\eta(u_\varepsilon)-I'_\eta(u_\varepsilon)g(u_\varepsilon))\varphi dq,\tag{23}$$

and

$$\int_0^T\int_{\varepsilon}^{1} k'_\varepsilon(x)(\Phi_\eta(u_\varepsilon)-I'_\eta(u_\varepsilon)g(u_\varepsilon))\varphi dq=\int_0^T\int_{\varepsilon}^{1} k'(x)(\Phi_\eta(u_\varepsilon)-I'_\eta(u_\varepsilon)g(u_\varepsilon))\varphi dq.\tag{24}$$

In addition, by referring to the definition of Φ_η and I_η,

$$\begin{aligned}&\int_0^T\int_{-\varepsilon}^{\varepsilon} k'_\varepsilon(x)(\Phi_\eta(u_\varepsilon)-I'_\eta(u_\varepsilon)g(u_\varepsilon))\varphi dq=-\int_0^T\int_{-\varepsilon}^{\varepsilon} k'_\varepsilon g(\kappa)sgn_\eta(u_\varepsilon-\kappa)\varphi dq\\ &+\int_0^T\int_{-\varepsilon}^{\varepsilon} k'_\varepsilon\varphi\Big(\int_\kappa^{u_\varepsilon} sgn_\eta(\tau-\kappa)g'(\tau)d\tau-(g(u_\varepsilon)-g(\kappa))sgn_\eta(u_\varepsilon-\kappa)\Big)dq.\end{aligned}$$

We look for a majoration of the right-hand side of this equality. First,

$$\left|\int_0^T\int_{-\varepsilon}^{\varepsilon} k_\varepsilon' g(\kappa) sgn_\eta(u_\varepsilon-\kappa)\varphi dq\right| \le |g(\kappa)| \int_0^T\int_{-\varepsilon}^{\varepsilon} |k_\varepsilon'|\varphi dq.$$

Now we turn on to the estimate of the term $|D(u_\varepsilon)|$ where:

$$D(u_\varepsilon)=\int_\kappa^{u_\varepsilon} sgn_\eta(\tau-\kappa)g'(\tau)d\tau-(g(u_\varepsilon)-g(\kappa))sgn_\eta(u_\varepsilon-\kappa).$$

More precisely, $|D(u_\varepsilon)|\le C_g\eta$ with $C_g=2M_g$. Indeed, let us fix η and κ. Then *a.e.* on Q:

- if $u_\varepsilon\ge\kappa+\eta$,
$$D(u_\varepsilon)=\int_\kappa^{\eta+\kappa}\frac{\tau-\kappa}{\eta}g'(\tau)d\tau+\int_{\kappa+\eta}^{u_\varepsilon} g'(\tau)d\tau-(g(u_\varepsilon)-g(\kappa))$$
and
$$|D(u_\varepsilon)|\le \eta M_g+|g(\kappa+\eta)-g(\kappa)|,$$
because $0\le\frac{\tau-\kappa}{\eta}\le 1$. Then we use the Lipschitz condition for g.
- if $\kappa-\eta\le u_\varepsilon\le\kappa+\eta$,
$$D(u_\varepsilon)=\int_\kappa^{u_\varepsilon}\frac{\tau-\kappa}{\eta}g'(\tau)d\tau-\frac{u_\varepsilon-\kappa}{\eta}(g(u_\varepsilon)-g(\kappa))$$
and
$$|D(u_\varepsilon)|\le M_g|u_\varepsilon-\kappa|+M_g\frac{|u_\varepsilon-\kappa|^2}{\eta}\le 2M_g\eta.$$
- if $u_\varepsilon\le\kappa-\eta$,
$$D(u_\varepsilon)=\int_{u_\varepsilon}^{\kappa-\eta} g'(\tau)d\tau+\int_{\kappa-\eta}^{\kappa}\frac{\tau-\kappa}{\eta}g'(\tau)d\tau+g(u_\varepsilon)-g(\kappa),$$
and $|D(u_\varepsilon)|\le C_g\eta$, as in the first case.

Eventually we deduce that the term:

$$\int_0^T\int_{-\varepsilon}^{\varepsilon} k_\varepsilon'\varphi\Big(\int_\kappa^{u_\varepsilon} sgn_\eta(\tau-\kappa)g'(\tau)d\tau-(g(u_\varepsilon)-g(\kappa))sgn_\eta(u_\varepsilon-\kappa)\Big)dq$$

is bounded by:

$$C_g\eta\int_0^T\int_{-\varepsilon}^{\varepsilon}|k_\varepsilon'|\varphi dxdt,$$

and

$$\int_0^T\int_{-\varepsilon}^{\varepsilon} k_\varepsilon'(x)(\Phi_\eta(u_\varepsilon)-I_\eta'(u_\varepsilon)g(u_\varepsilon))\varphi dq\le (C_g\eta+|g(\kappa)|)\int_0^T\int_{-\varepsilon}^{\varepsilon}|k_\varepsilon'|\varphi dxdt. \tag{25}$$

As k_ε is monotone on $[-\varepsilon,\varepsilon]$, for ε small enough, $|k_\varepsilon'|=sgn(k_R-k_L)k_\varepsilon'$. So,

$$\int_0^T\int_{-\varepsilon}^{\varepsilon}|k_\varepsilon'|\varphi dq=sgn(k_R-k_L)\int_0^T\int_{-\varepsilon}^{\varepsilon}k_\varepsilon'\varphi dq$$

Then we integrate by parts to obtain:

$$\begin{aligned}\int_0^T\int_{-\varepsilon}^{\varepsilon}|k_\varepsilon'|\varphi dq=&-sgn(k_R-k_L)\int_0^T\int_{-\varepsilon}^{\varepsilon}k_\varepsilon\varphi_x dq\\&+sgn(k_R-k_L)\int_0^T(k(\varepsilon)\varphi(t,\varepsilon)-k(-\varepsilon)\varphi(t,-\varepsilon))dt\,.\end{aligned} \tag{26}$$

Finally, from (21), (22), (23), (24), (25) and (26), for any positive η and ε, we have:

$$\begin{aligned}&-sgn(k_R-k_L)(C_g\eta+|g(\kappa)|)\int_0^T\int_{-\varepsilon}^{\varepsilon}k_\varepsilon\varphi_x dxdt+\int_Q k_\varepsilon(x)\Phi_\eta(u_\varepsilon)\varphi_x dxdt\\&+\int_0^T\int_{-1}^{-\varepsilon}k'(x)(\Phi_\eta(u_\varepsilon)-I'_\eta(u_\varepsilon)g(u_\varepsilon))\varphi dq+\int_\Omega I_\eta(u_0)\varphi(0,x)dx\\&+\int_0^T\int_{\varepsilon}^{1}k'(x)(\Phi_\eta(u_\varepsilon)-I'_\eta(u_\varepsilon)g(u_\varepsilon))\varphi dq+\int_Q I_\eta(u_\varepsilon)\varphi_t dxdt\\&+sgn(k_R-k_L)(C_g\eta+|g(\kappa)|)\int_0^T(k(\varepsilon)\varphi(t,\varepsilon)-k(-\varepsilon)\varphi(t,-\varepsilon))dt\geq 0.\end{aligned}$$

We take now the ε-limit. Clearly, because $(u_\varepsilon)_\varepsilon$ goes to u in $L^1(Q)$ and since I_η and Φ_η are Lipschitzian,

$$\lim_{\varepsilon\to0^+}\int_Q(I_\eta(u_\varepsilon)\varphi_t+k_\varepsilon(x)\Phi_\eta(u_\varepsilon)\varphi_x)dq=\int_Q(I_\eta(u)\varphi_t+k(x)\Phi_\eta(u)\varphi_x)dq.$$

Thanks to the definition of k_L and k_R and to the continuity of φ,

$$\lim_{\varepsilon\to0^+}\int_0^T(k(\varepsilon)\varphi(t,\varepsilon)-k(-\varepsilon)\varphi(t,-\varepsilon))dt=(k_R-k_L)\int_0^T\varphi(t,0)dt.$$

Moreover, $k_\varepsilon\varphi_x$ being bounded independently with respect to ε,

$$\lim_{\varepsilon\to0^+}\int_0^T\int_{-\varepsilon}^{\varepsilon}k_\varepsilon\varphi_x dxdt=0.$$

So, for any positive η, the following inequality holds:

$$\begin{aligned}&\int_Q(I_\eta(u)\varphi_t+k(x)\Phi_\eta(u)\varphi_x)dxdt+\int_Q k'(x)(\Phi_\eta(u)-I'_\eta(u)g(u))\varphi dxdt\\&+\int_\Omega I_\eta(u_0)\varphi(0,x)dx+(|g(\kappa)|\,+\,C_g\eta)|k_R-k_L|\int_0^T\varphi(t,0)dt\geq 0.\end{aligned}\tag{27}$$

We take the limit with respect to η through the Lebesgue dominated convergence Theorem, providing that u fulfills (5).

Lastly, let us establish that u satisfies (6)-(7). To this purpose, we use the functions H_η and Q_η defined in [8] for any τ, $\kappa\in\mathbb{R}$, by:

$$H_\eta(\tau,\kappa)=\left((\text{dist}(\tau,I[0,\kappa]))^2+\eta^2\right)^{\frac{1}{2}}-\eta$$

and

$$Q_\eta(\tau,\kappa)=\int_\kappa^\tau\partial_1H_\eta(\lambda,\kappa)g'(\lambda)d\lambda$$

where $I[0,\kappa]$ denotes the closed interval bounded by 0 and κ. The sequence $(H_\eta,Q_\eta)_\eta$ converges uniformly to $(\text{dist}(\tau,I[0,\kappa]),\mathcal{G}(\tau,0,\kappa))$ where:

$$\mathcal{G}(\tau,0,\kappa)=\frac{1}{2}(\Phi(\tau,0)-\Phi(\kappa,0)+\Phi(\tau,\kappa)).$$

By taking in (20) the test-function $\partial_1H_\eta(u_{\varepsilon,\mu},\kappa)\varphi$, for any function $\varphi\in C_c^\infty(]0,T[$

$\times\overline{\Omega})$, and passing to the limit with respect to μ, we obtain for any positive η and ε the following inequality:

$$\begin{aligned}\int_Q H_\eta(u_\varepsilon,\kappa)\varphi_t dxdt + \int_Q k_\varepsilon Q_\eta(u_\varepsilon,\kappa)\varphi_x dxdt \\ \int_Q k'_\varepsilon(x)(Q_\eta(u_\varepsilon,\kappa) - \partial_1 H_\eta(u_\varepsilon,\kappa)g(u_\varepsilon))\varphi dxdt \ \geq \ 0.\end{aligned}$$

If we only consider functions φ vanishing in a neighborhood of $\{x_0 = 0\}$ containing $[-\varepsilon,\varepsilon]$ (that will not be restictive in the sequel), we can take the ε-limit without difficulty to obtain:

$$\begin{aligned}\int_Q H_\eta(u,\kappa)\varphi_t dxdt + \int_Q kQ_\eta(u,\kappa)\varphi_x dxdt \\ \int_Q k'(x)(Q_\eta(u,\kappa) - \partial_1 H_\eta(u,\kappa)g(u))\varphi dxdt \ \geq \ 0.\end{aligned} \tag{28}$$

Then, for $(t,x) \in]0,T[\times\overline{\Omega}$, we choose in (28) a sequence of test-functions defined by $\varphi_n(t,x) = \beta(t)\alpha_n(x)$ with $\beta \in C_c^\infty(]0,T[)$, $\beta \geq 0$, and $\alpha_n \in C_c^\infty(\overline{\Omega})$ such as $\alpha_n \geq 0$, $\alpha_n(x)0$ on $]-1,1-\frac{1}{n}[$, $\alpha_n(1) = 1$ and $\|\alpha'_n\|_\infty \leq n$. On the one hand, by reasoning as in [8] we make sure that

$$\lim_{n\to\infty}\int_{1-\frac{1}{n}}^{1}\int_0^T \alpha'_n(x)k(x)Q_\eta(u,\kappa)\beta(t)dtdx \text{ exists and is nonnegative.}$$

On the other hand by using the definition of u_1^τ,

$$\lim_{n\to\infty}\int_{1-\frac{1}{n}}^{1}\int_0^T \alpha'_n(x)k(x)Q_\eta(u,\kappa)\beta(t)dtdx = \int_0^T k(1)Q_\eta(u_1^\tau(t),\kappa)\beta(t)dt.$$

Finally, when η goes to 0^+,

$$\int_0^T k(1)\,G(u_1^\tau,0,\kappa)\beta(t)dt \geq 0.$$

To conclude we just emphasize that the previous inequality is equivalent for all κ in $I[0,u_1^\tau]$, to:

$$sgn(u_1^\tau)k(1)(g(u_1^\tau) - g(\kappa)) \geq 0,$$

that is namely (6) when κ is reduced to belong to $I(0,u_1^\tau)$.
In the same way, by choosing $\varphi(t,x) = \beta(t)\delta_n(x)$ in (28), with $\beta \in C_c^\infty(]0,T[)$, $\beta \geq 0$, and $\delta_n \in C_c^\infty(\overline{\Omega})$ such as $\delta_n \geq 0$, $\delta_n(x) = 0$ on $]-1+\frac{1}{n},1[$, $\delta_n(-1) = 1$ and $\|\delta'_n\|_\infty \leq n$, by using the definition of u_{-1}^τ, we establish (7).

5.2. Second Step: $u_0 \in L^\infty(\Omega)$

We use a mollification process to come back to the first step. Indeed, for $j \in \mathbb{N}^*$, we consider the sequence $(u_0^j)_j$ such that u_0^j belongs to $C_c^\infty(\Omega)$ and (u_0^j) tends to u_0 in $L^1(\Omega)$. We denote u^j the entropy solution to (1) associated with the initial condition u_0^j so that, for any j, u^j fulfills (27) and (28). The comparison result (13) ensures that the sequence $(u_j)_j$ is a Cauchy sequence in $L^1(Q)$ and so tends to a limit, denoted u. Then the j-limit in (27) and (28) warrants that u is an entropy solution to (1).

To conclude, we point out that (17) or (18) implies (10), so that:

Corollary 1. *Assume that (17) or (18) holds. Then (1) has an unique entropy solution.*

6. Generalisation

In this section we keep the same assumptions on g but we consider that k has a finite number of discontinuities. Let $D=\{1,...,n-1\}, n\neq 0, x_0=-1, x_n=1$. We suppose that:

$$k \text{ is discontinuous at } x_i, i\in D, \tag{29}$$

while $k_{|]x_i,x_{i+1}[}\in W^{1,+\infty}(]x_i,x_{i+1}[)$. Of course we need a new definition of an entropy solution which has to be equivalent to Definition 1 when D is reduced to one point. So we say that:

Definition 2. *Under (29), a function u of $L^\infty(Q)$ is an entropy solution to Problem (1) if u satisfies (6)-(7) and if, $\forall\kappa\in\mathbb{R}$, $\forall\varphi\in C_c^\infty([0,T[\times\Omega)$, $\varphi\geq 0$,*

$$\begin{cases} \int_Q (|u(t,x)-\kappa|\varphi_t(t,x)+k(x)\Phi(u,\kappa)\varphi_x(t,x))dxdt \\ \qquad -\int_Q k'(x)sgn(u-\kappa)g(\kappa)\varphi dxdt \\ +\int_\Omega |u_0-\kappa|\varphi(0,x)dx+\sum_{i\in D}|(k_i^+-k_i^-)g(\kappa)|\int_0^T \varphi(t,x_i)dt \quad \geq 0 , \end{cases} \tag{30}$$

where

$$k_i^+=\lim_{x\to x_i^+}k(x) \quad and \quad k_i^-=\lim_{x\to x_i^-}k(x)$$

We denote γu_i^+ and γu_i^- the strong traces in $L^\infty(]0,T[)$ at $\{x=x_i\}$. By using the same techniques as before we can state the following theorem:

Theorem 3. *Under (18), when k satisfies (29), there exists a unique entropy solution to (1). Moreover, at every point x_i, $i\in D$, u satisfies the Rankine-Hugoniot condition:*

$$k_i^+g(\gamma u_i^+)=k_i^-g(\gamma u_i^-)$$

In addition, we adapt (17) under the form:

$$\begin{cases} \text{for } i,j\in D, i\neq j, sgn(k_i^+-k_i^-)=sgn(k_j^+-k_j^-), \\ \exists\alpha\in\mathbb{R}^-,\ \forall x\leq\alpha,\ (k_1^--k_1^+)g(x)\geq 0, \\ \exists\beta\in\mathbb{R}^+,\ \forall x\geq\beta,\ (k_1^--k_1^+)g(x)\leq 0. \end{cases} \tag{31}$$

This condition is satisfied when g is strictly monotone and the next corollary holds:

Corollary 2. *Assume that (31) is satisfied. Then problem (1) has a unique entropy solution, in the sense of Definition 2.*

References

[1] F. Bachmann, Analysis of a scalar conservation law with a flux function with discontinuous coefficients, *Advances in Differential equations*, **9**:11-12, 1317-1338, 2004

[2] C. Bardos, A.Y Leroux, J.C Nedelec, First order quasilinear equations with boundary conditions, *Comm. in partial differential equations,* **4**, 1017-1034, 1979.

[3] S. Diehl, A conservation law with point source and discontinuous flux function modelling continuous sedimentation, *SIAM J. Math. Anal.*, **56**, 388-419, 1996.

[4] G. Gagneux, M. Madaune-Tort, Analyse mathématique de modèles nonlinéaires de l'ingénierie pétrolière, *Mathématiques et Applications*, **22**, Springer-Verlag, Berlin, 1996.

[5] T. Gimse, N.H. Risebro,Solution of the Cauchy problem for a conservation law with a discontinuous flux function, *SIAM J. Math. Anal.*, **23**, 635-648, 1992.

[6] E. Godlewski, P.A. Raviart, Hyperbolic systems of conservation laws, *Mathématiques et Applications*, **3-4**, Ellipses, Paris, 1991.

[7] S.N. Kruzkov, First-order quasilinear equations with several independent variables, *Mat. Sb.* **81**, 228-255, 1970.

[8] J. Màlek, J. Nečas, M. Rokyta, M. Růžička, Weak and mesure-valued solutions to evolutionary PDEs *Applied Mathematics and Mathematical Computation* , **13**, Chapman & Hall, London, 1996.

[9] E. Yu. Panov, Property of strong precompactness for bounded sets of measure valued solutions of a first-order quasilinear equation, Sbornik: *Mathematics*, **190**:3, 427-446, 1999

[10] E. Yu. Panov, Existence of strong traces for generalized solutions of multidimensional scalar conservation laws, *Journal of Hyperbolic Differential Equations* , **2**:4, 885-908, 2005.

[11] N. Seguin, J. Vovelle, Analysis and approximation of a scalar conservation law with a flux fonction with discontinuous coefficients, *Math. Models Methods Appl. Sci.* **13**:2, 221-257, 2003.

[12] J.D. Towers, Convergence of a difference scheme for conservation laws with a discontinuous flux, *SIAM J. Numer. Anal.* **38**, 681-698,2000.

[13] A. Vasseur,Strong traces for solutions of multidimensional scalar conservation laws, *Arch. Rational Mech. Anal.* **160**, 181-193, 2001.

In: Research on Evolution Equation Compendium. Volume 2 ISBN: 978-1-60456-752-6
Editor: Gaston M. N'Guerekata

A New Exact Solution to the Delayed Diffusion Equation

P.M. Jordan*
Code 7181, Naval Research Laboratory,
Stennis Space Center, MS 39529, U.S.A.

Abstract

A new analytical solution to the one-dimensional delayed diffusion equation is derived in terms of the Lambert W-function. A new critical value of the thermal lag time parameter is also noted. The effects of varying this parameter are examined using numerical methods.

AMS Subject Classification: Primary 80A20

Key Words: Delayed diffusion equation; Fourier's law; thermal relaxation time; Lambert W-function

1. Introduction

According to Fourier's law, heat conduction in a homogeneous and isotropic solid body is described by the flux relation [1]

$$\mathbf{q} = -K\nabla\theta, \tag{1.1}$$

where $\mathbf{q}$, $K(>0)$, and $\theta = \theta(\mathbf{r}, \tau)$ are the heat flux vector, the (constant) thermal conductivity, and absolute temperature, respectively, and where $\mathbf{r}$ is the position vector and τ denotes the time. When combined with the conservation of energy law

$$\rho c_p \frac{\partial\theta}{\partial\tau} + \nabla\cdot\mathbf{q} = 0, \tag{1.2}$$

where ρ is the mass density and c_p is the specific heat at constant pressure, and assuming no heat sources/sinks are present, Eq. (1.1) results in the heat transport equation

$$\frac{\partial\theta}{\partial\tau} - \kappa\nabla^2\theta = 0, \tag{1.3}$$

where $\kappa = K/(\rho c_p)$ is the thermal diffusivity. Since Eq. (1.3), the well-known diffusion equation, is a parabolic PDE, we see that Fourier's law implies that a thermal disturbance at any point in a material body will be felt instantly, but unequally, at all other points of

*E-mail address: pjordan@nrlssc.navy.mil. The author thanks Prof. R. E. Mickens for the many enlightening discussions and his helpful suggestions. This work was supported by ONR/NRL funding (PE 061153N)

the body; in other word, Fourier's law predicts that thermal signals propagate with infinite speed, and thus conflicts with the requirements of causality. This drawback of Fourier's law, which Maxwell (by 1871) was clearly aware of, contrary to the claim made in Ref. [2], is most apparent under low temperature and/or high heat-flux conditions [3, 4].

To correct this unrealistic feature, which has come to be known as the "paradox of heat conduction," various modifications to Eq. (1.1) have been proposed, the best known of these being the Maxwell–Cattaneo (MC) law [2, 3]

$$\left\{1+\lambda_0\frac{\partial}{\partial\tau}\right\}\mathbf{q} = -K\nabla\theta. \tag{1.4}$$

Here, the thermal relaxation constant $\lambda_0(>0)$, the value which has been experimentally determined for a number of materials [3, 5–7], represents the time lag required to establish steady thermal conduction in a volume element once a temperature gradient has been imposed across it. Now the MC law, which has received a great deal of attention in the context of generalized thermoelasticity (see, e.g., [3, 7]), gives rise to the dissipative, hyperbolic heat transport equation

$$\frac{\partial\theta}{\partial\tau}+\lambda_0\frac{\partial^2\theta}{\partial\tau^2}-\kappa\nabla^2\theta = 0, \tag{1.5}$$

which is a special case of the telegraph equation known as the damped wave equation, or DWE for short (see, e.g., Refs. [8–10] and those therein). Thus, according to the MC law, heat conduction occurs via the propagation of damped thermal waves with (finite) propagation speed $c_0 = \sqrt{\kappa/\lambda_0}$, a phenomenon which has come to be known as "second sound" [2, 3, 11].

As noted by Tzou [12], the expression on the LHS of Eq. (1.4) consists of the first two terms of the Taylor series expansion of the LHS of the delayed Fourier flux relation

$$\mathbf{q}(\mathbf{r},\tau+\lambda_0) = -K\nabla\theta(\mathbf{r},\tau). \tag{1.6}$$

According to Eq. (1.6), a variation in the temperature gradient at time τ is "felt" by the heat flux vector at the later time $\tau+\lambda_0$. [Here, we note that the $\lambda_0\to 0$ limiting case of both Eqs. (1.4) and (1.6) is Fourier's law.] Under this flux law, and again assuming that no heat sources/sinks are present, heat transport is governed by the delayed diffusion equation

$$\frac{\partial\theta}{\partial\tau}(\mathbf{r},t+\lambda_0)-\kappa\nabla^2\theta = 0, \tag{1.7}$$

which we observe is simply Eq. (1.3) with time delay λ_0 included in the term on the LHS.

The primary aim of this work is to derive a new form of the solution to the one-dimensional (1D) case of Eq. (1.7) in the context of a well known initial–boundary value problem (IBVP). In the process, a new critical value of the lag time parameter is identified. Additionally, we numerically examine the effects of varying the lag time parameter and compare the maxima of the solutions of Eqs. (1.3) and (1.7) in the 1D case. To this end, the present communication is arranged as follows. In section 2, we formulate the IBVP and give a known exact solution. In section 3, we derive a new form of this solution in terms of the Lambert W-function. In section 4, numerical work is presented and discussed. And finally, in section 5, a summary of our findings is given.

2. Mathematical Analysis

2.1. Problem Formulation

Consider a slender, homogeneous, thermally conducting solid rod of constant cross-sectional area and constant thermal diffusivity κ. The rod occupies the open interval $(0,\ell)$ along the χ-axis of a Cartesian coordinate system. Let $\theta = \theta(\chi,\tau)$ denote the rod's absolute temperature, $\theta_0 \sin[\pi\chi/\ell]$ the rod's initial temperature (of course $\theta_0 > 0$), and for simplicity we assume that the temperature at both ends is maintained at zero. Furthermore, let the lateral face of the rod be fully insulated. Assuming that heat conduction within the rod is governed by Eq. (1.6), the mathematical model of this physical system consists of the following IBVP:

$$\theta_\tau(\chi,\tau+\lambda_0) = \kappa\theta_{\chi\chi}, \quad (\chi,\tau) \in (0,\ell)\times(0,\infty); \tag{2.1a}$$

$$\theta(0,\tau) = 0, \quad \theta(\ell,\tau) = 0, \quad \tau > 0; \tag{2.1b}$$

$$\theta(\chi,\tau) = \theta_0 \sin[\pi\chi/\ell], \quad (\chi,\tau) \in (0,\ell)\times[-\lambda_0,0]. \tag{2.1c}$$

Employing the following nondimensional quantities:

$$u = \theta/\theta_0, \qquad x = \chi/\ell, \qquad t = \tau(\kappa/\ell^2), \tag{2.2}$$

we recast our IBVP in dimensionless form as

$$u_t(x,t+\tau_0) = u_{xx}(x,t), \quad (x,t) \in (0,1)\times(0,\infty); \tag{2.3a}$$

$$u(0,t) = 0, \quad u(1,t) = 0, \quad t > 0; \tag{2.3b}$$

$$u(x,t) = \sin(\pi x), \quad (x,t) \in (0,1)\times[-\tau_0,0]. \tag{2.3c}$$

Here, $\tau_0 = \lambda_0(\kappa/\ell^2)$ is the (strictly positive) dimensionless thermal lag time.

2.2. Derivation of a Known Solution

Following [13], we assume $u(x,t) = X(x)T(t)$ and separate variables. Omitting the details, it is easy to show that

$$X_n(x) = \begin{cases} \sin(\pi x), & n = 1, \\ 0, & n = 2,3,4,\ldots, \end{cases} \tag{2.4}$$

and that T must satisfy the delay ODE (see Refs. [14–16] and those therein)

$$T'(t+\tau_0) + \pi^2 T(t) = 0. \tag{2.5}$$

Here, we note that the initial condition given in Eq. (2.3 c) has reduced to

$$T(t) = 1 \quad \text{for} \quad t \in [-\tau_0, 0]. \tag{2.6}$$

Next, we apply the Laplace transform $\mathcal{L}[\cdot]$ to Eq. (2.5). After using Eq. (2.6) and then solving the resulting subsidiary equation, the transform domain solution is found to be [16]

$$\bar{T}(s) = \frac{1}{s} - \frac{\pi^2}{s(s+\pi^2 e^{-s\tau_0})}, \tag{2.7}$$

where $\bar{T}(s) \equiv \mathcal{L}[T(t)]$ and s is the transform parameter. The inverse of the first term on the RHS of Eq. (2.7) is, of course, trivial. Obtaining the inverse of the second term, however, is somewhat more difficult.

As noted in [16], the easiest way to accomplish this is by first expanding the second term in Eq. (2.7) in powers of $1/s$. Doing so, and then simplifying, yields

$$\bar{T}(s) = \frac{1}{s} + \frac{1}{s}\sum_{m=0}^{\infty}\left(\frac{-\pi^2}{s}\right)^{m+1}\exp[-s(m\tau_0)]. \tag{2.8}$$

Here, we call attention to the fact that the $1/s$ factor outside the summation sign was inadvertently omitted in [13, Eq. (18)]. On inverting this series term-by-term using a table of inverses along with the properties of the Laplace transform (see, e.g., Ref. [16]), we obtain, after incorporating the contribution from the first term on the RHS of Eq. (2.8) and then multiplying by $\sin(\pi x)$, the closed-form solution

$$u(x,t) = \left\{\sum_{m=0}^{\lfloor t/\tau_0\rfloor+1}(-\tau_0\pi^2)^m\frac{[t/\tau_0-(m-1)]^m}{m!}\right\}\sin(\pi x) \qquad (t>0), \tag{2.9}$$

where $\lfloor\cdot\rfloor$ denotes the greatest integer (or floor) function. Equation (2.9) is the solution given by Su *et al.* [13].

2.3. Case of Zero Delay: Diffusion Equation Solution

Finally, we note that the Fourier-based solution,

$$u(x,t) = \exp(-\pi^2 t)\sin(\pi x) \qquad (t>0), \tag{2.10}$$

is easily determined by setting $\tau_0 \equiv 0$ in IBVP (2.3) and then using separation of variables.

3. Solution in Terms of the Lambert W-function

3.1. The Inversion Integral

In this section an alternative form of Eq. (2.9) will be derived. To this end, we observe that by the Laplace inversion theorem [17],

$$T(t) = 1 - \frac{\pi}{2\mathrm{i}}\int_{\gamma-\mathrm{i}\infty}^{\gamma+\mathrm{i}\infty}\frac{\mathrm{e}^{st}\,\mathrm{d}s}{s(s+\pi^2\mathrm{e}^{-s\tau_0})} \qquad (t>0), \tag{3.1}$$

where the constant $\gamma(>0)$ is chosen so that the line $s=\gamma$ lies to the right of all the singularities of the integrand. Omitting the details, it is not difficult to establish that the integral in Eq. (3.1) can be evaluated using the residue theorem (see, e.g., [17]); in particular, we note that the singularities of the integrand consist only of poles [16]. Determining these poles is, of course, equivalent to finding the roots of the transcendental equation

$$s(s+\pi^2\mathrm{e}^{-s\tau_0}) = 0. \tag{3.2}$$

Note that all roots of this equation are simple ones unless $\tau_0 = \tau^\star$, where $\tau^\star \equiv (\pi^2 e)^{-1}$, in which case $s = -1/\tau^\star$ is a double root. Clearly, Eq. (3.2) is satisfied either when $s = 0$ or when

$$s\tau_0 \exp(s\tau_0) = -\pi^2 \tau_0. \tag{3.3}$$

Let us now introduce the Lambert W-function [15, 18, 19], which is defined to be the solution of the equation $W(\zeta)\exp[W(\zeta)] = \zeta$. Using this function, the roots of Eq. (3.3) can be expressed in closed form. Specifically, we have

$$s = s_r \equiv \tau_0^{-1} W_r(-\pi^2 \tau_0) \qquad (r = 0, \pm 1, \pm 2, \ldots), \tag{3.4}$$

for $\tau_0 \neq \tau^\star$, and

$$s = s_{-1} = s_0 \equiv -1/\tau^\star \quad \text{and} \quad s = s_r \equiv \pi^2 e W_r(-e^{-1}) \quad (r = 1, \pm 2, \pm 3, \ldots), \tag{3.5}$$

when $\tau_0 = \tau^\star$. Here, $W_r(\cdot)$ denotes the rth ($r \in \mathbb{Z}$) branch of the W-function [15, 18] with $W_0(\cdot)$, also written as simply $W(\cdot)$ when there is no possibility for confusion, denoting the principal branch. Since the early 1990s, the W-function has appeared in a number of papers from various fields (see, e.g., Refs. [18–20] and those therein). It should be noted, however, that W is *not* a function, strictly speaking, since there are *two* distinct real values of $W(\zeta)$ for every $\zeta \in (-e^{-1}, 0)$.

3.2. Exact Time Domain Solution

On calculating and summing the residues of the integrand at each of the poles, the integral in Eq. (3.1) is readily evaluated in terms of an infinite series. Consequently, an exact expression for $T(t)$ is obtained and we find, after simplifying and multiplying by sin (πx), that the solution to IBVP (2.3) can also be expressed as

$$u(x,t) = \sin(\pi x) \begin{cases} T_1(t), & \tau_0 < \tau^\star \\ T_2(t), & \tau_0 = \tau^\star \end{cases} \qquad (t > 0), \tag{3.6}$$

where

$$T_1(t) = -\pi^2 \tau_0 \left\{ \frac{\exp\left[(t/\tau_0) W_0(-\pi^2\tau_0)\right]}{W_0(-\pi^2\tau_0)[1 + W_0(-\pi^2\tau_0)]} + \sum_{m=1}^{\infty} \left[\frac{\exp\left[(t/\tau_0) W_{-m}(-\pi^2\tau_0)\right]}{W_{-m}(-\pi^2\tau_0)[1 + W_{-m}(-\pi^2\tau_0)]} + \frac{\exp\left[(t/\tau_0) W_m(-\pi^2\tau_0)\right]}{W_m(-\pi^2\tau_0)[1 + W_m(-\pi^2\tau_0)]} \right] \right\} \tag{3.7}$$

and

$$T_2(t) = e^{-1} \left\{ \frac{8}{3} \exp(-t/\tau^\star) \left(1 + \frac{3t}{4\tau^\star}\right) - \frac{\exp\left[(t/\tau^\star) W_1(-e^{-1})\right]}{W_1(-e^{-1})[1 + W_1(-e^{-1})]} - \sum_{m=2}^{\infty} \left[\frac{\exp\left[(t/\tau^\star) W_{-m}(-e^{-1})\right]}{W_{-m}(-e^{-1})[1 + W_{-m}(-e^{-1})]} + \frac{\exp\left[(t/\tau^\star) W_m(-e^{-1})\right]}{W_m(-e^{-1})[1 + W_m(-e^{-1})]} \right] \right\}. \tag{3.8}$$

Here, we observe that the restriction $\tau_0 \in (0, \tau^\star]$ is necessary since u assumes negative values when $\tau_0 > \tau^\star$, in contradiction to the fact that u is the absolute temperature. We also observe that $\text{Re}[W_r(-\pi^2\tau_0)]$, where $r \in \mathbb{Z}$ and $\text{Re}[\cdot]$ denotes the real part of a complex quantity, is always negative, meaning that $T_{1,2} \to 0$ as $t \to \infty$, when $\tau_0 \in (0, \tau^\star]$. And it should be noted that solving Eq. (2.5) in terms of the W-function appears to have been first discussed in Ref. [15].

3.3. Analytical Results

Using the fact that [18]

$$W_0(\zeta) \approx \zeta - \zeta^2, \tag{3.9}$$

where $|\zeta| \ll e^{-1}$ is sufficiently small, it can be shown that

$$\sin(\pi x) \lim_{\tau_0 \to 0} T_1(t) = \exp(-\pi^2 t)\sin(\pi x); \tag{3.10}$$

i.e., we can recover the Fourier-based solution [Eq. (2.10)] by letting $\tau_0 \to 0$ in Eq. (3.6).

Based on the results given in [14], it is possible to obtain the large-t asymptotic expressions

$$u(x,t) \sim \sin(\pi x) \begin{cases} \dfrac{-\pi^2\tau_0 \exp\left[(t/\tau_0)W_0(-\pi^2\tau_0)\right]}{W_0(-\pi^2\tau_0)[1 + W_0(-\pi^2\tau_0)]}, & \tau_0 < \tau^\star \\ \dfrac{8}{3e}\left(1 + \dfrac{3t}{4\tau^\star}\right)\exp(-t/\tau^\star), & \tau_0 = \tau^\star \end{cases} \quad (t \to \infty). \tag{3.11}$$

Here, we should mention that the W-function is *not* explicitly discussed in [14]. Also, for a more detailed treatment of the asymptotic analysis of delay equations, see Ref. [21] and those therein.

4. Numerical Results

Since the solutions given in Eqs. (2.9) and (2.10) have been numerically compared/contrasted in [13], and Eqs. (2.9) and (3.6) are equivalent, there is no need to reproduce these graphs, which plotted u vs. x for fixed values of t, again here. Instead, our numerical work, which was carried out using the software package MATHEMATICA (Version 5.0), wherein $W_r(\cdot)$ is implemented as **ProductLog**$[r, \cdot]$, focuses on understanding the behavior of the difference function $\Delta(t)$, which we define below, as the lag time is varied, including the case of the critical value $\tau_0 = \tau^\star = (\pi^2 e)^{-1} \approx 0.0373$.

To this end, we have plotted in Fig. 1 the difference function

$$\Delta(t) \equiv \exp(-\pi^2 t) - \begin{cases} T_1(t), & \tau_0 < \tau^\star \\ T_2(t), & \tau_0 = \tau^\star \end{cases} \quad (t > 0), \tag{4.1}$$

which corresponds to subtracting Eq. (3.6) from (2.10) for fixed $x = 1/2$. Here, we see that $\max[\Delta(t)](\tau_0)$, which is a global maximum, is a bounded, increasing function of τ_0, as is the corresponding stationary point $t_C(\tau_0)$. In addition, it is clear that

$$\exp(-\pi^2 t) > \begin{cases} T_1(t), & \tau_0 < \tau^\star \\ T_2(t), & \tau_0 = \tau^\star \end{cases} \qquad (t > 0), \tag{4.2}$$

which is in agreement with the findings of [13]. Figure 1 also indicates that, after achieving its maximum, $\Delta(t)$ rapidly tends to zero as $t \longrightarrow \infty$, and that the rate of its decay is a decreasing function of τ_0.

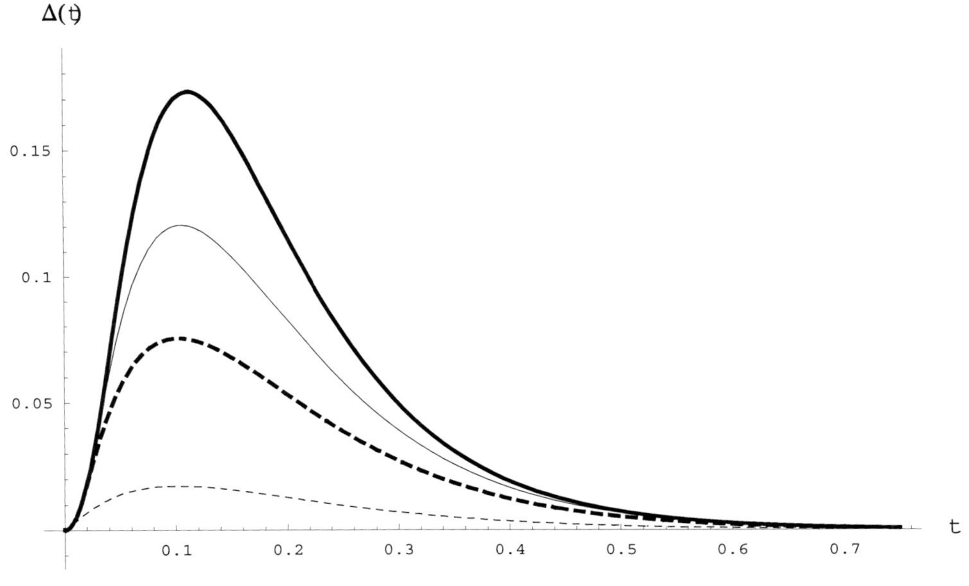

Figure 1. $\Delta(t)$ vs. t. Bold: $\tau_0 = \tau^\star$. Thin-solid: $\tau_0 = (0.75)\tau^\star$. Bold-broken: $\tau_0 = (0.5)\tau^\star$. Thin-broken: $\tau_0 = (0.125)\tau^\star$.

5. Summary

We have presented a new exact solution to the delayed diffusion equation in terms of the Lambert W-function. We have found that $\tau^\star = (\pi^2 e)^{-1}$ is a critical value of the lag time associated with this new solution and that

$$\tau_c < \tau^\star, \tag{5.1}$$

where $\tau_c \equiv (2\pi)^{-2} \approx 0.02533$ is the lag time value above which the solution of the corresponding IBVP involving the (1D) DWE can become negative [9, 10].

In addition, we plotted the difference function Δ for several values of $\tau_0 \leq \tau^\star$. We found that for $t > 0$, the maximum temperature predicated by Eq. (2.10), the Fourier-based solution, is always greater than that predicted by Eq. (3.6) and that the value of Δ rapidly tends to zero as $t \longrightarrow \infty$. Our numerical studies also indicated that there is a critical value of time, $t = t_C(\tau_0)$, at which this difference is greatest and that both max $[\Delta](\tau_0)$ and $t_C(\tau_0)$ are bounded, increasing functions of the lag time τ_0, meaning that both $\max[\Delta](\tau_0)$ and $t_C(\tau_0)$ are at their maximum when $\tau_0 = \tau^\star$.

Finally, in the course of this analysis the following (possibly new) identity was noticed:

$$-1 = \sum_{m=-\infty}^{+\infty} \frac{\eta}{W_m(-\eta)[1+W_m(-\eta)]} \qquad (\eta \neq 0, \mathrm{e}^{-1}). \tag{5.2}$$

References

[1] H. S. Carslaw and J. C. Jaeger, *Conduction of Heat in Solids*, 2nd edn., Oxford University Press, New York, 1959.

[2] D. D. Joseph and L. Preziosi, Heat waves. *Rev. Mod. Phys.* **61** (1989), 41–73; 62 (1990), 375–391.

[3] D. S. Chandrasekharaiah, Thermoelasticity with second sound: A review. *Appl. Mech. Rev.* **39** (1986), 355–376.

[4] W. Dreyer and H. Struchtrup, Heat pulse experiments revisited. *Cont. Mech. Thermodyn.* **5** (1993), 3–50.

[5] G. Caviglia, A. Morro, and B. Straughan, Thermoelasticity at cryogenic temperatures. *Int. J. Non-Linear Mech.* **27** (1992), 251–263.

[6] P. J. Antaki, Hotter than you think, *Machine Design*, July 13 (1995), 116–118.

[7] D. S. Chandrasekharaiah, Hyperbolic thermoelasticity: A review of recent literature. *Appl. Mech. Rev.* **51** (1998), 705–729.

[8] A. C. King, D. J. Needham, and N. H. Scott, The effects of weak hyperbolicity on the diffusion of heat. *Proc. Roy. Soc. A* **454** (1998), 1659–1679.

[9] R. E. Mickens and P. M. Jordan, A positivity-preserving nonstandard finite difference scheme for the damped wave equation. *Num. Meth. Partial Diff. Eq.* **20** (2004), 639–649.

[10] R. E. Mickens and P. M. Jordan, A new positivity-preserving nonstandard finite difference scheme for the DWE. *Num. Meth. Partial Diff. Eq.* **21** (2005), 976–985.

[11] C. I. Christov and P. M. Jordan, Heat conduction paradox involving second-sound propagation in moving media. *Phys. Rev. Lett.* **94** (2005), 154301.

[12] D. Y. Tzou, *Macro- to Microscale Heat Transfer: The Lagging Behavior*. Taylor & Francis, Washington, DC, 1997.

[13] S. Su *et al.*, Comparison of the solutions of a phase-lagging heat transport equation and damped wave equation. *Int. J. Heat Mass Transfer* **48** (2005), 2233–2241.

[14] B. K. Drives and R. D. Driver, Simplicity of solutions of $x'(t) = bx(t-1)$. *J. Math. Anal. Appl.* **157** (1991), 591–608.

[15] D. J. Jeffrey, D. E. G. Hare, and R. M. Corless, Unwinding the branches of the Lambert *W* function. *Math. Scientist* **21** (1996), 1–7.

[16] D. Zwillinger, *Handbook of Differential Equations*, 3rd edn., Academic Press, San Diego, CA, 1997.

[17] H. S. Carslaw and J. C. Jaeger, *Operational Methods in Applied Mathematics*. Dover, New York, 1963.

[18] R. M. Corless *et al.*, On the Lambert *W* function. *Advs. Comput. Math.* **5** (1996), 329–359.

[19] B. Hayes, Why *W*? *Amer. Scientist* **93** (2005), 104–108.

[20] P. M. Jordan, Finite-amplitude acoustic traveling waves in a fluid that saturates a porous medium: Acceleration wave formation. *Phys. Lett. A* **355** (2006), 216–221.

[21] A. C. Fowler, Asymptotic methods for delay equations. *J. Engng. Math.* **53** (2005), 271–290.

In: Research on Evolution Equation Compendium. Volume 2 ISBN: 978-1-60456-752-6
Editor: Gaston M. N'Guerekata

LEVEL SETS FOR REACTION DIFFUSION EQUATIONS

*Amin Boumenir**
Department of Mathematics,
University of West Georgia, Carrollton, GA 30118 USA

Abstract

We are concerned with the shape of the level sets of solution of reaction diffusion equations. Using the maximum principle we find sufficient conditions for their concavity.

1. Introduction

Due to the recent advances and applications of level sets in the area of image processing, numerical analysis, and differential equations, [1, 2, 4, 5, 6], we would like to investigate the shape of level sets of classical reaction diffusion equations defined by

$$\begin{cases} \frac{\partial}{\partial t}u = \Delta u + f(u) & x \in \Omega \Subset \mathbb{R}^m \quad t > 0 \\ u(x,t) = 0 & x \in \partial\Omega,\ t > 0 \\ u(x,0) = \varphi(x) & x \in \Omega \end{cases} \tag{1}$$

where $m \geq 1$. The nonlinearity f and initial condition φ are assumed to satisfy the conditions

[A] $f \in C^2$, $f(u) \geq 0,\ f'(u) > 0,$ and $f''(u) < 0$

[B] $\varphi \in C^2,\ \varphi \geq 0$ and $-f(\varphi) < \Delta\varphi < 0.$

Denote by $Q = \Omega \times [0,\ \infty)$ and let ∂Q represent its boundary. We would to study the shape of the level sets of the solution of (1) which are surfaces defined by the points $(x,\ t(x,n)) \in Q$ such that u takes the value n, i.e.

$$\sum_n := \{(x,t(x,n)) \in Q,\ u(x,t(x,n)) = n\}. \tag{2}$$

It is obvious that the geometry of the set $\{\Sigma_n\}_{n\in\mathbb{R}}$, which is a subset of $\mathbb{R}^m$ would fully describe the solution $(x,t,u(x,t))$ by level surfaces in $\mathbb{R}^{m+1}$. We are interested in finding sufficient conditions for the solution $u(.,t)$ to be, for example, concave down and also for Σ_n to be concave up. The level lines provide a phase space type analysis for the solution. For example, they cannot intersect except at a singularity and also play the role of characteristics, as they channel information. Since the gradient ∇u is orthogonal to Σ_n, i.e. $\nabla u \perp \Sigma_n$

*E-mail address: boumenir@westga.edu

and level sets method are powerful tools for Hamilton-Jacobi equations $u_t = H(\nabla u)$. However, in the case of reaction diffusion equations such as (1), the gradient is not available, unless you solve for the solution, which makes the standard computational machinery of the level set method difficult if not impossible to apply. Nevertheless, they still provide interesting geometrical insights for the solution especially when singularities such as blow-up for (1) occur. Condition [A] includes the interesting sub-critical nonlinearity that can be found in [3]. For example blow-up when $f(u) = |u|^{p-1}u$, occurs for

$$1 < p < \infty \quad \text{for } m = 1,2 \ \text{ and } 1 < p < \frac{m+2}{m-2} \qquad \text{for } \ m \geq 3.$$

Thus condition [A] which allows for $1 < p < 2$, includes the important case of blow-up of positive solutions. We first describe the case $m = 1$.

2. The One Dimensional Case

Equation (1) is then given by

$$\begin{cases} \frac{\partial}{\partial t}u = \frac{\partial^2}{\partial x^2}u + f(u) \quad x \in [-a,a] \ \ t > 0 \\ u(\pm a,t) = 0, \\ u(x,0) = \varphi(x) \geq 0 \ \ x \in [-a,a] \end{cases} \tag{3}$$

Proposition 1. *Assume that conditions [A] and [B] hold then the solution $u(.,t)$, when it exists, is positive and concave down*

Proof. First the solution exists locally and is positive. This follows from the maximum principle since $\frac{\partial}{\partial t}u - \frac{\partial^2}{\partial x^2}u = f(u) \geq 0$ and on the boundary $u \geq 0$ when $(x,t) \in \partial Q$. Next we prove that $u(.,t)$ is concave down. Since f is smooth, $u_{xx} = \frac{\partial^2}{\partial x^2}u$, satisfies

$$\begin{cases} \frac{\partial}{\partial t}u_{xx} - \frac{\partial^2}{\partial x^2}u_{xx} - f'(u)u_{xx} = f''(u)u_x^2 < 0 \quad x \in [-a,a], \, t > 0 \\ u_{xx}(\pm a,t) = \frac{\partial}{\partial t}u(\pm a,t) - f(u) < 0, \, t > 0 \\ u_{xx}(x,0) = \varphi''(x) < 0 \quad x \in [-a,a] \end{cases} \tag{4}$$

The maximum principle then implies that $u_{xx} < 0$ inside Q and thus the solution $u(.,t)$ is concave down. □.

Proposition 2. *Assume that conditions [A] and [B] hold. Then each level lines $t(x,n)$ defined by (2) has one minimum. If $n \leq \max\varphi$ then $\min_{-a\leq x\leq a}(t(x,n)) = 0$ while $n > \max\varphi$ then $\min_{-a\leq x\leq a} t(x,n) = \tau(n)$ where*

$$\max_{-a\leq x\leq a} u(x,\tau(n)) = n.$$

Proof. From the concavity of the solution, it follows that the solution $u(.,t)$, which is positive, has one maximum for each $t > 0$, say at $m(t)$,

$$u(x,t) < u(m(t),t) \qquad \text{for } x \neq m(t)$$

and clearly

$$\begin{cases} u_x(x,t) > 0 \text{ if } & -a < x < m(t) \\ u_x(x,t) < 0 \text{ if } & m(t) > x > a. \end{cases}$$

From (3) it follows that $\frac{\partial}{\partial t}u_t - \frac{\partial^2}{\partial x^2}u_t - f'(u)u_t = 0$ while $u_t(x,0) = \varphi''(x) + f(\varphi) > 0$ for $x \in [-a,a]$ and $u(\pm a,t) = 0$ for $t > 0$. Thus $u_t(x,t) \geq 0$ if $(x,t) \in \partial Q$, and by the maximum principle we have

$$u_t > 0 \text{ for } (x,t) \in Q. \tag{5}$$

Then by differentiating implicitly $u(x,t(x,n)) = n$ with respect to x, $t(x,n)$ satisfies a differential equation

$$t'(x,n) = -\frac{u_x(x,t(x,n))}{u_t(x,t(x,n))} \tag{6}$$

Thus from (6), the sign of u_x and u_t, we deduce that

$$t'(x,n) < 0 \text{ if } -a < x < m(t) \text{ while } t'(x,n) > 0 \text{ if } m(t) < x < a. \tag{7}$$

Each level lines $t(x,n)$ then must attain its minimum value which is a function of n. If we denote by $m_0 = \max_{-a\leq x\leq a} \varphi$, then for $n \leq m_0$ we have $\min_{-a\leq x\leq a} t(x,n) = 0$ since $u(x,0) \leq m_0$. If $n > m_0$, then $\tau(n) = \min_{-a\leq x\leq a}(t(x,n)) > 0$, otherwise $t(x,n)$ would intersect $t(x,m_0)$ which is impossible. Let now $\tau(n)$ be the first time where $\max_{-a\leq x\leq a} u(x,t) = u(x,\tau(n)) = n$, i.e. $u(x,t) < n$ for $t < \tau(n)$. Then $t(x,\tau(n) \geq \tau(n)$, otherwise $u(x,t(x,\tau(n)) < n$ would not hold. Since it is the first time then it is the minimum $\min_{-a\leq x\leq a}(t(x),n) = \tau(n) = \max_{-a\leq x\leq a} u(x,\tau(n))$. □

Proposition 3. *Assume that conditions [A] and [B] hold, and* $\varphi^{(4)} + 2f'(\varphi)\varphi'' + f''(\varphi)\varphi'^2 + f'(\varphi)f(\varphi) \leq 0$. *If* $t'(x,n).u_{tx} < 0$ *for* $(x,t) \in Q$ *then* $t''(x,n) > 0$, *i.e.* $t(x,n)$ *is concave up.*

Proof. For any fixed n, differentiate (6) with respect to x

$$u_t(x,t)t'' = -u_{xx}(x,t) - u_{tt}(x,t)t'^2 - 2u_{tx}(x,t)t'. \tag{8}$$

We already have $u_t > 0$ while $u_{xx} < 0$. Similarly, as done for u_{xx}, we can also show $u_{tt} < 0$ in Q if $u_{tt} \leq 0$ holds on ∂Q. Indeed, we already have

$$(u_{tt})_t - (u_{tt})_{xx} - f'(u)u_{tt} = f''(u)u_t^2 < 0.$$

Since u_{tt} satisfies the boundary condition $u_{tt}(\pm a,t) = 0$ and from $u_t = u_{xx} + f(u)$ on $t = 0$

$$\begin{aligned} u_{tt}(x,0) &= u_{xxt} + f'(u)u_t \\ &= \varphi^{(4)} + 2f'(\varphi)\varphi'' + f''(\varphi)\varphi'^2 + f'(\varphi)f(\varphi) \leq 0. \end{aligned}$$

By the maximum principle, we have $u_{tt}(x,t) < 0$ in Q. It remains to see that if $t'(x,n).u_{tx} < 0$ then (8) implies $t'' > 0$. □

The condition $t'.u_{tx} < 0$ is not needed when we are close to the boundary $x = \pm a$, and $t > 0$.

Corollary 4. *Assume that conditions [A] and [B] hold, and* $\varphi^{(4)}+2f'(\varphi)\varphi''+f''(\varphi)\varphi'^2+f'(\varphi)f(\varphi)\leq 0$. *If* $x\to\pm a$, *then the level sets are concave up.*

Proof. Clearly by the previous proposition and (7), we only need to show that $u_{xt}>0$ when $x\to -a^+$ and $u_{xt}<0$ when $x\to a^-$. Observe that from $u(-a,t)=0$ it follows $u_t(-a,t)=0$ and by (5) $u_t(x,t)>0$, we deduce that for $x\to -a^+$, $u_t(x,t)$ is increasing in x, and so $u_{xt}(x,t)>0$. Similarly if $x\to a^-$, then $u_{xt}<0$. By (7) and in both cases, $x\to\pm a$, we have $u_{xt}t'<0$. The concavity follows from (8). □

Corollary 5. *Assume that conditions [A] and [B] hold, and the solution converges to its steady state as* $t\to\infty$, *then, the level lines are concave up, i.e.* $t''(x,n)>0$.

Proof. Consider the region where the level curve is decreasing, see (7), and denote its inverse by $x(t,n)$ say, i.e. $u(x(t,n),t)=n$. The analogs of (6) and (8) are

$$u_x(x(t,n),t)\,x'(t,n)+u_t(x(t,n),t) = 0. \tag{9}$$

$$u_{xx}x'^2+u_x.x''+2u_{tx}x'+u_{tt} = 0. \tag{10}$$

Since $x'(t,n)<0$, the function $x(t,n)$, which is decreasing and bounded from below $x(t,n)>-a$, must have a limit as $t\to\infty$. Let $\lim_{t\to\infty}x(t,n)=\rho(n)$. Thus by continuity of the solution we have $\lim_{t\to\infty}u(\rho(n),t)=\lim_{t\to\infty}u(x(t,n),t)$ and so u_t, u_{tt} and $u_{xt}\to 0$ as $t\to\infty$, and so by (10) we have

$$u_{xx}x'^2+u_x.x''=o(1). \tag{11}$$

Since $u_{xx}<0$ while $u_x>0$ it follows that $x''>0$ and the level curve is concave. For the second branch of $t(x,n)$, when $t'(x,n)>0$, concavity means that $x''(t,n)<0$, which follows from (11). Thus by combining both branches of $x(t)$ we obtain $t''(x)>0$ and so $t(x,n)$ is concave up. □

3. The n-Dimensional Case

In the n-dimensional case, $\dim(Q)=m+1$, the dimension of the level surface, $\dim\Sigma_n=m\geq 2$ and thus we need m parameters for its representation,

$$\Sigma_n:=\{(x,t(x,n)):\ u(x,t(x,n))=n\}.$$

From the smoothness of the solution, we have

$$\begin{aligned}\frac{\partial}{\partial t}\frac{\partial u}{\partial x_i} &= \Delta\frac{\partial u}{\partial x_i}+f'(u)\frac{\partial u}{\partial x_i}\\ \frac{\partial}{\partial t}\frac{\partial^2 u}{\partial x_i^2} &= \Delta\frac{\partial^2 u}{\partial x_i^2}+f'(u)\frac{\partial^2 u}{\partial x_i^2}+f''(u)\left(\frac{\partial u}{\partial x_i}\right)^2\end{aligned}$$

and Δu satisfies the equation

$$\begin{cases}\frac{\partial}{\partial t}\Delta u=\Delta^2 u+f'(u)\Delta u+f''(u)\,||\nabla u||^2\\ \Delta u=-f(u)<0\ \ x\in\partial\Omega,\ t>0\\ \Delta u(x,0)=\Delta\varphi(x)<0\quad x\in\Omega.\end{cases}$$

By condition $[A]$, $f''(u) < 0$, and so

$$\frac{\partial}{\partial t}\Delta u - \Delta^2 u - f'(u)\Delta u \leq 0$$

which, by the maximum principle, implies that

$$\Delta u < 0 \text{ for } (x,t) \in Q.$$

Thus the solution $u(.,t)$ is concave down and obviously positive.

Proposition 6. *Assume that conditions $[A]$ and $[B]$ hold, then the solution $u(.,t)$ is positive and concave down.*

We next establish $\Delta t(x,n) > 0$, which means that the level surfaces are concave up. Using implicit differentiation

$$u(x_1, x_2, ..., x_n, t(x,n)) = n$$

we deduce

$$\Delta u + 2\nabla u_t.\nabla t + u_{tt}\,||\nabla t||^2 + u_t.\Delta t = 0.$$

Using similar arguments as in proposition 3, we have

Proposition 7. *Assume that conditions [A] and [B] hold, and $\Delta^2\varphi + 2f'(\varphi)\Delta\varphi + f''(\varphi)\,||\nabla\varphi||^2 + f'(\varphi)f(\varphi) \leq 0$. If for $(x,t) \in Q$ we have $\nabla u_t.\nabla t < 0$ then $\Delta t(x,n) > 0$, i.e. $t(x,n)$ is concave up.*

In dimension one, the concavity of the solution gives an insight on how a single blow-up point develops. Clearly the shape of level lines also gives a hint, in the n-dimensional case, on the geometry of the blow-up set which is still an open question.

References

[1] T. Chan and Vese, L. Active contour without edges, *IEEE trans. on Image processing* **10**, 266-277, (2001).

[2] S. Osher and R. Fedkiw, Level Set Methods and Dynamic Implicit Surfaces, *Applied Math. Sciences*, **153**, Springer, (2003).

[3] Y. Giga, S. Matsui, and S. Sasayama. Blow up rate for semilinear heat equations with subcritical nonlinearity. *Indiana Univ. Math. J.* **53**, 2, 483–514,(2004).

[4] G. Sapiro, *Geometric Partial Differential equations and Image Analysis* , Cambridge U. press, (2001).

[5] J. Sethian, Curvature and the evolution of fronts, *Comm.in.Math.Physics.* **101**, 487-499, (1985).

[6] J. Sethian, Level Set Methods and Fast Marching Methods: Evolving Interfaces in Computational Geometry, Fluid Mechanics, *Computer Vision and Materials Science,* Cambridge University Press, (1999).

In: Research on Evolution Equation Compendium. Volume 2 ISBN: 978-1-60456-752-6
Editor: Gaston M. N'Guerekata

On the Almost Periodicity of the Superposition of Functions

Stanisław Stoiński*
Faculty of Mathematics and Computer Science
Adam Mickiewicz University, Umultowska 87,
61-614 Poznań, Poland

Abstract

In this note we present some theorems on the superposition of a function which satisfies the Hölder condition and an (NS^p)-almost periodic (a.p.) function, as well an S^p-a.p. function. Moreover, we prove a theorem on the superposition of a continuous function and an (NH)-a.p. function. In the following, we give a theorem on the superposition of a differentiable function and a V_p-a.p. function. Finally, we prove a theorem on the superposition of a differentiable function and an $(NC^{(1)})$-a.p. function.

2000 Mathematics Subject Classification: 42A75 .

Key words and phrases: Almost periodic function, superposition, Hausdorff metric, variation.

1. Introduction

The continuous functions which are almost periodic (a.p. for short) with respect to the uniform metric, i.e. the uniformly a.p. functions, was defined and studied by H. Bohr (see [5]) in 1925 and later. V.V. Stepanov, H. Weyl and A.S. Besicovitch had studied the properties of functions a.p. with respect to integral metrics (see [4] - [6], [9]). B.M. Levitan had considered some kinde of generalization of Bohr's a.p. function, namely N-a.p. functions (see [9]). However functions a.p. with respect to the Hausdorff metric (see [8]) (which was studied by B. Sendov and others , see [10]), so-called H-a.p. functions was defined and investigated by S. Stoiński (see [11]-[14]) and a little later by A.S. Djafarov and G.M. Gasanov (see [7]). The definition and the properties of functions a.p. in variation (V_p-a.p. for short) can be found for example in [15] and [16].

It is well known (see [9], p.21) that if f is a uniformly continuous function on the set Y_g of values of a function g, where g is a uniformly a.p., i.e. B-a.p. function, then the superposition $f \circ g$ is B-a.p. The superpositions of different a.p. functions and continuous or uniformly continous functions, was considered by M. Adamczak and S. Stoiński in [2], [18] and [19].

*E-mail address: stoi@amu.edu.pl

This paper is concerned with the superposition of such functions as: Stepanov a.p., NH-a.p. or V_p-a.p. and $(NC^{(1)})$-a.p. functions and continuous or differentiable functions. The results of this type have application in the field of differential equations.

2. Preliminaries

Let F_Δ denote the class of closed and bounded with respect to the y axis point sets on the Oxy plane, such that their respective projections on the x axis are identical with the interval Δ (bounded or unbounded), and such that the intersection of every straight line $x = x_0$, $x_0 \in \Delta$, and $F \in F_\Delta$ is a closed interval. If $A, B \in F_\Delta$, then the Hausdorff distance between A and B we call the following number

$$r_\Delta(A,B) = \max(\sup_{X\in A}\inf_{Y\in B}\|X-Y\|_0, \sup_{X\in B}\inf_{Y\in A}\|X-Y\|_0),$$

where

$$\|X-Y\|_0 = \|X(x_1,y_1) - Y(x_2,y_2)\|_0 = \max(|x_1-x_2|, |y_1-y_2|)).$$

The functional $r_\Delta : F_\Delta \times F_\Delta \to \mathbb{R}_+$ is a metric in F_Δ. The following lemma is true (see [10]):

Lemma 1 . *Suppose that $A, B \in F_\Delta$. If:*
(a) for an arbitrary $X \in A$ there exists $Y \in B$ such that $\|X-Y\|_0 \leq \delta$
and
(b) for an arbitrary $X \in B$ there exists $Y \in A$ such that $\|X-Y\|_0 \leq \delta$,
then $r_\Delta(A,B) \leq \delta$.

It is easily seen that if $r_\Delta(A,B) \leq \delta$, then the conditions (a) and (b) from Lemma 1 hold.

By the complete graph $\bar{f}$ of the bounded function $f : \Delta \to \mathbb{R}$, we call a set of points (x,y), for which $x \in \Delta$, $I_f(x) \leq y \leq S_f(x)$, where I_f, S_f is the lower and upper Baire's function respectively, both with respect to the function f:

$$I_f(x) = \lim_{\delta\to 0}\inf_{|x'-x|\leq\delta} f(x') \qquad S_f(x) = \lim_{\delta\to 0}\sup_{|x'-x|\leq\delta} f(x').$$

The Hausdorff distance between two functions f and g, both defined and bounded on Δ, is defined in the following way: it is the Hausdorff distance between their respective complete graphs, i.e. $r_\Delta(f,g) = r_\Delta(\bar{f},\bar{g})$. Let $r_\mathbb{R} = r$.

A set $E \subset \mathbb{R}$ is called relatively dense iff there exists a positive number l such that in every open interval $(\omega, \omega + l)$, $\omega \in \mathbb{R}$, there is at least one element of the set E.

Let $f : \mathbb{R} \to \mathbb{R}$ be a bounded function. If for $\varepsilon > 0$ there is $r(f_\tau, f) \leq \varepsilon$, where $f_\tau(x) \equiv f(x+\tau)$,then the number $\tau \in \mathbb{R}$ is called an (H,ε)- almost period ((H,ε) -a.p.) of the function f. Let us denote the set of (H,ε)-a.p. of f by $E_H\{\varepsilon; f\}$. A bounded function $f : \mathbb{R} \to \mathbb{R}$ is called H-almost periodic (H-a.p.) iff for each $\varepsilon > 0$ the set $E_H\{\varepsilon; f\}$ is relatively dense. Let $\widetilde{H}$ be the space of H-a.p. functions. Examples, elementary properties of H-a.p. functions and the connection between H-a.p. functions and almost periodic functions of other types can be found in [11]-[14].

For $f:\mathbb{R}\to\mathbb{R}$ let us write

$$V_p(f) = \sup_{x\in\mathbb{R}} (|f(x)| + V_p(f;x)),$$

where

$$V_p(f;x) = \sup_{\Pi_x}\{\sum_{i=0}^{n-1} |f(t_{i+1}) - f(t_i)|^p\}^{\frac{1}{p}}, \qquad 1 \le p < \infty$$

Π_x is the partition $x-1 = t_0 < t_1 < t_2 < \ldots < t_n = x+1$ of the interval $\langle x-1, x+1\rangle$, $x \in \mathbb{R}$.

Let us denote

$$(BV_p)_{loc} = \{f:\mathbb{R}\to\mathbb{R} : \ V_p(f;x) < \infty \text{ for every } x \in \mathbb{R}\}.$$

Let $f \in (BV_p)_{loc}$. If for $\varepsilon > 0$ there is $V_p(f - f_\tau) \le \varepsilon$, then the number $\tau \in \mathbb{R}$ is called a (V_p, ε)-almost period of the function f. A continuous function $f \in (BV_p)_{loc}$ is called almost periodic in the p-th variation (V_p-a.p.) iff for each $\varepsilon > 0$ the set $E_v^p\{\varepsilon, f\} = \{\tau \in \mathbb{R} : V_p(f - f_\tau) \le \varepsilon\}$ is relatively dense. For $p = 1$ we obtain a V-a.p. function. Let $\widetilde{V}_p$ be the space of V_p-a.p. functions. Elementary properties of functions almost periodic in variation can be found in [15] and [16].

Let us define for $f \in C^{(n)}(\mathbb{R})$ the functional

$$D^{(n)}(f) = \sup_{x\in\mathbb{R}} (|f(x)| + \sum_{i=1}^{n} |f^{(i)}(x)|).$$

A function $f \in C^{(n)}(\mathbb{R})$ is called $C^{(n)}$-almost periodic ($C^{(n)}$a.p.) iff for each $\varepsilon > 0$ the set $E^{(n)}\{\varepsilon, f\} = \{\tau \in \mathbb{R} : D^{(n)}(f - f_\tau) \le \varepsilon\}$ is relatively dense. By $\widetilde{C^{(n)}}$ we denote the space of $C^{(n)}$-a.p. functions. Basic properties and examples of $C^{(n)}$-a.p. functions we can find in [1].

Let $f:\mathbb{R}\to\mathbb{R}$ be a continuous function. If for $N > 0$, $\varepsilon > 0$ there is

$$D_N(f - f_{\pm\tau}) = \max\{|f(x) - f_{\pm\tau}(x)| : -N \le x \le N\} \le \varepsilon,$$

then the number $\tau \in \mathbb{R}$ is called an (N, ε)-almost period ((N, ε)-a.p.) of the function f. Let $NE\{\varepsilon; f\}$ denote the set of (N, ε)-almost periods of f. A continuous function $f:\mathbb{R}\to\mathbb{R}$ is called N-almost periodic (N-a.p.) iff there exists a uniformly a.p. function φ (see [9]), a dominant function of f, such that for any two positive numbers N, ε there exists a $\delta > 0$ such that $E\{\delta; \varphi\} \subset NE\{\varepsilon; f\}$, where $E\{\delta; \varphi\} = \{\tau \in \mathbb{R} : \sup\{|\varphi(x) - \varphi_\tau(x)| : x \in \mathbb{R}\} \le \varepsilon\}$. By $\widetilde{N}$ we denote the space of N-a.p. functions. The above definitions were introduced by B. M. Levitan (see [9]).

The number $\tau \in \mathbb{R}$ is called an (NH, ε)-almost period ((NH, ε)-a.p.) of the bounded on every closed interval function $f:\mathbb{R}\to\mathbb{R}$ iff for two positive numbers N, ε we have $r_{\langle -N,N\rangle}(f, f_\tau) \le \varepsilon$. Let us denote by $NE_H\{\varepsilon; f\}$ the set of (NH, ε)-almost periods of f. A bounded on every closed interval function $f:\mathbb{R}\to\mathbb{R}$ is called (NH)-almost periodic ((NH)-a.p.) iff there exists an H-a.p. function φ, a dominant function of f, such that for any two positive numbers N, ε there exists a $\delta > 0$ such that $E_H\{\delta; \varphi\} \subset NE_H\{\delta; f\}$ (see [19]).

The number $\tau \in \mathbb{R}$ is called an (NS^p, ε)- almost period $((NS^p, \varepsilon)$-a.p.) of the function $f \in L^p_{loc}$, where $1 \le p < \infty$, iff for two positive numbers N, ε we have

$$(ND_{S^p})(f - f_\tau) = \max_{-N \le x \le N} (\int_{x-1}^{x+1} |f(u) - f_\tau(u)|^p du)^{\frac{1}{p}} \le \varepsilon.$$

Let us denote by $NE_{S^p}\{\varepsilon; f\}$ the set of all (NS^p, ε)-almost periods of f. A function $f \in L^p_{loc}$ is called (NS^p)-almost periodic ((NS^p)-a.p.) iff there exists an S^p-a.p. function φ (see [9]), a dominant function of f, such that for any two positive numbers N, ε there exists a $\delta > 0$ such that $E_{S^p}\{\delta; \varphi\} \subset NE_{S^p}\{\varepsilon; f\}$, where

$$E_{S^p}\{\delta; \varphi\} = \{\tau \in \mathbb{R} : \sup_{x \in \mathbb{R}} (\int_{x-1}^{x+1} |\varphi(u) - \varphi_\tau(u)|^p du)^{\frac{1}{p}} \le \varepsilon\}$$

(see [17]).

Let us write for $f \in C^{(n)}(\mathbb{R})$

$$(ND^{(n)})(f) = \max_{-N \le x \le N} (|f(x)| + \sum_{i=1}^{n} |f^{(i)}(x)|) \qquad \text{for} \quad N > 0.$$

The number $\tau \in \mathbb{R}$ is called an $((ND^{(n)}), \varepsilon)$-almost period $(((ND^{(n)}), \varepsilon)$-a.p.) of the function $f \in C^{(n)}(\mathbb{R})$, where $\varepsilon > 0$, iff $(ND^{(n)})(f - f_\tau) \le \varepsilon$. Let $NE^{(n)}\{\varepsilon; f\}$ denote the set of $((ND^{(n)}), \varepsilon)$-almost periods of f. A function $f \in C^{(n)}(\mathbb{R})$ is called $(NC^{(n)})$-almost periodic ($(NC^{(n)})$-a.p.) iff there exists a $C^{(n)}$-a.p. function φ, a dominant function of f, such that for any two positive numbers N, ε there exists a $\delta > 0$ such that $E^{(n)}\{\delta; \varphi\} \subset NE^{(n)}\{\varepsilon; f\}$ (see [3]).

3. Main Results

The following theorem, which concerns (NS^p)-a.p. functions, is true:

Theorem 1 . *If a function $f : \mathbb{R} \to \mathbb{R}$ satisfies the Hölder condition with the exponent $0 < \alpha \le 1$ and with the constant $M > 0$, a function g is $(NS^{p\alpha})$-a.p., where $p \ge \frac{1}{\alpha}$, then the superposition $f \circ g$ is (NS^p)-a.p.*

Proof. Since f satisfies the Hölder condition and $g \in L^{p\alpha}_{loc}$, so for an arbitrary interval $< a, b > \subset \mathbb{R}$ the following estimation

$$\int_a^b |f(g(u))|^p du \le 2^{p(1+\alpha)} M^p \int_a^b |g(u)|^{p\alpha} du + c,$$

where c is a certain constant, holds. Hence, it follows that $f \circ g \in L^p_{loc}$.

Because g is $(NS^{p\alpha})$-a.p., there exists an $S^{p\alpha}$-a.p. function φ, a dominant function of g, such that for any two positive numbers N, ε there exists a $\delta > 0$ such that $E_{S^{p\alpha}}\{\delta; \varphi\} \subset NE_{S^{p\alpha}}\{(\varepsilon/M)^{1/\alpha}; g\}$. Hence, for every $x \in \langle -N; N \rangle$ and $\tau \in E_{S^{p\alpha}}\{\delta; \varphi\}$ the following estimation

$$\Big(\int_{x-1}^{x+1} |f(g(u)) - f(g_\tau(u))|^p du\Big)^{\frac{1}{p}} \le M\Big(\sup_{-N \le x \le N} (\int_{x-1}^{x+1} |g(u) - g_\tau(u)|^{\alpha p} du)^{\frac{1}{\alpha p}}\Big)^{\alpha} \le \varepsilon,$$

holds, i.e. $\tau \in NE_{S^p}\{\varepsilon; f\circ g\}$, and so φ is a dominant function of $f\circ g$, too. Therefore, the superposition $f\circ g$ is (NS^p)-a.p. ■

We prove likewise the following:

Theorem 2 . *If a function f satisfies the Hölder condition with the exponent $0<\alpha\leq 1$, a function g is $S^{p\alpha}$-a.p., where $p\geq\frac{1}{\alpha}$, then the superposition $f\circ g$ is S^p-a.p.*

It is known (see [19]) that the superposition of a continuous function and a continuous H-a.p. function is (NH)-a.p. Now, we will prove the following theorem:

Theorem 3 . *If f is a continuous function on the set Y_g of values of a function g, where g is a continuous (NH)-a.p. function, then the superposition $f\circ g$ is (NH)-a.p.*

Proof. Fix arbitrarily $N>0$, $\varepsilon>0$. Let us denote $m_N=\min\{g(x): -N\leq x\leq N\}$, $M_N=\max\{g(x): -N\leq x\leq N\}$. By the Darboux property it follows that $\{g(x): -N\leq x\leq N\}=\langle m_N; M_N\rangle$.

Because g is (NH)-a.p., there exists an H-a.p. function φ, a dominant function of g, such that for any two positive numbers $N>0$, $\varepsilon^{'}>0$ there exists a $\delta^{'}>0$ such that $E_H\{\delta^{'};\varphi\}\subset NE_H\{\varepsilon^{'};g\}$. Hence, for $\tau\in E_H\{\delta^{'};\varphi\}$ we have

$$r_{\langle -N,N\rangle}(g_\tau,g)\leq\varepsilon^{'}.$$

Therefore, for $x\in\langle -N;N\rangle$ and $\tau\in E_H\{\delta^{'};\varphi\}$ we obtain

$$g_\tau(x)=g(x+\vartheta_1^{'}(x,\tau)\varepsilon^{'})+\vartheta_2^{'}(x,\tau)\varepsilon^{'}$$

and

$$g(x)=g_\tau(x+\vartheta_1^{''}(x,\tau)\varepsilon^{'})+\vartheta_2^{''}(x,\tau)\varepsilon^{'},$$

where $|\vartheta_i^{'}(x,\tau)|\leq 1$, $|\vartheta_i^{''}(x,\tau)|\leq 1$, $i=1,2$, $x+\vartheta_1^{'}\varepsilon^{'}, x+\vartheta_1^{''}\varepsilon^{'}\in\langle -N,N\rangle$.

If $m_N=\inf\{g(x): x\in\mathbb{R}\}$ and $M_N=\sup\{g(x): x\in\mathbb{R}\}$, then f is a uniformly continuous function on the set Y_g. Let $0<\varepsilon'<\min(\varepsilon,\delta)$, where a $\delta=\delta(\varepsilon)>0$ is a number which characterizes the uniform continuity of the function f on the closed interval Y_g, and $\tau\in E_H\{\delta';\varphi\}$. For an arbitrary $X=X(x,y)\in\overline{(f\circ g)_\tau}$ there exists $Y=Y(x+\vartheta_1^{'}\varepsilon^{'}, f(g(x+\vartheta_1^{'}\varepsilon')))\in\overline{f\circ g}$, where $x\in\langle -N,N\rangle$, such that $\|X-Y\|_0\leq\varepsilon$. Analogously, for an arbitrary $X=X(x,y)\in\overline{f\circ g}$ there exists $Y=Y(x+\vartheta_1^{''}\varepsilon^{'}, f(g_\tau(x+\vartheta_1^{''}\varepsilon^{'})))\in\overline{(f\circ g)_\tau}$, where $x\in\langle -N,N\rangle$, such that $\|X-Y\|_0\leq\varepsilon$. Hence, from the above estimates we get by Lemma 1

$$r_{\langle -N,N\rangle}((f\circ g)_\tau, f\circ g)\leq\varepsilon,$$

and so the superposition $f\circ g$ is an (NH)-a.p. function, because every (H,δ')-a.p. of φ is an (NH,ε)-a.p. of $f\circ g$, i.e. φ is a dominant function of $f\circ g$.

In the following we assume that $\langle m_N, M_N\rangle\subsetneq Y_g$. If Y_g is a bounded interval, i.e. there exist the constans $m=\inf\{g(x): x\in\mathbb{R}\}$ and $M=\sup\{g(x): x\in\mathbb{R}\}$, then we construct the closed interval

$$I_N=\begin{cases}\langle m_N-a_N, M_N+a_N\rangle, & \text{where } 0<a_N=\frac{1}{2}\min\{m_N-m, M-M_N\},\\ & \text{if } m_N>m,\ M_N<M,\\ \langle m_N, M_N+a_N\rangle, & \text{where } 0<a_N=\frac{1}{2}(M-M_N),\\ & \text{if } m_N=m,\ M_N<M,\\ \langle m_N-a_N, M_N\rangle, & \text{where } 0<a_N=\frac{1}{2}(m_N-m),\\ & \text{if } m_N>m,\ M_N=M.\end{cases}$$

In case when Y_g is an unbounded interval, then we construct the closed interval

$$I'_N = \begin{cases} \langle m_N - 1, M_N + 1\rangle, & \text{if} \quad Y_g = \mathbb{R}, \\ \langle m_N - 1, M_N + a_N\rangle, & \text{where} \quad 0 < a_N = \frac{1}{2}(M - M_N), \\ & \text{if} \quad M_N < M, \ \inf Y_g = -\infty, \\ \langle m_N - 1, M_N\rangle, & \text{if} \quad M_N = M, \ \inf Y_g = -\infty, \\ \langle m_N - a_N, M_N + 1\rangle, & \text{where} \quad 0 < a_N = \frac{1}{2}(m_N - m), \\ & \text{if} \quad m_N > m, \ \sup Y_g = +\infty, \\ \langle m_N, M_N + 1\rangle, & \text{if} \quad m_N = m, \ \sup Y_g = +\infty. \end{cases}$$

The function f is uniformly continuous on the closed interval I_N (or I'_N). Let $0 < \delta = \delta(N,\varepsilon) \le \min(1, a_N)$ be a number which characterizes the uniform continuity of a function f on the interval I_N (or I'_N), $0 < \varepsilon' < \min(\varepsilon, \delta)$ and $\tau \in E_H\{\delta'; \varphi\}$, where δ' is defined as in the first part of the proof.

Then for an arbitrary $X = X(x,y) \in \overline{(f\circ g)_\tau}$, where $x \in \langle -N, N\rangle$, there exists $Y = Y(x + \vartheta'_1\varepsilon', f(g(x+\vartheta'_1\varepsilon'))) \in \overline{f\circ g}$ such that $\|X - Y\|_0 \le \varepsilon'$. Inversely, for an arbitrary $X = X(x,y) \in \overline{f\circ g}$, where $x \in \langle -N, N\rangle$, there exists $Y = Y(x+\vartheta''_1\varepsilon', f(g_\tau(x+\vartheta''_1\varepsilon'))) \in \overline{(f\circ g)_\tau}$ such that $\|X - Y\|_0 \le \varepsilon$. From the above estimates we get by Lemma 1

$$r_{\langle -N,N\rangle}((f\circ g)_\tau, f\circ g) \le \varepsilon.$$

Hence, the superposition $f\circ g$ is an (NH)-a.p. function, because every (H,δ')-a.p. of φ is an $(NH,\varepsilon) - a.p.$ of $f\circ g$, i.e. φ is a dominant function of $f\circ g$. ■

It is known (see [19]), that if the derivative f' of a function $f\colon Y_g \to \mathbb{R}$ is uniformly continuous and bounded on Y_g, where g is a B-a.p. function such that the derivative g' is uniformly continuous, then the superposition $f\circ g$ is V-a.p.

The following theorem, which concerns V_p-a.p. functions, is true:

Theorem 4 . *If the derivative f' of a function $f\colon Y_g \to \mathbb{R}$ is uniformly continuous and bounded on the set Y_g of values of a function g, where g is a V_p-a.p. function, $1 \le p < \infty$, then the superposition $f\circ g$ is V_p-a.p.*

Proof. Because f' is bounded on Y_g, there exists a constant $M_1 > 0$ such that $|f'(u)| \le M_1$ for every $u \in Y_g$. The V_p-a.p. function g is V_p-bounded (see [16]), and so there exists a constant $M_2 > 0$ such that $V_p(g;x) \le M_2$ for every $x \in \mathbb{R}$. Since the derivative f' is bounded on Y_g, so f is uniformly continuous on Y_g. It is known that g is a V_p-a.p. function, and so g is uniformly a.p. Hence, g is uniformly continuous on $\mathbb{R}$ (see [9]). Therefore, for an arbitrary $\varepsilon > 0$ there exist a positive numbers δ_1, δ_2, δ_3 such that

$$|f'(u') - f'(u'')| < \frac{\varepsilon}{6M_2} \quad \text{for} \quad |u' - u''| < \delta_1, \quad u', u'' \in Y_g,$$

and

$$|g(t') - g(t'')| < \frac{\delta_1}{3} \quad \text{for} \quad |t' - t''| < \delta_2, \quad t', t'' \in \mathbb{R},$$

$$|f(u') - f(u'')| < \frac{\varepsilon}{3} \quad \text{for} \quad |u' - u''| < \delta_3, \quad u', u'' \in Y_g, \tag{1}$$

where $0 < \delta_3 < \min(\frac{\delta_1}{3}, \frac{\varepsilon}{6M_1}, \frac{\varepsilon}{3M_1(2n_0^{1/p}+1)})$, $n_0 = [\frac{2}{\delta_2}]+1$. Since $g \in (BV_p)_{loc}$, so for an arbitrary partition $\Pi_x : x-1 = t_0 < t_1 < ... < t_n = x+1$, where $x \in \mathbb{R}$, we have by the Lagrange theorem

$$\{\sum_{i=0}^{n-1} |(f \circ g)(t_{i+1}) - (f \circ g)(t_i)|^p\}^{\frac{1}{p}}$$

$$= \{\sum_{i=0}^{n-1} (|f'(u_i')||g(t_{i+1}) - g(t_i)|)^p\}^{\frac{1}{p}}$$

$$\leq M_1\{\sum_{i=0}^{n-1} |g(t_{i+1}) - g(t_i)|^p\}^{\frac{1}{p}},$$

where $u_i' \in (\min(g(t_i), g(t_{i+1})), \max(g(t_i), g(t_{i+1})))$. Hence, we obtain for every $x \in \mathbb{R}$

$$V_p(f \circ g; x) \leq M_1 V_p(g, x) \leq M_1 M_2 < \infty,$$

and so $f \circ g \in (BV_p)_{loc}$.

Since g is a uniformly a.p. function, so for $\tau \in E_v^p\{\delta_3; g\}$ we have $|g(t+\tau) - g(t)| \leq \delta_3$ for every $t \in \mathbb{R}$. From (1) we conclude that

$$|f(g(t+\tau)) - f(g(t))| < \frac{\varepsilon}{3}, \qquad t \in \mathbb{R}. \tag{2}$$

Because g is V_p-a.p., for $\tau \in E_v^p\{\delta_3; g\}$ we have for every partition Π_x, $x \in \mathbb{R}$, where the partition Π_x is defined as in the first part of the proof,

$$\{\sum_{i=0}^{n-1} |[g(t_{i+1}) - g_\tau(t_{i+1})] - [g(t_i) - g_\tau(t_i)]|^p\}^{\frac{1}{p}} \leq \delta_3. \tag{3}$$

Now, we present the estimate

$$V_p(f \circ g - (f \circ g)_\tau; x), \qquad x \in \mathbb{R},$$

for $\tau \in E_v^p\{\delta_3; g\}$.

We choose the subsequences $(t_{i_k'})$ and $(t_{i_k''})$ of the sequence (t_i), $i = 0, 1, 2, ..., n-1$, such that

$$|t_{i_k'+1} - t_{i_k'}| < \delta_2 \qquad \text{for} \quad i_k' \in \{0, 1, 2, ... n-1\}, \qquad k = 0, 1, 2 ..., n' - 1,$$

and

$$|t_{i_k''+1} - t_{i_k''}| \geq \delta_2 \qquad \text{for} \quad i_k'' \in \{0, 1, 2, ... n-1\}, \qquad k = 0, 1, 2 ..., n'' - 1.$$

Let us write

$$\sigma_p(f \circ g - (f \circ g)_\tau; \Pi_x) = \{\sum_{i=0}^{n-1} |[f(g(t_{i+1})) - f(g_\tau(t_{i+1}))] - [f(g(t_i)) - f(g_\tau(t_i))]|^p\}^{\frac{1}{p}}$$

$$= \{\sum_{k=0}^{n'-1} |[f(g(t_{i_k'+1})) - f(g_\tau(t_{i_k'+1}))] - [f(g(t_{i_k'})) - f(g_\tau(t_{i_k'}))]|^p$$

$$+\sum_{k=0}^{n''-1}|[f(g(t_{i''_k+1}))-f(g_\tau(t_{i''_k+1}))]-[f(g(t_{i''_k}))-f(g_\tau(t_{i''_k}))]|^p\}^{\frac{1}{p}}$$

$$\leq \{\sum_{k=0}^{n'-1}A^p_{i'_k}\}^{\frac{1}{p}}+\{\sum_{k=0}^{n''-1}A^p_{i''_k}\}^{\frac{1}{p}},$$

where

$$A_l=|[f(g(t_{l+1}))-f(g_\tau(t_{l+1}))]-[f(g(t_l))-f(g_\tau(t_l))]|$$

for $l=i'_k$ or $l=i''_k$.

By the Lagrange theorem it follows that there exists

$$u'_{i'_k}\in(\min(g(t_{i'_k}),g(t_{i'_k+1})),\ \max(g(t_{i'_k}),g(t_{i'_k+1}))),$$

$$u''_{i'_k}\in(\min(g_\tau(t_{i'_k}),g_\tau(t_{i'_k+1})),\ \max(g_\tau(t_{i'_k}),g_\tau(t_{i'_k+1})))$$

for which

$$A^p_{i'_k}\leq(|f'(u'_{i'_k})-f'(u''_{i'_k})||g_\tau(t_{i'_k+1})-g_\tau(t_{i'_k})|+|[g(t_{i'_k+1})-g_\tau(t_{i'_k+1})]-[g(t_{i'_k})-g_\tau(t_{i'_k})]||f'(u_{i'_k})|)^p,$$

and there exists

$$u'_l\in(\min(g(t_l),g_\tau(t_l)),\max(g(t_l),g_\tau(t_l))),\quad l=i''_k,i''_k+1,$$

for which

$$A^p_{i''_k}\leq(|f'(u'_{i''_k+1})-f'(u'_{i''_k})||g(t_{i''_k})-g_\tau(t_{i''_k})|$$

$$+|[g(t_{i''_k+1})-g_\tau(t_{i''_k+1})]-[g(t_{i''_k})-g_\tau(t_{i''_k})]||f'(u'_{i''_k+1})|)^p.$$

Using Minkowski's inequality, we obtain

$$\{\sum_{k=0}^{n'-1}A^p_{i'_k}\}^{\frac{1}{p}}\leq\{\sum_{k=0}^{n'-1}(|f'(u'_{i'_k})-f'(u''_{i'_k})||g_\tau(t_{i'_k+1})-g_\tau(t_{i'_k})|)^p\}^{\frac{1}{p}}$$

$$+\{\sum_{k=0}^{n'-1}(|[g(t_{i'_k+1})-g_\tau(t_{i'_k+1})]-[g(t_{i'_k})-g_\tau(t_{i'_k})]||f'(u_{i'_k})|)^p\}^{\frac{1}{p}}.$$

For $\tau\in E^p_\nu\{\delta_3;g\}$ we have $|u'_{i'_k}-u''_{i'_k}|<\delta_1$, and so

$$|f'(u'_{i'_k})-f'(u''_{i'_k})|<\frac{\varepsilon}{6M_2},\qquad k=0,1,2,\ldots n'-1,$$

and, using the estimation (3), we obtain

$$\{\sum_{k=0}^{n'-1}A^p_{i'_k}\}^{\frac{1}{p}}<\frac{\varepsilon}{3}.$$

In the following we have for $\tau\in E^p_\nu\{\delta_3;g\}$

$$\{\sum_{k=0}^{n''-1}A^p_{i''_k}\}^{\frac{1}{p}}\leq\{\sum_{k=0}^{n''-1}(|f'(u'_{i''_k+1})-f'(u'_{i''_k})||g(t_{i''_k})-g_\tau(t_{i''_k})|)^p\}^{\frac{1}{p}}$$

$$+\{\sum_{k=0}^{n''-1}(|[g(t_{i''_k+1})-g_\tau(t_{i''_k+1})]-[g(t_{i''_k})-g_\tau(t_{i''_k})]||f'(u'_{i''_k+1})|)^p\}^{\frac{1}{p}}$$

$$\leq M_1\delta_3(2n_0^{\frac{1}{p}}+1),$$

where $n_0=[2/\delta_2]+1$. Therefore, we obtain

$$\sigma_p(f\circ g-(f\circ g)_\tau;\Pi_x)<\frac{\varepsilon}{3}+M_1\delta_3(2n_0^{\frac{1}{p}}+1)<\frac{2}{3}\varepsilon. \quad (4)$$

From (2) and (4) it follows that

$$V_p(f\circ g-(f\circ g)_\tau)\leq\varepsilon$$

for $\tau\in E_v^p\{\delta_3;g\}$, i.e. the superposition $f\circ g$ is V_p-a.p. ■

It is known (see [2]), that if the derivative f' of a function $f\colon Y_g\to\mathbb{R}$ is uniformly continuous and bounded on Y_g, where g is a $C^{(1)}$-a.p. function, then the superposition $f\circ g$ is $C^{(1)}$-a.p.

Now, we will prove the following theorem:

Theorem 5 . *If the derivative f' of a function $f\colon Y_g\to\mathbb{R}$ is continuous on the set Y_g of values of a function g, where g is an $(NC^{(1)})$-a.p. function, then the superposition $f\circ g$ is $(NC^{(1)})$-a.p.*

Proof. It is known (see [3]) that since g is an $(NC^{(1)})$-a.p. function, so g and g' are N-a.p. Because the superposition of a continuous function and an N-a.p. function is an N-a.p. function (see [19]), the superpositions $f\circ g$ and $f'\circ g$ are N-a.p. For every $x\in\mathbb{R}$ we have $(f\circ g)'(x)=f'(u)\cdot g'(x)$ where $u=g(x)$, and hence the product $f'(g)\cdot g'$ is N-a.p. (see [9], p. 146). It is known (see [3]) that then $f\circ g$ is an $(NC^{(1)})$-a.p. function. ■

Acknowledgment

I would like to express my gratitude to the Reviewer for his invaluable critical remarks and suggestions.

References

[1] M. Adamczak, $C^{(n)}$-almost periodic functions, *Commentat. Math.* **37** (1997), 1–12.

[2] M. Adamczak, On compactness of $C^{(n)}$-almost periodic functions, *Demonstr. Math.* Vol.32 No 2 (1999), 385–393.

[3] M. Adamczak and S. Stoiński, On the $(NC^{(n)})$-almost periodic functions, Function Spaces, *Proceedings of the Sixth Conference, Wrocław 2001*, World Scientific Publishing Co. Pte. Ltd. (2003), 39–48.

[4] A. S. Besicovitch, *Almost periodic functions*, Cambridge 1932.

[5] H. Bohr, *Zur Theorie der fastperiodischen Funktionen*, I Teil, *Acta Math.* **45** (1925), 29–127; II Teil, *Acta Math.* **46** (1925), 101–214.

[6] C. Corduneanu, *Almost Periodic Functions*, Chelsea Publishing Co., New York, 1989.

[7] A. S. Djafarov and G. M. Gasanov, Almost periodic functions of the relatively Hausdorff metric and some of their properties, *Izv. Acad. Azerb. Sci. Ser. Sci. Phys. Tech. Math.* **1** (1977), 57–62 (in Russian).

[8] F. Hausdorff, *Grundzüge der Mengenlehre*, Leipzig 1914.

[9] B. M. Levitan, *Almost periodic functions*, Moscow 1953 (in Russian) .

[10] B. Sendov and B. Penkov, ε-entropy and ε-capacity of the space of continuous functions, *Izv. Math. Inst. Acad. Bulg. Sci.* **6** (1962), 27–50 (in Bulgarian).

[11] S. Stoiński, H-almost periodic functions, *Funct. Approximatio Comment. Math.* **1** (1974), 113–122.

[12] S. Stoiński, A connection between H-almost periodic functions and almost periodic functions of other types, *Funct. Approximatio Comment. Math.* **3** (1976), 205–223.

[13] S. Stoiński, On modulus of non-monotonicity of an H-almost periodic function, *Funct. Approximatio Comment. Math.* **4** (1976), 85–90.

[14] S. Stoiński, On superposition of the Bohr's almost periodic functions, *Funct. Approximatio Comment. Math.* **5** (1977), 85–89.

[15] S. Stoiński, Real-valued functions almost periodic in variation, *Funct. Approximatio Comment. Math.* **22** (1993), 141–148.

[16] S. Stoiński, Real-valued functions almost periodic in p-variation, *Fasc. Math.* **26** (1996), 155–162.

[17] S. Stoiński, N-almost periodic in the sense of variation functions, *Commentat. Math.* **41** (2001), 195–202.

[18] S. Stoiński, A note on N-almost periodic functions and (NI)-almost periodic functions, *Commentat. Math.* **44**(2) (2004), 199–204.

[19] S. Stoiński, On superposition of almost periodic functions, *International Journal of Evolution Equations* Vol.1 Number 2 (2005), 145–151.

In: Research on Evolution Equation Compendium. Volume 2 ISBN: 978-1-60456-752-6
Editor: Gaston M. N'Guerekata

TIME PERIODIC SOLUTIONS FOR QUASIGEOSTROPHIC MOTION AND THEIR STABILITY

Mei-Qin Zhan*
Department of Mathematics & Statistics
University of North Florida
Jacksonville, FL 32224, USA

Abstract

In this article, we study the quasigeostrophic equation, which is a prototypical geophysical fluid model. We will show the existence of time-periodic solutions for any the Coriolis parameter and the Ekman dissipation constant with nonhomogeneous boundary conditions, which improves the results obtained in [4] and [13].

1991 Mathematics Subject Classification: 35B05, 35B30, 35B65

Key words and phrases: quasigeostrophic fluid model, time-periodic motion, positive solution, monotonicity

1. Introduction

The two dimensional barotropic quasigeostrophic equation

$$\Delta\psi_t + J(\psi, \Delta\psi) + \beta\psi_x = \nu\Delta^2\psi - r\Delta\psi + f(x,y,t), \tag{1.1}$$

where $\psi(x,y,t)$ is the stream function, $\beta > 0$ is the meridional gradient of the Coriolis parameter, $\nu > 0$ is the viscous dissipation constant, $r > 0$ is the Ekman dissipation constant, and $f(x,y,t)$ is the wind forcing. Moreover, $\Delta = \partial_{xx} + \partial_{yy}$ is the Laplacian operator in the x,y plane, $x, y \in D$ an arbitrary bounded planar domain with piecewise smooth boundary, and $J(f,g) = f_x g_y - f_y g_x$ is the Jacobi operator.

This equation is derived as an approximation of the rotating shallow water equations by the conventional asymptotic expansion in small Rossby number. Schochet ([9]) has recently shown that quasigeostrophy is a valid approximation of the rotating shallow water equations in the limit of zero Rossby number, i.e. at asymptotically high rotation rate.

Assume ψ is C^2, let $\omega = \Delta\psi$, then the quasigeostrophic equation becomes the following system of equations,

$$\begin{aligned} &\omega_t + J(\psi, \omega) + \beta\psi_x = \nu\Delta\omega - r\omega + f(x,y,t) \\ &\Delta\psi = \omega, \end{aligned} \tag{1.2}$$

*E-mail address: mzhan@unf.edu

with the following boundary and initial conditions, which satisfies appropriate compatible conditions,

$$\begin{array}{lll}\psi(x,y,t) & = g_1(x,y,t) & \text{on } \partial D,\\ \omega(x,y,t) & = g_2(x,y,t) & \text{on } \partial D\\ \omega(x,y,0) & = \omega_0(x,y). & \end{array} \tag{1.3}$$

The time periodic solution of (1.2) has been studied by many authors. In [4], the author studied the quasigeostrophic model (1.2) with the homogeneous boundary conditions, $g_1(x,y,t) = g_2(x,y,t) = 0$. He showed that (1.2) admits at least one time periodic solution under a rather restrictive technique condition that the Coriolis parameter and the Ekman dissipation constant satisfy $r + \frac{\pi\nu}{|D|} > \frac{1}{2}\beta(\frac{|D|}{\pi} + 1)$, here $|D|$ is the area of domain D. In [13], under the same condition on r and β, the author shows that the when $f(x,y,t)$ is positive and periodic in time, the barotropic quasigeostrophic equation admits a time-periodic positive solution. Moreover, the author also showed that the solution is stable when its ω_x is bounded in L^∞ norm.

In this article, by using the method of upper and lower solutions and its associated monotone iteration, we will show that the equation (1.2) admits at least one time-periodic for any the Coriolis parameter and the Ekman dissipation constant with no homogeneous boundary conditions. Thus we are able to improve the result of [4] and [13].

2. The Main Results

In this section we will state and prove our main results.

We first give a sketch proof of the following existence of classical solution to the quasi-geostrophic equations (1.2)-(1.3). Let $\overline{D} = D \cup \partial D$, and $x = C^{1,2}([0,\infty) \times \overline{D}) \cap C([0,\infty))$.

Lemma 2.1. *The quasigeostrophic equations (1.2)-(1.3) have an unique classical smooth solution* $\omega, \phi \in x$ *when* $f(x,y,t) \in C([0,\infty), C^\alpha(\overline{D}))$, $g_1(x,y,t), g_2(x,y,t) \in C([0,\infty)C^\alpha(\partial D))$, *and* $\omega_0(x,y) \in C^{2,\alpha}(\overline{D})$.

Proof. First we transform the equations so that the boundary conditions become non-slippery. Let ω^* and ψ^* be the classical solutions of the following two linear equations,

$$\begin{array}{l}\omega_t^* = \nu\Delta\omega^* - r\omega^*\\ \omega^*(x,y,t) = g_2(x,y,t) \quad \text{on } \partial D\\ \omega(x,y,0) = 0.\end{array} \tag{2.4}$$

and

$$\begin{array}{l}\Delta\psi^* = \omega^*,\\ \psi^*(x,y,t) = g_1(x,y,t) \quad \text{on } \partial D,\end{array} \tag{2.5}$$

Set $\overline{\omega} = \omega - \omega^*$ and $\overline{\psi} = \psi - \psi^*$, we have the following equations,

$$\begin{array}{l}\overline{\omega}_t + J(\overline{\psi},\overline{\omega}) + \beta\overline{\psi}_x + \beta\psi_x^* + J(\overline{\psi},\omega^*) + J(\psi^*,\overline{\omega})\\ = \nu\Delta\overline{\omega} - r\overline{\omega} + f(x,y,t) + J(\omega^*,\psi^*)\\ \Delta\overline{\psi} = \overline{\omega},\end{array} \tag{2.6}$$

with the following non-slippery boundary conditions and initial condition

$$\begin{aligned} \overline{\psi}(x,y,t) &= 0 \quad \text{on } \partial D, \\ \overline{\omega}(x,y,t) &= 0 \quad \text{on } \partial D \\ \overline{\omega}(x,y,0) &= \omega_0(x,y). \end{aligned} \tag{2.7}$$

The equations (2.6-2.7) have the same nonlinear term as the vorticity form of the 2-D Navier-Stokes equations, so the methods in [12, 3, 6, 8] can be applied to obtain the existence and uniqueness of the classical solutions.

The main idea uses the nonlinear Galerkin method. By projecting $\overline{\omega}, \omega^*, \psi^*, \overline{f} = f(x,y,t) + J(\omega^*, \psi^*)$ onto a finite-dimensional subspace(expanded by finite-Fourier series). Let P^n be the projection operator and $G^n = P^n G$ etc to denote the projection of any function G into the subspace, the equations (2.6) become a system of ODEs

$$\begin{aligned} &\overline{\omega}^n_t + P^n J(\overline{\psi}, \overline{\omega}) + \beta \overline{\psi}^n_x + +\beta \psi^{*n}_x + P^n J(\overline{\psi}^n, \omega^{*n}) + P^n J(\psi^{*n}, \overline{\omega}^n) \\ &= \nu \Delta \overline{\omega}^n - r\overline{\omega} + f^n(x,y,t) + P^n J(\omega^{*n}, \psi^{*n}) \\ &\Delta \overline{\psi}^n = \overline{\omega}^n, \end{aligned} \tag{2.8}$$

which can be solved globally and to obtain the approximate solutions. To prove the convergence of the approximate solutions we need the following energy estimates.

First, multiply both sides of equation for $\overline{\omega}$ in (2.6) by $\overline{\omega}$ and integrate by part, notice

$$\int_D J(\overline{\psi}, \overline{\omega}) \overline{\omega} dxdy = 0$$

and the Poincaré inequality,

$$|\beta \int_D \overline{\psi}_x \overline{\omega} dxdy| \le \frac{1}{2} \beta (\frac{|D|}{\pi} + 1) \int_D \overline{\omega}^2 dxdy$$

we have

$$\frac{1}{2} \frac{d}{dt} ||\overline{\omega}||^2 + C_1 || \bigtriangledown \overline{\omega} ||^2 \le C_2(t) \int_D \overline{\omega}^2 dxdy + C_3(t) \tag{2.9}$$

Second, from the equation for $\overline{\psi}$ in (2.6) we have

$$\int_D |\Delta \overline{\psi}|^2 dxdy \le 2 \int_D |\overline{\omega}|^2 dxdy$$

Notice that

$$\begin{aligned} &| \int_D J(\overline{\psi}, \overline{\omega}) \Delta \overline{\omega} dxdy | \\ &\le C(\varepsilon) || \bigtriangledown \overline{\omega} ||^2_{L^4(D)}(t) || \bigtriangledown \overline{\omega} ||^2_{L^4(D)}(t) + \varepsilon ||\Delta \overline{\omega}||^2 \end{aligned} \tag{2.10}$$

and using this and Sobolev embedding, we know that $|| \bigtriangledown \overline{\omega} ||^2_{L^4(D)}(t) || \bigtriangledown \overline{\omega} ||^2_{L^4(D)}$ is in $L^1([0,T))$ for any $T > 0$.

Finally, multiply both sides of equation for $\overline{\omega}$ in (2.6) by $\Delta \overline{\omega}$ and integrate by part, using (2.10) we have

$$\begin{aligned} \tfrac{1}{2} \tfrac{d}{dt} || \bigtriangledown \overline{\omega} ||^2 + C_4 ||\Delta \overline{\omega}||^2 &\le C_5 || \bigtriangledown \overline{\omega} ||^2_{L^4(D)}(t) || \bigtriangledown \overline{\omega} ||^2_{L^4(D)}(t) \\ &+ C_6(t) || \bigtriangledown \overline{\omega} ||^2 + C_7(t) ||\overline{\omega}||^2 + C_8(t) \end{aligned} \tag{2.11}$$

Here $C_i(t), i = 1, \cdots 8$ denote various bounded or integrable functions in t that do not depend on $\overline{\psi}$, and $\overline{\omega}$. Inequalities (2.9-2.11) enable us to pass to limit to obtain strong solutions in Sobolev space $C^1([0,T), H^1(\overline{D})) \cap L^2([0,T), H^2(\overline{D}))$.

Then regularity results on the parabolic and elliptic equations, one has the classical smooth solutions. The uniqueness is a consequence of the comparison principle. □

The next result we want to present is the following comparison principle, this result indicates that the quasigeostrophic equation enjoys a comparison principle such that two solutions with different initial data and external forces can be compared.

Theorem 2.2. *Suppose that $f^{(i)}(x,y,t) \in C^\alpha(\overline{D})$, $g_1^{(i)}(x,y,t), g_2^{(i)}(x,y,t) \in C^\alpha(\partial D)$ $i = 1,2$ and $f^{(1)}(x,y,t) \geq f^{(2)}(x,y,t)$, $g_2^{(1)}(x,y,t) \geq g_2^{(2)}(x,y,t)$, $\omega_0^{(1)}(x,y) \geq \omega_0^{(2)}(x,y)$, for all x,y,t.*

Let $\omega^{(i)}(x,y,t) \in x$, $\psi^{(i)}(x,y,t) \in x$, $i = 1,2$, be the solutions of

$$\begin{aligned} &\omega_t^{(i)} + J(\psi^{(i)}, \omega^{(i)}) + \beta\psi_x^{(i)} = \nu\Delta\omega^{(i)} - r\omega^{(i)} + f^{(i)}(x,y,t) \\ &\Delta\psi^{(i)} = \omega^{(i)}, \end{aligned} \tag{2.12}$$

with boundary and initial conditions,

$$\begin{aligned} \psi^{(i)}(x,y,t) &= g_1^{(i)}(x,y,t) \quad \text{on } \partial D, \\ \omega^{(i)}(x,y,t) &= g_2^{(i)}(x,y,t) \quad \text{on } \partial D \\ \omega^{(i)}(x,y,0) &= \omega_0^{(i)}(x,y). \end{aligned} \tag{2.13}$$

We have $\omega^{(1)}(x,y,t) \geq \omega^{(2)}(x,y,t)$ for all x,y,t.

Proof. Let

$$\begin{aligned} \delta\omega(x,y,t) &= \omega^{(1)}(x,y,t) - \omega^{(2)}(x,y,t), \\ \delta\psi(x,y,t) &= \psi^{(1)}(x,y,t) - \psi^{(2)}(x,y,t), \\ \delta f(x,y,t) &= f^{(1)}(x,y,t) - f^{(2)}(x,y,t), \\ \delta g_1(x,y,t) &= g_1^{(1)}(x,y,t) - g_1^{(2)}(x,y,t), \\ \delta g_2(x,y,t) &= g_2^{(1)}(x,y,t) - g_2^{(2)}(x,y,t), \\ \delta\omega_0(x,y) &= \omega_0^{(1)}(x,y) - \omega_0^{(2)}(x,y), \end{aligned} \tag{2.14}$$

Using the equations (2.12) and (2.13) we have the following equations for the differences defined in (2.14)

$$\begin{aligned} &\delta\omega_t + J(\psi^{(1)}, \delta\omega) + J(\delta\psi, \omega^{(2)}) + \beta\delta\psi_x = \nu\Delta\delta\omega - r\delta\omega + \delta f(x,y,t) \\ &\Delta\delta\psi = \delta\omega, \end{aligned} \tag{2.15}$$

with boundary and initial conditions,

$$\begin{aligned} \delta\psi(x,y,t) &= \delta g_1(x,y,t) \quad \text{on } \partial D, \\ \delta\omega(x,y,t) &= \delta g_2(x,y,t) \quad \text{on } \partial D \\ \delta\omega(x,y,0) &= \delta\omega_0(x,y). \end{aligned} \tag{2.16}$$

Now let $u(x,y,t) = e^{-\lambda t}\delta\omega(x,y,t)$, and $\Psi(x,y,t) = e^{-\lambda t}\delta\psi(x,y,t)$, then

$$u_t = -\lambda u + e^{-\lambda t}\frac{d}{dt}\delta\omega(x,y,t),$$

so

$$u_t + \lambda u = e^{-\lambda t} \frac{d}{dt} \delta\omega(x, y, t).$$

Multiply both sides of equations in (2.15) and (2.16) by $e^{-\lambda t}$ we have

$$\begin{aligned} & u_t + \lambda u + J(\psi^{(1)}, u) + J(\Psi, \omega^{(2)}) + \beta\Psi_x \\ & \quad = \nu\Delta u - ru + e^{-\lambda t}\delta f(x, y, t) \\ & \Delta\Psi = u. \end{aligned} \tag{2.17}$$

with boundary and initial conditions,

$$\begin{aligned} \Psi(x,y,t) &= e^{-\lambda t}\delta g_1(x,y,t) \quad \text{on } \partial D, \\ u(x,y,t) &= e^{-\lambda t}\delta g_2(x,y,t) \quad \text{on } \partial D \\ u(x,y,0) &= u_0(x,y). \end{aligned} \tag{2.18}$$

Now if $\delta\omega(x_0, y_0, t_0) < 0$, at some point interior point $(x_0, y_0, t_0) \in (0, T) \times D$. We have u reaches its negative minimum at some interior $(x_1, y_1, t_1) \in (0, T) \times D$. So

$$u_t(x_1, y_1, t_1) = u_x(x_1, y_1, t_1) = u_y(x_1, y_1, t_1) = 0,$$

this indicates that, at (x_1, y_1, t_1),

$$J(\psi^{(1)}, u) = 0.$$

Furthermore, we have

$$\Delta u(x_1, y_1, t_1) \geq 0.$$

Using the equation (2.17) for u, we have for all $t \geq 0$,

$$\begin{aligned} & \lambda u(x_1, y_1, t_1) + J(\Psi, \omega^{(2)})(x_1, y_1, t_1) + \beta\Psi_x(x_1, y_1, t_1) \\ & = \nu\Delta u(x_1, y_1, t_1) - ru(x_1, y_1, t_1) + e^{-\lambda t}\delta f(x_1, y_1, t_1) \\ & \geq -ru(x_1, y_1, t_1) + e^{-\lambda t}\delta f(x_1, y_1, t_1) \end{aligned} \tag{2.19}$$

Notice that Ψ satisfied the following second-order linear elliptic equation,

$$\begin{aligned} & \Delta\Psi = u \\ & \Psi(x,y,t) \quad = e^{-\lambda t}\delta g_1(x,y,t) \quad \text{on } \partial D, \end{aligned} \tag{2.20}$$

so we have,

$$|\Psi_x(x_1, y_1, t_1)| + |\Psi_y(x_1, y_1, t_1)| \leq C(|u|(t_1) + |e^{-\lambda t_1}\delta g_2|(t_1)), \tag{2.21}$$

here $|h|(t)$ denotes the maximum value of any function $h(x, y, t)$ over D for each fixed t.

It is easy to see that

$$|J(\Psi, \omega^{(2)}) + \beta\Psi_x| \leq C(|\Psi_x|(t) + |\Psi_y|(t)), \tag{2.22}$$

Combine (2.21) and (2.22), we have

$$|J(\Psi, \omega^{(2)}) + \beta\Psi_x| \leq C(|u|(t_1) + |e^{-\lambda t_1}\delta g_2|(t_1)) \tag{2.23}$$

Since $u(x_1,y_1,t_1) < 0$ we can choose λ large enough in (2.19) so that, using the estimate (2.23),

$$0 > \lambda u(x_1,y_1,t_1) + \beta C(|u|(t_1) + |e^{-\lambda t_1}\delta g_2|) + ru(x_1,y_1,t_1),$$

which contradicts with

$$\lambda u(x_1,y_1,t_1) + \beta C(|u|(t_1) + |e^{-\lambda t_1}\delta g_2|(t_1)) + ru(x_1,y_1,t_1) \geq e^{-\lambda t}\delta f(x_1,y_1,t_1) \geq 0.$$

Therefore u, and hence $\delta\omega$ have to be nonnegative. This proves $\omega^{(1)}(x,y,t) \geq \omega^{(2)}(x,y,t)$ for all x,y,t. □

Corollary 2.3. *Suppose $f(x,y,t), g_2(x,y,t)$, and ω_0 is nonnegative, then the solution $\omega(x,y,t)$ to equation (1.2-1.3) is nonnegative for all time $t > 0$.*

Proof. Choose $f^{(2)}(x,y,t) \equiv g_1^{(2)}(x,y,t) \equiv g_2^{(2)}(x,y,t) \equiv 0$ in the Theorem 2.2, we have the result. □

Remark 2.4. *Let $L(\omega) = J(\psi,\omega) + \beta\psi_x - \nu\Delta\omega + r\omega$, this result shows that L is a monotone operator. So it is possible to prove existence of positive time-periodic solutions and stability of time-periodic and almost time-periodic solutions.*

Hereafter, we require that $f(x,y,t) \in C^{\alpha}(\overline{D})$, $g_1(x,y,t), g_2(x,y,t) \in C^{\alpha}(\partial D)$ are all time periodic functions, and ∂D is of class $C^{1+\alpha}$.

We define the upper and lower solutions for the quasigeostrophic equations below,

Definition 2.5. *A function $\overline{u} = (\overline{\omega},\overline{\psi})$ is called an upper solution of (1.2) in x if*

$$\begin{aligned}
&\overline{\omega}_t + J(\overline{\psi},\overline{\omega}) + \beta\overline{\psi}_x - \nu\Delta\overline{\omega} + r\overline{\omega} \geq f(x,y,t)\\
&\Delta\overline{\psi} = \overline{\omega},\\
&\overline{\psi}(x,y,t) \geq g_1(x,y,t) \quad on\ \partial D,\\
&\overline{\omega}(x,y,t) \geq g_2(x,y,t) \quad on\ \partial D\\
&\overline{\omega}(x,y,0) \geq \overline{\omega}(x,y,T).
\end{aligned} \tag{2.24}$$

Similarly, $\underline{u} = (\underline{\omega},\underline{\psi})$ is called a lower solution if it satisfies the inequalities in 2.24 in reversed order.

Lemma 2.6. *Suppose that*

(1) $f(x,y,t), g_2(x,y,t)$, and $g_1(x,y,t)$ are positive and

(2) there exists a $M > 0$ such that, $\forall(x,y,t) \in [0,T] \times \overline{D}$,

$$M > max\{\frac{f(x,y,t)}{r}, g_1(x,y,t), g_2(x,y,t)\},$$

we have $\underline{u}_0 = (0,0)$ is a lower solution of (1.2)-(1.3) and $\underline{u}_M = (M, M(\frac{1}{2}y^2+1))$ is an upper solution of (1.2)-(1.3).

Proof. It is easy to see that $\underline{u}_0 = (0,0)$ is a lower solution of (1.2)-(1.3). To show that $\underline{u}_M = (M, M(\frac{1}{2}y^2+1))$ is an upper solution of (1.2)-(1.3) one just needs to notice that, for $\overline{\psi} = M(\frac{1}{2}y^2+1)$ we have $\Delta\overline{\psi} = M$ and $\overline{\psi}_x = 0$. □

Let $J \equiv \{u = (\omega, \psi) \in x\,; 0 \leq \omega \leq M, \Delta\psi = \omega\}$, where M is the constant specified in Lemma 2.6. A pair of upper and lower solutions $\overline{u} = (\overline{\omega}, \overline{\psi}), \underline{u} = (\underline{\omega}, \underline{\psi}) \in J$ are said to be ordered in J if $\overline{\omega} \geq \underline{\omega}$ on $\overline{D}$.

Theorem 2.7. *Under assumption of Lemma 2.6, the quasigeostrophic equations (1.2)-(1.3) have both maximal time-periodic solution $\overline{u}$ and a minimal time-periodic solution $\underline{u}$ that satisfies,*

$$0 \leq \underline{u} \leq \overline{u}.$$

Here we say that $\overline{u}$ is maximal and $\underline{u}$ is minimal in the sense that, for any periodic solution $u \in J$, we have

$$\underline{u} \leq u \leq \overline{u}.$$

Proof. Starting from either $u^{(0)} = \underline{u}_0$ or $u^{(0)} = \overline{u}_M$ as an initial iteration we construct a sequence $\{u^{(m)} = (\omega^{(m)}, \psi^{(m)}\}, m = 1, 2, 3, \cdots$ by solving

$$\begin{aligned}
&\omega_t^{(m)} + J(\psi^{(m)}, \omega^{(m)}) + \beta\psi_x^{(m)} = \nu\Delta\omega^{(m)} - r\omega^{(m)} + f(x,y,t)\\
&\Delta\psi^{(m)} = \omega^{(m)},\\
&\psi^{(m)}(x,y,t) = g_1(x,y,t) \quad \text{on } \partial D,\\
&\omega^{(m)}(x,y,t) = g_2(x,y,t) \quad \text{on } \partial D\\
&\omega^{(m)}(x,y,0) = \omega^{(m-1)}(x,y,T).
\end{aligned} \tag{2.25}$$

Denote the sequence by $\{\overline{u}^{(m)}\}$ when $u^{(0)} = \underline{u}_0$ and $\{\overline{u}^{(m)}\}$ when $u^{(0)} = \overline{u}_M$.
By Theorem 2.2 and the fact that $\underline{u}_0$ is lower solution, we have

$$\underline{u}_0 \leq \underline{u}^{(1)}.$$

Similarly, the fact that $\overline{u}_M$ is an upper solution and $\overline{u}^{(1)}(x,y,0) = \overline{u}_M(x,y,T) = \overline{u}_M(x,y,0) = (M, M(\frac{1}{2}y^2 + 1))$ indicates that

$$\overline{u}^{(1)} \leq \overline{u}_M.$$

Since $\overline{u}^{(1)}$ and $\underline{u}^{(1)}$ satisfy the same equation with $\overline{u}^{(1)}(x,y,0) = (M, M(\frac{1}{2}y^2 + 1)) > \underline{u}^{(1)}(x,y,0) = (0,0)$, using Theorem 2.2, we have

$$\underline{u}^{(1)} \leq \overline{u}^{(1)}.$$

Inductively, for all $m \geq 1$ we have,

$$\underline{u}_0 \leq \underline{u}^{(m)} \leq \underline{u}^{(m+1)} \leq \overline{u}^{(m+1)} \leq \overline{u}^{(m)} \leq \overline{u}_M. \tag{2.26}$$

Hence $0 \leq \underline{\omega}^{(m)} \leq \overline{\omega}^{(m)}(x,y,t) \leq M$ for all m.
From $0 \leq |\omega^{(m)}(x,y,t)| \leq M$ for all m, and

$$\Delta\psi^{(m)} = \omega^{(m)},$$

we have

$$\int_{\overline{D}} |\Delta\psi^{(m)}|^2 \, dxdy = \int_{\overline{D}} \omega^{(m)} \Delta\psi^{(m)} \, dxdy \leq \varepsilon \int_{\overline{D}} |\Delta\psi^{(m)}|^2 \, dxdy + \frac{2M}{\varepsilon},$$

So $\{\psi^{(m)}(t)\} \subset H^2(\overline{D})$ is a bounded sequence. By Sobolev embedding we have $\{\psi^{(m)}(t)\} \subset C^{1,\alpha}(\overline{D})$ is a bounded sequence.

Applied the regularity results of the linear parabolic equations we can prove that $\{\psi^{(m)}\}, \{\omega^{(m)}\}$ are bounded sequences in x , and that $0 \leq |\omega^{(m)}(x,y,t)| \leq M$ for all m.

Combine this with the monotonicity of (2.26), we have

$\overline{u} = \lim_{m\to\infty} \overline{u}^{(m)} \in X$ and $\underline{u} = \lim_{m\to\infty} \underline{u}^{(m)} \in X$ are two classical solutions to (2.25) and $\overline{u}, \underline{u} \in J$.

To show that both $\overline{u}$ and $\underline{u}$ are time-periodic solutions, we need to show that $\omega(x,y,t) = \omega(x,y,t+T)$ for all $t > 0$, here we use $\omega(x,y,t)$ to denote either $\overline{\omega}(x,y,t)$ or $\underline{\omega}(x,y,t)$.

Let $\delta\omega(x,y,t) = \omega(x,y,t) - \omega(x,y,t+T)$ we have

$$\begin{aligned} &\delta\omega_t + J(\psi, \delta\omega) + J(\delta\psi, \omega) + \beta\delta\psi_x = \nu\Delta\delta\omega - r\delta\omega \\ &\Delta\delta\psi = \delta\omega, \end{aligned} \tag{2.27}$$

with boundary and initial conditions,

$$\begin{aligned} \delta\psi(x,y,t) &= 0 \quad \text{on } \partial D, \\ \delta\omega(x,y,t) &= 0 \quad \text{on } \partial D \\ \delta\omega(x,y,0) &= 0, \end{aligned} \tag{2.28}$$

Now proceed as in Theorem 2.2, we have $\delta\omega \geq 0$, replace $\delta\omega$ by $-\delta\omega$ we can get $\delta\omega \leq 0$ we have $\delta\omega(x,y,t) \equiv 0$. So $\omega(x,y,t)$ is periodic function in t with period T, hence $\psi(x,y,t)$ is periodic function in t with period T.

To prove that $\overline{u}$ is maximal and $\underline{u}$ is minimal in the sense that, for any periodic solution $u \in J$, we have

$$\underline{u} \leq u \leq \overline{u}.$$

To this end, one just need to see that we can regard u as both upper and lower solutions at the same time and we will have $u \geq \underline{u}^{(m)} \geq \underline{u}_0$ we have $u \geq \underline{u}$. Similarly, we have $u \leq \overline{u}$. □

Remark 2.8. *Notice, we just need to assume that $f(x,y,t), g_1(x,y,t)$, and $g_2(x,y,t)$ are bounded, we require that they are non-negative here just to show that the sign of the periodic solution agrees with the sign of $f(x,y,t), g_1(x,y,t)$, and $g_2(x,y,t)$. When $|f(x,y,t)| + |g_1(x,y,t)| + |g_2(x,y,t)| \leq M$, then $\underline{u}_0 = (-M, -M(\frac{1}{2}y^2+1))$ is the lower solution. In this case we still able to prove the existence of time-periodic solution without impose any conditions on the Coriolis parameter r and the Ekman dissipation constant β.*

It turns out that, under appropriate conditions, the time-periodic solution of (1.2) is stable in the sense stated in the following theorem:

Theorem 2.9. *Suppose that $f(x,y,t) \geq 0$, and there exists a $b < r$ such that*

$$|y\omega_x^*| \leq b,$$

where (ω^*, ψ^*) is the time-periodic solution to the equation

$$\begin{aligned} &\omega_t + J(\psi, \omega) + \beta\psi_x = \nu\Delta\omega - r\omega + f(x,y,t) \\ &\Delta\psi = \omega, \end{aligned} \tag{2.29}$$

with boundary and initial conditions,

$$\begin{aligned} \psi(x,y,t) &= g_1(x,y,t) \quad on\ \partial D, \\ \omega(x,y,t) &= g_2(x,y,t) \quad on\ \partial D \\ \omega(x,y,0) &= \omega_0(x,y). \end{aligned} \tag{2.30}$$

We have for any $\omega_0(x,y)$

$$\omega^*(x,y,0) - \delta_0 b \le \omega_0(x,y) \le \omega^*(x,y,0) + \delta_0 b,$$

if $(\omega(x,y,t),\psi(x,y,t))$ *is solution to the 2.29 with* $\omega(x,y,0) = \omega_0(x,y)$, *then*

$$|\omega(x,y,t) - \omega^*(x,y,t)| \le \delta_0 e^{(b-r)t}.$$

and

$$|\psi(x,y,t) - \psi^*(x,y,t)| \le C\delta_0 e^{(b-r)t},$$

where C depends on Ω *and* $g_1(x,y,t)$.

Proof. Let $\underline{\omega} = \omega^* - \delta_0 e^{(b-r)t}$, $\underline{\psi} = \psi^* - \frac{1}{2}y^2\delta_0 e^{(b-r)t}$.

It is easy to see that

$$\begin{aligned} &\Delta\underline{\psi} = \Delta(\psi^* - \tfrac{1}{2}y^2\delta_0 e^{(b-r)t}) = \Delta\psi^* - \delta_0 e^{(b-r)t} = \underline{\omega} \\ &\underline{\psi}(x,y,t) = g_1(x,y,t) - \delta_0 e^{(b-r)t} \le g_1(x,y,t) \quad \text{on } \partial D, \\ &\underline{\omega}(x,y,t) = g_2(x,y,t) - \delta_0 e^{(b-r)t} \le g_2(x,y,t) \quad \text{on } \partial D \\ &\underline{\omega}(x,y,0) = \omega^*(x,y,0) - \delta_0 = \omega^*(x,y,T) - \delta_0 \\ &\le \omega^*(x,y,T) - \delta_0 e^{(b-r)T} = \underline{\omega}(x,y,T). \end{aligned} \tag{2.31}$$

Now, for $\underline{\omega}$, we have

$$\begin{aligned} &\underline{\omega}_t + J(\underline{\psi},\underline{\omega}) + \beta\underline{\psi}_x - \nu\Delta\underline{\omega} + r\underline{\omega} \\ &= \omega_t^* + J(\psi^*,\omega^*) + \beta\psi_x^* - \nu\Delta\omega^* - r\omega^* \\ &\quad + y\delta_0 e^{(b-r)t}\omega_x^* + r\delta_0 e^{(b-r)t} - (b-r)\delta_0 e^{(b-r)t} \\ &= f(x,y,t) + y\delta_0 e^{(b-r)t}\omega_x^* - r\delta_0 e^{(b-r)t} - (b-r)\delta_0 e^{(b-r)t} \\ &\le f(x,y,t). \end{aligned} \tag{2.32}$$

Similarly, let $\overline{\omega} = \omega^* + \delta_0 e^{(b-r)t}, \overline{\psi} = \psi^* + \frac{1}{2}y^2\delta_0 e^{(b-r)t})$, we can show that

$$\begin{aligned} &\Delta\overline{\psi} = \Delta(\psi^* - \tfrac{1}{2}y^2\delta_0 e^{(b-r)t}) \\ &= \Delta\psi^* - \delta_0 e^{(b-r)t} = \overline{\omega} \\ &\overline{\psi}(x,y,t) = g_1(x,y,t) + \delta_0 e^{(b-r)t} \ge g_1(x,y,t) \quad \text{on } \partial D, \\ &\overline{\omega}(x,y,t) = g_2(x,y,t) + \delta_0 e^{(b-r)t} \ge g_2(x,y,t) \quad \text{on } \partial D \\ &\overline{\omega}(x,y,0) = \omega^*(x,y,0) + \delta_0 = \omega^*(x,y,T) + \delta_0 \\ &\ge \omega^*(x,y,T) + \delta_0 e^{(b-r)T} = \overline{\omega}(x,y,T). \end{aligned} \tag{2.33}$$

Now, for $\overline{\omega}$, we have

$$\begin{aligned} &\overline{\omega}_t + J(\overline{\psi},\overline{\omega}) + \beta\overline{\psi}_x - \nu\Delta\overline{\omega} + r\overline{\omega} \\ &= \omega_t^* + J(\psi^*,\omega^*) + \beta\psi_x^* - \nu\Delta\omega^* - r\omega^* \\ &\quad + y\delta_0 e^{(b-r)t}\omega_x^* + r\delta_0 e^{(b-r)t} + (b-r)\delta_0 e^{(b-r)t} \\ &= f(x,y,t) + y\delta_0 e^{(b-r)t}\omega_x^* - r\delta_0 e^{(b-r)t} + (b-r)\delta_0 e^{(b-r)t} \\ &\ge f(x,y,t). \end{aligned} \tag{2.34}$$

Using the comparison result of Theorem 2.2, we have that

$$\omega^* - \delta_0 e^{(b-r)t} \leq \omega \leq \omega^* + \delta_0 e^{(b-r)t}.$$

Elementary result on the elliptic equations show that

$$|\psi(x,y,t) - \psi^*(x,y,t)| \leq C\delta_0 e^{(b-r)t},$$

with C depends on Ω and $g_1(x,y,t)$. □

Remark 2.10. *Note that if $f(x,y,t) = 0$ and $g_1(x,y,t) = g_2(x,y,t) = 0$ we have the trivial time-period solution, $\omega(x,y,t) = \psi(x,y,t) = 0$. So the condition in the Theorem 2.9 is satisfied and we thus show that when the initial data is small, the solution will decay exponentially.*

Corollary 2.11. *For any $|\omega_0(x,y)| \leq \delta_0 b$, if $(\omega(x,y,t), \psi(x,y,t))$ is solution to the equation,*

$$\begin{aligned} &\omega_t + J(\psi,\omega) + \beta\psi_x = \nu\Delta\omega - r\omega \\ &\Delta\psi = \omega, \end{aligned}$$

with boundary and initial conditions,

$$\begin{aligned} \psi(x,y,t) &= 0 \quad on\ \partial D, \\ \omega(x,y,t) &= 0 \quad on\ \partial D \\ \omega(x,y,0) &= \omega_0(x,y). \end{aligned}$$

we have

$$|\omega(x,y,t)| \leq \delta_0 e^{(b-r)t}.$$

and

$$|\psi(x,y,t)| \leq C\delta_0 e^{(b-r)t},$$

where C depends only on Ω.

Proof. From the remark we see that $\omega(x,y,t) = \psi(x,y,t) = 0$ are periodic solution, from Theorem 2.9, we have the decaying result. □

3. Conclusion

The quasigeostrophic equation studied here models the geophysical fluid dynamics in the case of of infinite Rossby deformation radius and with flat bottom and is applicable to planetary-scale solutions. The situation with infinite Rossby deformation radius is equivalent to the rigid-lid approximation. We have shown that the nonlinear operator defined by quasigeostrophic equation is monotonic in the sense that positive solutions exist when the initial and boundary data is positive. Our result does not impose the conditions on β and r,. On the other hand the monotonicity of the operator $L(\omega) = J(\psi,\omega) + \beta\psi_x - \nu\Delta\omega + r\omega$, will certain enable one to obtain more properties of the long term dynamical behavior of the solutions, including the stability of the time-periodic solutions.

References

[1] A. J. Bourgeois and J. T. Beale; Validity of the quasigeostrophic model for large-scale flow in the atmosphere and ocean, *SIAM, J. Math. Anal.* **25**(1994), 1023-1068.

[2] B. Cushman-Roisin; *Introduction to Geophysical Dynamics,* Prentice Hall, 1994.

[3] Chae, Dongho; Lee, Jihoon Global well-posedness in the super-critical dissipative quasi-geostrophic equations. *Comm. Math. Phys.* **233** (2003), no. 2, 297–311.

[4] J. Duan; Time-Periodic Quasigeostrophic Motion under Dissipation and Forcing, *Appl. Math. Comput.*, **102**(1999), 121-127.

[5] D. Gilbarg and N. S. Trudinger; *Elliptic Partial Differential Equations of Second Order,* 2nd edition, Springer-Verlag, 1983.

[6] Constantin, Peter; Cordoba, Diego; Wu, Jiahong On the critical dissipative quasi-geostrophic equation. Dedicated to Professors Ciprian Foias and Roger Temam (Bloomington, IN, 2000). *Indiana Univ. Math. J.* **50** (2001), Special Issue, 97–107.

[7] A. Majda, and S. Wang; The Selective Decay Principle for Barostropic Geophysical Flows, *MAA.* **8** (2001), no. 4, 579–594.

[8] Mattingly, J. C. and Sinai, Ya. G., An elementary proof of the existence and uniqueness theorem for the Navier-Stokes equations, *Commun. Contemp. Math.,* vol. 1, no. 4, pp. 497–516, 1999.

[9] S. H. Schochet; Singular limits in bounded domains for quasilinear symetric hperbolic systems having a vorticity equation, *J. Diff. Eqns*, **68**(1987), 400-428

[10] R. Temam, *Infinite-Dimensional Dynamical Systems in Mechanics and Physics*, Springer-Verlag, (2000).

[11] G. Wolansky; Existence, uniqueness and stability of stationary barotropic flow with forcing and dissipation, *Comm. Pure Appl. Math.*, **41**(1988),19-46.

[12] J. Wu; The 2D dissipative quasi-geostrophic equation, *Appl. Math. Lett.* **15** (2002), no. 8, 925–930.

[13] M. Zhan; *On Quasigeostrophic Motion Under Dissipation and Forcing*, submitted.

In: Research on Evolution Equation Compendium. Volume 2 ISBN: 978-1-60456-752-6
Editor: Gaston M. N'Guerekata

JACOBIAN FEEDBACK LOOPS ANALYSIS II: STABILITY AND INSTABILITY

BOURAMA TONI
DEPARTMENT OF MATHEMATICS & COMPUTER SCIENCE
VIRGINIA STATE UNIVERSITY, PETERSBURG VA 23806
E-MAIL ADDRESS: BTONI@VSU.EDU

ABSTRACT. We investigate the loop stability conditions of differential systems, that is, the conditions of invariance of the Jacobian spectrum under any variation of entries that leaves unchanged the loop structure. We use the dynamical properties of Jacobian Loops described by the products of the matrix entries under cyclic permutations of the indices. It appears that all k-order Feedback Loops given by the union of disjoint simple loops involving k variables, must be positive for any asymptotic stable behavior. We also conjecture that the Loop structure requires a negative Feedback Loop of the system dimension for the onset of chaotic behavior.

1. INTRODUCTION

Consider the autonomous differential system

$$\begin{aligned} \dot{x}_i(t) = \frac{dx_i(t)}{dt} &= F_i(x_1, x_2, \cdots, x_n; k_1, k_2, \cdots, k_N) \\ x = (x_1, \cdots, x_n) \in \mathbb{R}^n, &\quad \mathcal{K} = (k_1, k_2, \cdots, k_N) \in \mathbb{R}^N, \end{aligned} \tag{1.1}$$

describing a dynamical system with phase space in $\mathbb{R}^n$, and the parameter/control space in $\mathbb{R}^N$. The component functions F_i, $i = 1, 2, \cdots, n$ of $F(x, \mathcal{K})$ are assumed to be at least $C^1(U)$, that is, differentiable along with their first partial derivatives on U an open set of $\mathbb{R}^n$. The partial order relation $x \leq y \iff x_i \leq y_i, \quad i = 1, ..., n$ defines the vector order in $\mathbb{R}^n$. The Jacobian matrix at $\bar{x} = (\bar{x}_1, \ldots, \bar{x}_n)$ is given by

$$J(\bar{x}) = DF(\bar{x}) = \left[\frac{\partial(F_1, \cdots, F_n)}{\partial(x_1, \cdots, x_n)}(\bar{x})\right] = \left[\frac{\partial F_i}{\partial x_j}(\bar{x})\right]_{1 \leq i,j \leq n} = [J_{ij}]_{1 \leq i,j \leq n}, \tag{1.2}$$

and in general depends on the state variables, except for linear systems [1,9].

The relation $\dot{x}_i(t) = F_i(x_1, \cdots, x_j, \cdots, x_n, \mathcal{K})$ actually shows how the rate of change in variable x_i is dependent on changes in any given variable x_j. Therefore the Jacobian entry $\frac{\partial F_i}{\partial x_j}(x) = J_{ij}$, for $1 \leq i, j \leq n$ describes the interaction between the variables x_i and x_j, as positive (respectively negative, no) interaction for $J_{ij} > 0$ (respectively $J_{i,j} < 0$, $J_{i,j} = 0$).

Dynamical systems theory is mostly based on quantitative values of the Jacobian entries. But for some systems, mainly in Biosciences such as Biology and Biochemistry, and in Economics, the relevant informations are of qualitative nature, that is, the sign $s_{i,j}$ of the nonzero Jacobian entries, $s_{i,j} \in \{+, -\}$, also called logical Jacobian entries. Combination of

1991 *Mathematics Subject Classification.* 34C,58F,92B,93D.
Key words and phrases. Jacobian Feedback loops, loop equivalence, stability.

the sign entries $s_{i,j}$ describe the Jacobian Feedback loops whose definition and dynamical roles are given below. The theory of Jacobian loops is therefore the analysis of the dynamics (simple and complex) using solely the loop-pattern Jacobian matrix, that is, even when only the signs, not the magnitudes of the Jacobian terms, are known. As such it is certainly an efficient tool in surveying dynamical response of models simulating physico-chemical, biological and economical systems by stressing qualitative understanding as the primary goal rather than numerical prediction.

In a series of studies we intend to show that the Jacobian Loop analysis is an efficient tool in the qualitative study of differential systems, in particular, for differential models where informations about the underlying dynamics often reside in the rules of construct of the system and not in the absolute values. See for instance [2,6]. Important questions to be addressed include the charaterization of Jacobian loops and their combinations promoting stability/instablity, exchange of stability and Hopf bifurcation, as well as the onset of chaotic behavior in evolution equations.

1.1. Jacobian Loops: Definitions and Notations.

For $\mathcal{I}_n$ the set of indices $1, \cdots, n$ we denote by $I_k = \{i_1, \cdots, i_k\}$ an ordered subset of k different elements of $\mathcal{I}_n$ and by $\tilde{I}_k = \pi_k(I_k) = \{j_1, \cdots, j_k\}$, with $\pi_k \in \Xi_k$ a permutation of I_k. Recall $Card(\Xi_k) = k!$, i.e., there are $k!$ permutations. Every permutation π_k may be factored into ν disjoint circular (cyclic) permutations $\sigma_i, i = 1, \cdots, \nu$, that is, $\pi_k = \sigma_1\sigma_2\cdots\sigma_\nu$. The signature of π_k, denoted $sg(\pi_k)$, is $(-1)^\eta$, η the number of inversions in π_k, that is, the number of pairs (j_m, j_n) with $j_m > j_n$ while $i_m < i_n$, for $j_m = \pi_k(i_m)$, and $j_n = \pi_k(i_n)$. The permutation π_k, is even (resp. odd) for an even (resp. odd) η. There are exactly $\frac{k!}{2}$ even and exactly $\frac{k!}{2}$ odd permutations in Ξ_k. We denote Ξ_k^c (resp. Ξ_k^e, Ξ_k^o) the subset of circular (resp. even, odd) permutations. [3]. The set Ξ_n is the classic symmetric group of permutations on the set of indices $\mathcal{I}_n$.

We have the following defining concepts.

The set of nonzero terms J_{ij}, $i \in I_k$, and $j \in \tilde{I}_k$, describes a *Jacobian loop* associated to the nonzero product

$$P(\pi_k, J) := \prod_{l=1}^{l=k} J_{i_l\pi_k(i_l)} = J_{i_1\pi_k(i_1)}J_{i_2\pi_k(i_2)}\cdots J_{i_k\pi_k(i_k)} \tag{1.3}$$

called a *loop product.* The loop is called a $k-$order *simple Jacobian loop* L_k when the permutation π_k is a $k-$cycle$= (i_1, i_2, \cdots, i_k)$ with the loop product

$$p_k = J_{i_1i_2}J_{i_2i_3}\cdots J_{i_{k-1}i_k}J_{i_ki_1}. \tag{1.4}$$

Its sign $sgn(L_k)$ is that of the loop product $p_k = P(L_k) := P(\pi_k, J)$. Its length or dimension $l(L_k) = k$ is the number of loop factors $J_{i_l\pi_k(i_l)}$ involved, as well as the number of related system variables x_i.

Therefore

(1) A simple Jacobian loop L_k is positive (resp. negative) for an even (resp. odd) number of negative loop factors $J_{i_l\pi_k(i_l)}$ in the loop product $P(L_k)$.
(2) The *loop graph* $\mathbb{L}_k$ associated with the loop L_k consists of k distinct vertices given by the system variables $x_i, i = 1, \ldots, k$ and k edges $E_{ij} = (x_i, x_j, s_{ij})$ directed from j to i where $s_{ij} = sign(J_{ij})$ denotes the nature of the interaction between the variable x_j and x_i.
(3) A positive (resp. negative) loop involving the variables $x_1, x_2, \ldots, x_k$ is also conveniently denoted $L^+_{x_1x_2\cdots x_k}$ (resp. $L^-_{x_1x_2\cdots x_k}$).

A non-circular permutation π_k yields a *union* of simple Jacobian loops, called a composite loop $\mathcal{L}_k^\nu = \cup_{i=1}^{i=\nu} L_i = (L_1, \ldots, L_\nu)$ of dimension $l(\mathcal{L}_k^\nu) = k$ given by the sum of the lengths of its ν simple components loops, i.e., $k = \sum_{i=1}^{i=\nu} d(L_i) = 1 + \cdots + \nu$.

A proper composite loop $\mathcal{L}_k^\nu$ of resonance (ν, k) is a disjoint union of ν simple loops of total length k, that is, the component loops do not share a vertex. We denote P_k^ν the loop product of a composite loop $\mathcal{L}_k^\nu$. The sign of a composite loop $\mathcal{L}_k^\nu$ is the sign of P_k^ν, or equivalently, $sign(\mathcal{L}_k^\nu) = \prod_{l=1}^{l=\nu} sign(L_i) = (-1)^{\nu_-}$, where ν_- is the number of negative simple loops in $\mathcal{L}_k^\nu$. $\chi_k^\nu = (-1)^{\nu+1}$ is the characteristic of the proper composite loop of resonance (ν, k). Therefore a $k-$order proper composite loop has a negative (resp. positive) resonance, i.e., a negative (resp. positive) characteristic for ν even (resp. odd).

Definition 1.1. *A $k-$order Feedback F_k is defined by*

$$F_k = \sum_{all \nu} (-1)^{\nu+1} P_k^\nu,$$

P_k^ν the loop product of the proper composite loop $\mathcal{L}_k^\nu$.

Consequently, (see also [8])

(1) A composite loop $\mathcal{L}_k^\nu$ is positive (resp. negative) for an even (resp. odd) number of its negative simple loops.

(2) A proper composite loop $\mathcal{L}_k^\nu$ with all component simple loops negative has a negative resonance in the Feedback F_k as defined above.

The *loop structure (or qualitative structure)*, denoted $\mathbb{L}_\mathcal{R}$, corresponding to the region $\mathcal{R}$ in the phase space or to a sign-pattern is the set of all Jacobian loops (simple and proper composite) along with their signs. The Jacobian Loop analysis required first the determination of the loop structure associated to the system in a given region of the phase space, either from the signed entries of a Jacobian matrix evaluated at equilibria or constant at some parameter values or solely from the logical evaluation of the interaction between the variables in terms of positive/negative.

Given a general n square matrix $A \in \mathbb{M}_n(\mathbb{R})$, $A = [a_{ij}]_{1 \le i,j \le n}$ the sign-pattern matrix $A_l = [s_{ij}]_{1 \le i,j \le n}, s_{ij} \in \{+, -\}$ is also called the *logical matrix.*

Definitions 1.2.

(1) *A matrix $B = [b_{ij}]_{1 \le i,j \le n}$ is logically or sign equivalent to $A = [a_{ij}]_{1 \le i,j \le n}$ in the region $\mathcal{R}$ if B has the same sign pattern as A, i.e., $A_l = B_l$. We denote $A \odot B$.*

(2) *A matrix $B = [b_{ij}]_{1 \le i,j \le n}$ is loop equivalent to $A = [a_{ij}]_{1 \le i,j \le n}$ in the region $\mathcal{R}$, if B yields the same loop structure $\mathbb{L}_\mathcal{R}$ as A. We denote $A \circlearrowleft B$.*

$\langle A \rangle$ denotes the logical equivalence class of matrix A represented by the *logical matrix A_l*. The logical equivalence class $\langle \rangle$ is actally represented by a $n \times n$ array of nonzero $s_{ij} = +, -$ entries such that, for any matrix $A = [a_{ij}] \in \langle \rangle$, $sign(a_{ij}) = s_{ij}$.

$\rangle A \langle$ denotes the loop equivalence class of matrix A. The class $\rangle\langle = \mathbb{L}$ is a loop structure represented by a set of simple signed loops, such that any matrix $A \in \rangle\langle$ has the loop structure $\mathbb{L}$.

Note that the loop equivalence class is the loop structure uncovered from a representative logical matrix. Therefore if $A \circlearrowleft B$ then they are sign-patterned in a way to provide the same loop structure, though they are not necessarily logically equivalent in the sense $sign(a_{ij}) = sign(b_{ij})$. Thus matrix operations such as negation, transposition, permutations, and signature similarity allow to derive one sign pattern from the other. Therefore

(1) Logical equivalence obviously implies loop equivalence but not inversely.

(2) Given two matrices A and B loop equivalent, the logical class $\langle B\rangle$ may be obtained from that of A by some combination of negation, transposition, permutation, and signature similarity.

Clearly, If $A \circlearrowleft B$ therefore the spectrum $\sigma(A)$ is the same as $\sigma(B)$.

Next, to adress the loop stability analysis, we recall how the role of loops is highly determined by their relation to the characteristic equation. (see also [8])

2. Loops and Jacobian Spectrum

For a matrix A, Jacobian or otherwise, given by $A = [a_{ij}]_{1\le i,j\le n}$ the characteristic polynomial is defined by the monic polynomial

$$\mathcal{C}_A(\lambda) = |\lambda I - A| = \lambda^n + c_1\lambda^{n-1} + \cdots + c_k\lambda^{n-k} + \cdots + c_{n-1}\lambda + c_n. \tag{2.1}$$

From Linear Algebra the coefficients may be expressed as

$$c_k = coefficient(\lambda^{n-k}) = \sum(-1)^k m_k, \quad k = 0, \cdots, n-1, \tag{2.2}$$

where the sum extends over all kth order principal minors m_k of A. For instance we have

$$\begin{aligned} c_n &= (-1)^n det(A) = (-1)^n|A|, \quad \text{for } k = n, \\ c_1 &= -\sum A_{ii} = -Tr(A), \quad \text{where Tr(A) is the trace of } A. \end{aligned} \tag{2.3}$$

From the theory of determinant and permutations we may write

$$m_k = \sum_{\pi_k\in\Xi_k} (-1)^\eta \prod_{i_l\in\mathcal{I}_k} A_{i_l\pi_k(i_l)} = \sum_{all\nu}(-1)^{k-\nu}\prod_{i=1}^{i=\nu} P(\sigma_i, A) = \sum_{all\nu}(-1)^{k-\nu}P(\mathcal{L}_k^\nu). \tag{2.4}$$

where the permutation $\pi_k \in \Xi_k$ of the indices $1 \le i_1 < i_2 < \cdots < i_k \le n$ factors into the cyclic permutations $(\sigma_1, \cdots, \sigma_\nu)$ yielding the proper composite loop $\mathcal{L}_k^\nu = (\sigma_1, \cdots, \sigma_\nu)$ with loop product $P(\mathcal{L}_k^\nu) = P_k^\nu$ as defined in section 1. Consequently we obtain an expression of the characteristic coefficients c_k in terms of the proper composite loops. Importantly it was proved in [8]

Lemma 2.1. *The kth order Feedback $\mathcal{F}_k$ involving all the proper composite loops $\mathcal{L}_k^\nu$ with $\nu = 1, \cdots, k$ may be expressed in terms of the coefficients of the characteristic polynomial by*

$$\mathcal{F}_k = c_k = \sum_{all\nu}(-1)^{\nu+1}P_k^\nu. \tag{2.5}$$

Note that:

(1) The $k-$ order Feedback $\mathcal{F}_k$ being the $k-$order coefficient of characteristic polynomial entails that the loop factors $A_{i\sigma(i)}$ defined in section 1. are the only Jacobian entries contributing to the characteristic equation, and therefore, influence directly the eigenvalues of the matrix, and consequently the dynamics.
(2) From the standard theory of equations it is also known that the coefficients $c_k = \mathcal{F}_k$ are related to the eigenvalues λ_i in a systematic way by the following Viete formulas:

(See [3,5]).

$$\begin{aligned}
c_1 = F_1 &= -(\lambda_1 + \cdots + \lambda_n) = -\sum_{i=1}^{i=1} A_{ii} = -Trace(A). \\
c_2 = F_2 &= \sum_{i,j=1,i<j} \lambda_i\lambda_j = \sum_{i,j=1,i<j} (A_{ii}A_{jj} - A_{ij}A_{ji}) \\
&= \lambda_1\lambda_2 + \lambda_1\lambda_3 + \cdots + \lambda_{n-1}\lambda_n. \\
c_3 = F_3 &= -\sum_{i,j,k=1,i<j<k} \lambda_i\lambda_j\lambda_k \\
&= -(\lambda_1\lambda_2\lambda_3 + \lambda_1\lambda_2\lambda_4 + \cdots + \lambda_{n-2}\lambda_{n-1}\lambda_n). \\
&\cdots \\
c_n = F_n &= (-1)^n\lambda_1\lambda_2\cdots\lambda_n = (-1)^n det(A).
\end{aligned} \tag{2.6}$$

It is also known that, given a n-degree polynomial with real coefficients

$$p(z) := z^n + a_1 z^{n-1} + \cdots + a_k z^{n-k} + \cdots + a_{n-1}z + a_n,$$

a necessary condition for $p(z)$ to have all its zeros z_* with negative real part, i.e., $p(z)$ is a *strongly stable* polynomial, is that all the coefficients $a_k > 0, k = 1, \cdots, n$. Indeed, if all real parts are negative then we have either the form

$$p(z) = \prod(z - (\alpha + i\beta)) = \prod(z^2 - 2\alpha z + \alpha^2 + \beta^2) = \prod(z^2 + az + b), \quad a > 0, \quad b > 0,$$

or

$$p(z) = \prod(z - \alpha) = \prod(z + a), \quad a > 0.$$

By successive multiplication we necessarily obtain $p(z)$ with $a_k > 0$. This immediately implies (see also [8])

Theorem 2.2. *A necessary condition to have all eigenvalues with negative real parts $\mathcal{R}_e < 0$ is that all **kth** order Feedback $\mathcal{F}_k$ must be positive.*

For the sake of completeness we recall The following results from [8] along with their proofs. They are used in the loop stability analysis.

Lemma 2.3. *A proper composite loop $\mathcal{L}_k^\nu$ with all ν components simple loops negative has a negative resonance in the **kth** order Feedback, that is, its characteristic χ_k^ν is negative.*

Proof. Indeed the term $(-1)^{\nu+1}P(\mathcal{L}_k^\nu)$ in the Feedback $\mathcal{F}_k$ has the sign $(-1)^\nu(-1)^{\nu+1} = -1$. Hence the claim. □

Theorem 2.4. *If there is no proper composite loop $\mathcal{L}_k$ of dimension $k \leq n$, then the characteristic coefficient $c_k = 0$.*

Moreover at least one proper composite loop $\mathcal{L}_n$ of the system dimension is necessary to have a nonsingular jacobian matrix.

Proof. From formulas (2.4), (2.5) the characteristic coefficient c_k can be written as

$$c_k = \mathcal{F}_k = \sum(-1)^\nu \prod_{l=1}^{l=r_1} J_{i_l\pi_k(i_l)} \prod_{l=r_1+1}^{l=r_2} A_{i_l\pi_k(i_l)} \cdots \prod_{l=r_{\nu-1}}^{l=r_\nu} A_{i_l\pi_k(i_l)}. \tag{2.7}$$

Terms in the expression of c_k with one cyclic permutation correspond to $r_1 = k$, those with two cyclic permutations correspond to $r_1 < k$, $r_2 = k - r_1$, and so on. Therefore, if there is no proper composite loop $\mathcal{L}_k$ of dimension k, then each term of the sum is zero. Hence the claim. For $k = n$ the system dimension, formulas (2.5)-(2.7) yield clearly $Det(A) = |A| = 0$. □

We say that the logical equivalent class $\langle A \rangle$ or the loop equivalence class $\rangle A \langle$ is *logically nondegenerate* if every matrix in the class is nonsingular in the sense $|A|$ is nonzero. We prove

Theorem 2.5. *If the loop structure $\mathbb{L}$ does contains a composite loop $\mathcal{L}_n$ of the dimension of the system, and all such loop $\mathcal{L}_n$ have the same sign, then the corresponding Jacobian determinant $|A|$ is nonzero.*

Proof. Indeed suppose all the composite loops $\mathcal{L}_n$ of the dimension of the system have the same sign. Then c_n, consisting of nonzero terms of the same sign, is therefore nonzero. Consequently, the Jacobian determinant is nonzero.

Moreover c_n is positive (resp. negative) if all $\mathcal{L}_n$ have an odd (resp. even) number ν_o (resp. ν_e) of simple loops L_i. □

We also prove

Theorem 2.6. *A positive simple loop in the loop equivalence class is a necessary condition for the Jacobian matrix to have a positive real eigenvalue.*

Proof. Recall the characteristic coefficients derived from (2.4), (2.5), (2.6), that is,

$$c_k = \sum_{\mathcal{L}_k=(L_1,\cdots,L_\nu)} (-1)^\nu P(\sigma_1, A) \cdot P(\sigma_2, A) \cdot \;\cdots\; \cdot P(\sigma_\nu, A), \tag{2.8}$$

where the simple Jacobian loops $L_i, i = 1, \cdots, \nu$ are defined by the cyclic permutations $\sigma_i, i = 1, \cdots, \nu$. Now assume that the region has a negative loop equivalence class, i.e., there is no positive simple loop in its loop structure. So every simple loop L_i defined by $\sigma_i \in \Xi_k^c$ is negative. Therefore the corresponding nonzero loop product $P(\sigma_i, A)$ is also negative. Then for a composite loop $\mathcal{L}_k = (L_1, \cdots, L_\nu)$ we have

$$sign(-1)^\nu P(\sigma_1, A) \cdot P(\sigma_2, A) \cdot \;\cdots\; \cdot P(\sigma_\nu, A) = sign((-1)^{2\nu}) = +. \tag{2.9}$$

Thus all the characteristic coefficients c_k are positive. This entails a characteristic polynomial of degree n with only positive coefficients. By Descartes' rules of sign it cannot have a positive real root. Hence the claim. □

3. Loop Stability

The loop stability refers to the invariance of the jacobian spectrum under any variation of entries that leave unchanged its loop structure. Jacobian loops and their combinations provide valuable information about the stability of a system even when only the signs, not the magnitudes of the Jacobian terms, are known.

Recall that Λ_A denotes the spectrum of matrix A, i.e., the set of all eigenvalues λ of A or zeros of the characteristic polynomial $\mathcal{C}_A(\lambda)$.

We summarize the classic characterizations of stability from Routh-Hurwitz and Lyapunov theories. For more details see [1,3,5].

Linear Stability Criteria. *The necessary and sufficient conditions to have all real parts negative are given by:*

(1) (**Lyapunov**) *There exist a positive definite symmetric matrix Q such that $QA + A^tQ$ is a negative definite matrix.*

(2) (**Routh-Hurwitz**) *All the Hurwitz determinants H_i are positive, where*

$$H_1 = c_1,$$
$$\vdots$$
$$H_n = \begin{vmatrix} c_1 & c_3 & c_5 & \cdots & c_{2n-1} \\ 1 & c_2 & c_4 & \cdots & c_{2n-2} \\ 0 & c_1 & c_3 & \cdots & c_{2n-3} \\ \vdots & \vdots & \vdots & \cdots & \vdots \\ 0 & 0 & 0 & \cdots & c_n \end{vmatrix} = c_n H_{n-1}, \quad c_j = 0, \quad j > n. \tag{3-1}$$

(3) (**Liénard-Chipart**) $c_k > 0$, *for all k, and alternate Hurtwitz determinants up to order n are positive.*

Definition 3.1.

(1) *The matrix A is stable (resp. asymptotically stable) if its characteristic polynomial $\mathcal{C}_A(\lambda)$ is stable (resp. strongly stable), that is, for every eigenvalue λ of A we have $\mathcal{R}_e(\lambda) \leq 0$ (resp. < 0).*

(2) *A is unstable if it is not stable. In other words, there is at least one eigenvalue λ such that $\mathcal{R}_e(\lambda) > 0$.*

(3) *The matrix A is of saddle-type if $\mathcal{R}_e(\lambda) > 0$ for some eigenvalues, and $\mathcal{R}_e(\lambda) < 0$ for the remainings.*

Therefore a sink equilibrium has a asymptotically stable Jacobian, whereas a source has an unstable Jacobian.

Moreover, if $\mathcal{R}_e(\lambda) > 0$ for every $\lambda \in \Lambda_A$ the instability is said to be *strong*, e.g., a source is strongly unstable. A *weak instability* is characterized by $\mathcal{R}_e(\lambda) > 0$, for some eigenvalues, and $\mathcal{R}_e(\lambda) = 0$ for the remainings.

Recall that, as a function of the matrix, Λ_A is neither additive nor multiplicative. Moreover we have the followings:

(1) For a nonsingular matrix A, i.e., $|A| = det(A) \neq 0$, $\Lambda_{A^{-1}} = \{\frac{1}{\lambda}, \quad \lambda \in \Lambda_A\}$. So A and A^{-1} must be stable simultaneously.

(2) $\Lambda_A = \Lambda_{A^t}$, A^t denotes the transpose of matrix A. Thus A and A^t must be stable simultaneously.

Definition 3.2.

(1) *The matrix A is loop stable (resp. asymptotically stable) if every matrix in the loop equivalence class $\rangle A \langle$ is stable (resp. asymptotically stable) .*

(2) *A is loop unstable if every matrix in the loop equivalence class $\rangle A \langle$ is unstable.*

We note that a loop equivalence class $\rangle A \langle$ that is not stable is not necessarily unstable. Instead it does have at least one matrix that is unstable, i.e., with an eigenvalue of positive real part in its spectrum.

The loop analysis is addressed here for irreducible matrices. Recall a matrix $A = [a_{ij}]_{1 \leq i,j \leq n}$ is *irreducible or indecomposable* if there is no simultaneous row-and-column permutation $P_r - P_c$ such that A is similar to

$$P_r A P_c = \begin{pmatrix} B & O \\ C & D \end{pmatrix}. \tag{3-2}$$

where P_r and P_c are respectively the row and column permutation matrices, and B, C are respectively a $p \times p$ and a $q \times q$ block such $p+q = n$, and O a $p \times q$ block of zeros. By a Laplace decomposition the spectrum of A is given by $\Lambda_A = \Lambda_B + \Lambda_D$, thus reducing its eigenvalue analysis to that of the individual diagonal block of lower dimension. Therefore we assume all matrices are irreducible without loss of generality. Actually a $P_r - P_c$ permutation amounts to a renumbering of the system variables, and renumbering should certainly not affect the properties of the system in general, and its asymptotic behavior in particular.

Theorem 2.2 in section 2. yields in the light of the Linear Stability criteria

Theorem 3.4. *All kth order Feedback F_k positive is a necessary condition for a stable loop equivalence class.*

Moreover we obtain

Theorem 3.5. *If the loop structure $\mathbb{L}$ has a positive simple loop L_k then the corresponding loop equivalence class cannot be stable.*

Proof. Indeed it suffices to construct a representative matrix with k eigenvalues arbitrarily close to the k-th roots of unity by continuity, and one of them is a simple positive eigenvalue. Therefore the matrix is unstable. Hence the claim. □

Similarly one can prove

Theorem 3.6. *The loop equivalence class is unstable if it has a composite loop $\mathcal{L}_n$ of the dimension of the system, positive for n even, and negative for n odd.*

Proof. From section 1. for any matrix A in such a loop structure, the presence of a composite loop of the dimension of the system ensures a nonzero determinant such that the characteristic coefficient $c_n = F_n = \mathcal{C}_A(0)$ is > 0 for n even, and < 0 for n odd. This entails the characteristic polynomial of any matrix in the class has a positive root. Hence the claim. □

The above Linear Stability criteria yields the following condition in terms of Jacobian loops

Theorem 3.7. *If an irreducible Jacobian matrix J is loop stable then the loop structure $\mathbb{L}$ has at least one simple loop L_1, and at least one proper composite loop $\mathcal{L}_n$ of the system dimension, and all simple Jacobian loops $L_k, k \leq 2$ are negative.*

We therefore note that negative loops promote stability whereas positive loops promote instability. A positive proper composite loop contributes a negative term to the characteristic polynomial, whereas a negative proper composite contributes a positive term.

Now assume that for the parameter vertor $\mathcal{K} = (k_1, \ldots, k_N)$ the Jacobian matrix is loop stable for the parameter value $\mathcal{K} = \mathcal{K}_*$, and loop unstable for $\mathcal{K} \neq \mathcal{K}_*$. The system then undergoes an exchange of loop stability at $\mathcal{K}_*$. Hopf bifurcation is known to ensure the existence of limit cycles or sustained oscillations. A system admits a Hopf bifurcation only if the corresponding Jacobian matrix admits and exchange of stability, that is,

Lemma 3.8. *In order to admit a Hopf bifurcation, the Jacobian matrix can neither be loop stable nor loop unstable throughout the entire region of the phase space.*

3.1. Loop Analysis in the Plane.

For a 2-dimensional system, we prove the following results.

Theoreom 3.9. *Any loop structure in the plane consisting of two 1-loop L_1 of opposite signs and a negative 2-loop cannot be loop stable or loop unstable.*

The proof is based on the following lemma:

Lemma 3.10. *The logical equivalence class given by*

$$\langle\rangle = \begin{pmatrix} + & - \\ + & - \end{pmatrix}, \tag{3-3}$$

cannot be not stable and cannot be not unstable, that is, there is a matrix $A \in \langle\rangle$ such A is unstable or stable.

Proof. The equivalence class above is the so-called 1-striped sign pattern; as such, given any monic quadratic polynomial $q(x) = x^2 + bx + c$, there is matrix M whose characteristic polynomial $\mathcal{C}_A(\lambda) = q(\lambda)$. Therefore there is certainly one whose spectrum contains an eigenvalue with a positive (resp. negative) real part. Hence the claim. □

Proof of Theorem 3.9. The proof of thereom 3.9 follows immediately; indeed by combinations such as negation, transposition, permutation, signature similarity, one can construct a matrix M with the loop structure of two 1-loop of opposite signs and a negative 2-loop in a such way that the matrix M is in class of the type in lemma 3.10. □

Based on the Linear Stability Theory, we can classify the loop equivalence classes in the plane as follows:

Theorem 3.11.

(1) *The loop structure consists of only a negative 2-loop L_2^-: The dynamic if that of a linear center, that is, a family of periodic orbits surrounding the origin.*
(2) *The loop structure $\mathbb{L}$ consists of a negative 2-loop L_2^- and two 1-loop of same sign; the dynamic is that of a focus point, that is, the presence of sustained oscillations around the origin. If the system is bounded then there exists a limit cycle surrounding the origin.*
(3) *The loop structure consists of only two 1-loops of opposite sign (resp. same sign). Then the origin is a saddle point (resp. a node stable for negative 1-loops, and unstable for positive 1-loops).*
(4) *These loop equivalence classes are the only loop equivalence classes in the plane.*

Proof. Immediate from Linear stability theory. Indeed every 2×2 matrix is amenable, via nonsingular linear transformations to one of the forms:

$$\begin{pmatrix} a & 0 \\ 0 & b \end{pmatrix}, \quad \begin{pmatrix} a & b \\ -b & a \end{pmatrix}. \tag{3-4}$$

□

Therefore we have

Corrollary 3.12. *A negative 2-loop L_2^- is a necessary condition for any periodic behavior such as a center, a focus, or a limit cycle, i.e., an isolated periodic orbit.*

3.2. Loop Analysis for 3-dimensional System.

Similarity to the 2-dimensional case, we prove

Theorem 3.13. *Any 3-dimensional loop structure containing the two negative 2-loops* L_{xy}^- *and* L_{yz}^-*, the positive 2-loop* L_{xz}^+*, in the absence of any positive 1-loop* L_y^+ *and negative 1-loops* L_x^- *and* L_z^- *is destabilized by the two positive 3-loops whenever they are present.*

Proof. Indeed such a loop may be represented by a logical class whose is similar by the usual operations of signatures and permutations, to

$$\begin{pmatrix} +/0 & - & + \\ + & -/0 & - \\ + & + & +/0 \end{pmatrix}. \tag{3-5}$$

One can easily construct a matrix in this class with a positive eigenvalue. □

Theorem 3.14. *For a 3-dimensional sytem the following four loop equivalence classes cannot be stable.*

(1) *The loop structure* $\langle\rangle$ *consists of two 1-loops* $L_1^\pm$ *of opposite signs, and two negative 2-loops* L_2^-*, no 3-loop.*
(2) *The loop structure* $\langle\rangle$ *consists of one positive 1-loop* L_x^-*, and two negative 1-loops* L_1^-*, one negative 2-loop* L_2^-*, and one positive 2-loop* L_2^+*, but no 3-loop.*
(3) *The loop structure* $\langle\rangle$ *consists of two 1-loops* $L_1^\pm$ *of opposite signs, and one negative 2-loop* L_2^-*, and a positive 3-loop* L_{xyz}^+*.*
(4) *The loop structure* $\langle\rangle$ *consists of two 1-loops* $L_1^\pm$ *of opposite signs, one negative 2-loop* L_2^-*, one positive 2-loop* L_2^+*, and a negative 3-loop* L_{xyz}^-*.*
(5) *These loop classes cannot be unstable, and are the only ones that cannot be stable and unstable.*
(6) *These equivalence classes are the ones necessary for any Hopf bifurcation ensuring the existence of limit cycles.*

Proof. We consider in each case a representative logical equivalence class, that is, a sign-pattern with such a loop pattern, and then use some results for advanced matrix theory on the so-called arbitrary spectrally sign pattern. By signatures and permatutions, a representative sign-pattern of the classes above is similar respectively the following sign-patterns well-known to be spectrally arbitrary, that is, they contains a matrix whose spectrum has a positive eigenvalue, root of an arbitrary monic cubic polynomial.

(1) The first case corresponds to the sign-pattern:

$$\begin{pmatrix} + & - & 0 \\ + & 0 & - \\ 0 & + & - \end{pmatrix} \tag{3-6}$$

(2) Case Two:

$$\begin{pmatrix} + & - & + \\ + & - & 0 \\ + & 0 & - \end{pmatrix} \tag{3-7}$$

(3) Case three:

$$\begin{pmatrix} + & - & 0 \\ + & 0 & - \\ + & 0 & - \end{pmatrix} \tag{3-8}$$

(4) Case four:

$$\begin{pmatrix} + & + & - \\ + & 0 & - \\ + & 0 & - \end{pmatrix} \tag{3-9}$$

□

The previous results entail the following conjecture illustrated in the jacobian loop analysis of the well-known Lorenz and Rossler systems, paradigms of chaotic dynamics. It is a step toward the necessary and/or sufficient conditions in terms of Feedback Loops for the onset of chaos.

Conjecture 3.15. *The previous loop equivalence classes are some of the necessary loop structures for the onset of any chaotic behavior in a 3-dimensional system.*

This conjecture is a step toward the determination of necessary and/or sufficient conditions in terms of Feedback Loops for the onset of chaos, in particular in Lorenz and Rossler systems. These systems are respectively:

3.3. The Lorenz System.

The Lorenz system is described by

$$\begin{aligned} \dot{x} &= \sigma(y - x) \\ \dot{y} &= \rho x - y - xz \\ \dot{z} &= -\beta z + xy \end{aligned} \tag{3-10}$$

Its equilibria are the origin $O = (0,0,0)$ for all values of the parameter ρ, and $E_{\pm} = (\pm\sqrt{\beta(\rho-1)}, \pm\sqrt{\beta(\rho-1)}, \rho - 1)$ appearing at $\rho > 1$ and symmetric with respect to the z-axis.

A Lorenz logical Jacobian class in the positive quadrant is represented by

$$\begin{pmatrix} - & + & 0 \\ s_{21} & - & \mp \\ \pm & \pm & - \end{pmatrix} \tag{3-11}$$

whereas along the z-axis we have

$$\begin{pmatrix} - & + & 0 \\ s_{21} & - & 0 \\ 0 & 0 & - \end{pmatrix} \tag{3-12}$$

where s_{21} is respectively $(+)$, (0), and $(-)$ under, in, and above the plane $z = \rho$. This implies a variation of a loop structure in these regions.

3.4. Rossler System.

The Rossler system is given by

$$\begin{aligned} \dot{x} &= -y - z \\ \dot{y} &= x + ay \\ \dot{z} &= bx + xz - cz \end{aligned} \tag{3-13}$$

For a wide range of the parameters a, b, c the system exhibits two unstable equilibria de type *saddle-focus* periodically repulsive (resp. attractive) in a plane while attractive (resp. repulsive) along a normal direction.

For the positive parameters a Rossler logical Jacobian class is given by

$$\begin{pmatrix} 0 & - & - \\ + & + & 0 \\ + & 0 & - \end{pmatrix}, \tag{3-14}$$

that is, the loop structure contains two negative 2-loops L^-_{xy} and L^-_{xz} along with two 1-loops of opposite signs L^+_y and L^-_z. Note, as conjectureed, the presence of two proper 3-dimensional composite loops $\mathcal{L}^2_3$ given by the negative (L^+_y, L^-_{xz}) and the positive (L^-_z, L^-_{xy}).

References

[1] Bellman, R., *Introduction to Matrix Analysis*, Classics in Applied Matematics 19, 2nd Ed. SIAM, 1997.

[2] Eisenfeld, J. and De Lisi, C., *On conditions for qualitative instability of regulatory loops with applications to immunological control loops*, Mathematics and Computers in Biomathematical Applications, Elsevier, New York (1994), 39-53.

[3] Gantmacher, F.R., ***Théorie des Matrices***, vol. 1,2, Collection Universitaire de Mathématiques, Dunod Paris, 1966.

[4] Gouzé, J.L., *Positive and negative circuits in Dynamical Systems*, J.Biol.Syst. **6** (1998), 11-15.

[5] Lancaster, P. and Tismenetsky, M., *The Theory of Matrices with Applications*, Academic Press, 1985.

[6] Quirk, R. and Ruppert, R., *Qualitative economics and the stability of equilibrium*, Rev. Econ. Studies **32** (1965), 311-326.

[7] Thomas, R., *The role of Feedback Circuits: positive feedback circuits are a necessary condition for positive eigenvalues in the feedback matrix*, Ber.Bunzenges.Phys.Chem. **98** (1994), 1148-1151.

[8] Toni, B., *Jacobian Feedback Loop Analysis I*, International Journal of Evolution Equations. **1(4)** (2005).

[9] Verhulst, F., ***Nonlinear Differential Equations and Dynamical Systems***, Springer-Verlag Universitext, 1990.

In: Research on Evolution Equation Compendium. Volume 2 ISBN: 978-1-60456-752-6
Editor: Gaston M. N'Guerekata

Existence of Oscillating Solution for Nonlinear State-Dependent Delay Differential Equation*

***Zhixian Yu*† *and Rong Yuan*‡**
School of Mathematical Science, Beijing Normal University,
Beijing 100875, People's Republic of China

Abstract

In this paper sufficient conditions for oscillation of all solutions are given for the equations

$$\dot{x}(t)+\sum_{i=1}^{n}p_i(t)f(x(t-r_i(t)))=0,$$

and

$$\dot{x}(t)+\sum_{i=1}^{n}p_i(x(t))f(x(t-r_i(x(t))))=0,$$

where $r_i(\cdot),p_i(\cdot)\in C(\mathbb{R},\mathbb{R}^+)$.

1991 Mathematics Subject Classification: 34K11

Key words and phrases: Oscillating solution; state-dependent; delay equation.

1. Introduction

In this paper, we will discuss the oscillation of all solutions for the following delay equations

$$\dot{x}(t)+\sum_{i=1}^{n}p_i(t)f(x(t-r_i(t)))=0, \tag{1.1}$$

and

$$\dot{x}(t)+\sum_{i=1}^{n}p_i(x(t))f(x(t-r_i(x(t))))=0, \tag{1.2}$$

where $r_i(\cdot),p_i(\cdot)\in C(\mathbb{R},\mathbb{R}^+)$, $f(x)$ is continuous. As a usual, a solution of equation (1.1) (or (1.2)) is said to oscillate if it has arbitrarily large zeros.

*Supported by National Natural Science Foundation of China
†E-mail address: yzx3411422@163.com
‡E-mail address: ryuan@bnu.edu.cn

Existence of oscillating solutions has been extensively discussed in the literatures (see [1-14]) and the references therein. This paper is motivated by the papers [5, 9]. Li and Kuang [9] considered the following first-order nonlinear nonautonomous delay differential equation

$$x'(t)+p(t)f(x(t-\tau(t)))=0, \tag{1.3}$$

where $p\in C([0,\infty),[0,\infty))$, $\tau\in C([0,\infty),(0,\infty))$, $\lim_{t\to\infty}(t-\tau(t))=\infty$, $f\in C(\mathbb{R},\mathbb{R})$ and $uf(u)>0$ for $u\neq 0$, and proved that every solution of Eq.(1.3) is oscillatory.

In [5], Ladas and Stavroulakis discussed the oscillation of all solutions of differential equations with deviating arguments of the form

$$y'(t)+\sum_{i=1}^{n}p_iy(t-\tau_i)=0 \tag{1.4}$$

and

$$y'(t)-\sum_{i=1}^{n}p_iy(t+\tau_i)=0 \tag{1.5}$$

in the condition that

$$\left(\prod_{i}^{n}p_i\right)^{\frac{1}{n}}\left(\sum_{i=1}^{n}\tau_i\right)>\frac{1}{e},$$

or

$$\frac{1}{n}\left(\sum_{i=1}^{n}(p_i\tau_i)^{\frac{1}{2}}\right)^2>\frac{1}{e},$$

where $\tau_i\,(i=1,2,\cdots,n)$ are positive constants, $p_i\,(i=1,2,\cdots,n)$ are all positive constants.

Some authors (for example, [1, 3, 13]) considered the oscillation problem for the delay differential equation

$$x'(t)+\sum_{i=1}^{n}q_i(t)x(t-\tau_i(t))=0, \tag{1.6}$$

where p_i, τ_i are continuous and positive functions. We would like to point out that our conditions and approaches are different from those in [1, 3, 13]. When we are back to the equations (1.4) and (1.5), our conditions coincide with these in [5].

In the sequel, we assume the following conditions:

$(H0)$. $f(x)x>0$ for $x\neq 0$, $0<\liminf_{x\to 0}\frac{f(x)}{x}=\alpha\leq\infty$.

$(H1)$. For Eq.(1.1),

$$\lim_{t\to\infty}p_i(t)=p_i,\ \lim_{t\to\infty}r_i(t)=r_i,$$

where $r_i(t)$ is bounded, $r_i'(t)<1$ and p_i, r_i are positive constants, $i=1,\ 2,\ \ldots n$.

$(H1')$. For Eq.(1.2),

$$\lim_{x\to 0} p_i(x) = p_i, \quad \lim_{x\to 0} r_i(x) = r_i,$$

where $r_i(x)$ is bounded, $r'(x) > 0$, and p_i and r_i are positive constants, $i = 1, 2, \dots n$.

$(H2)$. $(\prod_i^n p_i)^{\frac{1}{n}} (\sum_{i=1}^n r_i) > \frac{1}{\alpha_0 e}$.

$(H3)$. $\frac{1}{n}\left(\sum_{i=1}^n (p_i r_i)^{\frac{1}{2}}\right)^2 > \frac{1}{\alpha_0 e}$, where α_0 is a finite positive number with $0 < \alpha_0 \le \alpha$.

Now we can formulate our main results as follows.

Theorem 1.1 *Assume that the conditions $(H0)$, $(H1)$, $(H2)$ hold, then all the solutions of Eq.(1.1) are oscillatory.*

Theorem 1.2 *Assume that the conditions $(H0)$, $(H1')$, $(H2)$ hold, then all the solutions of Eq.(1.2) are oscillatory.*

Theorem 1.3 *Assume that the conditions $(H0)$, $(H1)$, $(H3)$ hold, then all the solutions of Eq.(1.1) are oscillatory.*

Theorem 1.4 *Assume that the conditions $(H0)$, $(H1')$, $(H3)$ hold, then all the solutions of Eq.(1.2) are oscillatory.*

The remaining parts of this paper are organized as follows. In section 2 we will give the proof of Theorems 1.1 and 1.2. The proof of Theorems 1.3 and 1.4 will be given in section 3. In the last section we will provide some examples.

2. Proof of Theorems 1.1 and 1.2

We begin with a Lemma.

Lemma 2.1 *[5] Assume that*

$$g(b_1, b_2, \cdots, b_n) = \sum_{i=1}^n \log b_i - \left(\sum_{i=1}^n q_i b_i\right)\left(\sum_{i=1}^n r_i\right)$$

is nonnegative for $b_i \ge 1$, $i = 1, \cdots, n$, $q_i > 0$, and $r_i > 0$. Then $g(b_1, b_2, \cdots, b_n)$ has a maximum at

$$\left(\frac{1}{q_1 \sum_{i=1}^n r_i}, \frac{1}{q_2 \sum_{i=1}^n r_i}, \cdots, \frac{1}{q_n \sum_{i=1}^n r_i}\right).$$

Moreover,

$$\left(\prod_{i=1}^n q_i\right)^{\frac{1}{n}} \left(\sum_{i=1}^n r_i\right) \le \frac{1}{e}.$$

Proof of Lemma 2.1. The proof has been given in [5]. For the reader's convenience its proof is quoted here.

Now observe that

$$\frac{\partial g}{\partial b_i} = \frac{1}{b_i} - q_i\left(\sum_{i=1}^{n} r_i\right) = 0,$$

for

$$b_i = \frac{1}{q_i \sum_{i=1}^{n} r_i}, \quad i = 1, 2, \ldots, n.$$

At the critical point

$$\left(\frac{1}{q_1 \sum_{i=1}^{n} r_i}, \frac{1}{q_2 \sum_{i=1}^{n} r_i}, \cdots, \frac{1}{q_n \sum_{i=1}^{n} r_i}\right),$$

the function $g(b_1, b_2, \cdots, b_n)$ has a maximum because the quadratic form

$$\sum_{i,j=1}^{n} \frac{\partial^2 g}{\partial b_i \partial b_j} a_i a_j = -\sum_{i=1}^{n} \frac{a_i^2}{b_i^2}$$

is negative definite. Since $g(b_1, b_2, \cdots, b_n) \geq 0$, the maximum of $g(b_1, b_2, \cdots, b_n)$ at the critical point should be nonnegative. So we have

$$\sum_{i=1}^{n}\left(-\log\left(q_i \sum_{i=1}^{n} r_i\right)\right) \geq \left(\sum_{i=1}^{n} q_i\right)\left(\frac{1}{q_i \sum_{i=1}^{n} r_i}\right)\left(\sum_{i=1}^{n} r_i\right) = n,$$

that is,

$$\left(\prod_{i}^{n} q_i\right)^{\frac{1}{n}}\left(\sum_{i=1}^{n} r_i\right) \leq \frac{1}{e}.$$

This completes the proof. □

Proof of Theorem 1.1 In fact we only need to prove that Eq. (1.1) does not have an eventually positive solution $x(t)$ [we can prove in the similar method that Eq. (1.1) does not have an eventually negative solution $x(t)$]. By contradiction, then $x'(t) < 0$, since $x(t)$ is an eventually positive solution and condition $(H0)$. So $x(t)$ has a limitation as $t \to \infty$. We claim that

$$\lim_{t \to \infty} x(t) = l = 0.$$

Indeed, if $l \neq 0$, then we have $l > 0$ and according to conditions $(H0)$ and $(H1)$, the following inequality

$$\begin{aligned} 0 &= \lim_{t \to \infty} x'(t) \\ &= \lim_{t \to \infty}\left(-\sum_{i=1}^{n} p_i(t) f(x(t - r_i(t)))\right) \\ &= -\sum_{i=1}^{n} p_i f(l) < 0 \end{aligned}$$

holds, which implies that $l = 0$.

On the other hand, by $(H1)$, for arbitrary $\varepsilon > 0$, there is a large enough T such that if $t > T$,

$$p_i - \varepsilon < p_i(t) < p_i + \varepsilon, \quad r_i - \varepsilon < r_i(t) < r_i + \varepsilon \tag{2.7}$$

and

$$\frac{f(x(t))}{x(t)} > \alpha_0 - \varepsilon. \tag{2.8}$$

Set

$$\varphi_i(t) = \frac{x(t - r_i(t))}{x(t)}, \quad t > T, \; i = 1, 2, \ldots n$$

and

$$b_i = \liminf_{t \to \infty} \varphi_i(t), \quad i = 1, 2, \ldots n.$$

Then $\varphi_i(t) > 1$ when $t > T$ and $b_i \geq 1$, $i = 1, 2, \ldots, n$, since $x'(t) < 0$.
Dividing $x(t)$ as $t > T$ in the both sides of (1.1), we obtain that

$$\frac{x'(t)}{x(t)} + \sum_{i=1}^{n} p_i(t)\varphi_i(t)\frac{f(x(t - r_i(t)))}{x(t - r_i(t))} = 0, \quad t > T, \; i = 1, 2, \ldots, n. \tag{2.9}$$

Integrating (2.9) over the intervals $[t - r_k(t), t]$, $k = 1, 2, \ldots, n$, then we get

$$\log x(t) - \log x(t - r_k(t)) + \int_{t - r_k(t)}^{t} \sum_{i=1}^{n} p_i(s)\varphi_i(s)\frac{f(x(s - r_i(s)))}{x(s - r_i(s))}\, ds = 0,$$

that is,

$$\log \varphi_k(t) - \int_{t - r_k(t)}^{t} \sum_{i=1}^{n} p_i(s)\varphi_i(s)\frac{f(x(s - r_i(s)))}{x(s - r_i(s))}\, ds = 0, \quad k = 1, 2, \ldots, n. \tag{2.10}$$

Now we will consider two cases:
Case 1: All numbers b_i, $i = 1, 2, \ldots, n$ are finite.

By (2.10) and (2.8) for large enough t $(t > T)$, we have

$$\log \varphi_k(t) \geq (\alpha_0 - \varepsilon)\int_{t - r_k(t)}^{t} \sum_{i=1}^{n} p_i(s)\varphi_i(s)\, ds, \quad k = 1, 2, \ldots, n. \tag{2.11}$$

according to $(H0)$, $\lim_{t\to\infty} x(t) = 0$ and the boundedness of $r_i(t)$. Therefore, combining (2.7) and (2.11), we obtain

$$\log b_k \geq (\alpha_0 - \varepsilon)\sum_{i=1}^{n}(p_i - \varepsilon)(r_k - \varepsilon)b_i. \tag{2.12}$$

Letting $\varepsilon \to 0$, then we have

$$\log b_k \geq \alpha_0 \sum_{i=1}^{n} p_i r_k b_i, \quad k = 1, 2, \ldots n, \tag{2.13}$$

which implies that

$$\sum_{k=1}^{n} \log b_k \geq \alpha_0 \left(\sum_{i=1}^{n} p_i b_i \right) \left(\sum_{k=1}^{n} r_k \right). \tag{2.14}$$

By inequality (2.14) and Lemma 2.1 where we take $q_i = \alpha_0 p_i$, we have the following inequality

$$\sum_{k=1}^{n} \left(-\log \left(\alpha_0 p_k \sum_{i=1}^{n} r_i \right) \right) \geq \left(\sum_{k=1}^{n} \alpha_0 p_k \right) \left(\frac{1}{\alpha_0 p_k \sum_{i=1}^{n} r_i} \right) \left(\sum_{k=1}^{n} r_k \right) = n. \tag{2.15}$$

Thus,

$$\left(\prod_{i=1}^{n} \alpha_0 p_i \right) \left(\sum_{i=1}^{n} r_i \right)^n \leq \frac{1}{e^n},$$

that is,

$$\left(\prod_{i}^{n} p_i \right)^{\frac{1}{n}} \left(\sum_{i=1}^{n} r_i \right) \leq \frac{1}{\alpha_0 e},$$

which contradicts $(H2)$. Therefore Eq. (1.1) could not have an eventually positive solution in case 1.

Case 2: At least one of the numbers b_i, $i = 1, 2, \ldots n$ is infinite. Without loss of generalization, we assume that for some $i \in \{1, 2, \cdots, n\}$,

$$\lim_{t \to \infty} \frac{x(t - r_i(t))}{x(t)} = +\infty.$$

According to $(H0)$, for large enough t_1, we have that if $t > t_1$,

$$x'(t) + p_i(t) f(x(t - r_i(t))) < 0. \tag{2.16}$$

and

$$\frac{f(x(t))}{x(t)} > \alpha_0 - \varepsilon.$$

Thus, for any $t > t_1$, we have the following inequality

$$x'(t) + (\alpha_0 - \varepsilon) p_i(t) x(t - r_i(t)) < 0. \tag{2.17}$$

Since $\lim_{t \to \infty} p_i(t) = p_i$, we have

$$\lim_{t \to \infty} \int_{t - r_i(t)}^{t} p_i(s) ds = p_i r_i > 0.$$

So there exist $d > 0$ and $t_2 > 0$ such that

$$\int_{t-r_i(t)}^{t} p_i(s) \geq d \quad \texttt{for} t > t_2.$$

Then for any $t > t_2$, there is $\xi(t) > t$ such that

$$\int_{t}^{\xi(t)} p_i(s)ds = \frac{d}{2} \tag{2.18}$$

and

$$\int_{\delta_i(\xi(t))}^{t} p_i(s)ds \geq \frac{d}{2}. \tag{2.19}$$

where $\delta_i(t) = t - r_i(t)$. Let $t_3 = \max\{t_1, t_2\}$. For any $t > t_3$, integrating (2.17) over the interval $[t, \xi(t)]$ and $[\delta_i(\xi(t)), t]$ respectively, we find

$$x(\xi(t)) - x(t) + (\alpha_0 - \varepsilon)\int_{t}^{\xi(t)} p_i(s)x(\delta_i(s))ds < 0 \tag{2.20}$$

and

$$x(t) - x(\delta_i(\xi(t))) + (\alpha_0 - \varepsilon)\int_{\delta_i(\xi(t))}^{t} p_i(s)x(\delta_i(s))ds < 0. \tag{2.21}$$

From $x'(t) < 0$, $\delta_i'(t) = 1 - r_i'(t) > 0$, (2.20) and (2.21), we obtain

$$x(\xi(t)) - x(t) + (\alpha_0 - \varepsilon)\int_{t}^{\xi(t)} p_i(s)dsx(\delta_i(\xi(t))) < 0 \tag{2.22}$$

and

$$x(t) - x(\delta_i(\xi(t))) + (\alpha_0 - \varepsilon)\int_{\delta_i(\xi(t))}^{t} p_i(s)dsx(\delta_i(t)) < 0 \tag{2.23}$$

for $t > t_3$.

Since $x(t) > 0$, omitting the first terms in (2.22) and (2.23), we conclude

$$-x(t) + (\alpha_0 - \varepsilon)\int_{t}^{\xi(t)} p_i(s)dsx(\delta_i(\xi(t))) < 0 \tag{2.24}$$

and

$$-x(\delta_i(\xi(t))) + (\alpha_0 - \varepsilon)\int_{\delta_i(\xi(t))}^{t} p_i(s)dsx(\delta_i(t)) < 0 \tag{2.25}$$

for $t > t_3$.

From (2.18), (2.19), (2.24) and (2.25), it follows

$$-x(t) + (\alpha_0 - \varepsilon)\frac{d}{2}x(\delta_i(\xi(t))) < 0 \tag{2.26}$$

and

$$-x(\delta_i(\xi(t))) + (\alpha_0 - \varepsilon)\frac{d}{2}x(\delta_i(t)) < 0 \tag{2.27}$$

for $t > t_3$. From (2.26) and (2.27), we obtain

$$x(t) > (\alpha_0 - \varepsilon)\frac{d}{2}x(\delta_i(\xi(t))) > (\alpha_0 - \varepsilon)^2\frac{d^2}{4}x(\delta_i(t)).$$

Then

$$\frac{x(\delta_i(t))}{x(t)} < \frac{4}{(\alpha_0 - \varepsilon)^2 d^2} < \infty$$

which is a contradiction to the assumption of case 2. From the above two cases, we conclude that equation (1.1) could not have an eventually positive solution. This completes the proof. □

Proof of Theorem 1.2 In fact we only need to prove that Eq. (1.2) does not have an eventually positive solution $x(t)$ [we can similarly prove that Eq.(1.2) does not have an eventually negative solution $x(t)$]. By contradiction, then $x'(t) < 0$, since $x(t)$ is an eventually positive solution and condition $(H0)$. Thus, $\lim_{t\to}x(t)$ exists. We claim that

$$\lim_{t\longrightarrow\infty} x(t) = l = 0.$$

Indeed, if $l \neq 0$, then we have $l > 0$ and according to conditions $(H0)$ and $(H1')$, the following inequality

$$\begin{aligned} 0 &= \lim_{t\longrightarrow\infty} x'(t) \\ &= \lim_{t\longrightarrow\infty}\left(-\sum_{i=1}^{n} p_i(x(t))f(x(t - r_i(x(t))))\right) \\ &= -\sum_{i=1}^{n} p_i(l)f(l) < 0 \end{aligned}$$

holds, which implies that $l = 0$.

By using the same notations as in the proof of Theorem 1.1, we also consider two cases:

Case 1: All numbers b_i, $i = 1,2,\ldots,n$ are finite.

Case 2: At least one of the numbers b_i, $i = 1,2,\ldots,n$ is infinite.

For case 1, the proof is similar to the proof of Theorem 1.1. So we omit it. Thus we only need to consider case 2, that is, for some $i \in \{1,2,\cdots,n\}$,

$$\lim_{t\longrightarrow\infty}\frac{x(t - r_i(x(t)))}{x(t)} = +\infty.$$

From $(H0)$, for large enough t_1, we have

$$x'(t) + p_i(x(t))f(x(t - r_i(x(t)))) < 0 \tag{2.28}$$

and

$$\frac{f(x(t))}{x(t)} > \alpha_0 - \varepsilon$$

for $t > t_1$. Thus, for any $t > t_1$, we have the following inequality

$$x'(t) + (\alpha_0 - \varepsilon)p_i(x(t))x(t - r_i(x(t))) < 0. \tag{2.29}$$

Since $\lim_{t\to\infty} x(t) = 0$ and $\lim_{t\to\infty} p_i(x(t)) = p_i$, we have

$$\lim_{t\to\infty} \int_{t-r_i(x(t))}^{t} p_i(x(s))ds = p_i r_i > 0.$$

So there are $d > 0$ and $t_2 > 0$ such that

$$\int_{t-r_i(x(t))}^{t} p_i(x(s)) \geq d \quad \texttt{for} t > t_2.$$

Then for any $t > t_2$, there is $\xi(t) > t$ such that

$$\int_{t}^{\xi(t)} p_i(x(s))ds = \frac{d}{2} \tag{2.30}$$

and

$$\int_{\delta_i(\xi(t))}^{t} p_i(x(s))ds \geq \frac{d}{2}, \tag{2.31}$$

where $\delta_i(t) = t - r_i(x(t))$.

Claim: $\delta_i(t) > \delta_i(s)$ for $t > s$, where s is large enough. In fact, for $t > s$, s is large enough, we have

$$\begin{aligned} \delta_i(t) - \delta_i(s) &= t - r_i(x(t)) - [s - r_i(x(s))] \\ &= (t-s) - [r_i(x(t)) - r_i(x(s))] \\ &= (t-s) - r_i'(\theta)x'(\tau)(t-s) \\ &= (t-s)[1 - r_i'(\theta)x'(\tau)] > 0, \end{aligned}$$

where $x(t) < \theta < x(s)$ and $s < \tau < t$. The inequality of the fourth line follows from $r'(x) > 0$ and $x'(t) < 0$.

Let $t_3 = \max\{t_1, t_2\}$. For any $t > t_3$, integrating (2.29) over the interval $[t, \xi(t)]$ and $[\delta_i(\xi(t)), t]$ respectively, we find

$$x(\xi(t)) - x(t) + (\alpha_0 - \varepsilon) \int_{t}^{\xi(t)} p_i(x(s))x(\delta_i(\xi(s)))ds < 0 \tag{2.32}$$

and

$$x(t) - x(\delta_i(\xi(t))) + (\alpha_0 - \varepsilon) \int_{\delta_i(\xi(t))}^{t} p_i(x(s))x(\delta_i(s))ds < 0. \tag{2.33}$$

Combining our claim and (2.32), (2.33), we obtain

$$x(\xi(t)) - x(t) + (\alpha_0 - \varepsilon) \int_{t}^{\xi(t)} p_i(x(s))ds x(\delta_i(\xi(s))) < 0 \tag{2.34}$$

and

$$x(t)-x(\delta_i(\xi(t)))+(\alpha_0-\varepsilon)\int_{\delta_i(\xi(t))}^{t} p_i(x(s))ds\,x(\delta_i(t))<0. \tag{2.35}$$

for $t>t_3$.

Since $x(t)>0$, omitting the first terms in (2.34) and (2.35), we conclude

$$-x(t)+(\alpha_0-\varepsilon)\int_{t}^{\xi(t)} p_i(x(s))ds\,x(\delta_i(\xi(s)))<0 \tag{2.36}$$

and

$$-x(\delta_i(\xi(t)))+(\alpha_0-\varepsilon)\int_{\delta_i(\xi(t))}^{t} p_i(x(s))ds\,x(\delta_i(t))<0. \tag{2.37}$$

for $t>t_3$. Combining (2.30), (2.31, (2.36) and (2.37), we have

$$-x(t)+(\alpha_0-\varepsilon)\frac{d}{2}x(\delta_i(\xi(s)))<0 \tag{2.38}$$

and

$$-x(\delta_i(\xi(t)))+(\alpha_0-\varepsilon)\frac{d}{2}x(\delta_i(t))<0 \tag{2.39}$$

for $t>t_3$. From (2.38) and (2.39), it follows

$$x(t)>(\alpha_0-\varepsilon)\frac{d}{2}\delta_i(\xi(t))>(\alpha_0-\varepsilon)^2\frac{d^2}{4}x(\delta_i(t)).$$

Then

$$\frac{x(\delta_i(t))}{x(t)}<\frac{4}{(\alpha_0-\varepsilon)^2d^2}<\infty,$$

which contradicts to the assumption of case 2.

From the above two cases, we obtain that equation (1.2) could not have an eventually positive solution. This completes the proof. □

Remark: If we use the condition $\delta_i(t):=t-r_i(t)=\max_{[0,t]}\{s-r_i(s)\}$ instead of $r_i'(t)<1$, then Theorem 1.1 still holds. Theorem 1.2 also holds if we replace $r'(x)>0$ by $r'(x)$ bounded or $r'(x)>0$ for $|x|<\delta$ since $\lim_{t\to\infty}x'(t)=0$.

Corollary 2.1 *Let*

$(H0')$. $\frac{f(x)}{x}\geq\alpha_0$ *for* $x\neq 0$, *where* $0<\alpha_0<\infty$.

Then

(i) Assume that $(H0')$, $(H1)$, $(H2)$ *hold, each solution of (1.1) is oscillatory.*

(ii) Assume that $(H0')$, $(H1')$, $(H2)$ *hold, each solution of (1.2) is oscillatory.*

Proof. We will obtain this result in the similar way as in Theorem 1.1. In fact, note that we needn't use $\lim_{t\to\infty} x(t) = 0$ when $(H0')$ holds for equation (1.1). We will omit the details. □

Theorem 2.1 *Assume* $(H1)$, $(H2)$ *and* $(H0')$. *Each solution of the advanced delay differential equation*

$$\dot{x}(t) - \sum_{i=1}^{n} p_i(t) f(x(t + r_i(t))) = 0, \tag{2.40}$$

is oscillatory.

Proof. We needn't use $\lim_{t\to\infty} x(t) = 0$ when $(H0')$ holds for the advanced equations. Thus, we can prove Theorem 2.1 in view of the similar technique. □

Remark. We consider the advanced equation

$$\dot{x}(t) - \sum_{i=1}^{n} p_i(x(t)) f(x(t + r_i(x(t)))) = 0, \tag{2.41}$$

corresponding to(1.2). Note $x'(t) > 0$ when $x(t) > 0$ for (2.41). So we can't prove $\lim_{t\to\infty} x(t) = 0$ in the condition $(H0')$. Thus, we can't conclude if each solution of (2.41) is oscillatory in the similar method.

Corollary 2.2 *Let*

$(H2')$. $n\left(\prod_{i=1}^{n} p_i\right)^{\frac{1}{n}} r > \frac{1}{\alpha_0 e}$, *where* $r = \min\{r_1, r_2, \ldots, r_n\}$

$(H2'')$. $\left(\sum_{i=1}^{n} r_i\right) p_i > \frac{1}{\alpha_0 e}$, *where* $p = \min\{p_1, p_2, \ldots, p_n\}$.

(i) Assume that $(H0)$ *or* $(H0')$, $(H1)$, $(H2')$ *hold, each solution of (1.1) is oscillatory.*
(ii) Assume that $(H0)$ *or* $(H0')$, $(H1)$, $(H2'')$ *hold, each solution of (1.1) is oscillatory.*
(iii) Assume that $(H0)$ *or* $(H0')$, $(H1')$, $(H2')$ *hold, each solution of (1.2) is oscillatory.*
(iv) Assume that $(H0)$ *or* $(H0')$, $(H1')$, $(H2'')$ *hold, each solution of (1.2) is oscillatory.*

Corollary 2.3 *Assume* $(H0')$, $(H1)$, *in addition, one of the conditions* $(H2')$ *and* $(H2'')$. *Then all of solutions of the advanced equation (2.40) oscillate.*

3. Proof of Theorems 1.3 and 1.4

We only prove Theorem 1.3. Theorem 1.4 can be proved similarly. So we omit it. We begin with another lemma.

Lemma 3.1 *For* $t > 0$, *then* $\frac{\log t}{t} \le \frac{1}{e}$.

Proof. Let $f(t)=\frac{\log t}{t}$. Observe that $f'(t)=\frac{1-\log t}{t^2}=0$ for $t=e$. Since $f'(t)\geq 0$ for $0<t\leq e$ and $f'(t)<0$ for $t>e$, $f(t)$ has the maximum at $t=e$. This implies that $\frac{\log t}{t}\leq\frac{1}{e}$.

Proof of Theorem 1.3. Similar to the proof of Theorem 1.1, we only need to prove that Eq. (1.1) does not have an eventually positive solution $x(t)$ [we can prove in the similar method that Eq. (1.1) does not have an eventually negative solution $x(t)$]. By contradiction, we also consider two cases:

Case 1: All numbers $b_i, i=1,2,\ldots n$ are finite, where b_i is as in the proof of Theorem 1.1.

Case 2: At least one of the numbers b_i, $i=1,2,\ldots n$ is infinite.

We only consider case 1: $b_i<\infty$ [we can prove Theorems 1.3 and 1.4 in the similar to the method of Theorems 1.1 and 1.2 for $b_i=\infty$].

Under the the conditions $(H0)$ and $(H1)$, we must have (2.13). By Lemma 3.1, we obtain

$$\frac{\log b_k}{b_k}\leq\frac{1}{e}\quad for\quad b_k\geq 1. \tag{3.42}$$

Combining (2.13) and (3.42), we find

$$\frac{1}{e}\geq\alpha_0\sum_{i=1}^{n}p_i r_k\frac{b_i}{b_k}, \tag{3.43}$$

which implies that

$$\frac{n}{\alpha_0 e}\geq\sum_{i=1}^{n}p_i r_i+\sum_{\substack{i,k=1\\ i<k}}^{n}(p_i r_k\frac{b_i}{b_k}+p_k r_i\frac{b_k}{b_i}).$$

Now using the result that

$$p_i r_k\frac{b_i}{b_k}+p_k r_i\frac{b_k}{b_i}\geq 2(p_i p_k r_i r_j)^{\frac{1}{2}},$$

we have

$$\frac{n}{\alpha_0 e}\geq\sum_{i=1}^{n}p_i r_i+2\sum_{\substack{i,k=1\\ i<k}}^{n}(p_i p_k r_i r_k)^{\frac{1}{2}}=(\sum_{i=1}^{n}(p_i r_i)^{\frac{1}{2}})^2),$$

which contradicts $(H3)$. This completes the proof. □

Corollary 3.1 *(i) Assume that $(H0')$, $(H1)$, $(H3)$ hold, each solution of (1.1) is oscillatory.*
(ii) Assume that $(H0')$, $(H1')$, $(H3)$ hold, each solution of (1.2) is oscillatory.

Proof. We will obtain this result which is the similar to the proof of Theorem 1.3. □

Corollary 3.2 *Assume $(H0')$, $(H1)$ and $(H3)$, each solution of the advanced delay differential equations (2.40) is oscillatory.*

Proof. We obtain easily the corollary according to the process of theorems 1.1 and 1.3. □

Corollary 3.3 *Let*

$(H3')$. $p_i r_i > \frac{1}{\alpha_0 e}$, *for some* $i \in \{1,2,\dots n\}$.

$(H3'')$. $\frac{1}{n}(\sum_{i=1}^{n} r_i p_i) > \frac{1}{\alpha_0 e}$.

Then,

(i) Assume that $(H0)$ *or* $(H0')$, $(H1)$, $(H3')$ *hold, each solution of (1.1) is oscillatory.*

(ii) Assume that $(H0)$ *or* $(H0')$, $(H1)$, (H'') *hold, each solution of (1.1) is oscillatory.*

(iii) Assume that $(H0)$ *or* $(H0')$, $(H1')$, $(H3')$ *hold, each solution of (1.2) is oscillatory.*

(iv) Assume that $(H0)$ *or* $(H0')$, $(H1')$, $(H3'')$ *hold, each solution of (1.2) is oscillatory.*

Corollary 3.4 *Assume* $(H0')$, $(H1)$, *in addition, one of the conditions* $(H3')$ *and* $(H3'')$. *Then all of solutions of the advanced equation (2.40) oscillate.*

Remark 3.1 *When* $p_i(t)$, $r_i(t)$, $p_i(x)$ *and* $r_i(x)$ *are constants for* $i = 1,2,\dots,n$ *and add* $f(x) = x$, *it is obvious that Theorems 1.1, 1.2, 1.3 and 1.4 coincide with those of G.Ladas and I. P.Stavroulakis [5].*

4. Examples

Example 4.1. We consider

$$x'(t) + (1+\arctan^2 t)\arctan x(t-(3+\frac{1}{\pi}\arctan^2 t))) +$$
$$(2+\arctan^2 t))\arctan x(t-(1+\frac{1}{\pi}\arctan^2 t))) = 0,$$

where $p_1(t) = 1+\arctan^2 t, p_2(t) = 2+\arctan^2 t,\ r_1(t) = 3+\frac{1}{\pi}\arctan^2 t, r_2(t) = 1+\frac{1}{\pi}\arctan^2 t$ and $f(x) = \arctan x$. It is easily checked that all the conditions of Theorems 1.1 and 1.3 are satisfied for the above delay differential equation. In fact, $p_1 = 1+\frac{\pi^2}{4}$, $p_2 = 2+\frac{\pi^2}{4}$, $r_1 = 3+\frac{\pi}{4}$, $r_2 = 1+\frac{\pi}{4}$, $r'(t) < 1$, $\lim_{x\to 0}\frac{f(x)}{x} = 1 > 0$,

$$\left(\prod_{i=1}^{2} p_i\right)\left(\sum_{i=1}^{2} r_i\right) > 4 > \frac{1}{e}$$

and

$$\frac{1}{2}\left(\sum_{i=1}^{2}(r_i p_i)^{\frac{1}{2}}\right)^2 > \frac{5+2\sqrt{6}}{2} > \frac{1}{e}.$$

Example 4.2. We consider

$$x'(t) + (1+\arctan^2 x(t))\arctan(x(t-(3+\arctan^2 x(t))))$$
$$+(2+\arctan^2 x(t)\arctan(x(t-(1+2\arctan^2 x(t)))) = 0,$$

where $p_1(x) = 1+\arctan^2 x, p_2(x) = 2+\arctan^2 x,\ r_1(x) = 3+\arctan^2 x, r_2(x) = 1+2\arctan^2 x$ and $f(x) = \arctan x$. It is also obvious that all the conditions of Theorems 1.2 and 1.4 are satisfied. We omit the details.

References

[1] Y. Cheng, Oscillation in nonautonomous scalar differential equations with deviating arguments, *Proc. Amer. Math. Soc.* **110** (1990), 711-719.

[2] M. Elabbasy, A. S. Hegazi, S. H.Saker, Oscillation of solutions to delay differential equations with positive and negative coefficients. Electron. *Electron. J. Differential Equations* 2000, No. 13, 13 pp. (electronic).

[3] B.R. Hunt, J.A. Yorke,When all solutions of $x' = -\sum q_i(t)x(t - T_i(t))$ oscillate. *J. Differential Equations.* **53** (1984), 139-145.

[4] J. Jiang, Oscillation criteria for first order nonlinear delay differential equations. *Hiroshima Math. J.* **31** (2001), 467-476.

[5] G.Ladas and I. P.Stavroulakis, Oscillations caused by several retarded and advanced arguments. *J. Differential Equations.* **44** (1982), 134-152.

[6] G.Ladas, Y. G.Sficas and I. P.Stavroulakis, Necessary and sufficient conditions for oscillations. *Amer. Math. Monthly.* **90** (1983),637-640.

[7] B. Li, Oscillation of first order delay differential equations. *Proc. Amer. Math. Soc.* **12** (1996), 3729-3737.

[8] B. Li, Oscillations of delay differential equations with variable coefficients. *J. Math. Anal. Appl.* **192** (1995), 312-321.

[9] B. Li and Y. Kuang, Sharp conditions for oscillations in some nonlinear nonautonomous delay differential equations. *Nonlinear Anal.* **29** (1997), 1265-1276.

[10] Xianyi Li and Deming Zhu, Oscillation and nonoscillation of advanced differential equations with variable coefficients. *J. Math. Anal. Appl.* **269** (2002), 462-488.

[11] Y. Kuang, and H. L. Smith, Slowly oscillating periodic solutions of autonomous state-dependent delay equations. *Nonlinear Anal.* **19** (1992), 855-872.

[12] M. R. S.Kulenović, G.Ladas, A.Meimaridou, Necessary and sufficient condition for oscillations of neutral differential equations. *J. Austral. Math. Soc. Ser. B* **28** (1987), 362-375

[13] M. K. Kwong, Oscillation of first-order delay equations. *J. Math. Anal. Appl.* **156** (1991), 274-286.

[14] X. H.Tang, Oscillation of first order delay differential equations with distributed delay. *J. Math. Anal. Appl.* **289** (2004), 367-378.

In: Research on Evolution Equation Compendium. Volume 2 ISBN: 978-1-60456-752-6
Editor: Gaston M. N'Guerekata

SEMILINEAR ABSTRACT DIFFERENTIAL EQUATIONS WITH DEVIATED ARGUMENT

*Ciprian G. Gal**
Morgan State University, Department of Mathematical Sciences,
Baltimore, Md, 21251, U.S.A.

Abstract

In this paper we prove local and global existence results for semilinear differential equations with deviated argument in Banach spaces.

1991 Mathematics Subject Classification: 34G10.

Key words and phrases: mild solutions, semigroups of linear operators, semilinear differential equations with deviated arguments.

1. Introduction

In the very recent papers [2] and [3], the existence and uniqueness of almost automorphic mild solutions with values in Banach spaces X and in locally convex spaces, respectively, for the differential equation with deviated argument

$$u'(t) = Au(t) + f(t, u(t), u[\alpha(u(t), t)]),\ t \in \mathbb{R},$$

were proved, where A is the infinitesimal (bounded) generator of a C_0-semigroup of operators $(T(t))_{t\geq 0}$ on a Banach space, satisfying an additional exponential-type conditions of stability of the form $||T(t)|| \leq Ke^{\omega t}, \forall t \in \mathbb{R}_+$ with $K > 0, \omega < 0$ and $\alpha(y,t) : X \times \mathbb{R} \to \mathbb{R}$ is an almost automorphic function with respect to t and Lipschitz in y. These results generalize some known classical results in the literature [4], [5], [6].

The goal of the present note is to prove the existence and uniqueness of local and global solutions for the initial value problem with deviated argument

$$u'(t) = Au(t) + f(t, u(t), u[\alpha(u(t), t)]),\ t \in \mathbb{R}_+,\ u(0) = u_0,$$

but without to impose any additional assumption on the semigroup $(T(t))_{t\geq 0}$.

These results generalize in some cases (that is, when A is bounded), the classical ones with usual (non-deviated) argument in [1].

*E-mail address: cgal@morgan.edu

2. Basic Results

Let $(X, ||\cdot||)$ be a Banach space. We consider the nonlinear differential equation with deviated argument

$$u'(t) = Au(t) + f(t, u(t), u[\alpha(u(t), t)]),\ t \in \mathbb{R}_+,\ u(0) = u_0, \tag{2.1}$$

where A is a bounded linear operator in X that generates a (C_0) semigroup $\{T(t)\}_{t\in\mathbb{R}_+}$ on X. It is easy to prove that any solution of this problem has the form

$$u(t) = T(t)u_0 + \int_0^t T(t-s)[f(s, u(s), u(\alpha(u(s), s)))]ds, \tag{2.2}$$

and we refer to any continuous

$$u \in C_L(\mathbb{R}_+, X) = \{u(\cdot) \in C(\mathbb{R}_+, X) : ||u(t) - u(s)|| \le L|t-s|, \forall t, s \in \mathbb{R}_+\}$$

satisfying the relation (2.2) as a mild solution of the above problem . Obviously, because of the absence, in general, of its differentiability, a mild solution is not a strong solution of the problem (2.1). A mild solution is thus a kind of generalized solution.

We have the following local existence result for (2.1).

Theorem 2.1 *Let $u_0 \in X$ and let $f : \mathbb{R}_+ \times X \times \mathbb{R}_+ \to X$ be continuous in all of its variables and satisfy the Lipschitz condition: for each $\delta > 0$, there is a $K = K(\delta) > 0$ such that*

$$||f(t, u_1, v_1) - f(t, u_2, v_2)|| \le K(||u_1 - u_2|| + ||v_1 - v_2||), \forall u_1, u_2, v_1, v_2 \in X,$$

whenever $0 \le t \le \delta$. Moreover, assume $\alpha : X \times \mathbb{R}_+ \to \mathbb{R}_+$ is globally continuous on $X \times \mathbb{R}_+$ and satisfies $\alpha(\cdot, 0) = 0$ and for each $\delta > 0$, there is a $L_\alpha = L_\alpha(\delta) > 0$ such that

$$|\alpha(u,t) - \alpha(v,t)| \le L_\alpha ||u - v||, \forall u, v \in X, \text{ whenever } 0 \le t \le \delta.$$

Then, for $\tau > 0$ sufficiently small , (2.1) has a unique solution in

$$C_L([0,\tau], X) = \{u(\cdot) \in C([0,\tau], X) : ||u(t) - u(s)|| \le L|t-s|, \forall t, s \in [0,\tau]\},$$

for some suitable $L = L(\tau) > 0$.

Proof. The idea of proof is that of Theorem 2.4, pp. 88-89 in [1], modified correspondingly to the deviated argument.

Let $\tau > 0$ and consider $Y = C_L([0,\tau], X)$. Let E be a closed neighborhood of u_0 in X. Define S by

$$(Su)(t) = T(t)u_0 + \int_0^t T(t-s)[f(s, u(s), u(\alpha(u(s), s)))]ds, \text{ for } t \in [0,\tau] \tag{2.3}$$

and $u \in \mathcal{M} = \{v(\cdot) \in Y : v(0) = u_0,\ v[0,\tau] \subset E\}$. Note that $\mathcal{M}$ is a complete metric space and $Su \in C([0,\tau], X)$ if $u \in C([0,\tau], X)$. We show that $Su \in Y$, for a suitable constant $L = L(\tau) > 0$. Let $u_0 \in X$. Since A is bounded, then by semigroup theory $u(t) = T(t)u_0 \in C^1([0,\tau],X)$, so that it is enough to show that $v(t) = \int_0^t T(t-s)[f(s,u(s),u(\alpha(u(s),s)))]ds \in C^1([0,\tau],X)$. It will follow $Su \in C_L([0,\tau],X)$. We have

$$v'(t) = T(0)[f(t,u(t),u(\alpha(u(t),t)))] + Av(t),$$

since A is linear and closed. Let M,ω such that $\|T(t)\| \leq Me^{\omega t}$, for $t \in \mathbb{R}_+$. Using the assumptions on f, we have for $t \in [0,\tau]$:

$$||v'(t)|| \leq \sup_{t\in[0,\tau]} \|f(t,u(t),u(\alpha(u(t),t)))\| + \sup_{t\in[0,\tau]} \|Av(t)\| \tag{2.4}$$

$$\leq \sup_{t\in[0,\tau]} \|f(t,u(t),u(\alpha(u(t),t))) - f(0,u_0,u_0)\| + \sup_{t\in[0,\tau]} \|f(0,u_0,u_0)\|$$

$$+ \sup_{t\in[0,\tau]} \int_0^t ||AT(t-s)[f(s,u(s),u(\alpha(u(s),s)))]||ds,$$

$$\leq (1 + Me^{\omega\tau}\tau|||A|||) \sup_{t\in[0,\tau]} \|f(t,u_0,u_0)\|$$

$$+K(\tau) \sup_{t\in[0,\tau]} (\|u(t)-u_0\| + \|u(\alpha(u(t),t)) - u(\alpha(u_0,0))\|)(1 + Me^{\omega\tau}\tau|||A|||) := M_f(\tau),$$

which exists because $u \in C_L([0,\tau],X)$. Here $|||\cdot|||$ stands for the operator norm. Estimate (2.4) yields for $t \in [0,\tau]$, $\tau > 0$ sufficiently small,

$$\sup_{t\in[0,\tau]} \left\| \frac{d}{dt}[Su(t)] \right\| \leq \sup_{t\in[0,\tau]} \|T(t)Au_0\| + \sup_{t\in[0,\tau]} \left\|v'(t)\right\|$$

$$\leq Me^{\omega\tau}\|Au_0\| + M_f(\tau). \tag{2.5}$$

Therefore, by the mean value theorem in vector spaces, $Su \in C_L([0,\tau],X)$, if L is chosen such that

$$L(\tau) \geq Me^{\omega\tau}\|Au_0\| + M_f(\tau) \tag{2.6}$$

for sufficiently small $\tau < 1$. Thus, for such chosen $L(\tau)$, the map S is well defined.

Next, we show that $S(\mathcal{M}) \subset \mathcal{M}$. For this, we assume without loss of generality that E is bounded. Finally, we have

$$\|Su - u_0\|_Y = \sup_{t\in[0,\tau]} \|Su(t) - u_0\|$$

$$\leq \sup_{t\in[0,\tau]} \|T(t)u_0 - u_0\| + \sup_{t\in[0,\tau]} \|v(t)\| := J_1(\tau) + J_2(\tau).$$

While $J_1(\tau) \to 0$ as $\tau \to 0$, calculating as in (2.4) and using the assumptions on the functions f,α, we also obtain the following estimate for

$$J_2(\tau) \leq Me^{\omega\tau}\tau \sup_{t\in[0,\tau]} \|f(t,u(t),u(\alpha(u(t),t)))\|$$

$$\leq Me^{\omega\tau}\tau\left(\sup_{t\in[0,\tau]}\|f(t,u_0,u_0)\|+\sup_{t\in[0,\tau]}\|f(t,u(t),u(\alpha(u(t),t)))-f(0,u_0,u_0)\|\right)$$

$$\leq Me^{\omega\tau}\tau\left(\sup_{t\in[0,\tau]}\|f(t,u_0,u_0)\|+K(\tau)(1+L(\tau)L_\alpha(\tau))\sup_{t\in[0,\tau]}\|u(t)-u_0\|\right).$$

Thus, it is clear $J_2(\tau)\to 0$ as $\tau\to 0^+$. It follows that S maps bounded sets of $\mathcal{M}$ into the same bounded sets (of $\mathcal{M}$), provided that τ is sufficiently small.

It remains to show that S is a contraction. We have from (2.3) :

$$\|Su-Sw\|_Y=\sup_{t\in[0,\tau]}\|Su(t)-Sw(t)\|$$

$$\leq Me^{\omega\tau}K(\tau)\int_0^\tau(\|u(s)-w(s)\|+\|u(\alpha(u(s),s))-w(\alpha(w(s),s))\|)\,ds$$

$$\leq Me^{\omega\tau}K(\tau)\int_0^\tau(\|u(s)-w(s)\|+\|u(\alpha(u(s),s))-u(\alpha(w(s),s))\|$$

$$+\|u(\alpha(u(s),s))-w(\alpha(w(s),s))\|)\,ds$$

$$\leq Me^{\omega\tau}K(\tau)\int_0^\tau\left(2\sup_{0\leq r\leq s}\|u(r)-w(r)\|+L(\tau)\left|\alpha(u(s),s)-\alpha(w(s),s)\right|\right)ds$$

$$\leq Me^{\omega\tau}K(\tau)\int_0^\tau(L(\tau)L_\alpha(\tau)+2)\sup_{0\leq r\leq s}\|u(r)-w(r)\|\,ds$$

$$\leq Me^{\omega\tau}K(\tau)(L(\tau)L_\alpha(\tau)+2)\tau\|u-w\|_Y. \tag{2.7}$$

Here obviously we can assume, without loss of generality, that $K(\tau),L(\tau),L_\alpha(\tau)$ are nondecreasing as functions of τ, which implies that we have

$$Me^{\omega\tau}K(\tau)(L(\tau)L_\alpha(\tau)+2)\tau\to 0$$

as $\tau\to 0^+$. Therefore S is a contraction as desired, for τ sufficiently small, that is, for

$$Me^{\omega\tau}K(\tau)(L(\tau)L_\alpha(\tau)+2)\tau<1.$$

This implies

$$L(\tau)<\left[\frac{1}{\tau Me^{\omega\tau}K(\tau)}-2\right]\frac{1}{L_\alpha}. \tag{2.8}$$

Taking into account the inequality satisfied by $L(\tau)$ above in the proof, concerning the fact that S is a well defined map, it follows from (2.6) and (2.8) that the following inequality must hold:

$$Me^{\omega\tau}||Au_0||+M_f(\tau)<L(\tau)<\left[\frac{1}{\tau Me^{\omega\tau}K(\tau)}-2\right]\frac{1}{L_\alpha}. \tag{2.9}$$

Now, since $M_f(\tau)$ is non-decreasing as function of τ, we get that $Me^{\omega\tau}||Au_0||+M_f(\tau)\to M||Au_0||+M_f(0+)$, as $\tau\to 0^+$ and also $\left[\frac{1}{\tau Me^{\omega\tau}K(\tau)}-2\right]\frac{1}{L_\alpha}\to+\infty$ as $\tau\to 0^+$, which implies that for sufficiently small $\tau<1$, the constant $L(\tau)$ exists. Thus, for such a constant satisfying (2.9), S is a well defined map from $C_{L(\tau)}([0,\tau],X)$ to $C_{L(\tau)}([0,\tau],X)$ if τ is sufficiently small. The contraction property of S was verified in (2.7). Hence, the Banach fixed point theorem implies that the desired solution of (2.1) is the fixed point of S. The proof is complete. □

Furthermore, we have the following global existence theorem for problem (2.1).

Theorem 2.2 *We have the same assumptions as in* **Theorem 2.1.** *Then, the initial value problem* (2.1) *has a unique solution in* $C_L([0,\infty),X)\cap C^1((0,\infty),X)$.

Proof. The idea of proof is that of Theorem 2.5, pp. 89-90 in [1], modified correspondingly to the deviated argument.

Let S be as in (2.3). We claim that

$$\|S^n u(t)-S^n w(t)\| \le \frac{[Me^{\omega t}K(t)(L(t)L_\alpha(t)+2)t]^n}{n!}\sup_{s\in[0,t]}\|u(s)-w(s)\|, \tag{2.10}$$

for all $u,w\in C_L([0,t],X)$, for all $t>0$. Let us denote $M_S^n(t)$ by

$$M_S^n(t):=\frac{[Me^{\omega t}K(t)(L(t)L_\alpha(t)+2)]^n}{n!}>0,$$

for $t>0$, $n\in\mathbb{N}$. We may assume without loss of generality that

$$t\longmapsto K(t),L(t),L_\alpha(t),M_S^n(t)$$

(for all n) are monotone nondecreasing functions in $\mathbb{R}_+$. For $n=1$, (2.10) holds because of the estimate in (2.9). We prove our claim using induction on n. Assume (2.10) is true. We have

$$\begin{aligned}
&\left\|S^{n+1}u(t)-S^{n+1}w(t)\right\|\\
&=\left\|\int_0^t T(t-s)[f(s,S^n u(s),S^n u(\alpha(S^n u(s),s)))-f(s,S^n w(s),S^n w(\alpha(S^n w(s),s)))]ds\right\|\\
&\le Me^{\omega t}K(t)\int_0^t(\|S^n u(s)-S^n w(s)\|+\|S^n u(\alpha(S^n u(s),s))-S^n w(\alpha(S^n w(s),s))\|)ds\\
&\le Me^{\omega t}K(t)\int_0^t(\|S^n u(s)-S^n w(s)\|+\|S^n u(\alpha(S^n u(s),s))-S^n u(\alpha(S^n w(s),s))\|\\
&\qquad+\|S^n u(\alpha(S^n u(s),s))-S^n w(\alpha(S^n w(s),s))\|)ds\\
&\le Me^{\omega t}K(t)\int_0^t(M_S^n(s)s^n\|u-w\|_Y+L(s)|\alpha(S^n u(s),s)-\alpha(S^n w(s),s)|\\
&\qquad+M_S^n(s)s^n\|u-w\|_Y)ds\\
&\le Me^{\omega t}K(t)\int_0^t(2M_S^n(s)s^n+L(s)L_\alpha(s)M_S^n(s))s^n\|u-w\|_Y ds\\
&\le Me^{\omega t}K(t)M_S^n(t)(2+L(t)L_\alpha(t))\|u-w\|_Y\int_0^t s^n ds\le M_S^{n+1}t^{n+1}\|u-w\|_Y,
\end{aligned}$$

Thus, by induction, (2.10) holds for all $n \in \mathbb{N}$. Next, we let $t > 0$ be fixed, but otherwise arbitrary and choose n so large that

$$\frac{\left[Me^{\omega\tau}K(t)\left(L(t)L_{\alpha}(t)+2\right)t\right]^{n}}{n!} < 1.$$

Then by (2.10), S has a unique fixed point in $Y = C_L([0,t], X)$ and therefore the problem (2.1) has a unique solution in $C_L([0,t], X)$, where L is chosen as in (2.9) (see the proof of **Theorem 2.1**). The result follows since $t > 0$ is arbitrary. □

Remark. When the generator A is bounded, Theorem 2.1 generalizes Theorem 2.4, p. 88 in [1], while Theorem 2.2 generalizes Theorem 2.5, p. 88-89 in [1].

References

[1] J. A. Goldstein, *Semigroups of Linear Operators and Applications*, Oxford University Press, Oxford, 1985.

[2] C. G. Gal, Almost automorphic mild solutions to some semilinear abstract differential equations with deviated argument, *Journal of Integral Equations and Applications*, **17** (2005), no. 4, 391–396.

[3] C. G. Gal, Almost automorphic mild solutions to some semilinear abstract differential equations with deviated argument in *Fréchet spaces*, submitteed.

[4] C. G. Gal, S.G. Gal and G.M. N'Guerekata, Almost automorphic functions in Fréchet spaces and applications to differential equations, *Semigroup Forum*, **75** (2005), No. 2, 23-48.

[5] G. M. N'Guérékata, Almost Automorphic and Almost Periodic Functions in Abstract Spaces, *Kluwer Academic*/Plenum Publishers, New York, 2001.

[6] G. M. N'Guérékata, Existence and uniqueness of almost automorphic mild solutions to some semilinear abstract differential equations, *Semigroup Forum,* **69**(2004), No. 1, 80-86.

In: Research on Evolution Equation Compendium. Volume 2 ISBN: 978-1-60456-752-6
Editor: Gaston M. N'Guerekata

Singular Solutions of a Semi-Linear Elliptic Equation on Nonsmooth Domains

Lotfi Riahi*
Department of Mathematics, National Institute of Applied Sciences and Technology, Charguia 1, 1080, Tunis, Tunisia

Abstract

Under general conditions on the signed Radon measure μ we prove the existence of positive singular solutions for the problem $\Delta u+\mu u^p=0$, $p>1$ on bounded NTA-domains in $\mathbb{R}^n$, $n\geq 2$. These results extend the recent ones proved by some authors to more general classes of potentials and domains. A new proof based on a simple fixed point argument is also given.

(2000)AMS Subject Classifications: 35J60, 35K55.

Key words and phrases: semi-linear elliptic equation, singular solution, Kato class, nonsmooth domain.

1. Introduction

In this paper we consider the semi-linear elliptic problem

$$(\mathcal{P})\begin{cases}\Delta u(x)+\mu u^p(x)=0, & x\in D\setminus\{0\}\\ u(x)>0, & x\in D\setminus\{0\}\\ u(x)\sim\lambda G(x,0) & \text{for } x \text{ near } 0 \text{ and } \lambda>0 \text{ small}\\ u(x)=0, & x\in\partial D,\end{cases}$$

where Δ is the Laplacian operator, $p>1$, D is a bounded non-tangentially accessible (NTA)-domain (see definition in [7]) in $\mathbb{R}^n$, $n\geq 2$ containing 0; G denotes the Laplacian Green function with Dirichlet boundary condition on D and μ is a signed Radon measure on D. Recall that, in the special cases $\mu=V(x)dx$ and D a bounded Lipschitz domain, Zhang and Zhao [12] studied this problem in the dimension $n\geq 3$ and they proved the existence of infinitely many positive continuous solutions provided that the function $W:=G^{p-1}(.,0)V$ is in the well known Kato class K_n, i.e. W is a Borel measurable function satisfying

$$\lim_{r\to 0}\sup_{x\in D}\int_{D\cap(|x-y|<r)}\frac{|W(y)|}{|x-y|^{n-2}}dy=0.$$

*E-mail address: Lotfi.Riahi@fst.rnu.tn

The main tool in obtaining the result is the $3G$-Theorem proved in [5]. By the same arguments, as indicated in [12, p.779], the result also holds for the problem $\Delta u + f(.,u) = 0$ provided that $f(x, G(x,0))/G(x,0)$ is in the Kato class K_n. In the two dimensional case the same problem is studied in [8] on Dini-smooth Jordan domains under the condition $f(x, G(x,0))/G(x,0)$ in the new functional class $\mathcal{K}(D)$ defined by

$$\limsup_{r\to 0}\sup_{x\in D}\int_{D\cap(|x-y|<r)} \frac{\rho(y)}{\rho(x)} G(x,y)|W(y)|dy = 0,$$

where $\rho(x) = d(x, \partial D)$ is the distance from x to the boundary of D. Recall that, for $n = 2$, the Kato class K_2 (see [2] and [4]) is defined as the set of Borel measurable functions W satisfying:

$$\limsup_{r\to 0}\sup_{x\in D}\int_{D\cap(|x-y|<r)} \operatorname{Log}\Big(\frac{1}{|x-y|}\Big)|W(y)|dy = 0.$$

From the inequalities in [11] and [8], we know that $K_2 \subsetneq \mathcal{K}(D)$. The key in proving the result in the two dimensional case is the $3G$-Theorem established in [11].

In this paper there are two aims. The first is to prove the existence of continuous solutions for the problem $(\mathcal{P})$ when $G^{p-1}(.,0)\mu$ is in a general class of signed Radon measures denoted by $\mathcal{K}^{\,loc}(D)$ covering the classes K_n and $\mathcal{K}(D)$, where D is a bounded NTA-domain (see definition in [7]) containing 0 in $\mathbb{R}^n$, $n \geq 2$. The second aim is to present a new and short proof based on a more simple fixed point theorem than the Schauder theorem used in the first papers and we show the uniqueness of the solution. The idea also could be applied to solve other nonlinear elliptic problems. We also establish characterizations of the classes $\mathcal{K}^{\,loc}(D)$ and $\mathcal{K}(D)$ which allow the reader to see the kind of measures admitted for the class $\mathcal{K}^{\,loc}(D)$ and not for the class $\mathcal{K}(D)$. At the end of the paper we provide an interesting example of measures in $\mathcal{K}^{\,loc}(D)\setminus\mathcal{K}(D)$.

Throughout the paper the letter C will denote a positive constant which may vary in value from line to line.

To present the result of the paper, we should first introduce the class $\mathcal{K}^{\,loc}(D)$. Let $\varphi(x) = 1 \wedge G(x,0)$, for all $x \in D$. A singed Radon measure μ on D is said to be in the class $\mathcal{K}^{\,loc}(D)$ if it satisfies :

(i) $\displaystyle \|\mu\| := \sup_{x\in D}\int_D \frac{\varphi(y)}{\varphi(x)} G(x,y)|\mu|(dy) < \infty,$

(ii) For any compact subset $E \subset D$, $\displaystyle \limsup_{r\to 0}\sup_{x\in E}\int_{D\cap(|x-y|<r)} G(x,y)|\mu|(dy) = 0.$

The result of the paper is the following.

Theorem 1.1. *Let μ be a signed Radon measure on D such that $G^{p-1}(y,0)\mu(dy)$ is in the class $\mathcal{K}^{\,loc}(D)$. Then there exists a number $\lambda_0 > 0$ such that for all $\lambda \in]0, \lambda_0[$, there is a unique continuous solution of the problem $(\mathcal{P})$ satisfying*

$$\frac{\lambda}{2} G(x,0) \leq u(x) \leq \frac{3\lambda}{2} G(x,0), \quad \text{for all} \quad x \in D\setminus\{0\},$$

and

$$\lim_{x\to 0} \frac{u(x)}{G(x,0)} = \lambda.$$

Remark 1.2. The condition (i) is more weaker than the Green bounded potential condition generated by the Kato class K_n used in [12]. Moreover the local Kato condition (ii) is more weaker than the global Kato conditions used in [12] and [8] and is sufficient to prove the continuity of solutions as for the Schrödinger equation in [2], [3] and [10]. These facts show the advantage of our class $\mathcal{K}^{\text{loc}}(D)$ with regards to the classes K_n and $\mathcal{K}(D)$.

Remark 1.3. Note that the measure μ may be singular with respect to the Lebesgue measure. Moreover, if we denote by Lip, NTA and Jor_{ds}, the sets of bounded Lipschitz domains, NTA-domains, Dini-smooth Jordan domains, respectively; we have $Lip \subsetneq NTA$ for $n \geq 2$ and $Jor_{ds} \subsetneq Lip$ for $n = 2$. In particular our result extends the ones proved in [12] and [8] to a more general class of potentials and a more general class of domains.

Remark 1.4. Let us denote by $Unif$ and Jor, the sets of bounded uniform domains (see definition in [1]) in $\mathbb{R}^n$, $n \geq 2$ and bounded Jordan domains (see definition in [4]) in $\mathbb{R}^2$, respectively. We have $NTA \subsetneq Unif$ for $n \geq 2$, $NTA \nsubseteq Jor$ and $Jor \nsubseteq NTA$ for $n = 2$. In contrast to NTA-domains, the uniform and Jordan domains may be irregular for the Dirichlet problem and their Green functions vanish outside a polar set of ∂D. Replacing the condition "$u(x) = 0, x \in \partial D$" by "$u(x) = 0$ *outside a polar set on* ∂D" and using the same arguments, based on the $3G$-Theorems established in [6] and [9], we may prove the existence of positive continuous solutions for the problem $(\mathcal{P})$ on uniform domains for $n \geq 3$ and Jordan domains for $n = 2$.

2. Preliminaries and Known Results

In this section we need to recall and prove some results which are the basic tools to prove Theorem 1.1. In particular the following $3G$-Theorem is recently established for Lipschitz domains [10], for NTA-domains [6] in $\mathbb{R}^n$, $n \geq 3$ and for Jordan domains in $\mathbb{R}^2$ [9]. It is also true for NTA-domains in $\mathbb{R}^2$ by similar arguments as in [6].

Theorem 2.1. (3G-Theorem) *Let D be a bounded NTA-domain in $\mathbb{R}^n$, $n \geq 2$. Then, there exists a constant $C_0 = C_0(n,D) > 0$ such that for all $x, y, z \in D$,*

$$\frac{G(x,z)G(z,y)}{G(x,y)} \leq C_0 \left[\frac{\varphi(z)}{\varphi(x)} G(x,z) + \frac{\varphi(z)}{\varphi(y)} G(z,y) \right].$$

Corollary 2.2. *Let μ be in the class $\mathcal{K}^{\text{loc}}(D)$. Then, for any nonnegative superharmonic function s on D, we have*

$$\sup_{x \in D} \int_D \frac{s(y)}{s(x)} G(x,y) |\mu|(dy) \leq 2C_0 \|\mu\|. \tag{2.1}$$

In particular

$$\sup_{x \in D} \int_D G(x,y) |\mu|(dy) \leq 2C_0 \|\mu\|. \tag{2.2}$$

Proof. (2.1) holds easily by writing $s(x) = \sup_n \int_D G(x,y) f_n(y) dy$, for an increasing sequence $(f_n)_n$ of nonnegative Borel measurable functions on D, using the $3G$-Theorem and the Fubini's theorem. (2.2) holds by taking $s = 1$ in (2.1).

Lemma 2.3. *Let μ be in the class $\mathcal{K}^{loc}(D)$. Then, the function*

$$H(x) = \frac{1}{G(x,0)} \int_D G(x,y) G(y,0) |\mu|(dy)$$

is continuous on D and tends to 0 as $x \to 0$.

Proof. Note first that by taking $s = G(.,0)$ in (2.1), we have $H(x) \leq 2C_0 \|\mu\|$, and so H is a real finite valued function on D. Let $x_0 \in D$ and $\delta > 0$ so that $E = \overline{B}(x_0, \delta) \subset D$. We will prove the continuity at x_0. Since $\mu \in \mathcal{K}^{loc}(D)$, by the condition (ii), for $\varepsilon > 0$ there is $r \in]0, \delta[$ small such that

$$\sup_{x \in E} \int_{D \cap (|x-y| < r)} G(x,y) |\mu|(dy) < \varepsilon. \tag{2.3}$$

$$\begin{aligned} H(x) &= \frac{1}{G(x,0)} \int_D G(x,y) G(y,0) |\mu|(dy) \\ &= \frac{1}{G(x,0)} \Big[\int_{B(x_0, \frac{r}{2})} G(x,y) G(y,0) |\mu|(dy) + \int_{D \cap B^c(x_0, \frac{r}{2})} G(x,y) G(y,0) |\mu|(dy) \Big] \\ &:= H_1(x) + H_2(x). \end{aligned}$$

We will estimate $H_1(x)$, for $x \in B(x_0, \frac{r}{2})$. Let $M = \sup_{x \in E} \frac{1}{\varphi(x)}$. By the $3G$-Theorem, we have

$$\begin{aligned} H_1(x) &\leq C_0 \Big[\int_{B(x_0, \frac{r}{2})} \frac{\varphi(y)}{\varphi(x)} G(x,y) |\mu|(dy) + \int_{B(x_0, \frac{r}{2})} \frac{\varphi(y)}{\varphi(0)} G(y,0) |\mu|(dy) \Big] \\ &\leq MC_0 \Big[\int_{B(x_0, \frac{r}{2})} G(x,y) |\mu|(dy) + \int_{B(x_0, \frac{r}{2})} G(y,0) |\mu|(dy) \Big] \\ &\leq MC_0 \Big[\int_{D \cap B(x,r)} G(x,y) |\mu|(dy) + \int_{B(x_0, \frac{r}{2})} G(y,0) |\mu|(dy) \Big]. \end{aligned} \tag{2.4}$$

Clearly the local Kato condition (ii) yields $\lim_{r \to 0} \int_{B(x',r)} G(x,y) |\mu|(dy) = 0$ for all $x, x' \in D$ (see also Proposition 7.1 in [3]). In particular, for r small enough,

$$\int_{B(x_0, \frac{r}{2})} G(y,0) |\mu|(dy) < \varepsilon. \tag{2.5}$$

Combining (2.3)-(2.5), we obtain $H_1(x) \leq 2MC_0 \varepsilon$.

We will estimate $H_2(x)$. We have

$$G(x,y) \leq \begin{cases} \dfrac{\Gamma(n/2 - 1)}{4\pi^{\frac{n}{2}} |x-y|^{n-2}} & \text{if} \quad n \geq 3. \\ \dfrac{1}{2\pi} (1 \vee \mathrm{Log} \dfrac{1}{|x-y|}) & \text{if} \quad n = 2. \end{cases}$$

It follows that for $x \in B(x_0, \frac{r}{4})$ and $y \in D \cap B^c(x_0, \frac{r}{2})$, $G(x,y) \leq \frac{c_n}{r^{n-1}}$, where c_n is a constant depending only on n. Using this inequality, $\int_D G(y,0)|\mu|(dy) < \infty$ and the dominated convergence theorem we easily obtain the continuity of the function

$$x \to \int_{D \cap B^c(x_0, \frac{r}{2})} G(x,y) G(y,0) |\mu|(dy)$$

at x_0. Since $\frac{1}{G(x,0)}$ is continuous on D, $H_2(x)$ is continuous at x_0. Hence $H(x)$ is continuous at x_0. Moreover when $x_0 = 0$, we remark that for $x \in B(0, \frac{r}{4})$, $H_2(x) \leq \frac{2c_n C_0 \|\mu\|}{G(x,0)}$ and $H_1(x) \leq 2MC_0\varepsilon$. Thus $H(x) \to 0$ as $x \to 0$.

3. Proof of Theorem 1.1.

We will show that there exists a number $\lambda_0 > 0$ such that for all $\lambda \in]0, \lambda_0[$, there is a unique function $u \in C(D \setminus \{0\})$ satisfying, for all $x \in D \setminus \{0\}$,

$$u(x) = \lambda G(x,0) + \int_D G(x,y) u^p(y) \mu(dy)$$

and

$$\frac{\lambda}{2} G(x,0) \leq u(x) \leq \frac{3\lambda}{2} G(x,0).$$

Clearly if u satisfies the above conditions, then u is a distributional solution of the problem ($\mathcal{P}$). By writing $w(x) = \frac{u(x)}{G(x,0)}$ and $\sigma = G^{p-1}(.,0)\mu$ it then suffices to show that there exists a unique function $w \in C(D)$ satisfying

$$w(x) = \lambda + \frac{1}{G(x,0)} \int_D G(x,y) G(y,0) w^p(y) \sigma(dy)$$

and

$$\frac{\lambda}{2} \leq w(x) \leq \frac{3\lambda}{2}.$$

For $\lambda > 0$, let then consider the set

$$S = \left\{ w \in C(D) : \frac{\lambda}{2} \leq w \leq \frac{3\lambda}{2} \right\}.$$

S is a closed subset in the Banach space $(C(D), \|.\|_\infty)$, where $\|w\|_\infty = \sup_{x \in D} |w(x)|$. Let us define the operator T on S by

$$Tw(x) = \lambda + \frac{1}{G(x,0)} \int_D G(x,y) G(y,0) w^p(y) \sigma(dy).$$

By Lemma 2.3, for $w \in S$, $Tw \in C(D)$. Moreover, by the $3G$-Theorem, we have

$$\left| \frac{Tw(x)}{\lambda} - 1 \right| \leq \lambda^{p-1} 2^{p+1} C_0 \|\sigma\|.$$

Then by choosing $0<\lambda<(2^{p+2}C_0\|\sigma\|)^{-\frac{1}{p-1}}$, we get $\frac{\lambda}{2}\le Tw\le\frac{3\lambda}{2}$ and so $Tw\in S$. On the other hand, by the $3G$-Theorem, for $w, w'\in S$, we have,

$$\begin{aligned}\|Tw-Tw'\|_\infty &\le \sup_{x\in D}\left(\frac{1}{G(x,0)}\int_D G(x,y)G(y,0)|\sigma|(dy)\right)\|w^p-w'^p\|_\infty\\ &\le 2C_0\|\sigma\|\,\|w^p-w'^p\|_\infty\\ &\le 2^p p\lambda^{p-1}C_0\|\sigma\|\,\|w-w'\|_\infty\end{aligned}$$

For $0<\lambda<(2^p pC_0\|\sigma\|)^{-\frac{1}{p-1}}$ we have $2^p p\lambda^{p-1}C_0\|\sigma\|<1$. Hence for $0<\lambda<\lambda_0=(2^{p+2}pC_0\|\sigma\|)^{-\frac{1}{p-1}}$, the operator T is a $\frac{1}{4}$-Lipschitz mapping from S into itself; and so by the well known fixed point theorem there exists a unique $w\in S$ such that $Tw=w$.

4. The Classes $\mathcal{K}^{\mathrm{loc}}(D)$ and $\mathcal{K}(D)$.

In this section, we prove characterizations of the classes $\mathcal{K}^{\mathrm{loc}}(D)$ and $\mathcal{K}(D)$ which allow us to show that $\mathcal{K}^{\mathrm{loc}}(D)$ contains more signed Radon measures than the class $\mathcal{K}(D)$.

Proposition 4.1. *Let μ be a signed Radon measure on D. Then $\mu\in\mathcal{K}^{\mathrm{loc}}(D)$ if and only if the function $P_\mu(x)=\int_D\frac{\varphi(y)}{\varphi(x)}G(x,y)|\mu|(dy)$ is in $C_b(D)$. Here $C_b(D)$ is the set of bounded continuous functions on D.*

Proof. Let $\mu\in\mathcal{K}^{\mathrm{loc}}(D)$. Then $\sup_{x\in D}P_\mu(x)=\|\mu\|<\infty$ and so P_μ is bounded. We will show the continuity of P_μ on D. Since φ is continuous on D, it then suffices to show the continuity of $J(x)=\int_D\varphi(y)G(x,y)|\mu|(dy)$. For any $r>0$ small, we have

$$\begin{aligned}J(x) &= \int_{D\cap(|x-y|<r)}\varphi(y)G(x,y)|\mu|(dy)+\int_{D\cap(|x-y|\ge r)}\varphi(y)G(x,y)|\mu|(dy)\\ &:= p_r(x)+q_r(x).\end{aligned}$$

By using the fact that $1_{\{|x-y|\ge r\}}G(x,y)\le\frac{c_n}{r^{n-1}}$; $\int_D\varphi(y)|\mu|(dy)<2C_0\|\mu\|$ and the dominated convergence theorem, it follows that q_r is continuous on D.
Now let $x_0\in D$ and $\delta>0$ such that $E=\overline{B}(x_0,\delta)\subset D$. By the assumption (ii), for $\varepsilon>0$, there is $r\in]0,\delta[$ small such that

$$\sup_{x\in E}\int_{D\cap(|x-y|<r)}G(x,y)|\mu|(dy)<\varepsilon.$$

Then for $x\in B(x_0,\delta)$, we have

$$\begin{aligned}|J(x)-J(x_0)| &\le |p_r(x)-p_r(x_0)|+|q_r(x)-q_r(x_0)|\\ &\le 2\varepsilon+|q_r(x)-q_r(x_0)|,\end{aligned}$$

and so $J(x)\longrightarrow J(x_0)$ as $x\longrightarrow x_0$. Thus J is continuous on D.

Conversely assume that $P_\mu \in C_b(D)$. Then $\|\mu\| = \sup_{x\in D} P_\mu(x) < \infty$. As above for $r > 0$ small $\frac{p_r(x)}{\varphi(x)} = P_\mu(x) - \frac{q_r(x)}{\varphi(x)}$ with $\frac{q_r}{\varphi}$ is continuous on D, and so $\frac{p_r}{\varphi}$ is continuous on D. Since $\lim_{r\to 0}\frac{p_r(x)}{\varphi(x)} = 0$ for all $x \in D$, by the Dini-theorem, it follows that, for any compact subset $E \subset D$,

$$\limsup_{r\to 0}\sup_{x\in E}\frac{p_r(x)}{\varphi(x)} = 0$$

which means that μ satisfies (ii).

Now, for $n \geq 2$, let us define the class $\mathcal{K}(D)$ as the set of signed Radon measures μ on D satisfying :

$$\lim_{r\to 0}\sup_{x\in D}\int_{D\cap(|x-y|<r)}\frac{\varphi(y)}{\varphi(x)}G(x,y)|\mu|(dy) = 0.$$

This class is used in [8] in the special cases of Dini-smooth Jordan domains in $\mathbb{R}^2$ and $\mu = V(x)dx$. Clearly by the compactness of $\overline{D}$, if $\mu \in \mathcal{K}(D)$ then $\|\mu\| < \infty$. Hence μ satisfies (i) and (ii) and so $\mu \in \mathcal{K}^{loc}(D)$. Combining with Lemma 3.2 in [10] and Lemmas 3.5 and 3.6 in [9], we then have the inclusions: $K_n \subset \mathcal{K}(D) \subset \mathcal{K}^{loc}(D)$. For interesting examples of functions μ in $\mathcal{K}^{loc}(D)\setminus K_n$, we refer the reader to Corollary 4.8 in [10]. In what follows we will prove a characterization of the class $\mathcal{K}(D)$.

Proposition 4.2. *Let μ be a signed Radon measure on D. Then $\mu \in \mathcal{K}(D)$ if and only if the function $P_\mu(x) = \int_D \frac{\varphi(y)}{\varphi(x)}G(x,y)|\mu|(dy)$ is in $C(\overline{D})$. Here $C(\overline{D})$ is the set of functions continuous up to the boundary on D.*

Proof. Let $\mu \in \mathcal{K}(D)$. Then $\mu \in \mathcal{K}^{loc}(D)$ and so by Proposition 4.1, $P_\mu \in C_b(D)$. It remains to prove that P_μ is continuous up to the boundary ∂D. Let $x_0 \in \partial D$. For $\varepsilon > 0$, there is $r > 0$ small so that

$$\sup_{x\in D}\int_{D\cap(|x-y|<r)}\frac{\varphi(y)}{\varphi(x)}G(x,y)|\mu|(dy) < \varepsilon.$$

For $x \in D\cap B(x_0,\frac{r}{2})$, we then have

$$Q_1(x) := \int_{D\cap(|x_0-y|<\frac{r}{2})}\frac{\varphi(y)}{\varphi(x)}G(x,y)|\mu|(dy) < \varepsilon.$$

Moreover the function

$$Q_2(x) := \int_{D\cap(|x_0-y|\geq\frac{r}{2})}\frac{\varphi(y)}{\varphi(x)}G(x,y)|\mu|(dy)$$

is the quotient of $\int_{D\cap(|x_0-y|\geq\frac{r}{2})}\varphi(y)G(x,y)|\mu|(dy)$ and $\varphi(x)$ which are two nonnegative solutions on $D\cap B(x_0,\frac{r}{4})$ vanishing on $\partial D\cap B(x_0,\frac{r}{4})$ and so by Theorem 7.9 in [7], $Q_2(x)$ has a limit as $x \to x_0$. Thus $P_\mu(x) = Q_1(x) + Q_2(x)$ has a limit as $x \to x_0$. Thus $P_\mu \in C(\overline{D})$.

Conversely, let μ be a Borel measurable function such that $P_\mu \in C(\overline{D})$. Then $\|\mu\| = \sup_{x\in D} P_\mu(x) < \infty$. For $r > 0$ small, put

$$p_r(x) = \int_{D\cap(|x-y|<r)} \frac{\varphi(y)}{\varphi(x)} G(x,y)|\mu|(dy)$$

and

$$q_r(x) = \int_{D\cap(|x-y|\geq r)} \frac{\varphi(y)}{\varphi(x)} G(x,y)|\mu|(dy).$$

Let $x_0 \in \partial D$. By the boundary Harnack principle (Theorem 5.1 in [7]), there is $C > 0$ such that for all $x \in D\cap B(x_0, \frac{r}{2})$ and all $y \in D \cap B^c(x,r)$, $\dfrac{G(x,y)}{\varphi(x)} = \dfrac{G(x,y)}{G(x,0)} \leq C$. Moreover $\lim_{x\to x_0} \dfrac{G(x,y)}{G(x,0)} = M(y,x_0)$, where $M(.,x_0)$ is the Martin-Poisson kernel with pôle x_0 normalized at 0. Since $\int_D \varphi(y)|\mu|(dy) \leq 2C_0\|\mu\|$, by the dominated convergence theorem $\lim_{x\to x_0} q_r(x) = \int_{D\cap(|x_0-y|\geq r)} \varphi(y)M(y,x_0)|\mu|(dy)$. Hence q_r has a limit at any point $x_0 \in \partial D$ and so $q_r \in C(\overline{D})$. Thus $p_r = P_\mu - q_r \in C(\overline{D})$. Since $\lim_{r\to 0} p_r(x) = 0$, by the Dini-theorem, we have $\lim_{r\to 0}\sup_{x\in\overline{D}} p_r(x) = 0$ and so $\mu \in \mathcal{K}(D)$.

Example 4.3. We will present an interesting example of signed Radon measures in $\mathcal{K}^{\text{loc}}(D) \setminus \mathcal{K}(D)$ even in simple smooth domains. Let us consider $D = B(0,1)$ the unit ball in $\mathbb{R}^n$, $n \geq 2$. From the formula of the Green function or the estimates given in [4] and [8], we know that $\frac{G(x,y)}{\varphi(x)} \leq \frac{C}{|x-y|^{n-1}}$, for all $x,y \in D$. Fix a point $z_0 \in \partial B(0,1)$. For all integer $m \geq 1$, let us define the sequence of disjoint balls B_m in D by $B_m = B(y_m, r_m)$, where $y_m = (1-\frac{1}{4m})z_0$ and $r_m = \frac{1}{16m(m+1)}$. Choose $\alpha_m > 0$ such that $\frac{\alpha_m G(.,y_m)}{\varphi} \leq 2^{-m}$ on $D\setminus B_m$ (it suffices to take $\alpha_m \leq 2^{-m} r_m^{n-1} C^{-1}$), and define

$$p_m = \frac{\alpha_m G(.,y_m)}{\varphi} \wedge 2 \quad \text{and} \quad p = \sum_{m=1}^{\infty} p_m.$$

Clearly $p_m \in C(D)$ and for all integer $k_0 \geq 1$, the series $\sum_{m=1}^{\infty} p_m$ converges uniformly on $D \setminus \bigcup_{k=k_0}^{\infty} B_k$ and so $p \in C(D)$. Moreover $p \leq 3$. Then $p \in C_b(D)$. We will prove that $p \notin C(\overline{D})$. We have

$$\limsup_{y\to z_0} p(y) \geq \limsup_{m\to\infty} p_m(y_m) = 2.$$

On the other hand by choosing $z_m = y_m + 2r_m z_0$, we clearly have $z_m \notin \bigcup_{k=1}^{\infty} B_k$, $z_m \to z_0$ as $m \to \infty$ and $p(z_m) \leq 1$ which imply

$$\liminf_{y\to z_0} p(y) \leq 1.$$

Thus $p \in C_b(D) \setminus C(\overline{D})$. Note that $\varphi p = \sum_{m=1}^{\infty} \varphi p_m = \sum_{m=1}^{\infty} \alpha_m G(.,y) \wedge 2\varphi$ is a Δ-potential on D, and so there is a positive Radon measure σ on D such that

$$\varphi(x)p(x) = \int_D G(x,y)\sigma(dy), \forall x \in D.$$

We then have

$$p(x) = \int_D \frac{\varphi(y)}{\varphi(x)} G(x,y)\mu(dy), \forall x \in D,$$

where $\mu = \frac{1}{\varphi}\sigma$. From Propositions 4.1 and 4.2, it follows that $\mu \in \mathcal{K}^{\,loc}(D) \setminus \mathcal{K}\,(D)$.

Acknowledgment

I want to sincerely thank Professor Wolfhard Hansen for his help in the example 4.3 and Professor Kentaro Hirata for his interesting remarks and comments about the paper.

References

[1] H. Aikawa : Boundary Harnack principle and Martin boundary for a uniform domain, *J. Math. Soc. Japan*, **53**, 1 (2001), 119-145.

[2] M. Aizenman and B. Simon : Brownian motion and Harnack's inequality for Schrödinger operators, *Comm. Pure Appl. Math.*, **35** (1982), 209-271.

[3] A. Boukricha, W. Hansen and H. Hueber : Continuous solutions of the generalized Schrödinger equation and perturbation of harmonic spaces, *Exposition Math.*, **5** (1987), 97-135.

[4] K.L. Chung and Z. Zhao: *From Brownian motion to Schrödinger's equation*, Springer-Verlag, New York 1995.

[5] M. Cranston, E.B. Fabes and Z. Zhao: Conditional gauge and potential theory for the Schrödinger operator, *Trans. Amer. Math. Soc.*, **37**, 1 (1988), 171-194.

[6] W. Hansen : Uniform boundary Harnack principle and generalized triangle property, *J. Funct. Anal.*, **226** (2005), 452-484.

[7] D. Jerison and C. Kenig : Boundary behavior of harmonic functions in nontangentially accessible domains, *Advances in Math.*, **46** (1982), 80-147.

[8] H. Mâagli and L. Mâatoug : Singular solutions of a nonlinear equation in bounded domains of $\mathbb{R}^2$, *Jour. Math. Anal. Appl.*, **270** (2002), 230-246.

[9] L. Riahi: A $3G$-Theorem for Jordan domains in $\mathbb{R}^2$, *Colloquium Math.*, **101** (2004), 1-7.

[10] L. Riahi: The $3G$-inequality for general Schrödinger operators on Lipschitz-domains, *Manuscripta Math.*, **116** (2005), 211-227.

[11] M. Selmi: Inequalities for Green function in a Dini-Jordan domain in $\mathbb{R}^2$, *Potential Anal.*, **13** (2000), 81-102.

[12] Q.S. Zhang and Z. Zhao: Singular solutions of semilinear elliptic and parabolic equations, *Mathematische Annalen*, **310** (1998), 777-794.

In: Research on Evolution Equation Compendium. Volume 2 ISBN: 978-1-60456-752-6
Editor: Gaston M. N'Guerekata

EXISTENCE OF WEIGHTED PSEUDO ALMOST PERIODIC SOLUTIONS TO SOME NON-AUTONOMOUS DIFFERENTIAL EQUATIONS

*Toka Diagana**
Department of Mathematics, Howard University,
2441 6th Street N.W., Washington D.C. 20059, USA

Abstract

The paper studies the so-called weighted pseudo almost periodic functions recently introduced by the author. Properties of those weighted pseudo almost periodic functions are discussed including a composition result of weighted pseudo almost periodic functions. The obtained results, subsequently, are utilized to study the existence and uniqueness of a weighted pseudo almost periodic solutions to some non-autonomous abstract differential equations.

1991 Mathematics Subject Classification: 44A35; 42A85; 42A75

Key words and phrases: weight, pseudo almost periodic, weighted pseudo almost periodic, almost periodic, heat equation.

1. Introduction

In Diagana [4], a new generalization of Bohr almost periodic functions was introduced. Such a new concept is called *weighted* pseudo almost periodicity and implements in a natural fashion the notion of *pseudo* almost periodicity due Zhang [13, 14, 15]. To construct those new spaces, the main idea consists of enlarging the so-called *ergodic* component, utilized in the Zhang's definition of pseudo almost periodicity, with the help of a *weighted measure* $d\mu(x) = \rho(x)dx$, where $\rho : \mathbb{R} \mapsto (0,\infty)$ is a *locally integrable* function over $\mathbb{R}$, which is commonly called *weight*. Unlike [4], here we provide with all complete proofs. We basically take a closer look into properties of weighted pseudo almost periodic functions and study their relationship with the Zhang's pseudo almost periodicity. For that, we consider a *binary* equivalence relation, $\prec$, on $\mathbb{U}_\infty$, the set of weights, which enables us to reorganize those weights into different equivalence classes. Among other things, if two weights ρ_1 and ρ_2 are equivalent, that is, $\rho_1 \prec \rho_2$, then their corresponding weighted pseudo almost periodic spaces coincide. In particular, when a weight ρ is bounded with $\inf_{x\in\mathbb{R}} \rho(x) > 0$, it is then equivalent to the constant function 1, and hence the weighted pseudo almost periodic space with weight ρ coincides with the Zhang's spaces (Corollary 3.4). In addition to

*E-mail address: tdiagana@howard.edu

the above, a composition result of weighted pseudo almost periodic functions is obtained (Theorem 3.6).

Applications include the search of weighted pseudo almost periodic solutions to the abstract non-autonomous differential equation

$$u'(t) = A(t)u(t) + f(t,u(t)), \quad t \in \mathbb{R}, \tag{1.1}$$

where $A(t) : D(A(t)) \subset \mathbb{X} \mapsto \mathbb{X}$ is a family of densely defined closed linear operators on a common domain $\mathbb{D} = D(A(t))$, which is independent of t, and $f : \mathbb{R} \times \mathbb{X} \mapsto \mathbb{X}$ is a weighted pseudo almost periodic function in the first variable uniformly in the second one. To illustrate Theorem 4.3 we consider the existence and uniqueness of a weighted pseudo almost periodic solution to the one dimensional heat equation.

2. Preliminaries

Let $(\mathbb{X}, \|\cdot\|)$, $(\mathbb{Y}, \|\cdot\|_{\mathbb{Y}})$ be two Banach spaces. The collection of all bounded linear operators from $\mathbb{X}$ into $\mathbb{Y}$ with be denoted $B(\mathbb{X},\mathbb{Y})$. This is simply denoted $B(\mathbb{X})$ when $\mathbb{X} = \mathbb{Y}$.

Let $\mathbb{U}$ denote the collection of all functions (weights) $\rho : \mathbb{R} \mapsto (0,\infty)$, which are locally integrable over $\mathbb{R}$ such that $\rho(x) > 0$ for almost each $x \in \mathbb{R}$. From now on, if $\rho \in \mathbb{U}$ and for $T > 0$, we then set

$$\mathbf{m}(T,\rho) := \int_{-T}^{T} \rho(x)dx.$$

As in the particular case when $\rho(x) = 1$ for each $x \in \mathbb{R}$, we are exclusively interested in those weights, ρ, for which, $\lim_{T\to\infty} \mathbf{m}(T,\rho) = \infty$. Thus throughout the rest of the paper we consider weights defined by

$$\mathbb{U}_\infty := \left\{ \rho \in \mathbb{U} : \lim_{T\to\infty} \mathbf{m}(T,\rho) = \infty \right\}.$$

Of course, one can always consider those weights $\rho \in \mathbb{U}$ for which $\lim_{T\to\infty} \mathbf{m}(T,\rho) \neq \infty$. However, those cases do not correspond to the situations that we want to consider here, as we examine some generalizations of the Zhang's pseudo almost periodicity. In addition to the above, the current setting requires the space of weights $\mathbb{U}_B$ defined by

$$\mathbb{U}_B := \left\{ \rho \in \mathbb{U}_\infty : \ \rho \ \text{ is bounded with } \ \inf_{x\in\mathbb{R}} \rho(x) > 0 \right\}.$$

Obviously, $\mathbb{U}_B \subset \mathbb{U}_\infty \subset \mathbb{U}$, with strict inclusions.

Let $BC(\mathbb{R},\mathbb{X})$ (respectively, $BC(\mathbb{R}\times\mathbb{Y},\mathbb{X})$) denote the collection of all $\mathbb{X}$-valued bounded continuous functions (respectively, the class of jointly bounded continuous functions $F : \mathbb{R}\times\mathbb{Y} \mapsto \mathbb{X}$). The space $BC(\mathbb{R},\mathbb{X})$ equipped with its natural norm, that is, the sup norm defined by

$$\|u\|_\infty = \sup_{t\in\mathbb{R}} \|u(t)\|$$

is a Banach space. Furthermore, $C(\mathbb{R},\mathbb{Y})$ (respectively, $C(\mathbb{R}\times\mathbb{Y},\mathbb{X})$) denotes the class of continuous functions from $\mathbb{R}$ into $\mathbb{Y}$ (respectively, the class of jointly continuous functions $F : \mathbb{R}\times\mathbb{Y} \mapsto \mathbb{X}$).

Definition 2.1. A function $f \in C(\mathbb{R},\mathbb{X})$ is called (Bohr) almost periodic if for each $\varepsilon > 0$ there exists $l(\varepsilon) > 0$ such that every interval of length $l(\varepsilon)$ contains a number τ with the property that

$$\|f(t+\tau) - f(t)\| < \varepsilon \text{ for each } t \in \mathbb{R}.$$

The number τ above is called an ε-*translation* number of f, and the collection of all such functions will be denoted $AP(\mathbb{X})$.

Definition 2.2. A function $F \in C(\mathbb{R}\times\mathbb{Y},\mathbb{X})$ is called (Bohr) almost periodic in $t \in \mathbb{R}$ uniformly in $y \in \mathbb{Y}$ if for each $\varepsilon > 0$ and any compact $K \subset \mathbb{Y}$ there exists $l(\varepsilon)$ such that every interval of length $l(\varepsilon)$ contains a number τ with the property that

$$\|F(t+\tau,y) - F(t,y)\| < \varepsilon \text{ for each } t \in \mathbb{R},\ y \in K.$$

The collection of those functions is denoted by $AP(\mathbb{Y},\mathbb{X})$.

To introduce those weighted pseudo almost periodic functions, we need to define the "weighted ergodic" space $PAP_0(\mathbb{X},\rho)$. Weighted pseudo almost periodic functions will then appear as perturbations of almost periodic functions by elements of $PAP_0(\mathbb{X},\rho)$.

Let $\rho \in \mathbb{U}_\infty$. Define

$$PAP_0(\mathbb{X},\rho) := \left\{ f \in BC(\mathbb{R},\mathbb{X}) : \lim_{T\to\infty} \frac{1}{\mathbf{m}(T,\rho)} \int_{-T}^{T} \|f(\sigma)\|\,\rho(\sigma)\,d\sigma = 0 \right\}.$$

Obviously, when $\rho(x) = 1$ for each $x \in \mathbb{R}$, one retrieves the so-called ergodic space of Zhang, that is, $PAP_0(\mathbb{X})$, defined by

$$PAP_0(\mathbb{X}) := \left\{ f \in BC(\mathbb{R},\mathbb{X}) : \lim_{T\to\infty} \frac{1}{2T} \int_{-T}^{T} \|f(\sigma)\|\,d\sigma = 0 \right\}.$$

Clearly, the spaces $PAP_0(\mathbb{X},\rho)$ are richer than $PAP_0(\mathbb{X})$ and give rise to an enlarged space of pseudo almost periodic functions. In Corollary 3.4, some sufficient condition on the weight $\rho \in \mathbb{U}_\infty$ are given so that $PAP_0(\mathbb{X},\rho) = PAP_0(\mathbb{X})$.

In the same way, we define $PAP_0(\mathbb{Y},\mathbb{X},\rho)$ as the collection of jointly continuous functions $F : \mathbb{R}\times\mathbb{Y} \mapsto \mathbb{X}$ such that $F(\cdot,y)$ is bounded for each $y \in \mathbb{Y}$ and

$$\lim_{T\to\infty} \frac{1}{\mathbf{m}(T,\rho)} \left\{ \int_{-T}^{T} \|F(s,y)\|\,\rho(s)\,ds \right\} = 0$$

uniformly in $y \in \mathbb{Y}$.

We are now ready to define the notion of weighted pseudo almost periodicity.

Definition 2.3. Let $\rho \in \mathbb{U}_\infty$. A function $f \in BC(\mathbb{R},\mathbb{X})$ is called weighted pseudo almost periodic (or ρ-pseudo almost periodic) if it can be expressed as $f = g + \phi$, where $g \in AP(\mathbb{X})$ and $\phi \in PAP_0(\mathbb{X},\rho)$. The collection of such functions will be denoted by $PAP(\mathbb{X},\rho)$.

Remark 2.4. (i) The functions g and ϕ in Definition 2.3 are respectively called the *almost periodic* and the *weighted ergodic perturbation* components of f.

(ii) The space $PAP(\mathbb{X},\rho)$ is a closed subspace of $(BC(\mathbb{R},\mathbb{X}),\|\cdot\|_\infty)$. This yields $(PAP(\mathbb{X},\rho),\|\cdot\|_\infty)$ is a Banach space.

(iii) $PAP(\mathbb{X}) \subset \bigcup_{\rho\in\mathbb{U}} PAP(\mathbb{X},\rho)$.

(iv) If $\rho \in \mathbb{U}_B$, then the space $PAP(\mathbb{X},\rho)$ is translation invariant.

Proposition 2.5. *Fix $\rho \in \mathbb{U}_B$. Let $f \in PAP_0(\mathbb{R},\rho)$ and let $g \in L^1(\mathbb{R})$. Then $f*g$, the convolution of f and g on $\mathbb{R}$, belongs to $PAP_0(\mathbb{R},\rho)$.*

Proof. From $f \in PAP_0(\mathbb{R},\rho)$ and $g \in L^1(\mathbb{R})$ it is clear that $f*g \in BC(\mathbb{R})$. Now setting

$$J(T,\rho) := \frac{1}{\mathbf{m}(T,\rho)} \int_{-T}^{T} \int_{-\infty}^{+\infty} |f(t-s)|\rho(t)|g(s)|\, ds\, dt$$

it follows that

$$\begin{aligned}
\frac{1}{\mathbf{m}(T,\rho)} \int_{-T}^{T} |(f*g)(t)|\rho(t)dt &\le J(T,\rho)\\
&= \int_{-\infty}^{+\infty} |g(s)| \left(\frac{1}{\mathbf{m}(T,\rho)} \int_{-T}^{T} |f(t-s)|\rho(t)dt \right) ds\\
&= \int_{-\infty}^{+\infty} |g(s)|\phi_T(s)ds,
\end{aligned}$$

where $\phi_T(s) = \dfrac{1}{\mathbf{m}(T,\rho)} \displaystyle\int_{-T}^{T} |f(t-s)|\rho(t)dt$.

Since $PAP_0(\mathbb{R},\rho)$ is translation invariant it follows that $\phi_T(s) \mapsto 0$ as $T \mapsto \infty$. Next, using the boundedness of ϕ_T ($|\phi_T(s)| \le \|f\|_\infty$) and the fact that $g \in L^1(\mathbb{R})$, the Lebesgue dominated convergence theorem yields

$$\lim_{T\to\infty} \left\{ \int_{-\infty}^{+\infty} |g(s)|\phi_T(s)ds \right\} = 0,$$

and hence $f*g \in PAP_0(\mathbb{R},\rho)$. □

Similarly, it is clear that if $h \in AP(\mathbb{R})$ and $\psi \in L^1(\mathbb{R})$, then the convolution $h*\psi \in AP(\mathbb{R})$. Combining those results one obtains:

Proposition 2.6. *Fix $\rho \in \mathbb{U}_\infty$. Let $f \in PAP(\mathbb{R},\rho)$ and let $g \in L^1(\mathbb{R})$. Then $f*g$, the convolution of f and g on $\mathbb{R}$, belongs to $PAP(\mathbb{R},\rho)$.*

Example 2.7. Let $\rho \in \mathbb{U}_\infty$. Then the function T defined by

$$T(x) = \int_{-\infty}^{\infty} K(x-y)f(y)dy,$$

where $K \in L^1(\mathbb{R})$ and $f \in PAP(\mathbb{R},\rho)$. Then $T \in PAP(\mathbb{R},\rho)$, by Proposition 2.6.

Proposition 2.8. *Let* $\rho \in \mathbb{U}_\infty$ *and let* $p > 1$. *If* $|f|^p \in PAP_0(\mathbb{R})$, *then* $f \in PAP_0(\mathbb{R}, \rho)$ *provided*

$$L(p) = \sup_{T>0}\left[(2T)^{\frac{1}{p}} \cdot \frac{\left[\mathbf{m}(T, \rho^{\frac{p}{p-1}})\right]^{1-\frac{1}{p}}}{\mathbf{m}(T, \rho)}\right] < \infty.$$

Proof. It suffices to note that the following inequality holds

$$\begin{aligned}\frac{1}{\mathbf{m}(T,\rho)}\int_{-T}^{T}|f(t)|\rho(t)dt &\leq \frac{1}{\mathbf{m}(T,\rho)}\left(\int_{-T}^{T}|f(t)|^p dt\right)^{1/p}\\ &\times \left(\int_{-T}^{T}\rho^{\frac{p}{p-1}}(t)dt\right)^{1-\frac{1}{p}}\\ &\leq L(p)\cdot\left(\frac{1}{2T}\int_{-T}^{T}|f(t)|^p dt\right)^{1/p}.\end{aligned}$$

Therefore passing to the limit as $T \to \infty$ in the previous inequality yields the desired result. □

Definition 2.9. Fix $\rho \in \mathbb{U}_\infty$. A function $F \in C(\mathbb{R} \times \mathbb{Y}, \mathbb{X})$ is called weighted pseudo almost periodic (or ρ-pseudo almost periodic) in $t \in \mathbb{R}$ uniformly in $y \in \mathbb{Y}$ if it can be expressed as $F = G + \Phi$, where $G \in AP(\mathbb{Y}, \mathbb{X})$ and $\phi \in PAP_0(\mathbb{Y}, \mathbb{X}, \rho)$. The collection of such functions will be denoted by $PAP(\mathbb{Y}, \mathbb{X}, \rho)$.

3. Properties of Weighted Pseudo Almost Periodic Functions

Let $\rho \in \mathbb{U}_\infty$. If $f, g \in PAP(\mathbb{X}, \rho)$ and let $\lambda \in \mathbb{R}$, then $f + \lambda g$ is also in $PAP(\mathbb{X}, \rho)$. Moreover, if $|f(\cdot)|$ is not even and if

$$\sup_{x\in\mathbb{R}}\left\{\frac{\rho(-x)}{\rho(x)}\right\} = M(\rho) < \infty,$$

then the function $\tilde{f}(x) := f(-x)$ for $x \in \mathbb{R}$ is also in $PAP(\mathbb{X}, \rho)$. In particular, if ρ is even, then $\tilde{f}$ belongs to $PAP(\mathbb{X}, \rho)$.

Theorem 3.1. *Fix* $\rho \in \mathbb{U}_\infty$. *The decomposition of a weighted pseudo almost periodic function* $f = g + \phi$, *where* $g \in AP(\mathbb{X})$ *and* $\phi \in PAP_0(\mathbb{X}, \rho)$, *is unique.*

Proof. Let $f = g_1 + \phi_1$ where $g_1 \in AP(\mathbb{X})$ and $\phi_1 \in PAP_0(\mathbb{X}, \rho)$. Following along the same line as in the proof of [13, Lemma 1.3] it easily follows that

$$g_1(\mathbb{R}) \subset \overline{f(\mathbb{R})}.$$

Thus, if $f = g_2 + \phi_2$ where $g_2 \in AP(\mathbb{X})$ and $\phi_2 \in PAP_0(\mathbb{X}, \rho)$, then

$$0 = f - f = (g_1 - g_2) + (\phi_1 - \phi_2) \in PAP(\mathbb{X}, \rho),$$

where $(g_1 - g_2) \in AP(\mathbb{X})$ and $(\phi_1 - \phi_2) \in PAP_0(\mathbb{X}, \rho)$. Hence, using the argument above, it follows that $(g_1 - g_2)(\mathbb{R}) \subset \{0\}$, and therefore, $g_1 = g_2$ and $\phi_1 = \phi_2$. □

Definition 3.2. Let $\rho_1, \rho_2 \in \mathbb{U}_\infty$. One says that ρ_1 is equivalent to ρ_2 and denote it $\rho_1 \prec \rho_2$ if and only if $\rho := \dfrac{\rho_1}{\rho_2} \in \mathbb{U}_B$.

Examples of equivalent weights include ρ and $\dfrac{1}{\rho}$, where ρ is any arbitrary weight in $\mathbb{U}_\infty$.

Let $\rho_1, \rho_2, \rho_3 \in \mathbb{U}_\infty$. It is clear that $\rho_1 \prec \rho_1$ (reflexivity); if $\rho_1 \prec \rho_2$, then $\rho_2 \prec \rho_2$ (symmetry); and if $\rho_1 \prec \rho_2$ and $\rho_2 \prec \rho_3$, then $\rho_1 \prec \rho_3$ (transitivity). So, $\prec$ is a binary equivalence relation on $\mathbb{U}_\infty$. Thus the equivalence class of a given weight $\rho \in \mathbb{U}_\infty$ will then be denoted by

$$cl(\rho) = \{\varpi \in \mathbb{U}_\infty : \rho \prec \varpi\}.$$

It is then clear that $\mathbb{U}_\infty = \bigcup_{\rho \in \mathbb{U}_\infty} cl(\rho)$.

Theorem 3.3. *Let $\rho \in \mathbb{U}_\infty$. If $\rho_1, \rho_2 \in cl(\rho)$, then $PAP(\mathbb{X}, \rho_1) = PAP(\mathbb{X}, \rho_2)$.*

Proof. Since $\rho_1 \prec \rho_2$ there exist some constants $K, K' > 0$ such that $K'\rho_2(x) \leq \rho_1(x) \leq K\rho_2(x)$ for each $x \in \mathbb{R}$. In particular, for each $T > 0$,

$$\frac{1}{K.\mathbf{m}(T,\rho_2)} \leq \frac{1}{\mathbf{m}(T,\rho_1)} \leq \frac{1}{K'.\mathbf{m}(T,\rho_2)}.$$

Let $\phi \in PAP_0(\mathbb{X}, \rho_2)$. In view of the above it easily follows that

$$\begin{aligned}
\frac{1}{\mathbf{m}(T,\rho_1)} \int_{-T}^{T} \|\phi(s)\|\rho_1(s)ds &= \frac{1}{\mathbf{m}(T,\rho_1)} \int_{-T}^{T} \|\phi(s)\| \left(\frac{\rho_1}{\rho_2}\right)(s)\rho_2(s)ds \\
&\leq \frac{K}{\mathbf{m}(T,\rho_1)} \int_{-T}^{T} \|\phi(s)\|\rho_2(s)ds \\
&\leq \frac{K}{K'.\mathbf{m}(T,\rho_2)} \int_{-T}^{T} \|\phi(s)\|\rho_2(s)ds.
\end{aligned}$$

Consequently,

$$\lim_{T\mapsto\infty} \frac{1}{\mathbf{m}(T,\rho_1)} \int_{-T}^{T} \|\phi(s)\|\rho_1(s)ds = 0,$$

and hence $\phi \in PAP_0(\mathbb{X}, \rho_1)$, that is, $PAP_0(\mathbb{X}, \rho_2) \subset PAP_0(\mathbb{X}, \rho_1)$.

To obtain the other inclusion, one simply uses the fact that the equivalence relation $\prec$ is symmetric, that is, $\rho_2 \prec \rho_1$, too. □

An immediate consequence of Theorem 3.3 is that $PAP(\mathbb{X}, \rho) = PAP(\mathbb{X}, cl(\rho))$. This enables us to identify the Zhang's space $PAP(\mathbb{X})$ with a weighted pseudo almost periodic class $PAP(\mathbb{X}, \rho)$.

Corollary 3.4. *If $\rho \in \mathbb{U}_B$, then $PAP(\mathbb{X}, \rho) = PAP(\mathbb{X}, cl(1)) = PAP(\mathbb{X})$.*

Proof. It suffices to see that there exist $K, K' > 0$ with $K' \leq \rho(x) \leq K$ for each $x \in \mathbb{R}$, which yields $\rho \prec 1$. □

In view of the above, the proof of the next corollary is quite immediate.

Corollary 3.5. *If* $\rho_1 \prec \rho_2$*, then (1)* $PAP(\mathbb{X},\rho_1+\rho_2) = PAP(\mathbb{X},\rho_1) = PAP(\mathbb{X},\rho_2)$*, and (2)* $PAP(\mathbb{X},\frac{\rho_1}{\rho_2}) = PAP(\mathbb{X},cl(1)) = PAP(\mathbb{X})$.

The next theorem implements the theorem of composition of pseudo almost periodic functions given in [1].

Theorem 3.6. *Let* $\rho \in \mathbb{U}_\infty$ *and let* $f \in PAP(\mathbb{Y},\mathbb{X},\rho)$ *satisfying the Lipschitz condition*

$$\|f(t,u)-f(t,v)\| \leq L.\|u-v\|_{\mathbb{Y}} \text{ for all } u,v\in\mathbb{Y},\ t\in\mathbb{R}.$$

If $h \in PAP(\mathbb{Y},cl(\rho))$*, then* $f(\cdot,h(\cdot)) \in PAP(\mathbb{X},cl(\rho))$.

Prior to proving Theorem 3.6, we give the following immediate consequences.

Corollary 3.7. *Let* $\rho_1,\rho_2 \in \mathbb{U}_\infty$ *with* $\rho_2 \in cl(\rho_1)$*. Let* $f \in PAP(\mathbb{Y},\mathbb{X},cl(\rho_1))$ *satisfying the Lipschitz's condition*

$$\|f(t,u)-f(t,v)\| \leq L.\|u-v\|_{\mathbb{Y}} \text{ for allu}, v\in\mathbb{Y},\ t\in\mathbb{R}.$$

If $h \in PAP(\mathbb{Y},\rho_2)$*, then* $f(\cdot,h(\cdot)) \in PAP_0(\mathbb{X},cl(\rho_1))$.

Corollary 3.8. *Let* $\rho \in \mathbb{U}_B$*. Let* $f \in PAP(\mathbb{Y},\mathbb{X},cl(\rho))$ *satisfying the Lipschitz's condition*

$$\|f(t,u)-f(t,v)\| \leq L.\|u-v\|_{\mathbb{Y}} \text{ for allu}, v\in\mathbb{Y},\ t\in\mathbb{R}.$$

If $h \in PAP(\mathbb{Y},\rho)$*, then* $f(\cdot,h(\cdot)) \in PAP(\mathbb{X})$.

Proof. (Theorem 3.6) We adapt our proof to the one given in [1]. Write $f = g+\phi$ where $g \in AP(\mathbb{Y},\mathbb{X})$ and $\phi \in PAP_0(\mathbb{Y},\mathbb{X},\rho)$. Similarly, write $h = h_1+h_2$, where $h_1 \in AP(\mathbb{Y})$ and $h_2 \in PAP_0(\mathbb{Y},\rho)$. It is obvious to see that $f(\cdot,h(\cdot)) \in C(\mathbb{R},\mathbb{X})$ as the composition of continuous functions. Now decompose f as follows

$$\begin{aligned} f(\cdot,h(\cdot)) &= g(\cdot,h_1(\cdot))+f(\cdot,h(\cdot))-g(\cdot,h_1(\cdot)) \\ &= g(\cdot,h_1(\cdot))+f(\cdot,h(\cdot))-f(\cdot,h_1(\cdot))+\phi(\cdot,h_1(\cdot)). \end{aligned}$$

Using the theorem of composition of almost periodic functions, it is easy to see that $g(\cdot,h_1(\cdot)) \in AP(\mathbb{X})$. Now, set

$$F(\cdot) = f(\cdot,h(\cdot))-f(\cdot,h_1(\cdot)).$$

Clearly, $F \in PAP_0(\mathbb{X},\rho)$. Indeed, for $T > 0$,

$$\begin{aligned} \frac{1}{\mathbf{m}(T,\rho)}\int_{-T}^{T}\|F(s)\|\rho(s)ds &= \frac{1}{\mathbf{m}(T,\rho)}\int_{-T}^{T}\|f(s,h(s))-f(s,h_1(s))\|\rho(s)ds \\ &\leq \frac{L}{\mathbf{m}(T,\rho)}\int_{-T}^{T}\|h(s)-h_1(s)\|\rho(s)ds \\ &\leq \frac{L}{\mathbf{m}(T,\rho)}\int_{-T}^{T}\|h_2(s)\|\rho(s)ds, \end{aligned}$$

and hence

$$\lim_{T\to\infty}\frac{1}{\mathbf{m}(T,\rho)}\int_{-T}^{T}\|F(s)\|\rho(s)ds=0.$$

To complete the proof we have to show that

$$\lim_{T\to\infty}\frac{1}{\mathbf{m}(T,\rho)}\int_{-T}^{T}\|\phi(s,h_1(s))\|\rho(s)ds=0.$$

As $h_1\in AP(\mathbb{Y})$, $h_1(\mathbb{R})$ is relatively compact. Thus for each $\varepsilon>0$ there exists a finite number of open balls $B_k=B(x_k,\frac{\varepsilon}{3L})$, centered at $x_k\in h_1(\mathbb{R})$ with radius for instance $\frac{\varepsilon}{3L}$ with $h_1(\mathbb{R})\subset\bigcup_{k=1}^{m}B_k$. Thus for $1\le k\le m$, the set $U_k=\{t\in\mathbb{R}: h_1(t)\subset B_k\}$ is open and $\mathbb{R}=\bigcup_{k=1}^{m}U_k$. Now, set $V_k=U_k-\bigcup_{i=1}^{k-1}U_i$ and $V_1=U_1$. clearly, $V_i\bigcap V_j=\emptyset$ for all $i\neq j$.

Since $\phi\in PAP_0(\mathbb{Y},\mathbb{X},\rho)$ there exists $T_0>0$ such

$$\lim_{T\to\infty}\frac{1}{\mathbf{m}(T,\rho)}\int_{-T}^{T}\|\phi(s,x_k)\|\rho(s)ds<\frac{\varepsilon}{3m}\quad\text{for } T\ge T_0 \tag{3.1}$$

and $k\in\{1,2,...,m\}$.

Moreover, since g ($g\in AP(\mathbb{Y},\mathbb{X})$) is uniformly continuous in $\mathbb{R}\times\overline{h_1(\mathbb{R})}$, one has

$$\|g(t,x_k)-g(t,x)\|<\frac{\varepsilon}{3}\ \text{for } x\in B_k,\ k=1,2,..,m. \tag{3.2}$$

Using above and the following the decompositions $\phi(\cdot,h_1(\cdot))=f(\cdot,h_1(\cdot))-g(\cdot,h_1(\cdot))$ and $\phi(t,x_k)=f(t,x_k)-g(t,x_k)$ it follows that

$$\begin{aligned}
&\frac{1}{\mathbf{m}(T,\rho)}\int_{-T}^{T}\|\phi(s,h_1(s))\|\rho(s)ds\\
&= \frac{1}{\mathbf{m}(T,\rho)}\sum_{k=1}^{m}\int_{V_k\bigcap[-T,T]}\|\phi(s,h_1(s))\|\rho(s)ds\\
&\le \frac{1}{\mathbf{m}(T,\rho)}\sum_{k=1}^{m}\int_{V_k\bigcap[-T,T]}\|\phi(s,h_1(s))-\phi(s,x_k)\|\rho(s)ds\\
&+ \frac{1}{\mathbf{m}(T,\rho)}\sum_{k=1}^{m}\int_{V_k\bigcap[-T,T]}\|\phi(s,x_k)\|\rho(s)ds\\
&\le \frac{1}{\mathbf{m}(T,\rho)}\sum_{k=1}^{m}\int_{V_k\bigcap[-T,T]}\|f(s,h_1(s))-f(s,x_k)\|\rho(s)ds\\
&+ \frac{1}{\mathbf{m}(T,\rho)}\sum_{k=1}^{m}\int_{V_k\bigcap[-T,T]}\|\phi(s,x_k)\|\rho(s)ds
\end{aligned}$$

$$+ \quad \frac{1}{\mathbf{m}(T,\rho)} \sum_{k=1}^{m} \int_{V_k \cap [-T,T]} \|g(s,h_1(s)) - g(s,x_k)\| \rho(s) ds$$
$$+ \quad \frac{1}{\mathbf{m}(T,\rho)} \sum_{k=1}^{m} \int_{V_k \cap [-T,T]} L\|h_1(s) - x_k\|_{\mathbb{Y}} \rho(s) ds$$
$$+ \quad \frac{1}{\mathbf{m}(T,\rho)} \sum_{k=1}^{m} \int_{V_k \cap [-T,T]} \|g(s,h_1(s)) - g(s,x_k)\| \rho(s) ds$$
$$+ \quad \sum_{k=1}^{m} \frac{1}{\mathbf{m}(T,\rho)} \int_{V_k \cap [-T,T]} \|\phi(s,x_k)\| \rho(s) ds.$$

For each $s \in V_k \cap [-T,T]$, $h_1(s) \in B_k$ in the sense that $\|h_1(s) - x_k\|_{\mathbb{Y}} < \frac{\varepsilon}{3L}$ for $1 \leq k \leq m$. Clearly, from (3.1)-(3.2) it easily follows that

$$\frac{1}{\mathbf{m}(T,\rho)} \int_{-T}^{T} \|\phi(s,h_1(s))\| \rho(s) ds \leq \varepsilon$$

for $T \geq T_0$, and hence

$$\lim_{T\to\infty} \frac{1}{\mathbf{m}(T,\rho)} \int_{-T}^{T} \|\phi(s,h_1(s))\| \rho(s) ds = 0.$$

□

4. Weighted Pseudo Almost Periodic Solutions

Let $\rho \in \mathbb{U}_\infty$. This section is devoted to the search of a ρ-pseudo almost periodic solution to the abstract non-autonomous differential equation (1.1). For that, we suppose that $\{A(t) : t \in \mathbb{R}\}$ is a family of closed linear defined on a common domain $\mathbb{D}$, which is independent of t, and dense in $\mathbb{X}$.

Throughout the rest of the paper we suppose that the system

$$\begin{cases} u'(t) &= A(t)u(t), \ t \geq s, \\ u(s) &= x \in \mathbb{X}, \end{cases} \tag{4.1}$$

has an associated evolution family of operators $\{U(t,s) : t \geq s \text{ with } t,s \in \mathbb{R}\}$, which is uniformly asymptotically stable.

Definition 4.1. A family of linear operators $\{U(t,s) : t \geq s \text{ with } t,s \in \mathbb{R}\} \subset B(\mathbb{X})$ is called an evolution family of operators for (4.1) whenever the following conditions hold:

(a) $U(t,s)U(s,r) = U(t,r)$ for every $r \leq s \leq t$;

(b) for each $x \in \mathbb{X}$ the function $(t,s) \to U(t,s)x$ is continuous and $U(t,s) \in B(\mathbb{X},\mathbb{D})$ for every $t > s$; and

(c) the function $(s,t] \to B(\mathbb{X})$, $t \to U(t,s)$ is differentiable with

$$\frac{\partial}{\partial t} U(t,s) = A(t)U(t,s).$$

More details on evolution families can be found in the book by Lunardi [12].

To deal with the existence and uniqueness of a ρ-pseudo almost periodic solution to (1.1), we require the following additional assumptions:

(H.1) The function $f : \mathbb{R} \times \mathbb{X} \mapsto \mathbb{X}, (t,u) \mapsto f(t,u)$ is ρ-pseudo almost periodic in $t \in \mathbb{R}$ uniformly in $u \in \mathbb{X}$. Furthermore, f satisfies the Lipschitz's condition: there exists $K > 0$ such that

$$\|f(t,u) - f(t,v)\| \leq K . \|u - v\|_{\mathbb{Y}}, \quad u, v \in \mathbb{X}, \; t \in \mathbb{R}.$$

(H.2) The evolution family $U(t,s)$ is asymptotically stable. Namely, there exist some constants $M, \omega > 0$ such that

$$\|U(t,s)\| \leq Me^{-\omega(t-s)} \text{ for every } t \geq s.$$

(H.3) Let $\omega > 0$. Setting

$$P(\omega) := \sup_{T>0} \left\{ \int_{-T}^{T} e^{-\omega(T+t)} \rho(t) dt \right\},$$

then $P(\omega) < \infty$.

Remark 4.2. Note that there are weights $\rho \in \mathbb{U}_\infty$ for which (H.3) is not satisfied. For instance, consider the weights defined for each $\omega > 0$ by $\rho_\omega(t) = |t|e^{2\omega t}$ for $t \in \mathbb{R}$. It is then clear that $P(\omega) = \infty$.

Theorem 4.3. *Fix $\rho \in \mathbb{U}_\infty$. Suppose that assumptions (H.1)-(H.2)-(H.3) hold. Then (1.1) has a unique weighted pseudo almost periodic solution whenever $K < \dfrac{\omega}{M}$.*

Proof. Let u be a bounded solution of (1.1). It is well-known that it can be expressed as

$$(\mathbb{M}u)(t) := \int_{-\infty}^{t} U(t,s) f(s,u(s)) ds \text{ for each } t \in \mathbb{R}.$$

Let $u \in PAP(\mathbb{X}, \rho)$ and write $f = g + \phi$ where $g \in AP(\mathbb{Y}, \mathbb{X})$ and $\phi \in PAP_0(\mathbb{X}, \mathbb{X}, \rho)$. Using (H.1) and the composition result, Theorem 3.6 it follows that $f(\cdot, u(\cdot)) = g(\cdot, u(\cdot)) + \phi(\cdot, u(\cdot))$ where $g(\cdot, u(\cdot)) \in AP(\mathbb{X})$ and $\phi(\cdot, u(\cdot)) \in PAP_0(\mathbb{X}, \rho)$. Rewrite $(\mathbb{M}u)$ as $(\mathbb{M}u) = (\mathbb{M}_1 u) + (\mathbb{M}_2 u)$, where

$$(\mathbb{M}_1 u)(t) := \int_{-\infty}^{t} U(t,s) g(s,u(s)) ds \text{ for each } t \in \mathbb{R}, \text{ and}$$

$$(\mathbb{M}_2 u)(t) := \int_{-\infty}^{t} U(t,s) \phi(s,u(s)) ds \text{ for each } t \in \mathbb{R}.$$

Since $g(\cdot, u(\cdot)) \in AP(\mathbb{X})$, for each $\varepsilon > 0$, there exists $l(\varepsilon) > 0$ such that each every interval of length $l(\varepsilon)$ contains a τ such that

$$\|g(t+\tau, u(t+\tau)) - g(t, u(t))\| < \frac{\omega . \varepsilon}{M} \text{ for all } t \in \mathbb{R}.$$

Considering $(\mathbb{M}_1 u)(t+\tau) - (\mathbb{M}_1)(t)$ and using assumption (H.2) it easily follows that $\|(\mathbb{M}_1 u)(t+\tau) - (\mathbb{M}_1 u)(t)\| < \varepsilon$ for all $t \in \mathbb{R}$, and hence $(\mathbb{M}_1 u)(\cdot) \in AP(\mathbb{X})$. The next step consists of showing that

$$\lim_{T\to\infty} \frac{1}{\mathbf{m}(T,\rho)} \int_{-T}^{T} \|(\mathbb{M}_2 u)(t)\| \rho(t) dt = 0.$$

Now it follows from (H.2) that

$$\lim_{T\to\infty} \frac{1}{\mathbf{m}(T,\rho)} \int_{-T}^{T} \|(\mathbb{M}_2 u)(t)\| \rho(t) dt \le I(\rho) + J(\rho),$$

where

$$I(\rho) := \lim_{T\to\infty} \frac{M}{\mathbf{m}(T,\rho)} \int_{-T}^{T} \rho(t) dt \left(\int_{-T}^{t} e^{-\omega(t-s)} \|\phi(s,u(s))\| \right), \quad \text{and}$$

$$J(\rho) := \lim_{T\to\infty} \frac{M}{\mathbf{m}(T,\rho)} \int_{-T}^{T} \rho(t) dt \left(\int_{-\infty}^{-T} e^{-\omega(t-s)} \|\phi(s,u(s))\| \right).$$

Now

$$\begin{aligned}
I(\rho) &= \lim_{T\to\infty} \frac{M}{\mathbf{m}(T,\rho)} \int_{-T}^{T} \|\phi(t,u(t))\| \rho(t) dt \left(\int_{-T}^{t} e^{-\omega(t-s)} ds \right) \\
&= \lim_{T\to\infty} \frac{M}{\mathbf{m}(T,\rho)} \int_{-T}^{T} \|\phi(t,u(t))\| \rho(t) dt \left[\frac{1}{\omega} \left(1 - e^{-\omega(t+T)} \right) \right] \\
&\le \frac{M}{\omega} \cdot \lim_{T\to\infty} \frac{M}{\mathbf{m}(T,\rho)} \int_{-T}^{T} \|\phi(t,u(t))\| \rho(t) dt \\
&= 0,
\end{aligned}$$

by $\phi(\cdot,u(\cdot)) \in PAP_0(\mathbb{X},\rho)$.

Similarly,

$$\begin{aligned}
J(\rho) &= \lim_{T\to\infty} \frac{M}{\mathbf{m}(T,\rho)} \int_{-\infty}^{-T} e^{\omega s} \|\phi(s,u(s))\| ds \int_{-T}^{T} e^{-\omega t} \rho(t) dt \\
&\le \lim_{T\to\infty} \left\{ \frac{P(\omega,\rho) . M . \sup_{s\in\mathbb{R}} \|\phi(s,u(s))\|}{\omega . \mathbf{m}(T,\rho)} \right\} \\
&= 0,
\end{aligned}$$

by (H.3).

To complete the proof, we apply the fixed-point principle of Banach to the nonlinear operator $\mathbb{M}$. In view of the above, it is clear that $\mathbb{M}$ maps $PAP(\mathbb{X},\rho)$ into itself. Moreover, for all $u, v \in PAP(\mathbb{X},\rho)$, it easy to see that

$$\|(\mathbb{M}u) - (\mathbb{M}v)\|_\infty \le \frac{MK}{\omega} \|u - v\|_\infty,$$

and hence $\mathbb{M}$ has a unique fixed-point, which obviously is the unique ρ-pseudo almost periodic solution to (1.1).

□

Remark 4.4. In the particular case when $\rho \in \mathbb{U}_B$, that is, $PAP(\mathbb{X}, cl(\rho)) = PAP(\mathbb{X})$ by Corollary 3.4, we retrieve the "non-weighted" situation, since assumption (H. 3) is always achieved in that event. This means that weighted pseudo almost periodic functions are good generalizations of the Zhang's pseudo almost periodic functions.

5. Example

To illustrate Theorem 4.3 we consider the existence and uniqueness of weighted pseudo almost periodic solutions to the heat equation given by the system

$$\begin{aligned} \frac{\partial u}{\partial t}(t,x) &= \frac{\partial^2 u}{\partial x^2}(t,x) + Q(t,x)u(t,x) + \gamma . H(t)\sin(u(t,x)) && (5.1)\\ u(t,0) &= u(t,\pi) = 0, \ \ t \in \mathbb{R} && (5.2) \end{aligned}$$

where $H(t) = \sin t + \sin(\sqrt{2}t) + e^{-|t|}$ for each $t \in \mathbb{R}$, $Q : \mathbb{R} \times [0,\pi] \mapsto \mathbb{R}$ is a jointly measurable function, and $\gamma > 0$ is a constant.

For that, we suppose that $\mathbb{X} = (L^2[0,\pi], \|\cdot\|_2)$ and define

$$\begin{aligned} D(A) &= \{u \in L^2[0,\pi] : u'' \in L^2[0,\pi], \ \ u(0) = u(\pi) = 0\}\\ Au(\cdot) &= u''(\cdot), \ \ \forall u(\cdot) \in D(A). \end{aligned}$$

Note that the linear operator A defined above is the infinitesimal generator of an analytic semigroup $(T(t))_{t\geq 0}$ on $L^2[0,\pi]$ satisfying $\|T(t)\| \leq e^{-t}$ for each $t \geq 0$.

Define the class of operators $A(t)$ by setting

$$\begin{aligned} D(A(t)) &= D(A)\\ A(t)v(x) &= Av(x) + Q(t,x)v \ \text{ for each } \ v \in D(A(t)). \end{aligned}$$

By assuming that $x \mapsto Q(t,x)$ is continuous for each $t \in \mathbb{R}$ with $Q(t,x) \leq -\omega_0$ $(\omega_0 > 0)$ for all $t \in \mathbb{R}, x \in [0,\pi]$, one sees that the system

$$\left\{ \begin{array}{ll} u'(t) & = A(t)u(t), \ \ t \geq s,\\ u(s) & = y \in \mathbb{X}, \end{array} \right. \tag{5.3}$$

has an associated evolution family $(U(t,s))_{t\geq s}$ on $L^2[0,\pi]$, which can be explicitly expressed by

$$U(t,s)v = T(t-s)e^{\int_s^t Q(\tau,y)d\tau} v.$$

Consequently,

$$\|U(t,s)\| \leq e^{-(1+\omega_0)(t-s)} \ \text{ for every } \ t \geq s.$$

Let $N \in \mathbb{N}$. Throughout the rest of this example, we suppose that the weight ρ is given by:

$$\rho(t) = (1 + |t|^2)^N, \ \ t \in \mathbb{R}.$$

Setting $f(t,u(t,\cdot)) = \gamma . H(t)\sin u(t,\cdot)$, one can clearly see that f is 3γ-Lipschitz in the sense that,

$$\|f(t,u(t,\cdot)) - f(t,v(t,\cdot))\|_2 \leq 3\gamma . \|u(t,\cdot) - v(t,\cdot)\|_2$$

for all $u(t,\cdot), v(t,\cdot) \in L^2[0,\pi]$, $t \in \mathbb{R}$.

Furthermore, it can be easily checked that $H \in PAP(\mathbb{R}, (1+|t|^2)^N)$. Consequently, f is pseudo almost periodic in $t \in \mathbb{R}$ uniformly in the second variable. Note that (H.3) is satisfied.

The previous discussion can be formulated as follows.

Theorem 5.1. *The heat equation with Dirichlet condition in (5.1)-(5.2) has a unique weighted pseudo almost periodic solution whenever*

$$\frac{3\gamma}{1+\omega_0} < 1.$$

References

[1] B. Amir and L. Maniar, Composition of pseudo-almost periodic functions and Cauchy problems with operator of nondense domain. *Ann. Math. Blaise Pascal* **6** (1999), no. 1, pp. 1–11.

[2] D. Bugajewski and T. Diagana, C. M. Mahop, Asymptotic and Pseudo Almost Periodicity of the Convolution Operator and Applications to Differential and Integral Equations, *J. Anal. Appl.* **25** (2006), 327-340.

[3] C. Corduneanu, *Almost Periodic Functions*, 2nd Edition, Chelsea-New York, 1989.

[4] T. Diagana, Weighted Pseudo Almost Periodic Functions and Applications, *C. R. Acad. Sci. Paris, Ser I* **343** (2006), no. 10, 643-646.

[5] T. Diagana, Pseudo Almost Periodic Solutions to Some Differential Equations, *Nonlinear Anal* **60** (2005), no. 7, pp. 1277–1286.

[6] T. Diagana and E. Hernàndez M., Existence and Uniqueness of Pseudo Almost Periodic Solutions to Some Abstract Partial Neutral Functional-Differential Equations and Applications, *J. Math. Anal. Appl.* **327**(2007), no. 2, 776-791.

[7] T. Diagana, C. M. Mahop, and G. M. N'Guérékata, Pseudo Almost Periodic Solution to Some Semilinear Differential Equations, *Math. Comp. Modelling* **43** (2006), no. 1-2, pp. 89-96.

[8] T. Diagana, C. M. Mahop, G. M. N'Guérékata, and B. Toni, Existence and Uniqueness of Pseudo Almost Periodic Solutions to Some Classes of Semilinear Differential Equations and Applications. *Nonlinear Anal.* **64** (2006), no. 11, pp. 2442-2453.

[9] T. Diagana, Existence and Uniqueness of Pseudo Almost Periodic Solutions to Some Classes of Partial Evolution Equations. *Nonlinear Anal.* **66** (2007), no. 2, 384-395.

[10] T. Diagana, *An Introduction to Classical and p-adic Theory of Linear Operators and Applications*, Nova Science Publishers, New York, 2006.

[11] H. X. Li, F. L. Huang, and J. Y. Li, Composition of pseudo almost-periodic functions and semilinear differential equations. *J. Math. Anal. Appl* **255** (2001), no. 2, pp. 436–446.

[12] A. LUNARDI, *Analytic semigroups and optimal regularity in parabolic problems* , PNLDE Vol. 16, Birkhäauser Verlag, Basel, 1995.

[13] C. Y. Zhang, Pseudo Almost Periodic Solutions of Some Differential Equations. *J. Math. Anal. Appl* **181** (1994), no. 1, pp. 62–76.

[14] C. Y. Zhang, Pseudo Almost Periodic Solutions of Some Differential Equations. II. *J. Math. Anal. Appl* **192** (1995), no. 2, pp. 543–561.

[15] C. Y. Zhang, Integration of Vector-Valued Pseudo Almost Periodic Functions, *Proc. Amer. Math. Soc* **121** (1994), no. 1, pp. 167–174.

In: Research on Evolution Equation Compendium. Volume 2 ISBN: 978-1-60456-752-6
Editor: Gaston M. N'Guerekata

A KRASNOSELSKII-TYPE FIXED POINT THEOREM FOR MULTIFUNCTIONS DEFINED ON A HYPERCONVEX SPACE

Marcin Borkowski

Abstract

We present a fixed point theorem for a sum of two convex-valued multifunctions acting on a weakly compact, hyperconvex subset of a normed space. The theorem is a multivalued version of a result of D. Bugajewski.

1. Introduction

In [3], D. Bugajewski proved the following Krasnoselskii-type theorem in a hyperconvex setting.

Theorem 1. *Let K be a bounded hyperconvex subset of a normed space $(X, \|\cdot\|)$ such that $\lambda K \subset K$ for every $\lambda \in (0,1]$. Assume that*

1. *$f_1 : K \to X$ is nonexpansive;*
2. *$f_2 : K \to X$ is completely continuous;*
3. *$f_1(x) + f_2(y) \in K$ for any $x, y \in K$;*
4. *every sequence (x_n) such that $x_n \in K$ for $n \in \mathbb{N}$ and*

$$\lim_{n\to\infty} \big(x_n - f(x_n)\big) = 0,$$

where $f := f_1 + f_2$, has a limit point.

Then, f has a fixed point.

Recall that the assumption that $\lambda H \subset H$ can be released, as it was shown in [4].

Recently, M. Özdemir and S. Akbulut published the paper [6] with a multivalued version of Bugajewski's theorem. Unfortunately, their proof contains some errors. We will state a slightly different version of this theorem and then discuss the errors in [6].

2. Preliminaries

Let X be a metric space. By $\mathcal{B}(X)$ we denote the family of nonempty, bounded and closed subsets of X and by $\mathcal{K}(X)$ the set of nonempty compact subsets of X.

In what follows, we will use the symbol d_X for a metric in the space X and H_X for the Hausdorff metric in the hyperspace $\mathcal{B}(X)$; we will write d and H if the underlying space is obvious from the context.

By A' we will denote the complement of the subset A of some space X, i.e., the set $X \setminus A$.

Definition 1. We call a mapping $f\colon X \to Y$ between metric spaces *nonexpansive*, if $d(f(x), f(y)) \leq d(x,y)$ for each $x, y \in X$.

Definition 2. Let A be any subset of a metric space X. The *Kuratowski measure of noncompactness* of the set A, denoted by $\alpha(A)$, is the greatest lower bound of the numbers $\varepsilon > 0$ such that A can be covered by a finite family of sets of diameter not greater than ε. (We put $\alpha(A) = +\infty$ for unbounded sets.) A mapping $f\colon X \to Y$ between metric spaces is called *α-condensing* if $\alpha(f(A)) \leq \alpha(A)$ for each nonempty $A \subset X$ and $\alpha(f(A)) < \alpha(A)$ provided that $\alpha(A) > 0$.

The aforementioned notions can be easily transfered to the setting of multivalued mappings.

Definition 3. Let X and Y be metric spaces. A multifunction $F\colon X \to \mathcal{B}(Y)$ is called *nonexpansive*, if it is nonexpansive as a mapping from (X, d_X) to $(\mathcal{B}(Y), H_Y)$. A multifunction $F\colon X \to 2^Y \setminus \{\emptyset\}$ is called *α-condensing* if $\alpha(F(A)) \leq \alpha(A)$ for each nonempty $A \subset X$ and $\alpha(F(A)) < \alpha(A)$ whenever $\alpha(A) > 0$, where $F(A) := \bigcup_{a \in A} F(a)$.

We will use the following well-known facts about multifunctions (see, e.g., [5]): a nonexpansive, compact-valued multimap is continuous (in particular, upper semi-continuous); moreover, if F_1, F_2 are compact-valued and upper semi-continuous multifunctions defined on a subset of a normed space, then so is $F_1 + F_2$.

Theorem 2. *Let X and Y be metric spaces. Then, $F : X \to \mathcal{K}(Y)$ is upper semi-continuous if and only if for each sequence $(x_n) \subset X$ converging to some $x \in X$ and sequence $(y_n) \subset Y$ such that $y_n \in F(x_n)$, there is a subsequence of (y_n) converging to a point in $F(x)$.*

Definition 4. Let Z be a nonempty subset of a metric space X and let $F\colon Z \to 2^X \setminus \{\emptyset\}$ be a closed-valued multifunction. We call F *hemicompact* if each sequence $(x_n) \subset Z$ such that $\lim_{n\to\infty} d(x_n, F(x_n)) = 0$ has a convergent subsequence.

Definition 5. Let Z be a subset of a normed space X and let $F\colon Z \to \mathcal{K}(X)$. We call F *strongly continuous* if for each sequence $(x_n) \subset Z$ weakly convergent to $x \in Z$, every sequence $(y_n) \subset X$ such that $y_n \in F(x_n)$ for $n \in \mathbb{N}$ has a subsequence convergent in norm to some point in $F(x)$.

Fact 3 ([8, p. 31, Remark (iii)]). *Let Z be a weakly compact subset of a normed space X, and let $F : Z \to \mathcal{K}(X)$ be strongly continuous. Then $F(Z)$ is compact.*

Proof. Indeed, let $(y_n) \subset F(Z)$. For each $n \in \mathbb{N}$ choose $x_n \in Z$ such that $y_n \in F(x_n)$. We can assume that $x_n \rightharpoonup x \in Z$. This means that (y_n) has a subsequence convergent to some $y \in F(x) \subset F(Z)$. □

Fact 4 ([8, p. 31, Remark (v)]). *A strongly continuous map $F\colon Z \to \mathcal{K}(X)$ defined on a weakly compact subset Z of a normed space X is upper semi-continuous.*

Proof. Let $x_n \to x$ in X and $F(x) \subset V$, where $V \subset X$ is open. Assume $F(x_n) \cap V'$ for infintely many $n \in \mathbb{N}$. Passing to a subsequence we can assume that $F(x_n) \cap V' \neq \emptyset$ for each $n \in \mathbb{N}$. Choose $y_n \in F(x_n) \cap V'$; from strong continuity we can assume that $y_n \to y \in F(x)$. Then $y \in F(x) \cap V'$—contradiction. □

Now let us recall the definition of a hyperconvex space, introduced by Aronszajn and Panitchpakdi:

Definition 6 ([2]). We call a metric space X *hyperconvex*, if every collection of closed balls $\{\overline{B}(x_i, r_i)\}_{i \in I}$ in X such that $d(x_i, x_j) \leq r_i + r_j$ for any $i, j \in I$ has a nonempty intersection.

Note that if H is a hyperconvex subspace of a metric space X, then there exists a nonexpansive retraction of X onto H, i.e., a nonexpansive mapping $R\colon X \to H$ such that $R|_H$ is the identity map on H (see [2]).

We will also need a Darbo-type fixed-point theorem for multifunctions. In order to state it, we will introduce a so-called $\mathcal{U}_c^{\kappa}$ class of multivalued mappings.

First, if $\mathcal{X}$ is some class of multivalued mappings, we will denote by $\mathcal{X}(X,Y)$ the subclass of mappings in $\mathcal{X}$ acting on X and taking values in $2^Y \setminus \{\emptyset\}$, and by $\mathcal{X}_c$ the class of finite compositions of mappings from $\mathcal{X}$. By $\mathcal{U}$ we will denote a class of multivalued mappings such that each F in $\mathcal{U}_c$ is upper semi-continuous and compact-valued (in particular, all single-valued, continuous mappings belong to $\mathcal{U}$) and each $F \in \mathcal{U}_c(\overline{B}^n, \overline{B}^n)$ has a fixed point, where $\overline{B}^n$ is a closed n-dimensional Euclidean ball. Further, given a class $\mathcal{U}$, we will denote by $\mathcal{U}_c^{\kappa}(X,Y)$ the family of mappings $F\colon X \to 2^Y \setminus \{\emptyset\}$ such that for any $K \in \mathcal{K}(X)$, there exists an $F_K \in \mathcal{U}_c(X,Y)$ such that $F_K(x) \subset F(x)$ for each $x \in K$. We will also note that Kakutani multimappings (i.e., convex- and nonempty-valued, upper semi-continuous multimaps from a subset of a Hausdorff topological vector space to another such space) are an example of $\mathcal{U}_c^{\kappa}$ maps (see [7]).

Theorem 5 ([1, Theorem 2.4]). *Let X be a hyperconvex bounded metric subspace of a topological vector space and let $F \in \mathcal{U}_c^{\kappa}(X,X)$ be α-condensing. Then F has a fixed point.*

3. Main Result

Theorem 6. *Let us assume that H is a nonempty, hyperconvex and weakly compact subset of a normed space X, $F_1 : H \to \mathcal{K}(X)$ is strongly continuous, $F_2\colon H \to \mathcal{K}(X)$ is nonexpansive, $F := F_1 + F_2$ is hemicompact with convex compact values lying in H. Then, F has a fixed point.*

Proof. Let $R\colon X \to H$ be a nonexpansive retraction of X onto H. For each $n \in \mathbb{N}$, define a multifunction $G_n\colon H \to 2^H \setminus \{\emptyset\}$ by the formula $G_n := R((1 - \frac{1}{n})F)$. Obviously, the map

$F\colon H \to \mathcal{K}(X)$ is Kakutani, so it is a $\mathcal{U}_c^\kappa$ mapping. Now let $K \subset H$ be compact and let $F_K \in \mathcal{U}_c(H, \mathcal{K}(X))$ be such that $F_K(x) \subset F(x)$ for each $x \in H$. Since $\mathcal{U}_c$ is closed under finite superpositions, $R((1-\frac{1}{n})F_K) \in \mathcal{U}_c(H, \mathcal{K}(X))$ and $R((1-\frac{1}{n})F_K(x)) \subset G_n(x)$ for each $x \in K$. This shows that G_n is a $\mathcal{U}_c^\kappa$ mapping.

Let $A \subset H$. We have:

$$\begin{aligned}\alpha(G_n(A)) &= \alpha\big(R((1-\tfrac{1}{n})(F_1+F_2)(A))\big) \leq \\ &\leq \alpha((1-\tfrac{1}{n})F_1(A) + (1-\tfrac{1}{n})F_2(A)) \leq \\ &\leq (1-\tfrac{1}{n})\alpha(F_1(A)) + (1-\tfrac{1}{n})\alpha(F_2(A)) \\ &\leq (1-\tfrac{1}{n})\alpha(A).\end{aligned}$$

Since weakly compact sets are bounded, we can infer from Theorem 5 that each G_n has a fixed point x_n.

For each $n \in \mathbb{N}$, choose $x_n' \in (1-\frac{1}{n})F(x_n)$ such that $x_n = R(x_n')$. Since $\frac{n}{n-1}x_n' \in F(x_n) \subset H$, and in particular, the sequence (x_n') is bounded, we have:

$$\begin{aligned}d(x_n, F(x_n)) &\leq \|x_n - \tfrac{n}{n-1}x_n'\| = \\ &= \|R(x_n') - R(\tfrac{n}{n-1}x_n')\| \leq \\ &\leq \|(1-\tfrac{n}{n-1})x_n'\| = \\ &= \tfrac{1}{n-1}\|x_n'\| \to 0.\end{aligned}$$

That means that (passing to a subsequence if necessary) we can assume that $x_n \to x \in H$.

Choose sequences (y_n) and (z_n) such that $y_n \in F_1(x_n)$, $z_n \in F_2(x_n)$ and $x_n = R((1-\frac{1}{n})(y_n+z_n))$. From Theorem 2 we infer that $y_n \to y \in F_1(x)$ and $z_n \to z \in F_2(x)$ (again passing to subsequences if necessary). Of course, $y+z \in H$. We have:

$$x_n = R((1-\tfrac{1}{n})(y_n+z_n)) \to R(y+z) = y+z \in F_1(x)+F_2(x) = F(x)$$

and the proof is finished. □

4. Remarks

We will now turn our attention to the main result of the paper [6]. It states the following:

Conjecture. *Let H be a nonempty bounded hyperconvex subspace of a normed space X such that $\lambda H \subset H$ for each $\lambda \in (0,1]$. Assume that a multivalued mapping $F := F_1 + F_2\colon H \to \mathcal{K}(H)$ is hemicompact. Assume also that $F_1, F_2\colon H \to \mathcal{K}(X)$ are convex-valued, F_1 is strongly continuous, F_2 nonexpansive and $F_1(H)+F_2(H) \subset H$. Then, F has a fixed point.*

Unfortunately, the proof given therein contains a few errors.

1. The authors use the fact that the metric projection of X onto H is single-valued, continuous and nonexpansive. However, none of these statements need be true. Indeed, let $X := \mathbb{R}^2$ with the l_1-type norm and $H := \operatorname{conv}\{(0,0),(0,1)\} \cup \operatorname{conv}\{(0,0),(1,0)\}$;

then, the metric projection of X onto H fails to be unique at the point $(1,1)$; moreover, considering the points $a := (1-\varepsilon, 1)$ and $b := (1, 1-\varepsilon)$ for $\varepsilon \in (0,1)$ shows that it is not continuous (and hence also not nonexpansive). (Note that using a nonexpansive retraction instead of the metric projection solves both these problems.)

2. The authors use some properties of strongly continuous multifunctions, analogous to Facts 3 and 4, but without the assumption of weak compactness. The proofs of these facts, however, require that the domain be weakly compact, which went unnoticed in the quoted paper. (In fact, the authors replaced weak convergence by norm convergence in their definition of strong continuity, which resulted in a notion equivalent to upper semi-continuity. Obviously this rendered the analogue of Fact 4 true, but Fact 3 remains false with this new definition.)

3. The authors infer α-condensingness of a mapping from its nonexpansiveness; the example of the identity map shows that this inference is false.

4. In the course of the proof, the authors define multifunctions G_n ($n \in \mathbb{N}$) as a sum of certain two multivalued mappings; however, these mappings are defined on different domains, so this sum is ill-defined. (It turns out that if we consider this sum on the intersection of the domains of the summands, we arrive at the conclusion that using the metric projection, apart from being wrong, was unnecessary.)

In view of the above, the question whether the statement given is true remains open.

References

[1] R. P. Agarwal and D. O'Regan, *Hyperconvex spaces and fixed points*, *Georgian Math. Journal* **9** (2002), no. 2, 199–206.

[2] N. Aronszajn and P. Panitchpakdi, *Extension of uniformly continuous transformations and hyperconvex metric spaces*, *Pacific J. Math* **6** (1956), 405–439.

[3] D. Bugajewski, *Fixed-point theorems in hyperconvex spaces revisited*, *Math. Comput. Modelling* **32** (2000), no. 13, 1457–1461.

[4] D. Bugajewski and R. Espínola, *Remarks on some fixed point theorems for hyperconvex spaces and absolute retracts*, Function Spaces the Fifth Conference (*Lecture Notes in Pure and Applied Mathematics Series* **213**). Edited by H. Hudzik and L. Skrzypczak, 2000, pp. 85–92.

[5] S. Hu and N. S. Papageorgiou, *Handbook of Multivalued Analysis*, Vol. I: Theory, Kluwer Academic Publishers, Dordrecht/Boston/London, 1997.

[6] M. Özdemir and S. Akbulut, *A fixed point theorem for multivalued maps in hyperconvex spaces*, *Appl. Math. Comput.* **157** (2004), 637–642.

[7] D. O'Regan, *Admissible spaces and α-fixed points*, *Nonlinear Funct. Anal. & Appl.* **10** (2005), no. 1, 55–64.

[8] ———, *New fixed-points results for 1-set contractive set-valued maps*, *Computers Math. Applic.* **35** (1999), no. 4, 27–34.

INDEX

A

B

C

D

E

J

K

L

M

N

O

P

Q

R

S

T

U

V

Erratum for *Research on Evolution Equations Compendium, Volume 1*

Editor: Gaston M N'Guerekata

p. i-422:

Volume 2 should be Volume 1

p. 129:

Equations 1.1 and 1.2 should be:

$$\frac{\partial v}{\partial t} = \frac{\partial^2 v}{\partial x^2} + h(v) - w, \quad (1.1)$$

$$\frac{\partial w}{\partial t} = \epsilon(v - \gamma w) \qquad (1.2)$$

p. 183:

The last equation should be:

$$q(x) = 2\frac{d}{dx}H(x, x),$$